METHODS IN MOLECULAR BIOLOGY™



*Series Editor*
**John M. Walker**
**School of Life Sciences**
**University of Hertfordshire**
**Hatfield, Hertfordshire, AL10 9AB, UK**

For further volumes:
http://www.springer.com/series/7651

# Antiviral Methods and Protocols

**Second Edition**

Edited by

**Edwin Yunhao Gong**

*Janssen Infectious Diseases-Diagnostics BVBA,*
*Johnson & Johnson Corporation, Beerse, Belgium*

Humana Press

*Editor*
Edwin Yunhao Gong
Janssen Infectious Diseases-Diagnostics BVBA
Johnson & Johnson Corporation
Beerse, Belgium

ISSN 1064-3745 ISSN 1940-6029 (electronic)
ISBN 978-1-4939-6035-4 ISBN 978-1-62703-484-5 (eBook)
DOI 10.1007/978-1-62703-484-5
Springer New York Heidelberg Dordrecht London



*Cover illustration*: The blue background is the crystal violet staining of the whole cell monolayer. The white dots are virus particles (one white dot is derived from one virion). In the last well, only one virion is present at that drug concentration (the test compound at that concentration totally inhibits virus replication), and one can count the number of white dots to calculate the drug effect. This image represents duplicate wells for each drug test concentration.

Printed on acid-free paper

Humana Press is a brand of Springer
Springer is part of Springer Science+Business Media (www.springer.com)

# Preface

Viruses are the leading cause of disease and death worldwide, and antiviral drugs are one of the most effective ways, in addition to vaccines, to control and manage viral infections. Although antiviral drugs have been successfully developed for some viral diseases, there remains a clear unmet medical need to develop novel antiviral agents for the control and management of many viruses that currently have no or limited treatment options as well as a need to overcome the limitations associated with the existing antiviral drugs, such as adverse effects and emergence of drug-resistant mutations.

The first edition of *Antiviral Methods and Protocols* was published in the *Methods in Molecular Medicine* series (discontinued some years ago) in 2000. This new edition, published in the *Methods in Molecular Biology* series, reflects the major technical advances for the last 13 years in antiviral research and discovery. Although it is flagged as the second edition, all chapters are completely new, and none of the chapters is simply updated from the previous edition. This edition focuses on many important human viruses that cause major problems in public health, such as human immunodeficiency virus type 1 (HIV-1), hepatitis viruses (hepatitis B and C viruses), herpes viruses, and influenza virus. Many new chapters describing antiviral methods and protocols against respiratory viruses, such as human respiratory syncytial virus (4 chapters), and some important emerging viruses, such as dengue virus (7 chapters), West Nile virus (1 chapter), and chikungunya virus (2 chapters), are included. In addition, as the in vivo evaluation of antiviral efficacy and modes of action using animal models is very important to antiviral drug discovery and development, several chapters describing animal models for evaluation of antiviral agents in vivo are also presented.

Consistent with the approach of the *Methods in Molecular Biology* series, each chapter contains detailed cutting-edge research techniques that are currently used in the field of antiviral research. Many methods and protocols presented in this edition are fully validated, high-throughput antiviral assays that can be used for screening compound libraries in pharmaceutical companies and research institutions. Overall, this new edition of *Antiviral Methods and Protocols* will serve as a laboratory reference for pharmaceutical and academic biologists, medicinal chemists, and pharmacologists as well as for virologists in the field of antiviral research and drug discovery.

I would like to thank Professor John Walker, chief editor of the series, for his advice. I also wish to express my gratitude to the editors from Springer, especially Patrick Marton, David Casey, Anne Meagher and Paul Wehn, for making this new edition possible, and to Sundaramoorthy Balasubramanian, Project Manager at SPi Global, India, for his valuable contribution to edit this book, and to all the chapter authors for their valuable contributions to this book.

***Beerse, Belgium*** ***Edwin Yunhao Gong***

# Contents

## Contributors

SOFIA L. ALCARAZ-ESTRADA • *Division de Medicina Genomica, Centro Medico Nacional-ISSSTE, Mexico, DF, Mexico*

DALE L. BARNARD • *Department of Animal, Dairy, and Veterinary Sciences, Institute for Antiviral Research, Utah State University, Logan, UT, USA*

RALF BARTENSCHLAGER • *Department of Infectious Diseases, Molecular Virology, University of Heidelberg, Heidelberg, Germany*

TOMAS BERGSTRÖM • *Department of Clinical Virology, University of Gothenburg, Göteborg, Sweden*

JAN MARTIN BERKE • *Janssen Infectious Diseases-Diagnostics BVBA, Johnson & Johnson Corporation, Beerse, Belgium*

TIMOTHY M. BLOCK • *Institute for Biotechnology and Virology Research, Drexel University College of Medicine, Doylestown, PA, USA*

CHRISTOPHE BODENREIDER • *Novartis Institute for Tropical Diseases, Singapore*

JEAN-FRANÇOIS BONFANTI • *Janssen Research and Development, Val-de-Reuil, France*

NIGEL BOURNE • *Department of Pediatrics, University of Texas Medical Branch, Galveston, TX, USA*

LIEVE BUNKENS • *Janssen Infectious Diseases-Diagnostics BVBA, Johnson & Johnson Corporation, Beerse, Belgium*

DAWEI CAI • *Institute for Biotechnology and Virology Research, Drexel University College of Medicine, Doylestown, PA, USA*

REGINALD CLAYTON • *BioReliance Ltd., Glasgow, Scotland, UK*

ERNA CLEIREN • *Janssen Infectious Diseases-Diagnostics BVBA, Johnson & Johnson Corporation, Beerse, Belgium*

MARLEEN CLYNHENS • *Janssen Infectious Diseases-Diagnostics BVBA, Johnson & Johnson Corporation, Beerse, Belgium*

ASHLEY DAGLEY • *Department of Animal, Dairy and Veterinary Sciences, Institute for Antiviral Research, Utah State University, Logan, UT, USA*

GÉRY DAMS • *Janssen Infectious Diseases-Diagnostics BVBA, Johnson & Johnson Corporation, Beerse, Belgium*

BEN DE CLERCK • *Janssen Infectious Diseases-Diagnostics BVBA, Johnson & Johnson Corporation, Beerse, Belgium*

PASCALE DEHERTOGH • *Janssen Infectious Diseases-Diagnostics BVBA, Johnson & Johnson Corporation, Beerse, Belgium*

LEEN DELANG • *Rega Institute for Medical Research, Leuven, Belgium*

ALAIN DELOOF • *Janssen Infectious Diseases-Diagnostics BVBA, Johnson & Johnson Corporation, Beerse, Belgium*

BENOIT DEVOGELAERE • *Biocartis NV, Mechelen, Belgium*

KOEN DOCKX • *Janssen Infectious Diseases-Diagnostics BVBA, Johnson & Johnson Corporation, Beerse, Belgium*

GREGORY FANNING • *Janssen Infectious Diseases-Diagnostics BVBA, Johnson & Johnson Corporation, Beerse, Belgium*
ERIC FERRARI • *Department of Infectious Diseases, Merck Research Laboratories, Kenilworth, NJ, USA*
BART FEVERY • *Janssen Infectious Diseases-Diagnostics BVBA, Johnson & Johnson Corporation, Beerse, Belgium*
MAXIM FEYAERTS • *Janssen Infectious Diseases-Diagnostics BVBA, Johnson & Johnson Corporation, Beerse, Belgium*
KATLEEN FIERENS • *Janssen Infectious Diseases-Diagnostics BVBA, Johnson & Johnson Corporation, Beerse, Belgium*
WOLFGANG FISCHL • *Haplogen GmbH, Vienna, Austria*
PEGGY GELUYKENS • *Janssen Infectious Diseases-Diagnostics BVBA, Johnson & Johnson Corporation, Beerse, Belgium*
OLIVIA GOETHALS • *Janssen Infectious Diseases-Diagnostics BVBA, Johnson & Johnson Corporation, Beerse, Belgium*
EDWIN YUNHAO GONG • *Janssen Infectious Diseases-Diagnostics BVBA, Johnson & Johnson Corporation, Beerse, Belgium*
HAITAO GUO • *Institute for Biotechnology and Virology Research, Drexel University College of Medicine, Doylestown, PA, USA*
JU-TAO GUO • *Institute for Biotechnology and Virology Research, Drexel University College of Medicine, Doylestown, PA, USA*
EMMANUEL GUSTIN • *Janssen Infectious Diseases-Diagnostics BVBA, Johnson & Johnson Corporation, Beerse, Belgium*
STEPHANIE HAMILTON • *Biota Pharmaceuticals, Inc., Notting Hill, VIC, Australia*
PASCALE HOLEMANS • *Biocartis NV, Mechelen, Belgium*
LIESELOT HOUSPIE • *Rega Institute for Medical Research, University of Leuven, Leuven, Belgium*
HSUEH-CHENG HUANG • *Department of Infectious Diseases, Merck Research Laboratories, Kenilworth, NJ, USA*
GABRIELA ISPAS • *Janssen Infectious Diseases-Diagnostics BVBA, Johnson & Johnson Corporation, Beerse, Belgium*
TANIA IVENS • *Biocartis NV, Mechelen, Belgium*
DIRK JOCHMANS • *Rega Institute for Medical Research, University of Leuven, Leuven, Belgium*
JUSTIN G. JULANDER • *Department of Animal, Dairy and Veterinary Sciences, Institute for Antiviral Research, Utah State University, Logan, UT, USA*
HANNAH KENENS • *Janssen Infectious Diseases-Diagnostics BVBA, Johnson & Johnson Corporation, Beerse, Belgium*
ELS KEYAERTS • *Rega Institute for Medical Research, University of Leuven, Leuven, Belgium*
KATHRYN M. KITRINOS • *Gilead Sciences Inc., Foster City, CA, USA*
CHRISTIAN D. KLEIN • *Medicinal Chemistry, Institute of Pharmacy and Molecular Biotechnology (IPMB), Heidelberg University, Heidelberg, Germany*
DIANA KOLETZKI • *Janssen Diagnostics BVBA, Beerse, Belgium*
GUENTER KRAUS • *Janssen Infectious Diseases-Diagnostics BVBA, Johnson & Johnson Corporation, Beerse, Belgium*
LEEN KWANTEN • *Janssen Infectious Diseases-Diagnostics BVBA, Johnson & Johnson Corporation, Beerse, Belgium*

OLIVER LENZ • *Janssen Infectious Diseases-Diagnostics BVBA, Johnson & Johnson Corporation, Beerse, Belgium*
SIEW PHENG LIM • *Novartis Institute for Tropical Diseases, Singapore*
YANG LIU • *Gilead Sciences Inc., Foster City, CA, USA*
PEDRO LORY • *Janssen Infectious Diseases-Diagnostics BVBA, Johnson & Johnson Corporation, Beerse, Belgium*
ANNA LUNDIN • *Department of Clinical Virology, University of Gothenburg, Göteborg, Sweden*
ANGELA LUTTICK • *Biota Pharmaceuticals, Inc., Notting Hill, VIC, Australia*
PIET MAES • *Rega Institute for Medical Research, University of Leuven, Leuven, Belgium*
GEERT MEERSSEMAN • *Biocartis NV, Mechelen, Belgium*
JOHAN NEYTS • *Rega Institute for Medical Research, Leuven, Belgium*
MAINA NGOTHO • *Institute of Primate Research, Nairobi, Kenya*
HUI NIE • *Institute for Biotechnology and Virology Research, Drexel University College of Medicine, Doylestown, PA, USA*
CHRISTOPH NITSCHE • *Medicinal Chemistry, Institute of Pharmacy and Molecular Biotechnology (IPMB), Heidelberg University, Heidelberg, Germany*
RADHAKRISHNAN PADMANABHAN • *Department of Microbiology and Immunology, Georgetown University, Washington, DC, USA*
THERESA PATTERY • *Janssen Diagnostics BVBA, Beerse, Belgium*
ERIN DONOHUE REICHERT • *Defense Threat Reduction Agency, Fort Belvoir, VA, USA*
DIRK ROYMANS • *Janssen Infectious Diseases-Diagnostics BVBA, Johnson & Johnson Corporation, Beerse, Belgium*
WOUTER SCHUL • *Department of Infectious Diseases, Novartis Institutes for Biomedical Research, Emeryville, CA, USA*
PEI-YONG SHI • *Novartis Institute for Tropical Diseases, Singapore*
KENNY SIMMEN • *Janssen Infectious Diseases-Diagnostics BVBA, Johnson & Johnson Corporation, Beerse, Belgium*
DONALD F. SMEE • *Department of Animal, Dairy, and Veterinary Sciences, Institute for Antiviral Research, Utah State University, Logan, UT, USA*
ALEXANDRA SMETS • *Biocartis NV, Mechelen, Belgium*
LIESBET SMEULDERS • *Janssen Infectious Diseases-Diagnostics BVBA, Johnson & Johnson Corporation, Beerse, Belgium*
VEERLE SMITS • *Janssen Diagnostics BVBA, Beerse, Belgium*
HANS STEVENS • *Rega Institute for Medical Research, University of Leuven, Leuven, Belgium*
LIEVEN J. STUYVER • *Janssen Diagnostics BVBA, Beerse, Belgium*
EDWARD TRYBALA • *Department of Clinical Virology, University of Gothenburg, Göteborg, Sweden*
SIMON P. TUCKER • *Biota Pharmaceuticals, Inc., Notting Hill, VIC, Australia*
FRANCES VALENCIA • *Department of Experimental Pathology, University of Texas Medical Branch, Galveston, TX, USA*
KOEN VAN ACKER • *Biocartis NV, Mechelen, Belgium*
ELLEN VAN DAMME • *Janssen Infectious Diseases-Diagnostics BVBA, Johnson & Johnson Corporation, Beerse, Belgium*
JOKE VAN DEN BERG • *Janssen Infectious Diseases-Diagnostics BVBA, Johnson & Johnson Corporation, Beerse, Belgium*
PETER VAN DEN EEDE • *Janssen Diagnostics BVBA, Beerse, Belgium*

CHRISTEL VAN DEN EYNDE • *Janssen Infectious Diseases-Diagnostics BVBA, Johnson & Johnson Corporation, Beerse, Belgium*
MARIJKE VAN DER AUWERA • *Biocartis NV, Mechelen, Belgium*
LIESBET VAN DER HELM • *Janssen Infectious Diseases-Diagnostics BVBA, Johnson & Johnson Corporation, Beerse, Belgium*
BARBARA VAN KERCKHOVE • *Janssen Infectious Diseases-Diagnostics BVBA, Johnson & Johnson Corporation, Beerse, Belgium*
MARNIX VAN LOOCK • *Janssen Infectious Diseases-Diagnostics BVBA, Johnson & Johnson Corporation, Beerse, Belgium*
MARC VAN RANST • *Rega Institute for Medical Research, University of Leuven, Leuven, Belgium*
BERTRAND VAN SCHOUBROECK • *Janssen Infectious Diseases-Diagnostics BVBA, Johnson & Johnson Corporation, Beerse, Belgium*
LIESBETH VAN WESENBEECK • *Janssen Infectious Diseases-Diagnostics BVBA, Johnson & Johnson Corporation, Beerse, Belgium*
GENEVIÈVE VANDERCRUYSSEN • *Biocartis NV, Mechelen, Belgium*
LEEN VANHOOREN • *Janssen Diagnostics BVBA, Beerse, Belgium*
JANNICK VERBEECK • *KU Leuven, Leuven, Belgium*
INGE VEREYCKEN • *Janssen Infectious Diseases-Diagnostics BVBA, Johnson & Johnson Corporation, Beerse, Belgium*
ANN VERHEYEN • *Janssen Infectious Diseases-Diagnostics BVBA, Johnson & Johnson Corporation, Beerse, Belgium*
NICK VERHEYEN • *Janssen Infectious Diseases-Diagnostics BVBA, Johnson & Johnson Corporation, Beerse, Belgium*
YVAN VERLINDEN • *Janssen Infectious Diseases-Diagnostics BVBA, Johnson & Johnson Corporation, Beerse, Belgium*
RENÉ VERLOES • *Janssen Infectious Diseases-Diagnostics BVBA, Johnson & Johnson Corporation, Beerse, Belgium*
KATRIEN VERMEIREN • *Biocartis NV, Mechelen, Belgium*
RONALD L. VESELENAK • *Department of Microbiology and Immunology, University of Texas Medical Branch, Galveston, TX, USA*
LEEN VIJGEN • *Janssen Infectious Diseases-Diagnostics BVBA, Johnson & Johnson Corporation, Beerse, Belgium*
JORGE VILLACIAN • *Janssen Infectious Diseases-Diagnostics BVBA, Johnson & Johnson Corporation, Beerse, Belgium*
JOHAN VINGERHOETS • *Janssen Infectious Diseases-Diagnostics BVBA, Johnson & Johnson Corporation, Beerse, Belgium*
INGE VLIEGEN • *Rega Institute for Medical Research, Leuven, Belgium*
WILLIAM T. WINDSOR • *In Vitro Pharmacology, Merck Research Laboratories, Kenilworth, NJ, USA*
RAN YAN • *Institute for Biotechnology and Virology Research, Drexel University College of Medicine, Doylestown, PA, USA*
ANDY YIP • *Novartis Institute for Tropical Diseases, Singapore*
RUMIN ZHANG • *In Vitro Pharmacology, Merck Research Laboratories, Kenilworth, NJ, USA*

# Part I

## Human Immunodeficiency Virus Type 1 (HIV-1)

# Chapter 1

# A Fluorescence-Based High-Throughput Screening Assay to Identify HIV-1 Inhibitors

**Peggy Geluykens, Koen Van Acker, Johan Vingerhoets, Christel Van den Eynde, Marnix Van Loock, and Géry Dams**

## Abstract

Highly active antiretroviral therapy (HAART) dramatically increases the long-term survival rates of human immunodeficiency virus type 1 (HIV-1) infected patients. Yet, poor adherence to therapy, adverse effects and the occurrence of resistant viruses can compromise the efficacy of HAART regiments. Therefore, there remains a clear unmet medical need for novel drugs and treatment options. In this chapter, we describe an HIV-1 antiviral high-throughput screening assay based on an HIV-1 permissive T lymphoblastoid MT4 cell line, stably transfected with a construct carrying an HIV-1 long terminal repeat promoter driving the expression of a reporter gene (enhanced green fluorescent protein). This assay runs in a 384-well format and enables the identification of HIV-1 inhibitors during a high-throughput screening campaign. In parallel, a cytotoxicity assay is performed to evaluate the compound-related in vitro toxicity.

**Key words** HIV-1, High-throughput screening, Fluorescence-based assay, HIV-1 inhibitors

## 1 Introduction

In 2009, approximately 33 million people were living with HIV-1 (WHO, AIDS epidemic update, 2010). More importantly, it was reported for the first time that the spread of HIV-1 was halted and even reversed. This success can be contributed to the implementation of HAART [1]. Nevertheless, there remains a significant unmet need as the long-term success of HAART can be compromised by poor adherence to therapy, adverse effects and HIV-1's propensity to develop resistance against antiretroviral drugs, which limit treatment options for patients.

Cell-based antiviral assays are widely used to underpin the discovery of antiviral compounds and offer tractable methodologies for large scale operations such as high-throughput screening and smaller, routine monitoring of medicinal chemistry efforts during hit to lead and lead optimization processes [2]. Cell-based antiviral assays offer

Edwin Yunhao Gong (ed.), *Antiviral Methods and Protocols*, Methods in Molecular Biology, vol. 1030,
DOI 10.1007/978-1-62703-484-5_1, © Springer Science+Business Media, LLC 2013

an attractive way to identify HIV-1 inhibitors, as these "holistic" assays encompass, in one single test, the complete range of antiviral and cellular targets which are critical for viral replication. This enables targeting the virus in its natural cellular environment, whereas biochemical assays are mostly focused on one specific target and the active compounds in these assays are not necessarily able to enter cells due to their chemical properties. In the first generation cell-based assays, the antiviral activity of compounds was evaluated based on their inhibition of viral cytopathic effects (CPE) in cell culture. In these assays, an HIV-1-susceptible cell line is infected with the wild-type virus in the presence of a potential antiviral compound. When viral replication proceeds, the cells are killed by the virus, whereas antiviral active compounds protect the cells from virus-mediated cell death. In parallel, the cytotoxicity of the compounds is assessed on uninfected cells, which enables the determination of the compound's selectivity. However, observing virus-induced CPE requires several days (>5 days) of incubation, which makes the assay more prone to compound-related toxicity. Therefore, the tolerance towards toxic compounds is low, which compromises the ability to identify compounds with low selectivity indexes. Pauwels et al. [3] described such a CPE-based antiviral assay, which was based on the infection of human MT4 cells with laboratory-adapted HIV-1 strains. Due to the excellent replication rate of these HIV-1 strains in the MT4 cell line, HIV-1-mediated cell death can be detected with the vital dye 3-(4,5-dimethylthiazol-2-yl)-2,5-diphenyltetrazolium bromide (MTT), at 5 days post-infection [4]. The first HIV-1 non-nucleoside reverse transcriptase (RT) inhibitor was discovered with such an MT4/MTT assay [5].

Later on, the second generation antiviral screening assays were developed, in which target cells are equipped with a reporter gene to assess the inhibition of viral replication by small molecules. Upon HIV-1 infection and replication, the reporter gene is expressed in an HIV-1 dependent manner. Antiviral compounds which inhibit HIV-1 replication also inhibit the reporter gene expression. Throughout the years, several reporter genes have been used to develop screening assays, including chloramphenicol acetyltransferase, luciferase, green fluorescent protein, alkaline phosphatase, and β-galactosidase [6]. As described above, the selectivity of the compounds is determined by running a cytotoxicity test in parallel using mock-infected reporter cells in which the reporter gene expression is controlled by a constitutive promoter.

Here, we describe in details a cell-based reporter assay using stably transfected MT4 cells that carry an enhanced green fluorescent protein (EGFP) reporter gene under the control of the HIV-1 long terminal repeat (LTR) promoter [7]. These cells express basal levels of EGFP that can be easily detected by fluorescence microscopy or flow cytometry. During HIV-1 replication, the viral Tat protein is expressed and binds to the LTR, up-regulating the EGFP expression levels at least tenfold. Consequently, the infected cells can be discriminated from noninfected cells by their fluorescence intensity. Active antiviral compounds inhibit the HIV-1 replication cycle and as such will reduce the Tat-induced EGFP signal. EGFP expression levels can be quantified by automated fluorescence microscopy or flow cytometry. The EGFP-based assay has several advantages over the CPE assays. First, EGFP is an intrinsically fluorescent protein which does not require any substrate, making the assay homogeneous and noninvasive. Due to this mix-and-read principle, the readout is completely harmless to the cells and the same plate can be measured repeatedly, which introduces flexibility and reduces costs considerably. Second, the EGFP signal is generated before the MT4 cells die, allowing detection of viral infection in living cells which results in a shorter incubation time as the assay does not rely on life–death discrimination. Third, the microscopic readout allows for the convenient retrieval of images and raw data afterwards. One disadvantage of the EGFP-based assay, however, is that the intrinsically fluorescent compounds or serum components can interfere with the fluorescent readout of the assay.

## 2 Materials

### 2.1 Reagents

1. HIV-1 IIIB virus: kindly provided by Dr Guido van der Groen (Institute of Tropical Medicine, Antwerp, Belgium).
2. MT4 cells (human T lymphotropic virus type 1-transformed human T lymphoblastoid cell line): kindly provided by Dr Naoki Yamamoto (National Institute of Infectious Diseases, AIDS Research Center, Tokyo, Japan).
3. MT4-LTR-EGFP cells were obtained by transfecting MT4 cells with a selectable construct encompassing the HIV-1 LTR promoter driving the expression of EGFP and subsequent selection of stably transfected cells [7] (*see* **Note 1**).
4. MT4-CMV-EGFP cells were obtained by stably transforming MT4 cells with a CMV–EGFP construct [7].
5. RPMI/10 % FBS: RPMI-1640 with Ultra Glutamine and HEPES without phenol red (Lonza-BioWhittaker®), supplemented with 10 % fetal bovine serum (Sigma) and 0.02 mg/mL gentamicin (Gibco).
6. Geneticin® (G418; Gibco).

### 2.2 Consumables

1. Corning® 384 Well Flat Clear Bottom Black Polystyrene TC-Treated Microplates (Corning).
2. Corning® 96 Well Clear V-Bottom TC-Treated Microplates (Corning).
3. Tissue culture flasks.

### 2.3 Equipment

1. Z2™ Coulter counter®, Analyzer (Beckman Coulter).
2. Humidified 37 °C incubator (5 % $CO_2$).
3. Allegra® X-15R Centrifuge (Beckman Coulter).
4. Multidrop combi reagent dispenser with dispensing cassettes (Thermo Scientific).
5. Automated fluorescence microscope equipped with a 488 nm (blue) laser or a fluorescence spectrophotometer.

## 3 Methods

### 3.1 Preparation of Virus Stocks

1. Thaw a tube of IIIB virus in the incubator at 37 °C, e.g., a virus stock with a titer of $2.6 \times 10^6$ tissue culture infectious dose per mL ($TCID_{50}$/mL).
2. Pellet $3 \times 10^7$ MT4-LTR-EGFP cells by centrifuging for 5 min at $500 \times g$ in a 50 mL sterile tube.
3. Discard the supernatant and infect the cells with IIIB at a multiplicity of infection (MOI) of 0.01 $TCID_{50}$/cell. In this example, the virus titer is $2.6 \times 10^5$ $TCID_{50}$/mL, thus add 1.15 mL of virus stock on the pellet.
4. Mix the cell pellet and virus dilution by briefly shaking the tube and incubate for 60 min at 37 °C. Meanwhile, preheat a 175 $cm^2$ cell culture flask containing 200 mL RPMI/10 % FBS at 37 °C.
5. Resuspend the cells with virus in the 200 mL preheated RPMI/10 % FBS.
6. Incubate the culture flask in the incubator at 37 °C.
7. Microscopically assess CPE and GFP-signal on a daily basis until full CPE is observed, typically 5–7 days post-infection.
8. To harvest the virus, transfer the supernatant into a 500 mL sterile tube and centrifuge at $2{,}000 \times g$ and 4 °C for 10 min to pellet the cells.
9. Transfer the supernatant to a new 500 mL sterile tube.
10. Aliquot the viral supernatant into 1.5 mL or 4.5 mL cryovials (one extra 1.5 mL tube for titration, *see* Subheading 3.2) and store at −80 °C.

### 3.2 Titration of the Virus Stocks

1. Perform virus dilutions using a 96-well V-bottom plate in a following plate layout: column 1 is the medium control (MC), columns 2–11 are assigned to prepare a fivefold serial dilution of the virus, column 12 is the cell control (CC). To determine the virus titer ($TCID_{50}$/mL), the virus dilution series is repeated eight times (rows A to H). Therefore, dispense 100 μL of RPMI/10 % FBS into column 1 (MC) and columns 3–12 of a 96-well V-bottom plate using a multi-channel pipette or multidrop combi reagent dispenser.

2. Thaw virus at 37 °C. Add 125 μL undiluted virus to each well of column 2 with a multi-channel pipette.
3. Make a 1/5 serial dilution of the virus from columns 2 to 11 in the 96-well plate by pipetting 25 μL of column 2 into the 100 μL RPMI/10 % FBS of column 3. Mix thoroughly by pipetting up and down and transfer 25 μL to the next column. Repeat this through column 11 (*see* **Note 2**).
4. Prepare RPMI/10 % FBS containing 2 % DMSO, 4 mL for each 384-well plate.
5. Dispense 10 μL of RPMI/10 % FBS containing 2 % DMSO in an entire 384-well black plate using a multidrop combi reagent dispenser.
6. Transfer four times 15 μL from each well of the 96-well plates to each of the quadruplicate wells in the black 384-well plate (e.g., well A1 of the 96-well plate corresponds to wells A1, A2, B1, and B2 of the 384-well plate).
7. Prepare an MT4-LTR-EGFP cell suspension in RPMI/10 % FBS; 6 mL at $4 \times 10^5$ cells/mL for each 384-well black plate.
8. Dispense 15 μL of the MT4-LTR-EGFP cell suspension into columns 3–24 using a multidrop combi reagent dispenser (columns 3–22 contain the serial dilution of the virus stock, columns 23 and 24 are CC). In addition, dispense 15 μL of RPMI/10 % FBS into columns 1 and 2 (MC).
9. Incubate the 384-well plate for 3 days in a humidified incubator at 37 °C.
10. Afterwards, count the wells that show CPE or fluorescence for every virus dilution and calculate the titer ($TCID_{50}$/mL) according to the method of Reed and Muench [8].

### 3.3 HIV-1 Antiviral High-Throughput Screening Assay

#### 3.3.1 Preparation of the Compound Test Plates

1. Prepare fourfold serial dilutions of test compounds in RPMI/10 % FBS at 4× final concentrations and add 10 μL/well of each test concentration to the corresponding wells of columns 3–22 of a 384-well black plate (the DMSO concentration in each well is 2 %, *see* **Note 3**). Typically, we test four concentrations for each compound and 80 compounds per plate. For each set of test compounds, prepare duplicate plates: one for testing the antiviral activity and one for assessing the cytotoxicity. In our laboratory, the stock solutions of test compounds are made in 100 % DMSO and diluted to test concentrations using medium.
2. In the activity plates, dispense 10 μL of RPMI/10 % FBS containing 2 % DMSO in columns 1, 2, 23, and 24. In this plate format, columns 1 and 2 are used as virus control (VC; absence of compounds); columns 3–22 contain the serial dilutions of the test compounds; and columns 23 and 24 serve as cell control (CC; absence of virus and compounds).

3. In the toxicity plates, dispense 10 μL of RPMI/10 % FBS containing 2 % DMSO in columns 1, 2, 23, and 24 columns. In this plate format, columns 1 and 2 are used as CC, columns 3–22 contain the serial dilutions of the test compounds, and columns 23 and 24 serve as MC.

#### 3.3.2 Antiviral Testing

1. For activity testing, prepare 6 mL of MT4-LTR-EGFP cell suspension in RPMI/10 % FBS ($4 \times 10^5$ cells/mL) for every 384-well test plate. In parallel, prepare 12 mL of MT4-CMV-EGFP cell suspension in RPMI/10 % FBS ($2 \times 10^5$ cells/mL) for each toxicity 384-well plate.
2. Thaw the virus stock at 37 °C and prepare a virus dilution in RPMI/10 % FBS to generate an MOI of 0.0025 $TCID_{50}$/cell. Prepare 6 mL for each 384-well activity test plate; e.g., to obtain an MOI of 0.0025 using a virus stock with a titer of $3.33 \times 10^5$ $TCID_{50}$/mL, 3 μL virus stock per mL of virus dilution is required.
3. In each 384-well activity plate, dispense 15 μL of RPMI/10 % FBS into the wells of columns 23 and 24 using a multidrop combi reagent dispenser, and dispense 15 μL virus dilutions into the wells of columns 1–22. In addition, dispense 15 μL of the MT4-LTR-EGFP cell suspension (6,000 cells/well) into all wells of the plate.
4. In each 384-well toxicity plate, dispense 30 μL of RPMI/10 % FBS into the wells of columns 23 and 24 and dispense 30 μL of the MT4-CMV-EGFP cell suspension (6,000 cells/well) into the wells of column 1 until 22, using a multidrop combi reagent dispenser.
5. Place the plates in a humidified incubator at 37 °C and incubate for 3 days.
6. Three days post-infection, measure the EGFP fluorescence using an automatic fluorescence microscope at 488 nm (blue laser).

#### 3.3.3 Data Analysis

1. Calculate the dose response curves and determine the half maximal effective concentration ($EC_{50}$), which represents the concentration of a compound at which the virus replication is inhibited by 50 %, as measured by a 50 % reduction of the EGFP fluorescent intensity compared with the VC. The $EC_{50}$ is calculated using linear interpolation of a dose response curve. For each compound concentration the percentage inhibition is determined. The percent inhibition ($I$) for every test concentration is calculated using the following formula: $I = 100 \times ((S_{VC} - S_T)/(S_{VC} - S_{CC}))$; $S_T$, $S_{CC}$, and $S_{VC}$ are the amount of EGFP signal in the test compounds, CC, and VC, respectively. In parallel, determine also the half maximal cytotoxic concentration ($CC_{50}$), defined as the concentration required to reduce the EGFP fluorescent intensity by 50 %

compared to that of the untreated control wells. Both $EC_{50}$ and $CC_{50}$ for each compound can be calculated using computer software such as GraphPad Prism.

2. Determine the selectivity index (SI), defined as the ratio of the $CC_{50}$ to the $EC_{50}$. We use an $SI \geq 4$ as the definition of "hits".

## 4 Notes

1. The MT4-LTR-EGFP and the MT4-CMV-EGFP cell lines are maintained in RPMI/10 % FBS supplemented with 0.5 mg/mL geneticin for selective pressure. The cell lines are incubated in a humidified incubator at 37 °C. The cells are passaged every 3 or 4 days. For a 3- or 4-day cell culture, $7.5 \times 10^4$ cells/mL or $3.75 \times 10^4$ cells/mL are seeded in a T175 cell culture flask, respectively. The cells are cultured for maximum 30 passages. Prior to an experiment, all cell lines are cultured in medium without the geneticin selection agent.
2. To avoid carryover of virus, transfer the virus suspension from the highest concentration to the first dilution with minimal touching of the medium present in the well. Discard the tip and use a new tip to mix medium and virus before transferring the suspension to the next dilution. Use this procedure for every dilution step.
3. The final DMSO concentration in each well cannot exceed 0.5 % because of the toxicity to the cells. The volume of the compounds in a plate is 10 μL/well and is fourfold diluted with the cell suspension (and virus suspension) in a total volume of 40 μL. Therefore, the DMSO concentration in the compounds' dilutions is 2 % and should be the same in all wells. The cell and virus control wells should contain the same DMSO concentration as the compound wells.

### References

1. Pomerantz RJ, Horn DL (2003) Twenty years of therapy for HIV-1 infection. Nat Med 9:867–873
2. Westby M, Nakayama GR, Butler SL et al (2005) Cell-based and biochemical screening approaches for the discovery of novel HIV-1 inhibitors. Antiviral Res 67:121–140
3. Pauwels R, Balzarini J, Baba M et al (1988) Rapid and automated tetrazolium-based colorimetric assay for the detection of anti-HIV compounds. J Virol Methods 20:309–321
4. Pannecouque C, Daelemans D, De Clercq E (2008) Tetrazolium-based colorimetric assay for the detection of HIV replication inhibitors: revisited 20 years later. Nat Protoc 3: 427–434
5. Pauwels R, Andries K, Debyser Z et al (1992) Potent and highly selective human immunodeficiency virus type 1 (HIV-1) inhibition by a series of alfa-anilinophenylacetamide derivatives targeted at HIV-1 reverse transcriptase. Proc Natl Acad Sci USA 90:1711–1715
6. Naylor LH (1999) Reporter gene technology: the future looks bright. Biochem Pharmacol 58:749–757
7. Dirk Jochmans D, Deval J, Kesteleyn B et al (2006) Indolopyridones inhibit human immunodeficiency virus reverse transcriptase with a novel mechanism of action. J Virol 80:12283–12292
8. Reed LJ, Muench H (1938) A simple method of estimating fifty percent endpoints. Am J Hyg 27:493–497

# Chapter 2

# A Homogeneous Time-Resolved Fluorescence Assay to Identify Inhibitors of HIV-1 Fusion

Liesbet Smeulders, Lieve Bunkens, Inge Vereycken, Koen Van Acker, Pascale Holemans, Emmanuel Gustin, Marnix Van Loock, and Géry Dams

## Abstract

The human immunodeficiency virus type 1 (HIV-1) initiates infection through sequential interactions with CD4 and chemokine coreceptors unmasking the gp41 subunit of the viral envelope protein. Consequently, the N-terminal heptad repeats of gp41 form a trimeric coiled-coil groove in which the C-terminal heptad repeats collapse, generating a stable six-helix bundle. This brings the viral and cell membrane in close proximity enabling fusion and the release of viral genome in the cytosol of the host cell.

In this chapter, we describe a homogeneous time-resolved fluorescence assay to identify inhibitors of HIV-1 fusion, based on the ability of soluble peptides, derived from the N- and C-terminal domains of gp41, to form a stable six-helix bundle in vitro. Labeling of the peptides with allophycocyanin and the lanthanide europium results in a Föster resonance energy transfer (FRET) signal upon formation of the six-helix bundle. Compounds interfering with the six-helix bundle formation inhibit the HIV-1 fusion process and suppress the FRET signal.

**Key words** HIV-1, Gp41, Fusion inhibitor, Screening, FRET, HTRF

## 1 Introduction

HIV-1 capsid is enveloped with a lipid bilayer spiked with trimeric gp120–gp41 glycoprotein complexes [1, 2]. These spikes play a pivotal role in the HIV-1 entry process and are anchored in the viral membrane via the gp41 transmembrane protein to which the exterior gp120 subunits are non-covalently attached. The HIV-1 entry process is initiated upon binding of gp120 to the cellular CD4 receptor inducing conformational changes which in turn enables binding of gp120 to chemokine co-receptors. This latter interaction facilitates insertion of the N-terminal fusion domain of gp41 into the cell membrane. During this process, the N-terminal heptad repeat domains (HRN) form a trimeric coiled-coil structure. Next, the gp41 C-terminal heptad repeat domains (HRC) collapse into these hydrophobic grooves resulting in the formation

Edwin Yunhao Gong (ed.), *Antiviral Methods and Protocols*, Methods in Molecular Biology, vol. 1030,
DOI 10.1007/978-1-62703-484-5_2, © Springer Science+Business Media, LLC 2013

of a stable six-helix bundle, a trimer of heterodimers. Finally, this brings viral and cell membranes into close proximity promoting fusion and delivery of the viral capsid to the cytoplasm. Hence, interfering with six-helix bundle formation prevents the virus from entering the cell, precluding HIV-1 infection.

HRC-derived peptides are potent inhibitors of six-helix bundle formation with reported antiviral activities at nanomolar concentrations, e.g., T-20 [3], T-1249 [4], and C-34 [5, 6]. More importantly, the six-helix bundle is a validated target for HIV-1 antiviral drug development, as T-20 (enfuvirtide) was approved for clinical use [7]. Yet, a search for a small molecule fusion inhibitor is still justified as the T-20 peptide can be administered only by injection, is prone to proteolytic cleavage, and resistance does occur [8].

HRC-derived peptides not only are potent inhibitors but also enable the in vitro formation of six-helix bundles, when mixed together with the HRN-derived peptides. Therefore, combining synthetic peptides overlapping gp41 residues 546–581 (N36) and 628–655 (C34) offers an attractive approach to develop biochemical assays which enable identification of peptidomimetic or small molecule inhibitors of six-helix bundle formation [9–12].

Here, we describe a homogeneous time-resolved fluorescence (HTRF) assay to measure the inhibition of six-helix bundle formation, based on N36–C34 interactions (Fig. 1). To address aggregation issues of N36 in the absence of C34, the N-terminal part of N36 is fused to a soluble alpha-helical peptide derived from GCN4, denoted IQ [10, 13]. The resulting fusion peptide, IQN36, is labeled with the light emitting fluorophore allophycocyanin (APC) and the C34 peptide is labeled with the UV-excitable fluorophore europium (Eu). Formation of the six-helix bundle brings both fluorophores in close proximity enabling FRET between the UV-excitated Eu donor and APC acceptor, resulting in emission of red light. Since this technique allows detection of changes in the nanometer range, compounds that interfere with the binding of C34 to IQN36 will quantitatively suppress the UV-induced signal.

## 2 Materials

### 2.1 Peptides and Reagents

1. Biotin-labeled IQN36 (Table 1): IQN36 is synthesized using standard protocols (Abgent, San Diego) and biotinylated at the N-terminus of the IQ moiety (PerkinElmer). The stock solution is prepared in DMSO at a concentration of 2 mM and stored at 4 °C.
2. Europium labeled C34 (C34-Eu) (Table 1): C34 is synthesized using standard protocols (Abgent, San Diego) and Eu-labeled at its C-terminus (PerkinElmer). Spin down briefly the tube before reconstituting C34-Eu in HEPES buffer to a concentration of 100 μM. Mix by vortexing and store aliquots at −20 °C.

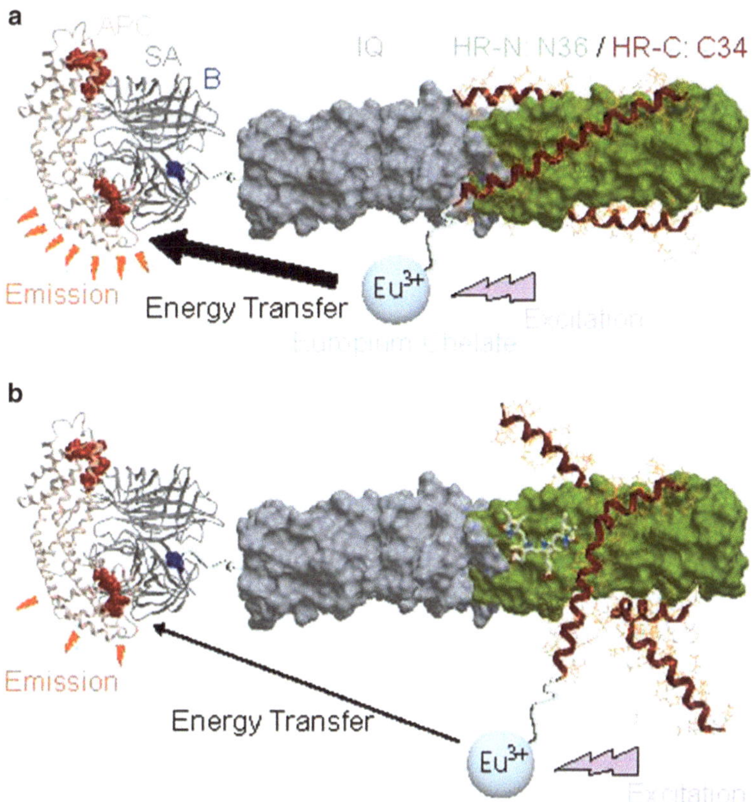


**Fig. 1** (**a**) Structure of the Eu/allophycocyanin (APC)-based assay. C34 and N36 are derived from the gp41 heptad-repeats, respectively, the C- and N-terminal. A chelate of the UV-excitable fluorescent donor $Eu^{3+}$ is bound to C34 via a short linker. N36 is fused with an IQ moiety for solubility reasons and labeled with the light-emitting acceptor APC via a biotin-streptavidin linker. IQN36-APC moieties trimerize to give rise to a coiled-coil structure. When C34 binds the hydrophobic grooves of the IQN36 trimer, both fluorophores come in close proximity, resulting in FRET and the emission of APC-fluorescence. *SA* streptavidin, *B* biotin. (**b**) Mechanism of FRET inhibition. Compounds that interfere with the binding of C34 to IQN36 will quantitatively suppress the UV-induced signal [14]. Reproduced from Dams et al. (2007) with permission from SAGE Publications

**Table 1**
**Amino acid sequence of IQN36 and C34**

| Peptide | Amino acid sequence |
|---|---|
| IQN36 | RMKQIEDKIEEIESKQKKIENEIARIKKLISGIVQQQNNLLRAIEAQQHLLQ LTVWGIKQLQARIL |
| C34 | WMEWDREINNYTSLIHSLIEESQNQQEKNEQELL |

3. Streptavidin conjugated to SureLight-Allophycocyanin (streptavidin-APC, PerkinElmer): reconstitute to 1 mL with deionized $H_2O$ and allow the vial to sit on ice for 20–30 min. Mix gently by vortexing or tapping the tube before use. The stock solution is stored at 4 °C and the concentration is batch dependent.
4. 100 mM HEPES buffer (pH 7.2): Weigh 23.83 g of HEPES and add PBS to a final volume of 1 L. Adjust the pH to 7.2 and sterilize the solution through a 0.22 μm filtration; store at 4 °C.

### 2.2 Consumables

1. White 384-well non-tissue culture treated plates (Costar, Corning).
2. Costar® 50 mL reagent reservoir (Costar, Corning).
3. Corning® 28 mm diameter syringe filter, 0.2 μm (Corning).

### 2.3 Equipment

1. Multidrop combi liquid dispenser with dispensing cassettes (Thermo Scientific).
2. Viewlux™ (PerkinElmer) or other fluorescent readers.

## 3 Methods

Carry out all procedures at room temperature unless otherwise specified.

### 3.1 Preparation of 384-Well Test Plates

Perform fourfold serial dilutions to prepare 4× final concentrations of test compounds in PBS and add 10 μL of each test concentration to each of the quadruple wells of a 384-well plate (final concentration of DMSO is 0.5 %, *see* **Note 1**). Typically, we test eight compounds per plate with nine concentrations of each compound. In this plate format, columns 1–16 contain the serial dilutions of the test compounds, columns 17–20 are used as positive control (no compounds) and columns 21–24 as negative control (no APC-labeled IQN36). In our laboratory, stock solutions of test compounds are made in DMSO and dispensed in 384-well plates using a 96-channel liquid handler.

### 3.2 Preparation of Working Solutions

All working solutions need to be freshly prepared and are diluted in 100 mM HEPES buffer (pH 7.2).

1. Prepare a 40 nM working solution of streptavidin-APC (6 mL per 384-well plate).
2. Prepare a 400 nM working solution of IQN36-biotin (5 mL per 384-well plate).
3. Prepare a 500 nM working solution of C34-Eu (5 mL per 384-well plate).

### 3.3 HTRF Assay

1. For each 384-well plate, prepare 1 mL of negative control mix by adding equal volumes of streptavidin-APC working solution and 100 mM HEPES buffer. Incubate the mix for 30 min at room temperature.
2. For each 384-well plate, prepare 10 mL of test mix by adding equal volumes of streptavidin-APC and IQN36-biotin working solutions. Incubate the mix for 30 min at room temperature.
3. Dispense 20 μL/well of the negative control mix to columns 21–24 (negative control wells) using a multidrop combi liquid dispenser (final concentration of streptavidin-APC is 10 nM).
4. Dispense 20 μL/well of the test mix to columns 1–20 (test wells and positive control wells; the final concentration of streptavidin-APC is 10 nM and the final concentration of IQN36-biotin is 100 nM) (*see* **Note 2**).
5. Incubate the plates at room temperature for 2 h (*see* **Note 3**).
6. Add 10 μL working solution of Eu-labeled C34 peptide to all wells (final concentration of C34-Eu is 125 nM) (*see* **Note 2**).
7. Incubate the plates at room temperature for 30 min.
8. Measure the FRET signal using a Viewlux™ reader with the following instrument settings: a standard filter set for the LANCE® assay (PerkinElmer) (a DUG11 (330/75) excitation filter, a 400LP dichroic mirror, a 618/8 emission filter for the fluorescence of the $Eu^{3+}$ donor, and a 671/8 emission filter for the fluorescence of the APC acceptor); a chopper wheel system provides a 50 μs delay after the excitation flash, to suppress the fast-decaying signals due to background fluorescence and direct excitation of the acceptor, and is followed by a 354 μs signal integration period [14].
9. Calculate the 50 % inhibitory concentration ($IC_{50}$) of the test compounds (*see* **Note 4**).

## 4 Notes

1. Although DMSO concentrations up to 5 % do not have a negative influence on the results, a final DMSO concentration of 0.5 % is typically applied.
2. Optimal concentrations of labeled peptides are selected from a three-dimensional analysis encompassing a concentration range of (a) streptavidin-APC, (b) IQN36-biotin, and (c) C34-Eu to generate robust dose–response curves, a high signal-to-background ratio, a good sensitivity, an acceptable $Z'$, and an economical use of C34-Eu [14]. The $Z'$-factor is a simple statistical parameter to evaluate the quality of high-throughput screening assays [15].

(a) The optimal concentration of streptavidin-APC is 10 nM as this yields the best signal-to-background ratio. Higher concentrations do not improve the signal-to-noise ratio, whereas concentrations below 1 nM compromise the acceptor signal.

(b) To optimize the IQN36-biotin concentrations, the trimerization level as well as the solubility of the peptide is considered. Under the experimental conditions of this assay, 100 nM yields the best results.

(c) The cost of the C34-Eu peptide is a point to consider when screening large numbers of compounds to generate millions of data points. Concentrations of 250 nM or more yield high signal-to-background ratios and a Z′ factor of 0.9 [14]. However, a C34-Eu concentration of 125 nM still results in a $Z'$ factor of 0.8, which is still acceptable for a biochemical high-throughput screening assay. In addition, 50 % reduction of C34-Eu usage positively influences the cost/benefit ratio of the assay.

3. Prior to the addition of C34-Eu, test compounds are preincubated with IQN36-APC for 2–4 h, allowing the inhibitor to bind the IQN36 coiled-coil without having to compete with C34-Eu after the stable six-helix bundle is formed.

4. The $IC_{50}$ is defined as the concentration of compound achieving 50 % inhibition of the FRET signal in compound-treated wells compared to the positive control. First, the percentage inhibition is determined for every compound concentration. The percent inhibition (I) is calculated using the following formula: $I = 100 \times (S_{PC} - S_T)/(S_{PC} - S_{NC})$; $S_T$, $S_{NC}$, and $S_{PC}$ are the amount of fluorescent signal in the test compounds, NC (negative control) and PC (positive control), respectively. The $IC_{50}$ is calculated using linear interpolation of a dose response curve. For each compound concentration, the median inhibition of the quadruple replicates is calculated. If the medium inhibition values of all tested concentration is higher or lower than 50 %, the lowest concentration is reported with an "<" sensor, or the highest concentration is reported with a ">" sensor, respectively. If the median inhibition value of a tested concentration is exactly 50 % this value is reported. In all other cases, the $IC_{50}$ value is calculated following the formula:

$$IC_{50} = \exp\left(\ln(C_h) - \ln(C_h / C_l) \times (I_h - 0.5) / (I_h - I_l)\right)$$

$C_h$ and $C_l$ are the compounds concentration of which the inhibition values ($I_h$ and $I_l$) are below and above 50 % inhibition, respectively. The formula uses a linear interpolation in the log-concentration domain. Calculations can be performed using computer software such as GraphPad Prism. All dose–response curves of active compounds are reviewed by the operator.

## Acknowledgement

Liesbet Smeulders and Lieve Bunkens contributed equally to the work.

### References

1. Eckert DM, Kim PS (2001) Mechanisms of viral membrane fusion and its inhibition. Annu Rev Biochem 70:777–810
2. Gallo SA, Finnegan CM, Viard M et al (2003) The HIV Env-mediated fusion reaction. Biochim Biophys Acta 1614:36–50
3. Wild CT, Shugars DC, Greenwell TK et al (1994) Peptides corresponding to a predictive alpha-helical domain of human immunodeficiency virus type 1 gp41 are potent inhibitors of virus infection. Proc Natl Acad Sci U S A 91:9770–9774
4. Greenberg ML, Davison D, Jin L et al (2002) In vitro antiviral activity of T-1249 a second generation fusion inhibitor. Antivir Ther 7: S10
5. Jiang S, Lin K, Strick N et al (1993) HIV-1 inhibition by a peptide. Nature 365:113
6. Weissenhorn W, Dessen A, Harrison SC et al (1997) Atomic structure of the ectodomain from HIV-1 gp41. Nature 387:426–430
7. Kilby JM, Hopkins S, Venetta TM et al (1998) Potent suppression of HIV-1 replication in humans by T-20, a peptide inhibitor of gp41-mediated virus entry. Nat Med 4:1302–1307
8. Joly V, Jidar K, Tatay M et al (2010) Enfuvirtide: from basic investigations to current clinical use. Expert Opin Pharmacother 11: 2701–2713
9. Chan DC, Fass D, Berger JM et al (1997) Core structure of gp41 from the HIV envelope glycoprotein. Cell 89:263–273
10. Lu M, Blacklow SC, Kim PS (1995) A trimeric structural domain of the HIV-1 transmembrane glycoprotein. Nat Struct Biol 2:1075–1082
11. Root MJ, Steger HK (2004) HIV-1 gp41 as a target for viral entry inhibition. Curr Pharm Des 10:1805–1825
12. Liu S, Jiang S (2004) High throughput screening and characterization of HIV-1 entry inhibitors targeting gp41: theories and techniques. Curr Pharm Des 10:1827–1843
13. Eckert DM, Malashkevich VN, Kim PS (1998) Crystal structure of GCN4-pIQI, a trimeric coiled coil with buried polar residues. J Mol Biol 284:859–865
14. Dams G, Van Acker K, Gustin E et al (2007) A time-resolved fluorescence assay to identify small-molecule inhibitors of HIV-1 fusion. J Biomol Screen 12:865–874
15. Zhang JH, Chung TD, Oldenburg KR (1999) A simple statistical parameter for use in evaluation and validation of high throughput screening assays. J Biomol Screen 4:67–73

# Chapter 3

# Identification of HIV-1 Reverse Transcriptase Inhibitors Using a Scintillation Proximity Assay

**Bertrand Van Schoubroeck, Marnix Van Loock, Tania Ivens, Pascale Dehertogh, Dirk Jochmans, and Géry Dams**


## Abstract

The human immunodeficiency virus type 1 (HIV-1) reverse transcriptase (RT) converts the viral single-stranded RNA into double-stranded DNA. The inhibition of reverse transcription in the viral life cycle has proven its efficacy as a clinically relevant antiviral target, but the appearance of resistance mutations remains a major cause of treatment failure and stresses the continuous need for new antiviral compounds. In this chapter, we describe an HIV-1 RT scintillation proximity assay (SPA) to identify inhibitors of the RT. The assay uses an RNA/DNA (poly(rA)/oligo(dT)) template/primer bound to SPA beads, which contain scintillant. Reverse transcriptase extends the primer by incorporating [$^3$H]dTTP and dTTP, which results in light production by the scintillant in the bead. Compounds that inhibit reverse transcriptase will prevent the incorporation of tritiated dTTP resulting in a reduction of emitted light compared to the untreated controls.


**Key words** HIV-1, Reverse Transcriptase, SPA technology, Antiviral inhibitors

## 1 Introduction

Upon entry into the host cell, the HIV-1 virus is uncoated and the viral RT converts the single-stranded RNA genome into double-stranded DNA (dsDNA). Within the context of the pre-integration complex, the dsDNA is translocated to the nucleus and integrates in the genome of the host cell [1]. The different steps of reverse transcription are well characterized and have been reviewed elsewhere [2]. The RT incorporates deoxyribonucleotide triphosphates (dNTPs) at the 3′ end of the primer, using the viral RNA as a template. The inhibition of this step in the viral life cycle played a pivotal role in the establishment of highly active antiretroviral therapy as RT inhibitors are included in the current standard of care [3]. Both nucleoside RT inhibitors (NRTIs) and non-nucleoside RT inhibitors (NNRTIs) have proven their clinical efficacy. NRTIs are phosphorylated in the host cell and the NRTI-triphosphate competes with the natural dNTPs for binding in the

Edwin Yunhao Gong (ed.), *Antiviral Methods and Protocols*, Methods in Molecular Biology, vol. 1030,
DOI 10.1007/978-1-62703-484-5_3, © Springer Science+Business Media, LLC 2013

active site. These compounds are chain terminators because they lack the 3′-OH group [4]. In contrast, NNRTIs are another chemical class of small molecules that bind to a pocket close to the active site of RT [5, 6]. The appearance of resistance mutations to these compounds stresses the continuous need for new compounds with another mechanism of action [7]. Recently, a novel type of RT inhibitor, called Indopy-1, has been described [8]. This class of compounds also targets the RT by binding to the active site of the enzyme; but unlike NRTIs, they do not require to be phosphorylated to block the DNA synthesis, although they share the same binding site [8].

SPA is a homogeneous technology based on the fact that in an aqueous environment weak β-emitters, like [$^3$H], need to be in close proximity of scintillant molecules before they can transfer their energy to the scintillant [9, 10]. This energy transfer results in the emission of light, which can be measured using a photomultiplier. The chemical incorporation of the scintillant into beads eliminated the requirement of separating bound from unbound radiolabeled molecules, resulting in a homogeneous (mix-and-read) assay [11]. The combination of these beads with the [$^3$H] has proven ideal because tritium has a very short path length of 1.5 μm, which requires that the tritium-labeled molecules interact with the coated beads in order to be in the vicinity of the scintillant to emit lights. Therefore, the energy of unbound [$^3$H] is dissipated in the solution, generating only low levels of background signal [10].

In this chapter, we describe an RT assay using SPA (Fig. 1) that makes use of a 5′-biotinylated 16 mer oligo(T) DNA primer annealed to a 300-base poly(rA) template. This RNA/DNA (poly(rA)/oligo(dT)) template/primer is bound to the SPA beads via a streptavidin/biotin linkage. In the presence of RT, [$^3$H] dTTP and dTTP are incorporated through primer extension. In the presence of RT inhibitors the incorporation of [$^3$H]dTTP and dTTP is prevented, resulting in a reduced signal compared with untreated controls.

## 2 Materials

### 2.1 Reagents

1. RPMI/10 % FBS: RPMI-1640 supplemented with 10 % heat-inactivated fetal bovine serum (FBS), 1 % L-glutamine and 0.02 mg/mL gentamicin (Gibco).
2. HIV-1 reverse transcriptase (GE Healthcare, 200 units/vial).
3. Quan-T-RT™ [$^3$H]-Reverse Transcriptase kit (GE Healthcare) (*see* **Note 1**).
4. Efavirenz, a marketed NNRTI [12, 13] as a reference compound.

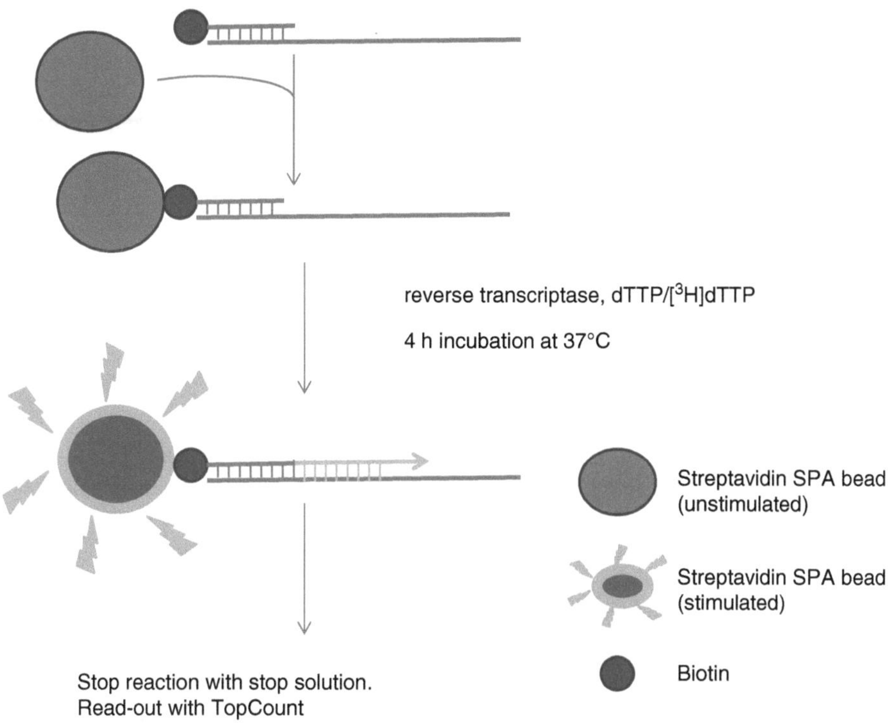


**Fig. 1** Principle of the HIV-1 RT assay using SPA. See text for details

### 2.2 Consumables

1. F96 MicroWell™ MaxiSorp™ plate (Nunc).
2. TopSeal-A 384, Clear Self-Adhesive Topseal for 384-well Microplates (PerkinElmer).
3. Corning® 96-Well Clear V-Bottom TC-Treated Microplates (Corning).

### 2.3 Equipment

1. Matrix Impact 250 12-channel pipettes (Thermo Scientific).
2. A humidified incubator at 37 °C.
3. TopCount 12 detector, 96 and 384 format (PerkinElmer).

## 3 Methods

### 3.1 Determination of the Optimal RT Concentration

For every new batch of RT, the optimal enzymatic concentration needs to be determined due to the batch-to-batch variations.

1. Prepare a twofold serial dilution of RT in RPMI/10 % FBS in a 96-well V-bottom plate, starting with a 1/150 dilution in well A1.
2. Prepare an assay mix (*see* Table 1) without enzyme and add 20 μL/well to row A of a F96 MicroWell™ MaxiSorp™ plate, which already contains 25 μL RPMI/10 % FBS supplemented with 2 % DMSO per well. Wells 1–9 contain the serial dilutions of the enzyme, and wells 10–12 serve as negative control (no enzyme).

**Table 1**
**Composition of the assay mix**

| Component | Volume/reaction (μL) |
|---|---|
| Primer/template | 5 |
| Buffer (5.25×) | 10 |
| TTP/[$^3$H]TTP (500 μCi/mL) | 0.5 |
| RPMI/10 % FBS | 4.5 |
| RT solution | 5 |

3. Add 5 μL of the RT serial dilution to wells 1–9 and add 5 μL of RPMI/10 % FBS to wells 10–12. This results in a final volume of 50 μL/well.
4. Incubate the plates for 4 h at 37 °C (*see* **Note 2**).
5. Add 100 μL of stop solution (provided in the kit) to each well of row A.
6. Seal the plates with Topseal-A 384 (*see* **Note 3**).
7. Measure the plates using the TopCount: 2 min delay time, exposure time = 1 min/well.
8. Determine the optimal RT concentration based on a concentration that gives 2,000–4,000 counts per minute (CPM) (*see* **Note 4**).

### 3.2 Scintillation Proximity Assay

1. Prepare fourfold serial dilutions of test compounds in RPMI/10 % FBS at 2× final concentrations and transfer 25 μL of each test concentration to a 96-well MaxiSorp™ plate (2 % DMSO) (*see* **Notes 5** and **6**). All compounds are tested in duplicate. Typically, we test eight compounds per plate with nine concentrations of each compound. In this plate format, columns 1–9 are assigned to the test compounds. On every test plate, columns 10 and 11 are the positive control wells (no compound present) and column 12 is the negative control wells (add 100 μL stop solution to the wells immediately after addition of the assay mix) (*see* **step 5**). Columns 10–12 contain 25 μL RPMI/10 % FBS supplemented with 2 % DMSO. Depending on the number of compounds the plate format can be changed, for example, four compounds per plate with each compound dilution in duplicate. We use Efavirenz as a positive control, which is a known NNRTI.
2. Based on the optimal RT concentration determined under Subheading 3.1, prepare a RT solution at 10× the final enzyme concentration in the assay by diluting the RT stock solution provided by the supplier. Typically, we dilute the RT stock solution 10,000 times in RPMI/10 % FBS.

3. Prepare the assay mix as described in Table 1 and keep on ice.
4. Add 25 μL of assay mix to all wells of the compound plate. The final volume in each well is 50 μL with a final DMSO concentration of 1 %.
5. Add 100 μL of stop solution to the negative control wells (column 12) to stop the reaction in these wells.
6. Incubate the plates at 37 °C for 4 h (*see* **Note 2**).
7. Add 100 μL of stop solution to each well of columns 1–11.
8. Seal the plates with Topseal-A 384 (*see* **Note 3**).
9. Read the plates using the Topcount: 2 min delay time, exposure time = 1 min/well.
10. Calculate the 50 % inhibitory concentration ($IC_{50}$) of test compounds. $IC_{50}$ is defined as the compound concentration that inhibits 50 % of the HIV-1 RT activity (reduction of CPM) compared with untreated (= without compound) controls. First, the % inhibition is calculated for every concentration using the following formula: % inhibition = $((C^+ - X)/(C^+ - C^-)) \times 100$ in which $X$ is the CPM measured at each compound concentration. $C^+$ is the median CPM of the positive control wells and $C^-$ is the median CPM of the negative control wells (background). For each compound concentration, the average inhibition of the duplicates is calculated. Next, calculate the ($IC_{50}$) for each compound using a linear interpolation of a dose–response curve using computer software such as GraphPad Prism.

## 4 Notes

1. Use the radioactive materials according to the local rules of good laboratory practice.
2. When dealing with edge effect, put the plates in a petri dish with a wet tissue or add an extra water reservoir.
3. Clean the top and bottom of each plate with a wet tissue to avoid electrostatic charges that can interfere with the measurement.
4. In our experience, a robust and reproducible biochemical assay was obtained when an RT concentration gives 2,000–4,000 CPM measured by a TopCount. Typically, we can dilute the RT stocks 10,000 times, but batch-to-batch variations need to be taken into account.
5. For longer storage, place the compound plates at −20 °C to avoid edge effects.
6. When nucleotide inhibitors are tested in this biochemical assay, triphosphorylated nucleotides should be used.

## References

1. Freed EO (2001) HIV-1 replication. Somat Cell Mol Genet 26:13–33
2. Jonckheere H, Anne J, De Clercq E (2000) The HIV-1 reverse transcription (RT) process as target for RT inhibitors. Med Res Rev 20:129–154
3. Sterne JAC, Hernán MA, Ledergerber B et al (2005) Long-term effectiveness of potent antiretroviral therapy in preventing AIDS and death: a prospective cohort study. Lancet 366:378–384
4. St Clair MH, Richards CA, Spector T et al (1987) 3′-Azido-3′-deoxythymidine triphosphate as an inhibitor and substrate of purified human immunodeficiency virus reverse transcriptase. Antimicrob Agents Chemother 31:1972–1977
5. Rittinger K, Divita G, Goody RS (1995) Human immunodeficiency virus reverse transcriptase substrate-induced conformational changes and the mechanism of inhibition by nonnucleoside inhibitors. Proc Natl Acad Sci U S A 92:8046–8049
6. Smerdon SJ, Jager J, Wang J et al (1994) Structure of the binding site for nonnucleoside inhibitors of the reverse transcriptase of human immunodeficiency virus type 1. Proc Natl Acad Sci U S A 91:3911–3915
7. Shafer RW, Schapiro JM (2008) HIV-1 drug resistance mutations: an updated framework for the second decade of HAART. AIDS Rev 10:67–84
8. Jochmans D, Deval J, Kesteleyn B et al (2006) Indolopyridones inhibit human immunodeficiency virus reverse transcriptase with a novel mechanism of action. J Virol 80: 12283–12292
9. Bosworth N, Towers P (1989) Scintillation proximity assay. Nature 341:167–168
10. Cook N, Harris A, Hopkins A, Hughes K (2002) Scintillation proximity assay (SPA) technology to study biomolecular interactions. In: Coligan JE et al (eds) Current protocols in protein science. Wiley, New York, pp 19.8.1–19.8.35
11. Berry J, Price-Jones M (2005) Measurement of radioligand binding by scintillation proximity assay. Methods Mol Biol 306:121–137
12. Balani SK, Kauffman LR, deLuna FA et al (1999) Nonlinear pharmacokinetics of efavirenz (DMP-266), a potent HIV-1 reverse transcriptase inhibitor, in rats and monkeys. Drug Metab Dispos 27:41–45
13. Young SD, Britcher SF, Tran LO et al (1995) L-743, 726 (DMP-266): a novel, highly potent nonnucleoside inhibitor of the human immunodeficiency virus type 1 reverse transcriptase. Antimicrob Agents Chemother 39:2602–2605

# Chapter 4

# Biochemical Screening Assays to Identify HIV-1 Integrase Inhibitors

**Marleen Clynhens, Alexandra Smets, Inge Vereycken, Marnix Van Loock, Reginald Clayton, Geert Meersseman, and Olivia Goethals**

## Abstract

Human immunodeficiency virus type 1 (HIV-1) integrase is, in addition to reverse transcriptase and protease, an important enzymatic target for antiretroviral drug development. Integrase plays a critical role in the HIV-1 life cycle coordinating the integration of the reverse-transcribed viral DNA into the host genome. This integration step is the net result of two consecutive integrase-related processes. First, integrase removes a dinucleotide from the 3′ viral DNA ends in a process called 3′-processing. Next, in a process called strand transfer, the viral DNA is integrated into the host genomic DNA. Early on, biochemical assays have played a critical role in understanding the function of HIV-1 integrase and the discovery of small-molecule inhibitors. In this chapter we describe two biochemical assays to identify inhibitors of the 3′-processing and strand transfer process of HIV-1 integrase.

**Key words** HIV-1, Integrase, Strand transfer, 3′-Processing, Real-time PCR

## 1 Introduction

The viral enzyme integrase (IN) is essential for the HIV replication cycle and is encoded by the *pol* gene. The full-length IN comprises 288 amino acids and consists of three domains. The N-terminal domain of the enzyme coordinates a zinc atom, promotes multimerization, and is essential for enzymatic activity [1]. The catalytic core domain (CCD) binds viral and chromosomal DNA, contains a catalytic triad (D,D-35-E motif), which coordinates two magnesium atoms in the active site, and is involved in IN multimerization [2]. The C-terminal domain binds the viral DNA proximal to the long terminal repeat (LTR) and is also involved in IN multimerization [3].

During the integration process, IN recognizes the LTR sequences and binds to both ends of the viral DNA. Within the cytoplasm of the host cells, the integrase removes a GT dinucleotide,

Edwin Yunhao Gong (ed.), *Antiviral Methods and Protocols*, Methods in Molecular Biology, vol. 1030,
DOI 10.1007/978-1-62703-484-5_4, © Springer Science+Business Media, LLC 2013

next to a conserved 3′CA sequence, from each viral cDNA 3′end in a process called 3′-processing.

The 3′-processed DNA is then transported to the nucleus, in the context of the pre-integration complex. Subsequently, IN initiates a strand transfer process by transferring the processed "donor" DNA into an "acceptor" strand of DNA, which in the context of a cellular infection represents the host cell chromosomes. Following strand transfer, host DNA repair enzymes repair the remaining gaps in the DNA to complete the integration process ensuring a stable association of viral genome with the host cell [4–6]. Both the 3′-processing and the strand transfer reaction can be recapitulated in vitro and form the basis of screening and profiling assays for identification of enzymatic inhibitors where double-stranded DNA is used as "acceptor" DNA.

Early on, biochemical assays facilitated the unraveling of the IN function [7, 8]. These assays used radioactively labeled LTR sequences to study IN functionality and 3′-processed or integrated products were identified using electrophoresis. Although these assays were very informative they were not amenable for high-throughput screening. The introduction of novel biochemical techniques resulted in an array of novel assays based on, e.g., surface plasmon resonance, anisotropy, scintillation proximity, and fluorescent/quenching assays [9–12]. These reductionistic but target-specific biochemical assays resulted in the discovery of small-molecule integrase strand transfer inhibitors [13–17]. In addition, these assays have also been used to delineate small molecules and determine the structure–activity relationship [18].

To investigate the 3′-processing activity of IN, an assay was developed based on the removal of the GT dinucleotide and disruption of a quenching probe (Q-probe) dependent on its proximity to a fluorophore (Fig. 1). Two oligonucleotides are hybridized to form a 20 bp (base pair) U5 LTR of human immunodeficiency virus type 1 (HIV-1). The sense strand oligo, containing the removable 3′GT dinucleotide, is labeled at the 3′end with an Alexa Fluor® 488 fluorophore. The antisense strand oligo is labeled at the 5′end with a quencher, dabcyl. Hybridization of both strands results in quenching of the signal produced by the fluorophore. In the presence of IN and the cofactor lens epithelium-derived growth factor (LEDGF/p75), the 3′-processing occurs where the alexa-labeled GT dinucleotide is removed from the Q-probe, resulting in an increase of fluorescent signal (Fig. 1). Inhibition of the 3′-processing by a compound results in reduction of the fluorescent signal in comparison with the positive control.

The Strand Transfer TaqMan (STT) assay is a biochemical assay combined with real-time PCR (rtPCR) using a TaqMan probe to identify compounds that inhibit the strand transfer

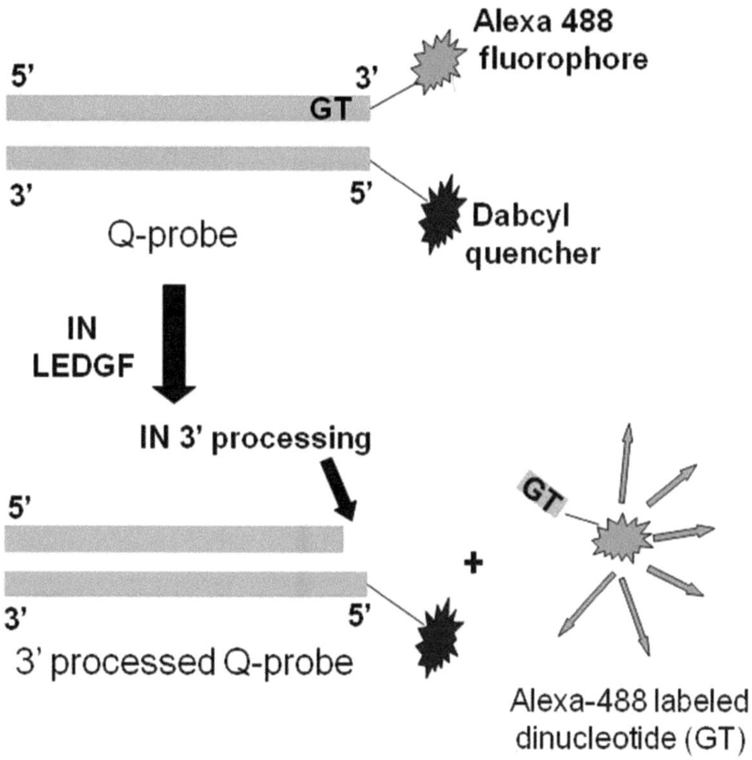


**Fig. 1** Principle of the 3′-processing assay. Two oligonucleotides are hybridized in a 1.5 ratio (dabcyl/alexa) to form the 20 bp U5 LTR of HIV-1. The sense strand oligo contains the removable 3′GT dinucleotide and is labeled with the Alexa Fluor® 488 fluorophore. The antisense strand oligo is 5′ labeled with dabcyl as a quencher. In the presence of IN and the cofactor LEDGF/p75, 3′-processing occurs and the alexa-labeled GT dinucleotide is released from the quencher, resulting in an increase of fluorescent signal

activity of HIV-1 IN. The probe, targeting a 20-nucleotide sequence within the LTR, is labeled with FAM at the 5′end and Black Hole Quencher® (BHQ) at the 3′end. The donor DNA, mimicking the U5 LTR end of the HIV-1 genomic DNA, contains a forward primer sequence and is pre-incubated with IN and LEDGF/p75 to reach steady state of 3′-processing. Subsequent incubation with target DNA containing the reverse primer sequence leads to complete strand transfer. As target DNA we opted for a plasmid carrying an HIV-1 genome to act as a DNA acceptor. A specific amplicon can be generated with the rtPCR when strand transfer has occurred, and therefore compounds that inhibit the strand transfer reaction will generate higher Ct values in comparison to the untreated positive control, enabling the identification of specific strand transfer inhibitors (Fig. 2).

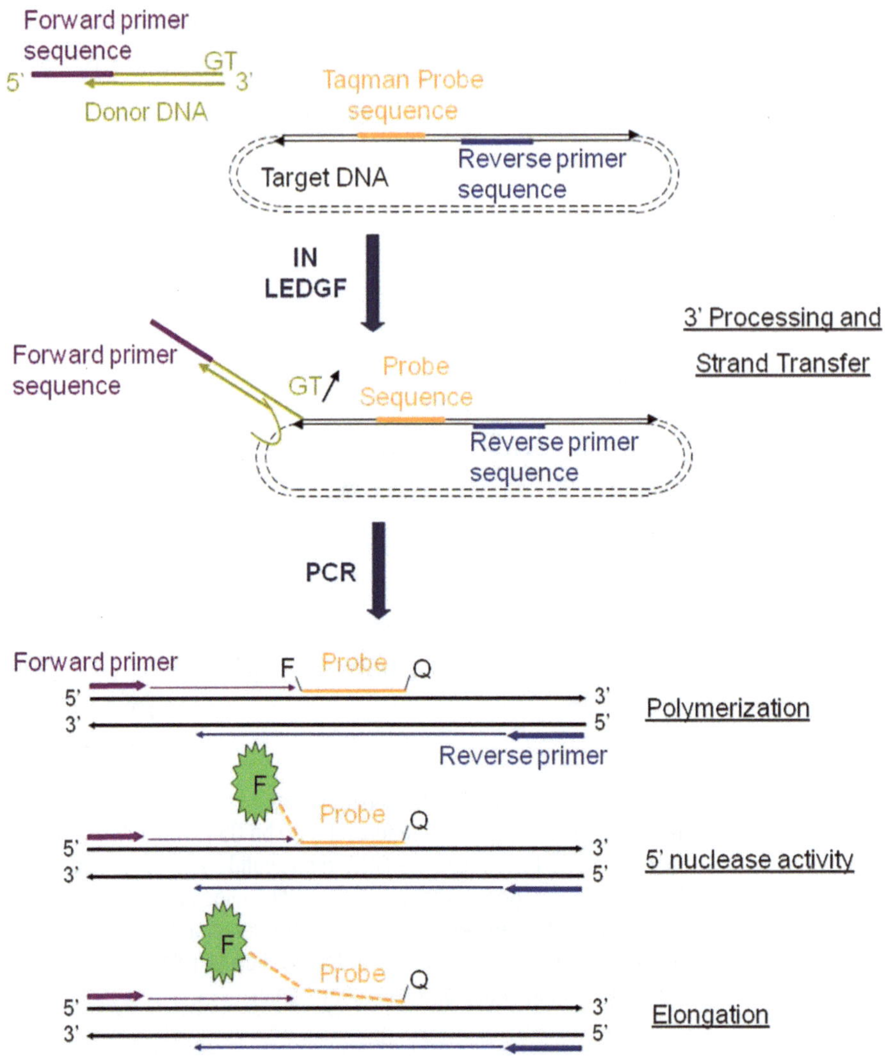


**Fig. 2** Principle of the STT assay. The target DNA is a pNL4.3 plasmid, containing the TaqMan probe sequence and the reverse primer sequence. The probe itself is labeled with FAM at the 5′end and Black Hole Quencher® (BHQ) at the 3′end. The donor DNA, mimicking a U5 LTR end of HIV-1, contains the forward primer sequence and is pre-incubated with IN and LEDGF/p75 to reach steady-state 3′-processing. Subsequently, incubation with the target DNA leads to strand transfer. During rtPCR, the fluorophore is separated from the quencher, which is reflected in a Ct value. Compounds that inhibit the strand transfer reaction show increased Cts compared to the Ct value of the positive control wells

## 2 Materials

### 2.1 Integrase Proteins

1. HIV-1 IN: Briefly, upon induction with isopropylthiogalactopyranoside (IPTG) a $HIS_6$-tagged IN is expressed from plasmid pCPH6P-HIV1-IN [19] in transformed Rosetta™ 2 competent cells (Novagen). Cells are harvested at 6 h post induction and lysed by sonication. Purified protein is obtained

by performing in succession: affinity chromatography, removal of the $HIS_6$-tag, ion exchange chromatography, and gel filtration chromatography. Aliquots of the purified protein are stored at −80 °C in a 50 mM Tris–HCl buffer (pH 7.4) supplemented with 7.5 mM CHAPS, 10 % glycerol, 5 mM dithiothreitol (DTT), and 1 M NaCl.

2. LEDGF/p75: Briefly, upon induction with IPTG, a $HIS_6$-tagged LEDGF/p75 is expressed from plasmid pCP6H75 [20] in transformed Arctic cells according to the manufacturer's instructions (Stratagene). Cells are harvested at 24 h post induction and lysed by sonication. Purified protein is obtained by performing in succession: affinity chromatography, removal of the $HIS_6$-tag, ion exchange chromatography, and gel filtration chromatography. Finally, the purified LEDGF/p75 is concentrated using an Amicon Ultra-15 filter. The purified protein is stored at −80 °C in a 50 mM Tris–HCl buffer (pH 7.4) supplemented with 10 % glycerol, 5 mM DTT, and 120 mM NaCl.

### 2.2 Oligos for the 3′-Processing Assay

1. INss (sense strand): 5′-TGTGGAAAATCTCTAGCAGT-3′-Alexa Fluor® 488 fluorophore.
2. INas (antisense strand): dabcyl-5′-ACTGCTAGAGATTTTCC ACA-3′.

### 2.3 Primers for the STT Assay

1. INdon1: 5′-ACTGCTAGAGATTTTCCACAC-3′ (100 μM stock solution).
2. INdon2: 5′-CCCTCAAGGACGTGTGGAAAATCTCTAGC AGT-3′ (100 μM stock solution).
3. INfwd (forward primer): 5′-CCCTCAAGGACGTGTG-3′ (100 μM stock solution).
4. INrev (reverse primer): 5′-CCCAGGCCACACCTC-3′ (100 μM stock solution).
5. INtm (TaqMan probe): FAM-5′-CTTTCCGCTGGGGACT TTCC-3′-BHQ (100 μM stock solution).

### 2.4 Reagents

1. 1 M 3-(*N*-morpholino) propanesulfonic acid (MOPS): Weigh 209 g MOPS and dissolve in deionized $H_2O$. Adjust the pH to 7.2 in a final volume of 1 L. Wrap in aluminum foil and store at 4 °C.
2. 1 M DTT: Dissolve 154.3 g DTT in deionized $H_2O$ to give a final volume of 1 L. Store in aliquots at −20 °C (*see* **Note 1**).
3. 1 M $MgCl_2$ (*see* **Note 1**).
4. 1 M L-glutamic acid monopotassium salt monohydrate: Dissolve 203 g L-glutamic acid monopotassium salt in deionized $H_2O$. Adjust to a final volume of 1 L. Store at room temperature (*see* **Note 1**).

5. 5 M NaCl.
6. 1 M Tris–HCl (pH 8).
7. 10 % SDS.
8. 1 M NaOH.
9. UltraPure™ DNase/RNase-free distilled water (Gibco).
10. Target DNA: pNL4.3 (store at −20 °C) [21].
11. TaqMan®Universal PCR Master Mix (2×), No AmpErase® UNG (Applied Biosystems).
12. Reference compounds:
    (a) Raltegravir (Shanghai Medicilon Inc): 20 mM stock solution in 100 % DMSO.
    (b) Elvitegravir (Selleck): 20 mM stock solution in 100 % DMSO.

### 2.5 Consumables

1. Costar® 384-well black flat bottom polystyrene non-treated microplates (Corning, Cat. No. 3573).
2. MicroAmp® Optical 384-Well Reaction Plate (Applied Biosystems, Cat. No. 4343370).
3. MicroAmp® Optical Adhesive Film (Applied Biosystems, Cat. No. 4311971).
4. 2 mL Eppendorf tubes.
5. Transparent 384-well plates.
6. Reagent reservoirs.

### 2.6 Equipment

1. Humidified 37 °C incubator.
2. Water-bath 37 °C.
3. Heating system 95 °C with removable blocks.
4. Multidrop combi reagent dispenser with dispensing cassettes (Thermo Scientific).
5. Multichannel pipettes.
6. ABI7900HT fast real-time PCR system (Applied Biosystems).
7. Viewlux™ (PerkinElmer) or other fluorescent reader.

## 3 Methods

### 3.1 Preparation of 384-Well Test Plates

For compound testing, prepare fourfold serial dilutions in assay buffer (*see* Subheading 3.2.2, **step 1**) at a 4× final concentration. Transfer 10 μL of each test concentration to the quadruple wells of a 384-well black plate (the DMSO concentration in each well is 2 %). In this format, we typically test eight compounds per plate with nine concentrations for each compound. On each test plate, columns 1–18 are assigned to test compounds, columns 19–22 are the positive control wells (2 % DMSO in the absence of compound), and columns 23–24 are the negative control wells (0.5 % SDS or 400 mM NaOH stop solution added to the wells prior to addition of IN in the 3′-processing assay and the STT assay, respectively). We use Raltegravir and Elvitegravir as positive dose–response controls. In our laboratory, stock solutions of test compounds are prepared in DMSO, diluted in the assay buffer, and dispensed in 384-well plates using a 96-channel liquid handler.

## 3.2 3′-Processing Assay

### 3.2.1 Preparation of the Q-Probe

Freshly prepare an annealing buffer: 10 mM Tris–HCl, 100 mM NaOH, adjust the pH to 8. Anneal INss and INas in a 1.5 ratio of dabcyl/alexa by mixing 500 μL INss, 750 μL INas, and 8.75 mL annealing buffer to obtain a final concentration of 5 μM quenched probe. Aliquot 1 mL of this solution into each 2 mL Eppendorf tube and place the tubes in a preheated heating block at 95 °C for 1 min. Remove the blocks from the apparatus and let the tubes cool down inside the block for 30 min. Store the Q-probe at −20 °C.

### 3.2.2 Performing the 3′-Processing Assay

1. Prepare 20 mL assay buffer per 384-well test plate. Make a solution of 25 mM MOPS (pH 7.2), 10 mM DTT, 10 mM $MgCl_2$, 15 mM L-glutamic acid monopotassium salt, and deionized $H_2O$. This assay buffer must be freshly prepared before each experiment (*see* **Note 2**).
2. Prepare the stop solution: 0.5 % SDS in deionized $H_2O$.
3. Dispense 10 μL of the 0.5 % SDS stop solution to columns 23–24 (negative control wells) of the test plates using a multi-drop combi reagent dispenser (*see* **Note 3**).
4. Preheat the plates in a humidified 37 °C incubator. Place the plates one by one, next to each other, to ensure a good temperature distribution over the plates.
5. Prepare 12 mL assay mix in assay buffer for each 384-well test plate at room temperature by adding the Q-probe to a final concentration of 50 nM, LEDGF/p75 to a final concentration of 100 nM, and IN to a final concentration of 150 nM. Avoid light exposure as much as possible by wrapping the assay mix with aluminum foil.
6. Place a new multidrop dispensing cassette, rinse the tubes with deionized $H_2O$, and add 30 μL of the assay mix to each well of the test plate (*see* **Note 4**).
7. Incubate the plates for 2 h in a humidified 37 °C incubator and place the plates one by one, next to each other. Shield the test plates from light.
8. After 2 h of incubation, stop the reaction by adding 10 μL of 0.5 % SDS stop solution to all wells, except the negative control wells (*see* **Note 3**).
9. Cover the plates with aluminum foil to protect them from light exposure.

10. Measure the fluorescence during 5 s on the Viewlux™ using an excitation filter of 480 nm and an emission filter of 540 nm.
11. Calculate the 50 % inhibitory concentration ($IC_{50}$) for each compound, using software such as GraphPad Prism. The $IC_{50}$ value is defined as the concentration of compound at which 50 % of the 3′-processing activity of IN is inhibited (reduction of fluorescence) compared to the positive control wells. The $IC_{50}$ is calculated using linear interpolation of a dose–response curve. For each compound concentration, the median inhibition of the quadruple replicates is calculated.

### 3.3 STT Assay

#### 3.3.1 Donor DNA Preparation

1. Prepare the annealing buffer: 100 mM NaCl and 10 mM Tris–HCl (pH 8) in deionized $H_2O$.
2. Create the donor DNA by annealing oligo INdon1 and oligo INdon2 to each other by adding 25 μL of INdon1 stock solution to 25 μL of INdon2 stock solution in a 2 mL Eppendorf tube (Fig. 2), and then add 450 μL of annealing buffer. For annealing of the oligos, preheat the heating block to 95 °C, then remove the blocks from the apparatus, place the tubes in the block, and let them cool down inside the block for 30 min. Store the tubes at −20 °C.

#### 3.3.2 Strand Transfer Reaction in 384-Well Plates

1. Prepare the assay buffer (*see* Subheading 3.2.2, **step 1**): Prepare 50 mL assay buffer per 384-well compound plate.
2. Prepare the STT stop solution: 400 mM NaOH in deionized $H_2O$.
3. Add 40 μL of NaOH stop solution per well to columns 23–24 (negative control wells) using a multidrop combi reagent dispenser (*see* **Note 5**).
4. Preheat the plates in a humidified 37 °C incubator as described in Subheading 3.2.2, **step 4**.
5. Prepare 1 mL of IN/LEDGF/donor DNA mix per 384-well compound plate containing 2.4 μM IN, 1.6 μM LEDGF, and 1.2 μM donor DNA in the assay buffer (*see* **Note 6**). Incubate this mix for 5 min at 37 °C in a water-bath. Then, put the IN/LEDGF/donor DNA mix on ice for 2 min.
6. Meanwhile, dilute the target DNA to a final concentration of 2.43 nM in assay buffer. Prepare 15 mL for each 384-well test plate.
7. Dilute the IN/LEDGF/donor DNA mix 22.4 times in assay buffer to a final concentration of 106.8 nM IN, 72 nM LEDGF, and 53.4 nM donor DNA, respectively. Prepare 15 mL of IN/LEDGF/donor dilution for each 384-well test plate.
8. Add 15 μL of diluted target DNA to all wells of a 384-well test plate with a multidrop combi reagent dispenser.

**Table 1**
**Components of the real-time PCR master mix**

| | Stock concentration | Final concentration | Volume/ reaction (μL) |
|---|---|---|---|
| DNase/RNase-free $H_2O$ | | | 2.46 |
| TaqMan®Universal PCR master mix | 2× | 1× | 5 |
| INfwd | 100 μM | 125 nM | 0.0125 |
| INrev | 100 μM | 125 nM | 0.0125 |
| INtm (probe) | 100 μM | 125 nM | 0.0125 |

9. Add 15 μL of IN/LEDGF/donor dilution to all wells of a 384-well test plate with a multidrop combi reagent dispenser (*see* **Note 7**).
10. Incubate the plates for 30 min in a humidified 37 °C incubator.
11. Stop the reaction by adding 40 μL of NaOH stop solution to all wells, except columns 23–24 (negative control wells) with a multidrop combi reagent dispenser (*see* **Notes 5** and **8**).
12. Before performing the rtPCR, a 32-fold dilution of the strand transfer reaction product is required. Therefore, add 10 μL of the test plate, using a multichannel pipette, to each well of a transparent 384-well plate, which already contains 70 μL DNase/RNase-free $H_2O$ per well (dilution plate 1: 1/8 dilution). Mix gently and transfer 10 μL of dilution plate 1 to a second transparent 384-well plate, which already contains 30 μL DNase/RNase-free $H_2O$ (dilution plate 2: 1/4 dilution).

#### 3.3.3 rtPCR Assay

1. Prepare 4.5 mL of rtPCR master mix for each 384-well plate as described in Table 1.
2. Dispense 7.5 μL of the rtPCR master mix into the wells of a MicroAmp® Optical 384-Well Reaction Plate using a multichannel pipette.
3. Transfer 2.5 μL of the diluted strand transfer reaction product of each well of dilution plate 2 (*see* Subheading 3.3.2, **step 12**) to the respective well of the ABI 384-well reaction plate using a multichannel.
4. Seal the plate with a MicroAmp® Optical Adhesive Film.
5. Place the sealed ABI 384-well reaction plate in the ABI 7900HT system and run the cycling program as described in Table 2.
6. Following amplification, record the Ct values for each condition.

**Table 2**
**Cycling parameters for the real-time PCR**

| Step | Temperature | Time | |
|---|---|---|---|
| Amplitaq Gold®activation | 95 °C | 15 min | |
| Denaturation | 94 °C | 15 s | 50 cycles |
| Annealing | 53 °C | 22 s | |

#### 3.3.4 Data Analysis

1. Calculate the inhibition using the Ct values for each compound concentration as follows:

$$\% \text{ inhibition} = 1 - (1 + PCR_{eff})^{-(Ct - medianCtpos)}$$

With PCReff = PCR efficiency = 1 and Ctpos = Ct of the positive control wells

Positive controls (no compound) should have a Ct of around 20 and negative controls (inactivated IN) should have a Ct higher than 35.

2. Generate a dose–response curve for each compound using the calculated % inhibition. Based on the dose–response curves, calculate the 50 % inhibitory concentration ($IC_{50}$) for each compound using computer software as GraphPad.

## 4 Notes

1. The assay buffer contains DTT as a reducing agent to prevent the formation of disulfide bonds. $Mg^{2+}$ is added as a cofactor in the active site and potassium is used as salt. The use of sodium chloride instead of potassium chloride is not recommended, since this would remove the $Mg^{2+}$ out of the active site.
2. Alternatively, it is possible to make a 10× assay buffer and store aliquots at −20 °C. Aliquots can only be used once after thawing and should only be used to prepare 1× assay buffer on the day of the experiment.
3. 0.5 % SDS denatures IN, which consequently fails to bind the Q-probe in the 3′-processing assay.
4. After dispensing the 0.5 % SDS stop solution to columns 23–24, it is important to switch to a new dispensing cassette, as residual traces of SDS can inhibit IN activity in the assay mix. In addition, once the assay mix is added to a test plate, the test plate should be placed immediately back into a humidified incubator at 37 °C to ensure optimal IN activity. Do not stack the plates. Process a maximum of 20 compound plates with one assay mix.

5. 400 mM NaOH inhibits the activity of IN.
6. For the preparation of the IN/LEDGF/donor DNA mix, thaw LEDGF and IN slowly on ice. First, mix IN and LEDGF together, and then add them to the donor DNA in the assay buffer, right before the incubation starts.
7. After addition of 15 μL of target DNA dilution and 15 μL of IN/LEDGF/donor DNA mix to 10 μL of compound in the wells, the final concentrations of IN, LEDGF, donor DNA, and target DNA in the wells are 40, 27, 20, and 1 nM, respectively.
8. To ensure optimal STT assay performance, new multidrop cassettes should be used for every solution that is dispensed: IN/LEDGF/donor DNA mix, target DNA dilution, and NaOH stop solution.

## References

1. Bushman FD, Engelman A, Palmer I et al (1993) Domains of the integrase protein of human immunodeficiency virus type 1 responsible for polynucleotidyl transfer and zinc binding. Proc Natl Acad Sci USA 90(8): 3428–3432
2. Asante-Appiah E, Skalka AM (1997) A metal-induced conformational change and activation of HIV-1 integrase. J Biol Chem 272(26): 16196–16205
3. Jenkins TM, Engelman A, Ghirlando R et al (1996) A soluble active mutant of HIV-1 integrase: involvement of both the core and carboxyl-terminal domains in multimerization. J Biol Chem 271(13):7712–7718
4. Asante-Appiah E, Skalka AM (1997) Molecular mechanisms in retrovirus DNA integration. Antiviral Res 36(3):139–156
5. Engelman A, Mizuuchi K, Craigie R (1991) HIV-1 DNA integration: mechanism of viral DNA cleavage and DNA strand transfer. Cell 67(6):1211–1221
6. Esposito D, Craigie R (1999) HIV integrase structure and function. Adv Virus Res 52: 319–333
7. Katz RA, Merkel G, Kulkosky J et al (1990) The avian retroviral IN protein is both necessary and sufficient for integrative recombination in vitro. Cell 63(1):87–95
8. Bushman RD, Craigie R (1991) Activities of human immunodeficiency virus (HIV) integration protein in vitro: specific cleavage and integration of HIV DNA. Proc Natl Acad Sci USA 88:1339–1343
9. Yi J, Asante-Appiah E, Skalka AM (1999) Divalent cations stimulate preferential recognition of a viral DNA end by HIV-1 integrase. Biochemistry 38:8458–8468
10. Deprez E, Tauc P, Leh H et al (2000) Oligomeric states of the HIV-1 integrase as measured by time-resolved fluorescence anisotropy. Biochemistry 39:9285–9294
11. He HQ, Ma XH, Liu B et al (2007) High-throughput real-time assay based on molecular beacons for HIV-1 integrase 3′-processing reaction. Acta Pharmacol Sin 28:811–817
12. Grobler JA, Stillmock KA, Hazuda D (2009) Scintillation proximity assays for mechanistic and pharmacological analyses of HIV-1 integration. Methods 47:249–253
13. John S, Fletcher TM III, Jonsson CB (2005) Development and application of a high-throughput screening assay for HIV-1 integrase enzyme activities. J Biomol Screen 10:606–614
14. Farnet CM, Wang B, Lipford JR et al (1996) Differential inhibition of HIV-1 preintegration complexes and purified integrase protein by small molecules. Proc Natl Acad Sci USA 93:9742–9747
15. Hansen MS, Smith GJ III, Kafri T et al (1999) Integration complexes derived from HIV vectors for rapid assays in vitro. Nat Biotechnol 17:578–582
16. Wang Y, Klock H, Yin H et al (2005) Homogeneous high-throughput screening assays for HIV-1 integrase 3beta-processing, strand transfer activities. J Biomol Screen 10:456–462
17. Hazuda DJ, Felock P, Witmer M et al (2000) Inhibitors of strand transfer that prevent integration and inhibit HIV-1 replication in cells. Science 287:646–650

18. Hazuda DJ, Anthony NJ, Gomez RP et al (2004) A naphthyridine carboxamide provides evidence for discordant resistance between mechanistically identical inhibitors of HIV-1 integrase. Proc Natl Acad Sci USA 101:11233–11238
19. Hare S, Shun M-C, Gupta SS et al (2009) A novel co-crystal structure affords the design of gain-of-function lentiviral integrase mutants in the presence of modified PSIP1/LEDGF/p75. PLoS Pathog 5(1):e1000259. doi:10.1371/journal.ppat.1000259
20. Cherepanov P, Maertens G, Proost P et al (2003) HIV-1 integrase forms stable tetramers and associates with LEDGF/p75 protein in human cells. J Biol Chem 278: 372–381
21. Adachi A, Gendelman HE, Koenig S et al (1986) Production of acquired immunodeficiency syndrome-associated retrovirus in human and nonhuman cells transfected with an infectious clone. J Virol 59: 284–291

# Chapter 5

# HIV-1 Genotyping of the Protease-Reverse Transcriptase and Integrase Genes to Detect Mutations That Confer Antiretroviral Resistance

**Peter Van den Eede, Liesbeth Van Wesenbeeck, Yvan Verlinden, Maxim Feyaerts, Veerle Smits, Ann Verheyen, Leen Vanhooren, Alain Deloof, Jorge Villacian, and Theresa Pattery**

## Abstract

Major advances in antiretroviral (ARV) therapy during the last decade have made HIV-1 infections a chronic, manageable disease. In spite of these significant advancements, ARV drug resistance remains a hurdle for HIV-infected patients who are committed to lifelong treatments. Several commercially marketed and/or laboratory-developed tests (LDT) are available to detect resistance-associated mutations (RAMs) in HIV-1, by genotyping. These genotyping tests mainly comprise polymerase chain reaction (PCR)-amplification and population, nucleotide sequencing (Sanger methodology) of a large part of the protease (PR), reverse transcriptase (RT), and integrase (IN) genes. In this chapter, we describe HIV-1 PR, RT, and IN genotyping on clinical samples (plasma), using the LDT methodology performed at Janssen Diagnostics BVBA, Belgium (JDx), where the PR–RT genotyping is used as input, to generate a CE-marked vircoTYPE™ HIV-1 report while the IN genotyping is performed as a research-use-only (RUO) assay. The complete HIV-1 PR gene (297 bp; 99 amino acids) and a large part of the RT gene (the first 1,200 bp; 400 amino acids) are amplified and sequenced as a single 1,497 bp fragment. Genotyping of the IN gene is performed by amplification and sequencing of the RT–IN region (the last 459 bp; 153 amino acids of RT with the complete 867 bp; 289 amino acids of IN). This methodology allows identification of nucleoside/nucleotide reverse transcriptase, non-nucleoside reverse transcriptase, protease, and integrase inhibitor (NRTI, NtRTI, NNRTI, PI, INI) RAMs in the PR–RT and IN genes, which allows to predict viral response against current ARV regimens.

**Key words** HIV-1, Genotyping, Resistance testing, Resistance-associated mutations, Reverse transcriptase inhibitors, Protease inhibitors, Integrase inhibitors, vircoTYPE™ HIV-1

## 1 Introduction

During the last decade, significant progress has been made in managing human immunodeficiency virus (HIV-1)-infected patients [1, 2]. In the 1990s, treatment strategies were challenged

Edwin Yunhao Gong (ed.), *Antiviral Methods and Protocols*, Methods in Molecular Biology, vol. 1030,
DOI 10.1007/978-1-62703-484-5_5, 

by the rapid evolution of viral resistance, which was considered an inevitable consequence of antiretroviral (ARV) therapy.

The introduction of combination therapies with effective ARV regimens combining different modes of action drastically improved the mortality and morbidity of HIV-1. Despite their success, patients are committed to lifelong adherence to ARV therapy. Due to the long-term therapy, side effects, non-adherence, and emergence of resistant viral strains are major concerns that challenge a sustained control of viral replication. From a historical viewpoint, drugs acting on PR and RT constitute the core of the ARV treatment, according to the World Health Organization (WHO) guidelines used in the management and treatment of HIV-1-infected patients. The NRTIs and NtRTIs compete with natural deoxyribonucleotides during the reverse transcription process and block DNA chain elongation, when inserted into the DNA strand. The NNRTIs block in a noncompetitive manner and prevent the reverse transcriptase from converting RNA to DNA. On the other hand, the PIs block the enzymatic activity by binding to the active site of the viral protease. Recently, other ARV drug classes using distinct mechanisms and targeting different steps of the viral life cycle like entry inhibitors, fusion inhibitors, and INI have entered into clinical practice with less frequent usage. The increasing number of ARV drugs and regimen options available today results in multiple resistant pathways, complicating the manual selection of an effective ARV therapy, especially when considering that patients are committed to lifelong treatment and can suffer from toxicities that limit their treatment options [3]. Therefore, resistance testing by genotyping is broadly recommended as a standard of care, to evaluate treatment options in both newly infected and treatment-experienced patients, who are failing their current treatment regimens [4, 5].

For HIV-1 resistance testing, several methods based on genotyping and/or phenotyping are currently available either as LDT or marketed devices, to assess the susceptibility of a specific HIV-1 virus to individual ARV compounds [6–9]. For phenotyping assays, the antiviral activity is a direct measure ($IC_{50}$-50 % inhibitory concentration of an ARV drug) of the growth of the virus, in the presence of the ARV compound, by comparing to a reference, wild-type strain [7, 9]. Although labor intensive and cumbersome, phenotyping is still the preferred choice for patients with complex mutational patterns, for new ARV compound profiling, or to assess the impact of newly identified mutations on ARVs. Today, genotyping assays using population-based Sanger sequencing to detect RAMs are popular, as they are rapid and cost-effective, when compared to phenotyping. The nucleotide sequence obtained from a clinical sample is aligned and analyzed using expert-defined rule-based systems (e.g., ANRS—France, Rega—Belgium, Stanford—USA) or data-driven algorithms like vircoTYPE™ HIV-1 (Janssen Diagnostics BVBA, Belgium) [6–8]. It is important to note that

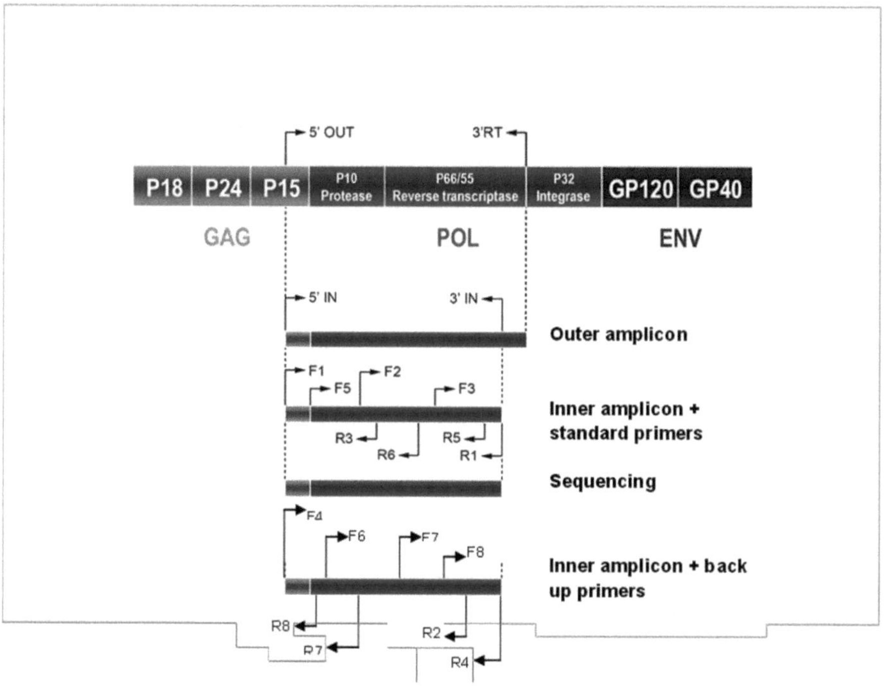


**Fig. 1** Overview of the primer annealing sites in HIV-1 PR–RT region for the different amplification or sequencing steps of the genotyping protocol

the data interpretation of genotyping results is complex and requires regular reviews and frequent updates. Historically, VirtualPhenotype™ was one of the first data-driven interpretation systems [9]. Today, genotyping with vircoTYPE™ HIV-1 allows calculation of the predictive phenotypic fold change (FC) in $IC_{50}$ and utilizes two clinical cutoffs (CCOs) to address the fact that resistance for most ARV drugs is not a black and white phenomenon, but a continuum. CCOs are defined as the baseline phenotypic fold change in $IC_{50}$ values above which an ARV-treated patient's virological response begins to decline (CCO1) or is almost lost (CCO2). The CCOs presented on the vircoTYPE™ HIV-1 report are derived from in vivo viral activity [7, 8].

In this chapter, we describe the LDT methodology of JDx to genotype the HIV-1 PR–RT domain (Fig. 1) and the RT–IN domain (Fig. 2) from viral RNA, which is derived from plasma samples of HIV-1-infected patients. The PR–RT nucleotide sequence is subsequently used as input to generate a vircoTYPE™ HIV-1 report [7]. Briefly, the methodology is as follows: Total viral RNA is extracted from plasma that is then reverse-transcribed to cDNA (RT-PCR) and the HIV-1 PR–RT or RT–IN region are subsequently amplified using specific PCR amplification primers. The amplified PCR products are purified and visualized by agarose gel electrophoresis to verify the amplification success rate (>95 % for samples with viral load ≥1,000 copies/mL). If the amplification

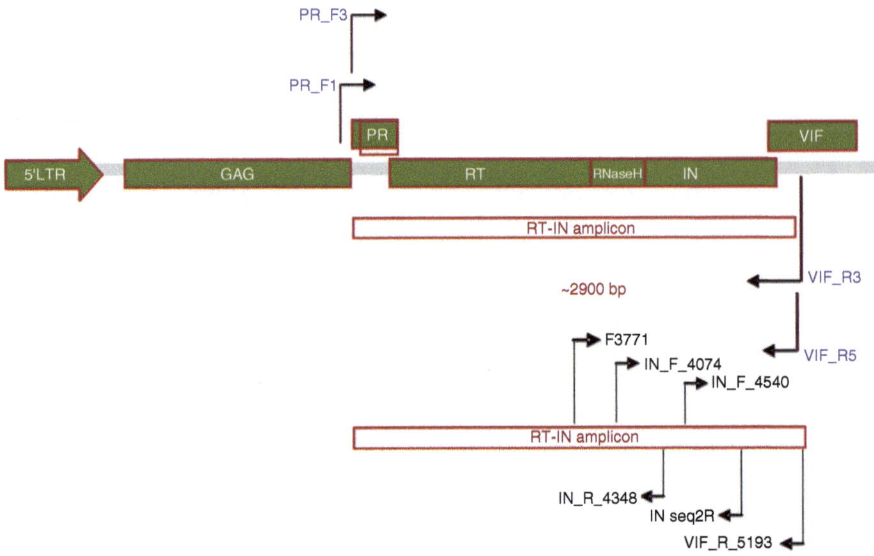


**Fig. 2** Overview of the primer annealing sites in HIV-1 RT–IN region for the different amplification or sequencing steps of the genotyping protocol

was unsuccessful, the procedure is repeated from plasma by performing an ultrasensitive viral RNA extraction. A standard set of primers (forward and reverse) is used to sequence the amplified PCR products on both the upper and lower strand (bidirectional sequencing). The nucleotide sequence obtained from the patient sample is aligned against the wild-type, reference HXB2 HIV-1 sequence (Genbank Accession number K03455) to construct a nucleotide contig. The PR–RT contig contains the complete 99 amino acids of the PR gene and the first 400 amino acids of the RT gene. The IN contig covers the complete 289 amino acids of the IN gene and is generated by assembling the RT–IN sequence that corresponds to the last 153 amino acids of RT gene, containing the 120 amino acids of the RNase H domain, and the complete IN gene. If nucleotide sequencing fails the first time, the process is repeated from the purified PCR-amplified fragment with a new set of primers (forward and reverse backup primers), to generate a nucleotide contig.

## 2 Materials

### 2.1 Total Viral RNA Extraction from Plasma

1. UltraPure™ DNase/RNase-free distilled water.
2. NucliSens® easyMAG® lysis buffer (bioMérieux, Cat. No. 280134).
3. NucliSens® easyMAG® extraction buffer 1 (bioMérieux, Cat. No. 280130).
4. NucliSens® easyMAG® extraction buffer 2 (bioMérieux, Cat. No. 280131).

5. NucliSens® easyMAG® extraction buffer 3 (bioMérieux, Cat. No. 280132).
6. NucliSens® easyMAG® magnetic silica (bioMérieux, Cat. No. 280133).
7. Aluminum sealing tape.
8. NucliSens® easyMAG® disposables: Aspirator disposables, sample strips, sterile tubes of 50 and 5 mL.
9. Sterile reagent reservoir.
10. NucliSens® easyMAG® instrument (bioMérieux, Cat. No. 280140).
11. Eppendorf centrifuge.
12. Control samples: (a) Negative control (NC): DNase/RNase-free water. (b) Positive control (PC): 600 μL of wild-type HIV-1 IIIB recombinant virus stock at a concentration of 10,000–20,000 copies/mL.

### 2.2 One-Step RT-PCR and Inner PCR

1. Deoxynucleotidetriphosphate (dNTP) mix: 10 mM for each dNTP (Promega, Cat. No. U1511).
2. SuperScript III One-Step RT-PCR System with Platinum Taq (Invitrogen, Cat. No. 12574-035).
3. SuperScript III One-Step RT-PCR System containing 2× Reaction Mix, and Buffer containing 0.4 mM of each dNTP, 2.4 mM $MgSO_4$ (Invitrogen, Cat. No. 12574-035).
4. Expand High Fidelity PCR system (Roche, Cat. No. 11759167103).
5. Expand High Fidelity Buffer (Roche, Cat. No. 11759167103).
6. Primers (*see* Tables 1 and 2).
7. 96-well PCR plates.
8. Sealing tape for 96-well plates.
9. Thermocycler: Gene Amp PCR System 9700 (PerkinElmer).

### 2.3 Amplicon Purification and Gel Analysis

1. QIAquick 96 PCR purification kit containing the extraction buffers PE, PM, and EB (Qiagen, Cat. No. 963141).
2. $QIA_{VAC}$ 96 well (Qiagen, Cat. No. 9014579).
3. V bottomed 96 well plates and cover (Costar).
4. Absolute ethanol.
5. Ultrapure™ 10× Tris–Acetate–EDTA (TAE) buffer, pH 8.3 (Invitrogen).
6. 6× DNA loading dye.
7. Generuler™ 1 kb Plus DNA ladder (Fermentas, Cat. No. SM1333).
8. Precast gels Latitude HT containing 0.5 μg/mL ethidium bromide (Lonza, Cat. No. 57234).

**Table 1**
**HIV-1 PR–RT amplification primers with their annealing position on the reference HXB2 strain and the concentration of the working solution with their final concentration in the PCR mix**

| Name | Sequence (5′ to 3′) | Position of primers HXB2 | Concentration working solution (μM) | Concentration PCR (μM) |
|---|---|---|---|---|
| 3′RT | CATTGCTCTCCAATTACTGTGATAT TTCTCATG | 4263–4295 | 20 | 0.2 |
| 5′OUT | GCCCCTAGGAAAAAGGGCTGTTGG | 2008–2031 | 20 | 0.2 |
| 3′IN | CATCTACATAGAAAGTTTCTGCTCC | 3855–3879 | 20 | 0.2 |
| 5′IN | CTAGGAAAAAGGGCTGTTGGAAATG | 2012–2036 | 20 | 0.2 |

**Table 2**
**HIV-1 RT–IN amplification primers with their annealing position on the reference HXB2 strain and the concentration of the working solution and their final concentration in the PCR mix**

| Name | Sequence (5′ to 3′) | Position of primers HXB2 | Concentration working solution (μM) | Concentration PCR (μM) |
|---|---|---|---|---|
| PR_F1 | CCCTCAAATCACTCTTTGGCAACGAC | 2260–2277 | 20 | 0.2 |
| PR_F3 | GCTCTATTAGATACAGGAGCAGATG | 2316–2340 | 20 | 0.2 |
| VIF_R3 | CTCCTGTATGCAGACCCCAATATG | 5243–5266 | 20 | 0.3 |
| VIF_R5 | GGGATGTGTACTTCTGAACTT | 5193–5213 | 20 | 0.3 |

9. Gel tray.
10. Gel electrophoresis system.
11. Power supply.
12. LAS 3000 mini gel imager (Fujifilm).
13. LAS 3000 mini gel imager software (Fujifilm).
14. Aide Image analysis software (Aidetek).

### 2.4 Sequencing Reaction

1. ABI Big Dye terminator V3.1 cycle sequencing kit (Applied Biosystems, Cat. No. 433745).
2. 2.5× dilution buffer: 40 mL 1 M Tris–HCl (pH 9.0), 1 mL of 1 M $MgCl_2$, 159 mL of UltraPure DNase/RNase-free distilled water.
3. Sequencing primers (*see* Tables 3 and 4).

### 2.5 Purification of Sequencing Reactions and Capillary Gel Electrophoresis

1. ABgene Thermofast 96 PCR plate.
2. DyeEx kit (Qiagen, Cat. No. 63183).
3. Clearseal 3730.

**Table 3**
**HIV-1 PR–RT sequencing primers**

| Name | Sequence (5′ to 3′) | Position of primers HXB2 | Concentration working solution (μM) |
|---|---|---|---|
| F1 | GAGAGCTTCAGGTTTGGGG | 2170–2188 | 4 |
| F2 | AATTGGGCCTGAAAATCC | 2696–2713 | 4 |
| F3 | CCTCCATTCCTTTGGATGGG | 3222–3241 | 4 |
| F5 | CACTCTTTGGCAACGACCC | 2261–2279 | 4 |
| R1 | CTCCCACTCAGGAATCC | 3778–3794 | 4 |
| R3 | CTTCCCAGAAGTCTTGAGTTC | 2797–2817 | 4 |
| R5 | GGGTCATAATACACTCCATG | 3492–3511 | 4 |
| R6 | GGAATATTGCTGGTGATCC | 3012–3030 | 4 |
| F4 | CAGACCAGAGCCAACAGCCCC | 2142–2162 | 4 |
| F6 | GGTACAGTATTAGTAGGACC | 2469–2488 | 4 |
| F7 | GTACTGGATGTGGGTGATGC | 2871–2890 | 4 |
| F8 | GTGGGAAAATTGAATTGGG | 3330–3348 | 4 |
| R2 | GTACTGTCCATTTATCAGG | 3255–3273 | 4 |
| R4 | CTAACTGGTACCATAATTTCACTAAGGGAGG | 3807–3837 | 4 |
| R7 | CATTGTTTAACTTTTGGGCC | 2601–2620 | 4 |
| R8 | GATAAAACCTCCAATTCC | 2397–2414 | 4 |

The first eight primers are the first-line sequencing primers and the last eight primers are the backup sequencing primers

**Table 4**
**HIV-1 RT–IN sequencing primers**

| Name | Sequence (5′ to 3′) | Position of primers HXB2 | Concentration working solution (μM) |
|---|---|---|---|
| F3771 | GCCACCTGGATTCCTGAGTG | 3771–3790 | 4 |
| IN_F_4074 | CAACCAGATAAAAGTGAATCAG | 4074–4095 | 4 |
| IN_F_4540 | TAGCAGGAAGATGGCCAGT | 4540–4558 | 4 |
| IN_R_4348 | CTCCTTTTAGCTGACATTTATCAC | 4348–4371 | 4 |
| Inseq2R | CTGCCATTTGTACTGCTGTC | 4748–4767 | 4 |
| VIF_R_5193 | ATGTGTACTTCTGAACTT | 5193–5210 | 4 |

4. Eppendorf 5810 centrifuge.
5. Holder for the ABI3730XL DNA analyzer.
6. TE buffer 10 mM Tris, adjusted to pH 8.0 with HCl, 1 mM EDTA (Ambion).
7. Xts-384 variable temperature thermal sealer.
8. HI-DI Formamide (Applied Biosystems).
9. Big Dye Terminator v3.1 Sequencing standard kit (Applied Biosystems, Cat. No. 4336921).
10. Clear Seal 3730 (Applied Biosystems).
11. 96-well PCR plate (ABgene).
12. ABI 3730XL Analyzer (Applied Biosystems).
13. Sequencher™ 4.1.4 for windows (Gene Codes Corp.).
14. BioEdit v7 (Ibis Biosciences).

## 3 Methods

### *3.1 Total Viral RNA Extraction*

#### *3.1.1 Preparation of Reagents*

1. Take the buffers out from 4 °C and ensure that they are at room temperature prior to use.
2. Invert the bottles containing the solution several times to ensure that all crystals are dissolved.
3. Thaw the samples and spin down before opening the tubes.
4. Switch on the NucliSens® easyMAG® instrument and computer (*see* **Note 1**).

#### *3.1.2 Extraction of Viral RNA*

A maximum of three, 8-well strips, equivalent to 22 plasma samples, 1 NC, and 1 PC can be processed in a single run (*see* **Note 2**).

1. Label the three strips before loading 256 μL of each plasma sample to the dedicated wells when the standard elution volume is 60 μL. If the extraction needs to be repeated, e.g., after the inner PCR step no PCR amplicon is observed, the ultrasensitive RNA extraction protocol requires 600 μL of plasma instead of 256 μL and adjust elution volume to 25 μL.
2. Place the strips in the NucliSens® easyMAG® instrument. Follow the instructions on the screen to run the NucliSens® easyMAG® instrument. Define extraction parameter: starting volumes, elution volume, type of samples, and sample ID. Check the lot number, volume, and expiry date for all reagent bottles (lysis buffer and extraction buffer 1, 2, and 3) and replenish or replace the bottles where required, before starting the lysis step.
3. Press run to start the lysis.
4. During the 10-min incubation of the lysis buffer, prepare the "premix": Vortex the NucliSens® easyMAG® magnetic silica

prior to use and dilute per eight samples 550 μL of silica in 550 μL of extraction buffer 3. Vortex the "premix" and dispense 125 μL to an 8-well microtiter plate strip.

5. Following completion of lysis, add 100 μL of "premix" to the sample strip containing the lysed samples and homogenize the mixtures. Resume the extraction and after approximately 40 min, the extraction is finished.
6. Transfer the extracted RNA elution (60 or 25 μL in case of repeat) to a 96-well PCR plate. Seal the plate with aluminum sealing tape and store the RNA samples at 4 °C until further use, within the same day. Immediately proceed to the one-step RT-PCR reaction. Following the start of the RT-PCR reaction, leftover RNA can be stored at −80 °C for a maximum period of 6 months.

### 3.2 One-Step RT-PCR and Inner PCR

#### 3.2.1 Primer Preparation

1. Dissolve the lyophilized primers for the one-step RT-PCR and inner PCR in Ultrapure™ DNase/RNase-free distilled water to prepare a stock solution of 100 μM and aliquot (*see* **Note 3**).
2. Prepare the working solution (20 μM) by diluting one aliquot of 100 μM stock solution in Ultrapure™ DNase/RNase-free distilled water. Keep one vial of each primer to prepare the master mix and store the remaining primer vials at −20 °C.

#### 3.2.2 One-Step RT-PCR (Outer PCR)

1. Calculate the volume of reagents needed based on the number of samples to be processed. The total amount of reactions to be prepared is equal to *X* samples + 8 negative controls spread over a full plate, 1 outer mix control, and 1 positive control (*see* **Note 4**). The total reaction volume is 35 μL for PR–RT and 25 μL for RT–IN. The reagent composition of the outer PCR reaction mix for PR–RT and RT–IN are summarized in Tables 5 and 6.
2. Thaw all reagents: UltraPure DNase/RNase-free distilled water and SuperScript III One-Step RT-PCR System 2× Reaction Mix, forward and reverse primers, excluding the enzyme SuperScript III One-Step RT-PCR System with Platinum Taq.
3. Vortex the reagents briefly, excluding the Taq enzyme and primers, and spin down.
4. Prepare the master mix by adding the correct volume of the reagents, with exception of the Taq enzyme, to a tube.
5. Mix by gently pipetting up and down or inverting the tube.
6. Take the Taq enzyme out of the freezer (−20 °C) at the time of use, and transfer the calculated volume of enzyme to the master mix. Mix by gently inverting the tube several times.

**Table 5**
**HIV-1 PR–RT one-step RT-PCR (outer PCR) reaction mix composition**

| Reagents | Volume per reaction (μL) | Concentration |
|---|---|---|
| RNase/DNase-free water | 6.1 | |
| SuperScript III One-Step RT-PCR System 2× reaction mix | 17.5 | 1× |
| Primer 5′OUT | 0.35 | 0.2 μM |
| Primer 3′RT | 0.35 | 0.2 μM |
| SuperScript III One-Step RT-PCR Platinum Taq | 0.7 | (Proprietary info manufacturer) |
| Sample | 10 | |
| Total volume | 35 | |

**Table 6**
**HIV-1 RT–IN one-step RT-PCR (outer PCR) reaction mix composition**

| Reagents | Volume per reaction (μL) | Concentration |
|---|---|---|
| RNase/DNase-free water | 6.5 | |
| SuperScript III One-Step RT-PCR System 2× reaction mix | 12.5 | 1× |
| Primer PR_F1 | 0.25 | 0.2 μM |
| Primer VIF_R3 | 0.25 | 0.2 μM |
| SuperScript III One-Step RT-PCR Platinum Taq | 0.5 | (Proprietary info manufacturer) |
| Sample | 5 | |
| Total volume | 25 | |

7. Dispense the master mix in a 96-well PCR plate (25 μL/well for PR–RT or 20 μL for RT–IN).
8. Prepare the sample and control sample RNA (*see* **Note 5**).
9. Add 10 μL (PR–RT) or 5 μL (RT–IN) of sample or positive control RNA to the corresponding reaction mix. Add 10 μL (PR–RT) or 5 μL (RT–IN) of Ultrapure™ DNase/RNase-free distilled water to the dedicated wells for outer PCR-negative controls and outer mix control.
10. Seal the plate with sealing tape.
11. Place the PCR plate in the thermocycler.

12. Conduct the PCR using the following reaction conditions:

    *PR–RT*: Initiate the reaction with a reverse transcription step for 30 min at 53 °C, followed by a denaturation step for 2 min at 94 °C. Continue with 40 cycles of denaturation at 94 °C for 15 s, annealing at 55 °C for 30 s, and elongation at 68 °C for 150 s. The final elongation step is at 68 °C for 7 min. The plates should be maintained at 12 °C until removal from the thermocycler.

    *RT–IN*: Initiate the reaction with the reverse transcription step of 30 min at 56 °C, followed by a denaturation step for 2 min at 94 °C. Continue with 40 cycles of denaturation at 94 °C for 15 s, annealing at 62 °C for 30 s, and elongation at 68 °C for 150 s. The final elongation is at 68 °C for 10 min. The plates should be maintained at 12 °C until removal from the thermocycler.

13. Proceed immediately with the inner PCR or store the plates at −20 °C.

#### 3.2.3 Inner PCR Reaction

1. Calculate the required reagent volumes. The total number of reactions is equal to the outer PCR plate plus 1 additional inner mix control (*see* **Note 4**). The total reaction volume for both PR–RT and RT–IN is 100 μL. The reagent composition of the inner PCR reaction mix for PR–RT and RT–IN are summarized in Tables 7 and 8.
2. Prepare the PCR reagents as described in Subheading 3.2.2, **steps 2** and **3**.
3. Prepare the master mix by adding the calculated volumes of reagents to a tube.
4. Mix by gently pipetting up and down or inverting the tube.

**Table 7**
**HIV-1 PR–RT inner PCR reaction mix composition**

| Reagents | Volume per reaction (μL) | Concentration |
|---|---|---|
| RNase/DNase-free water | 77.14 | |
| Expand high-fidelity buffer | 10 | 1× |
| dNTPs | 2 | 0.2 mM |
| Primer 5′IN | 1 | 0.2 μM |
| Primer 3′IN | 1 | 0.2 μM |
| Expand high-fidelity PCR system | 0.86 | 0.03 U/μL |
| DNA outer PCR | 8 | |
| Total volume | 100 | |

**Table 8**
**HIV-1 RT–IN inner PCR reaction mix composition**

| Reagents | Volume per reaction (μL) | Concentration |
|---|---|---|
| RNase/DNase-free water | 79.82 | |
| Expand high-fidelity buffer | 10 | 1× |
| dNTPs | 2 | 0.2 mM |
| Primer PR_F3 | 1.52 | 0.3 μM |
| Primer VIF_R5 | 1.52 | 0.3 μM |
| Expand high-fidelity PCR system | 1.14 | 0.04 U/μL |
| DNA outer PCR | 4 | |
| Total volume | 100 | |

5. Take the enzyme out from the freezer and immediately add the correct volume of enzyme to the master mix. Mix by gently inverting the tube several times.
6. Dispense the master mix in a 96-well PCR plate (92 μL/well for PR–RT or 96 μL for RT–IN).
7. Prepare the samples and control samples (*see* **Note 5**).
8. Add 8 μL of outer PCR for PR–RT and 4 μL of outer PCR for RT–IN product to the appropriate reaction mix. Add 8 μL (PR–RT) or 4 μL (RT–IN) of ultrapure DNase/RNase-free distilled water to outer PCR mix control.
9. Seal the plate.
10. Place the PCR plate in the thermocycler.
11. Start the PCR using the following reaction conditions:

    *PR–RT*: Initiate the reaction with a denaturation step of 2 min at 94 °C and then 30 cycles starting with a denaturation for 15 s at 94 °C followed by 30 s of annealing at 60 °C, and elongation step of 2 min at 72 °C. After the first ten cycles, the duration of the elongation step increases with 5 s for each cycle. Complete the reaction with a final extension of 7 min at 72 °C and keep the plates at 12 °C until removed from the cycler.

    *RT–IN*: Initiate the reaction with a denaturation step of 2 min at 94 °C and then 35 cycles starting with a denaturation step for 15 s at 94 °C, followed by 30 s of annealing at 60 °C, and an elongation step of 3 min at 68 °C. Complete the reaction with a final extension of 10 min at 68 °C and keep the plate at 12 °C until removed from the cycler.
12. Store the plates at −20 °C or immediately perform the purification.

### 3.3 Purification of the PCR-Amplified Product and Gel Electrophoresis

#### 3.3.1 Purification of PCR-Amplified Product

1. Prepare the QIAvac 96 and QIAquick 96-well plates and seal unused rows on the 96-well plates.
2. Prepare the Qiagen PE buffer by adding absolute ethanol and following the instructions on the bottle.
3. Transfer 100 μL of the inner PCR product from the PCR plate to the QIAquick 96-well plate and add 3 volumes (300 μL) of PM buffer. Switch on the vacuum source at a pressure of 0.2–0.6 bar for 3 min.
4. After the liquid has passed through the membrane, switch out the vacuum. Add 900 μL of PE buffer to the individual wells and switch on the vacuum again for 3 min at 0.2–0.6 bar. Repeat this wash step one more time.
5. When the PE buffer has passed through the membrane, apply a pressure of maximal 0.1 bar for an additional 10 min to dry the membrane.
6. Switch out the vacuum source and gently tap the top of the plate on a stack of tissues to remove all drops. Blot the nozzles of the plate with clean tissue to remove excess liquid.
7. Apply a vacuum pressure of 0.6 bar for 2 min to remove all ethanol from the membrane.
8. Place the QIAquick 96-well plate on the 96-well V-bottom plate.
9. Elute the DNA by applying 100 μL of EB buffer directly onto the center of each filter membrane.
10. Wait for 1 min and centrifuge at $317 \times g$ for 5 min to collect the DNA in the 96-well V-bottom plate.
11. Seal the plate with the purified inner PCR product.

#### 3.3.2 Gel Electrophoresis

1. Prepare the electrophoresis buffer: Take 200 mL of 10× TAE buffer and add to 1,800 mL of DNase/RNase-free water. Mix by inverting the bottle ten times.
2. Fill the electrophoresis reservoir with electrophoresis buffer.
3. Take the precast Latitude HT gel out of its cover and place the gel carefully in a tray (*see* **Note 6**).
4. Submerge the gel in the electrophoresis reservoir to fill the slots with electrophoresis buffer and take it back out to load the samples.
5. Dispense 1 μL/well of 6× DNA loading dye into U-bottom 96-well plate.
6. Add 5 μL of purified PCR product to the loading dye in the U-bottom 96-well plate (one sample/well) and mix by pipetting up and down. Aspirate 5 μL of the loading dye sample mix and load it on the gel. Ensure that all air gaps are removed from the slots. Leave some slots open for the molecular weight ladder.

7. Mix the ready-to-use Generuler 1 kb Plus DNA ladder and dispense 6 μL in the dedicated slots.
8. Place the loaded gel gently into the electrophoresis chamber.
9. Seal the purification plate containing the purified PCR product and store the plates at 4 °C.
10. Connect the power supply to the electrophoresis chamber and set the electrophoresis conditions at 150 V for 30 min.
11. When the electrophoresis is completed, start the Fuji Las-3000 mini gel imager and open the LAS-3000 gel imager software on the computer.
12. Wait until the message ready is shown and place the gel on the bottom tray of the imager.
13. Switch on the UV light.
14. Set the exposure using the following settings:
    Exposure type: "Precision"; Sensitivity/Resolution: "High"; Method fluorescence: "EtBr"; Tray position: "3" (max. 180 × 120 mm); Focusing brightness: "10"; Exposure time: Manual (1/8 to 1/60 s depending on the best exposure time as no PCR product or ladder are overexposed).
15. Take the picture. Switch off the UV light and discard the gel.
16. Verify that the amplicons are of expected length, by comparing with the standard molecular weight ladder (PR–RT = 1.8 kb and RT–IN = 2.9 kb).
17. Check that all positive and negative controls are positive and negative, respectively. If one of the controls does not meet the requirements, the experiment needs to be repeated from RNA extraction using a higher starting volume (600 μL) of plasma and a smaller elution volume (25 μL).

### 3.4 Sequencing Reaction

#### 3.4.1 Preparing the Sequencing Reaction

1. Thaw the working solution of the sequencing primers for the amplified product (eight primer pairs for PR–RT and six primer pairs for RT–IN, *see* also Tables 3 and 4) (*see* **Note 7**).
2. Spin the primers down before opening the tube.
3. Distribute 4 μL of each primer onto the ABgene plate within each row. Prepare the amount of primer plates that can be used immediately and corresponding to the number of samples that need to be sequenced (*see* **Note 8**).
4. Take the Big Dye Terminator mix out from −20 °C and place immediately on ice. Take 2.5× dilution buffer and the ultrapure DNase/RNase-free distilled water out of the fridge and bring to room temperature.
5. Prepare the sequencing master mixtures in a 12-well strip or a 96-well plate. Calculate the total number of sequencing reactions and required volume of the reagents. One sequencing reaction contains 3 μL of ultrapure DNase/RNase-free distilled

water, 2.5 μL 2.5× dilution buffer, and 1 μL of Big Dye Terminator (*see* **Note 4**).

6. Following use, place the Big Dye Terminator mix immediately at −20 °C.
7. Dispense the required volume of sample (1 μL of sample/ sequencing reaction) from the purification plate to the correct sequencing master mixtures prepared in the 12-well strip or the 96-well plate (sample 1 in master mix 1, sample 2 in master mix 2, and so forth).
8. Mix well by vortexing or inverting the strip/plates and spin down.
9. Distribute the sequencing reaction mix over the primer plates with master mix 1 in column 1, etc. and dispense 7.5 μL of sequencing reaction mix into each well.
10. Seal the plates and spin down.
11. Place the ABgene plates with the sequencing reaction in the thermocycler.
12. Start the sequencing reaction with denaturation at 96 °C for 10 s followed by annealing at 50 °C for 5 s and elongation at 60 °C for 4 min. Repeat this cycle 25 times, and store the plates at 12 °C until they are removed from the thermocycler.

*3.4.2 Purification of the Sequencing Reaction Product*

1. For each sequencing plate, a DyeEx 96 plate is required and the tape from the bottom of the plate should be removed.
2. Place the DyeEx 96 plate on top of a waste collection plate and carefully remove the tape on top of the plate (*see* **Note 9**).
3. Centrifuge the DyeEx 96 plate with the waste collection plate for 3 min at 1,000 × *g* (Eppendorf 5810) to remove the storage buffer in the DyeEx plate. Discard the flow-through (*see* **Note 10**).
4. Apply 300 μL of TE buffer on the gel-bed surface of each well of the DyeEx plate and repeat **step 3**.
5. Place the DyeEx 96 plate on top of an elution plate (96-well ABgene plate) using the adaptor.
6. Slowly apply 11.5 μL of sequencing sample and 11.5 μL of TE buffer on the gel-bed surface of each well.
7. Centrifuge for 3 min at 1,000 × *g*.
8. Place the ABgene plates containing the purified sequencing reaction products in the holders for the ABI3730XL DNA Analyzer.
9. Seal the ABgene plate containing the purified sequencing product with the clear seal 3730 using the Xts-384 variable temperature thermal sealer set at 170 °C for 2 s and seal the plates.
10. Proceed immediately with the capillary electrophoresis of the sequence products.

*3.4.3 Capillary Electrophoresis of Purified Sequencing Product*

1. Start up the computer.
2. Start up the ABI 3730XL DNA analyzer (Applied Biosystems).
3. Start the 3730 Data collection v3.0 software.
4. Open the stacker and place up to 16 plates into the stacker. Ensure that the plate's assembly fits in the stacker and close the stacker door.
5. Import the plate layout and start the run.
6. After the run is completed, view the run history file for unexpected errors.
7. Discard the plates.

*3.4.4 Analyzing Sequencing Data*

1. Import the sample sequences into the Sequencher™ 4.1.4 software.
2. Select all sequences.
3. Use the "Call secondary peak" command to remove all secondary peaks that are lower than 25 % of the highest peak at that position.
4. Trim the sequences using the following settings:

   Trim the 5′end, no more than 25 %, until the first 25 bases contain less than 1 ambiguity.

   At the 3′ starting from 100 bases from the 5′ trim, trim the first 25 bases containing more than 5 ambiguities.
5. Select "Trim checked items" to remove the areas marked in red leaving only the required sequences and close the window.
6. Import the relevant sequences from the samples to align the sequences of each sample with the HXB2 (*see* **Note 11**) reference viral sequence. Choose the "Import command" from the file menu and in the Sequence submenu, browse for the reference sequence file and click open.
7. Assemble the raw sequences to construct the contig. Make sure that the assembly parameters are set as follows: Select "Dirty data," with a minimum match of 85 % and with a minimum overlap of 20 bp.
8. Select "Assemble automatically" to generate the contig.
9. Open the assembled contig by double clicking and showing the overview of the sequence alignment.
10. Delete any sequence that overhangs the 3′ and 5′ ends of the reference.
11. Select the reference sequence and in the Contig menu choose "Remove selected sequences". The reference sequence will disappear from the assembled contig (*see* **Note 12**).
12. Visualize the amino acid sequences by clicking on the button in the lower right corner.

13. Select "Show chromatograms".
14. Look at the chromatogram for ambiguities (heterogeneities or points of contig disagree) (*see* **Note 13**).
15. If the point of ambiguity results in a correct consensus sequence, then move on to the next ambiguity.
16. If the ambiguity results in an incorrect consensus sequence, manually edit by using the following rules: Only change if the strands in both directions (forward and reverse) agree. When mixtures are observed and are not automatically called on both strands although they are visible (intensity below 25 % of highest peak) in all fragments covering that position, it may be manually edited if the secondary peak is higher than 25 % of the primary peak in at least one sequence fragment (*see* **Note 14**).
17. After editing, overview the contig with the overlapping fragments (*see* **Note 15**).
18. If no consensus sequence can be made because of high background or gaps, repeat the sequencing reaction using the backup primers for PR–RT (*see* Table 3) or the same primers for RT–IN. If there are upcoming inserts, frameshifts, or stop codon, restart the experiment again from RNA extraction.
19. Export the consensus sequence as a fasta file.

#### 3.4.5 BLAST Analysis (See **Note 16**)

1. Open Bioedit.
2. Go to the file menu and select "New alignment".
3. Go to the "File" menu and select "Import".
4. Select sequence alignment file.
5. Select the sequence files of interest and choose "Open".
6. Go to Accessory applications menu and select BLAST-Local Blast.
7. Select program "Blastn".
8. Select alignment database of sequences obtained during the last experiment of that day.
9. Select Tabular output and an expectation value of 1E-100.
10. Set the maximum number of reported hits and alignments to indicate the first five.
11. Press "Do search".
12. Save results and search for sequences with a similarity of 98 % or more with a different sample ID (*see* **Note 17**).
13. For all samples with a similarity of 98 % check whether the IDs of the involved samples are the same, if they are from the same patient, and/or the involved samples show concordance with their respective previous visits.
14. If the samples with a 98 % or more similarity do not follow any of the criteria mentioned above, repeat the analysis from RNA extraction.

## 4 Notes

1. Avoid magnetic fields (e.g., mobiles or cordless phones) near the NucliSens® easyMAG® as the electromagnetic waves may affect the extraction and cause malfunctioning of the machine.
2. All sample manipulations and buffer preparations are performed under a laminar flow hood.
3. Leave primers at room temperature for 15 min to ensure that all lyophilized primers are dissolved before continuing.
4. Count one or two samples extra to cope with pipetting errors.
5. Spin the DNA samples down prior to opening the plate or the tubes in order to avoid contaminations and unwanted spills.
6. Take the necessary precautions when handling the gel, as the gel and the pouch in the package contain ethidium bromide.
7. For PR–RT region, two sets of eight primers exist. When sequencing the sample for the first time, use the eight standard primers; when repeating the experiment due to errors in previous sequences obtained with the standard primers, use the backup primer set (*see* Table 3 and Fig. 1). For RT–IN, only one set of six primers (*see* Table 4) exists. In cases where the sequencing reaction needs to be repeated, the same set of primers is used again (Fig. 2).
8. Sequence only those samples for which a PCR amplicon of expected size (PR–RT = 1.8 kb and RT–IN = 2.9 kb) was observed on an agarose gel. A maximum of 12 samples can be loaded on one plate with one sample/column.
9. Cautiously remove the cover as the DyeEx 96 plate contains storage buffer.
10. The gel-bed surface in the wells may vary.
11. HXB2 (Genbank Accession number K03455) positions 2,253 to 3,749 bp for PR–RT and region 4,230 to 5,096 bp for IN.
12. In some cases, it may be unable to delete the reference strain and this means that there is a gap in the consensus sequence as result of the trimming process. However, the sequence data in that region may have clean and normal peaks of height and shape in the chromatogram, with minimal background. In this case the sequence may be elongated by clicking "revert to experimental data" until the gap is closed. Use the "Call secondary peak" as bases might be changed in mixtures and to remove all secondary peaks that are lower than 25 % of the highest peak at that position.
13. Use CTRL + N on the computer to go immediately to the point of ambiguity.

14. The editing of the sequences is under the condition that the background levels of the sequences are low.
15. If the contig contains regions that are covered by only one strand, the sequence data in these regions should have clean peaks of normal shape and signal height with minimal background level. If not, these sequences should not be exported and should be considered as a gap.
16. Perform a BLAST analysis on all sequences in an experiment daily and 6 monthly, to monitor potential cross contaminations between analyzed samples and/or potential sample mix-ups.
17. The laboratory standards for New York State Department of Health state that sequence homology above 98 % should be investigated for possible contamination, unless other criteria have been adapted by the assay manufacturer.

## Acknowledgements

The authors would like to thank the Diagnostic Lab and the R&D at Janssen Diagnostics BVBA for their contribution.

### References

1. Zolopa AR (2010) The evolution of HIV treatment guidelines: current state-of-the-art of ART. Antiviral Res 85(1):241–244
2. Stewart A, Chan Carusone S, To K et al (2012) Causes of death in HIV patients and the evolution of an AIDS hospice: 1988–2008. AIDS Res Treat 2012:390406
3. Lengauer T (2012) Bioinformatical assistance of selecting anti-HIV therapies: where do we stand? Intervirology 55:108–122
4. U.S. Department of Health and Human Services (2012) Panel on Antiretroviral Guidelines for Adults and Adolescents—A Working Group of the Office of AIDS Research Advisory Council (OARAC). http://aidsinfo.nih.gov/guidelines/html/1/adult-and-adolescent-treatment-guidelines/0/
5. EACS guidelines (2011) European Guidelines for treatment of HIV infected adults in Europe http://www.europeanaidsclinicalsociety.org/index.php?option=com_content&view=article&id=59&Itemid=41
6. Frentz D, Boucher CA, Assel M et al (2010) Comparison of HIV-1 genotypic resistance test interpretation systems in predicting virological outcomes over time. PLoS One 5(7): e11505
7. Pattery T, Verlinden Y, De Wolf H et al (2012) Development and performance of conventional HIV-1 phenotyping (Antivirogram®) and genotype-based calculated phenotyping assay (virco®TYPE HIV-1) on protease and reverse transcriptase genes to evaluate drug resistance. Intervirology 55(2):138–146
8. Van Houtte M, Picchio G, Van Der Borght K et al (2009) A comparison of HIV-1 drug susceptibility as provided by conventional phenotyping and by a phenotype prediction tool based on viral genotype. J Med Virol 81(10):1702–1709
9. Zazzi M, Incardona F, Rosen-zvi M et al (2012) Predicting response to antiretroviral treatment by machine learning: the EuResist project. Intervirology 55:123–127

# Part II

## Hepatitis Viruses

# Chapter 6

# In Vitro Kinetic Profiling of Hepatitis C Virus NS3 Protease Inhibitors by Progress Curve Analysis

**Rumin Zhang and William T. Windsor**

## Abstract

Kinetic profiling of drug binding to its target reveals important mechanistic parameters including drug–target residence time. In this chapter, we focus on global progress curve analysis as a convenient method for kinetic profiling. Detailed guidelines with pros and cons for various experimental designs and data analysis are provided. Kinetic profiling of Boceprevir and Telaprevir is illustrated.

**Key words** HCV NS3 protease, Binding kinetics, Kinetic profiling, Drug–target residence time, Global progress curve analysis, Boceprevir, Telaprevir

## 1 Introduction

The dual impact of binding kinetics on drug discovery and drug action has been emphasized in recent reviews [1–5]. Notably, the concept of drug–target residence time is gaining wide recognition [6, 7]. For nonhuman targets such as the nonstructural proteins in hepatitis C virus (HCV), a drug with a sufficiently long drug–target residence time will result in sustained target engagement between dosing and is expected to contribute toward greater clinical efficacy. A longer residence time will assure longer target engagement until target degradation and/or the next drug dose. Thus, identifying drug candidates with optimal drug–target residence time is crucial during drug discovery. It is well known that binding kinetics of such long residence time drugs, if ignored, can lead to substantial underestimation of the true potency [8]. There are many established ways to measure binding kinetics and the drug–target residence time [6, 9–11]. In this chapter, progress curve analysis is illustrated to profile critical binding kinetics for HCV NS3 inhibitors. The guidelines for experimental design and data analysis as detailed herein are generally applicable to other slow binding inhibitors. An in-depth study on many practical aspects of

Edwin Yunhao Gong (ed.), *Antiviral Methods and Protocols*, Methods in Molecular Biology, vol. 1030,
DOI 10.1007/978-1-62703-484-5_6, © Springer Science+Business Media, LLC 2013

enzymology is available to interested readers from several excellent reviews or books [12].

From the perspective of binding kinetics, enzyme inhibitors such as HCV NS3 inhibitors can be classified into four main categories (Table 1). When the binding equilibriums are quickly established within the time frame (usually seconds) of mixing assay reagents (enzyme, inhibitor, and substrate), then the binding is said to be fast. Fast binding inhibitors (FBI) exhibit a linear time course of inhibition, giving a time-independent half maximal inhibition concentration ($IC_{50}$) value. Fast binding is a relative term, as seconds can be relatively slow in millisecond time scale stopped flow experiments. Our discussion here assumes the standard time-frame of mixing assay reagents in seconds. Any time-dependent inhibition not attributable to substrate depletion, product inhibition and/or enzyme instability is indicative of slow binding kinetics. Slow binding inhibitors (SBI) exhibit curvilinear time course of inhibition, yielding time-dependent $IC_{50}$ values. The true inhibition potency can be substantially underestimated by the apparent potency (often measured as $IC_{50}$) when the drug–target residence time is much longer than the assay timeframe [8].

Under "tight binding condition" (i.e., when the apparent inhibition constant is comparable to or well below the active enzyme concentration used in the assay), enzyme will be quantitatively titrated by the inhibitor added and free inhibitor concentration will be significantly depleted before saturation binding. As a result, the $IC_{50}$ value obtained from dose-dependent inhibition curves will approach the practical limit of the assay imposed by the active enzyme concentration [12]. A tight binding condition is useful for accurate determination of active site concentration of the enzyme (thus active site titration). This condition poses serious problems, however, for inhibitor potency estimation and must be addressed with proper experimental design and data analysis.

Fast or slow binding, with or without tight binding, gives rise to four combinatorial classes of inhibitors, including FBI, SBI, fast and tight binding inhibitors (FTBI), slow and tight binding inhibitors (STBI) (Table 1). With respect to SBI or STBI, it is possible that slow binding occurs through one or two kinetically distinguishable steps [7]. In a 2-step reaction, the slow step may be before or after a fast step. The slow step may be due to a slow isomerization of the enzyme, the inhibitor or their initial binary complex, as well as due to a slow induced fit between enzyme and inhibitor, a slow covalent modification of the enzyme by a reversible inhibitor or an irreversible inactivator. The depicted 2-step binding model under Table 1 is the most common type. The HCV protease inhibitors Boceprevir and Telaprevir are examples of 2-step inhibitors involving a slow step that is reversible following a mechanism-based acylation [13, 14].

**Table 1**
**Kinetic classification of enzyme inhibitors**

| Classes | Description |
|---|---|
| FBI | Fast binding inhibitor, 1-step |
| FTBI | Fast and tight binding inhibitor, 1-step |
| SBI | Slow binding inhibitor, 1- or 2-step (see diagrams below) |
| STBI | Slow and tight binding inhibitor, 1- or 2-step |
| 1-step SBI model 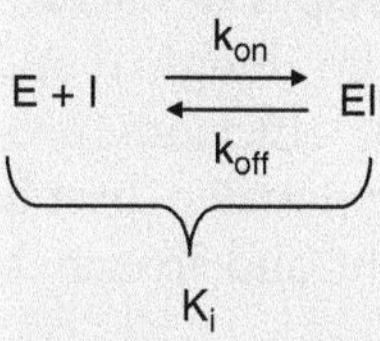  | 2-step SBI model 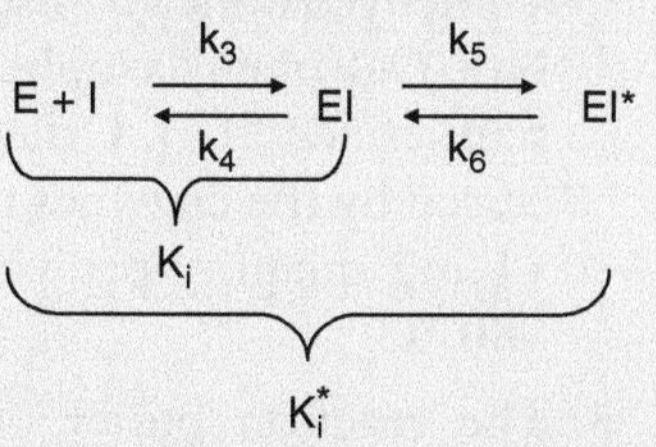  |
| $K_i = k_{off}/k_{on}$ | $K_i = k_4/k_3$; $K_i^* = K_i/(1 + k_5/k_6)$ |
| $v_s/v_0 = 1/(1+[I]/K_i)$ | $v_i/v_0 = 1/(1+[I]/K_i)$; $v_s/v_0 = 1/(1+[I]/K_i^*)$ |
| $k_{obs} = k_{off}(1+[I]/K_i)$ | $k_{obs} = k_6(v_i/v_s)$ |
| $\tau = 1/k_{off}$ | $\tau = 1/k_{off} = (k_4 + k_5 + k_6)/(k_4k_6) > 1/k_6$ |

Where, E is the enzyme, I is the inhibitor, EI is the initial binary binding complex, EI* is the isomerized or modified enzyme–inhibitor complex, $K_i$ and $K_i^*$ are, respectively, initial and overall inhibition constant, $k_{on}$ (or $k_3$) and $k_{off}$ (or $k_4$) are, respectively, association and dissociation rate constant of the first step, $k_5$ and $k_6$ are, respectively, forward and backward isomerization or modification rate constants of the second step, $v_0$, $v_i$, and $v_s$ are, respectively, the uninhibited (no compound control), initial and final (with compound) steady state velocities. Tau ($\tau$) is the residence time (the time it takes for an inhibitor to dissociate from ~63 % of the complex), simply determined by the reciprocal of the overall dissociation rate constant ($k_{off}$). Notice the overall dissociation rate constant ($k_{off}$) takes on different formulism for 1-step and 2-step SBI. Not shown in the diagrams are the binding reaction between substrate and enzyme (the corresponding association and dissociation rate constants are $k_1$ and $k_2$, respectively, in case readers wonder about their absence in the diagram). In the presence of substrate, the apparent inhibition constant ($K_i^{app}$) or apparent overall inhibition constant ($K_i^{*app}$) should be used in place of $K_i$ and $K_i^*$, respectively, in equations other than their own definitions containing microscopic rate constants. *See* Table 5 for the changes and also equations for additional models. Tight binding happens when active enzyme concentration is more than 10 % of $K_i^{app}$ or $K_i^{*app}$. The substrate effect on the apparent inhibition constant or $IC_{50}$ was reported by Cheng and Prusoff [27]. In the presence of a ***competitive*** substrate, $K_i^{app} = K_i(1+[S]/K_m)$, $K_i^{*app} = K_i^*(1+[S]/K_m)$. In the presence of an ***uncompetitive*** substrate, $K_i^{app} = K_i(1+K_m/[S])$, $K_i^{*app} = K_i^*(1+K_m/[S])$. Notice the reversal of [S] and $K_m$ ratio for competitive versus uncompetitive substrate, which makes the effect of substrate concentrations on both $K_i^{app}$ and $K_i^{*app}$ trending in opposite directions for competitive versus uncompetitive inhibitors. In the presence of a ***noncompetitive*** substrate, $K_i^{app} = K_i([S]+K_m)/([S]/\alpha + K_m)$, $K_i^{*app} = K_i^*([S]+K_m)/([S]/\alpha + K_m)$, where $\alpha$ is the ratio of equilibrium dissociation constant of the tertiary complex ESI over that of the binary complex EI. For pure noncompetitive inhibitors (i.e., $\alpha = 1$), there is no difference between the apparent and intrinsic equilibrium inhibition constants. For bi-substrate enzyme systems, the apparent Michaelis-Menten constant should be used (*see* ref. 27, 28 for details)

## 2 Materials

### 2.1 Reagents

1. NS4A(amino acids 21–32)-tethered single chain NS3 (amino acids 3–181) protease of various genotypes, via a GSGS tetrapeptide linker: prepared as described [15]. Store frozen at −80 °C.
2. A substrate peptide containing Europium fluorophore and QSY7quencher,Ac-Cys(Eu)-Asp-Asp-Met-Glu-Glu-Abu-[COO]-Ala-Ser-Ala-Lys(QSY7)-amide: custom ordered (PerkinElmer). Alternatively, a chromogenic substrate peptide containing a 4-phenylazophenyl ester, Ac-Asp-Thr-Glu-Asp-Val-Val-Pro-Nva-*O*-4-phenylazophenyl ester, can be custom designed and synthesized [16]. Other substrates that can be efficiently processed by the enzymes may also be used. Store powder (best in ~1 mg quantity per vial) protected from light and frozen at −80 °C.
3. The reaction buffer (Teknova, Cat. No. H1039): 25 mM MOPS, pH 6.5, 300 mM NaCl, 0.05 % lauryl maltopyranoside, 5 μM EDTA, 20 % glycerol. Store refrigerated.
4. Compounds are dissolved in neat DMSO solution, usually at 10 mM stock. Store sealed at room temperature under desiccation.

### 2.2 Consumables and Equipment

1. Black, opaque flat bottom, 384-well microtiter plates are used for fluorescent detection, while black, clear flat bottom, 384-well microtiter plates (preferably of minimal UV absorption) are used for UV/VIS spectrophotometric detection. Round bottom, 96- or 384-well polypropylene microtiter plates are used for compound dilutions in DMSO.
2. A plate reader such as PHERAstar® FS (BMG Labtech,) is used to monitor the progress of reactions. Other plate readers such as EnVision® (PerkinElmer), SpectraMax® (Molecular Devices) and Tecan® (Tecan, Switzerland) may also be used.

## 3 Methods

Progress curve analysis is our recommended method of choice for medium to high throughput kinetic profiling. Progress curve analysis does not require specialized instrumentation of label-free methods such as surface plasmon resonance. It can be conveniently performed in an ordinary biochemistry lab. Kinetic profiling using progress curve analysis, with or without preincubation (*see* **Note 1**), has been well documented in literature [16–21]. All assays are performed at 25 °C, unless otherwise noted. A 50 μL final reaction volume is assumed, but the volume may be scaled up or down.

### 3.1 Enzymatic Assay Optimization

As a prerequisite of progress curve analysis, a range-finding experiment should be performed by varying both enzyme and substrate concentrations to establish steady state enzyme kinetics conditions under which the time-dependent product formation for no inhibitor control reactions is linear within the intended time window of observation (usually ~2 h, sometimes longer for very slow binding inhibitors). If possible, kinetic parameters for substrate turnover, such as maximal velocity ($V_{max}$) and Michaelis-Menten constant ($K_m$), should be obtained.

1. Thaw an aliquot of the frozen enzyme stock on ice for at least 30 min. Warm up the frozen substrate powder and refrigerated reaction buffer to room temperature for at least 30 min.
2. Twofold serially dilute the thawed enzyme stock in reaction buffer at working stock concentrations of 2× the final enzyme concentrations (*see* **Note 2**). In our experience, the final concentrations of HCV NS3 protease should be in low nM to sub-nM range.
3. Dissolve substrate powder to a fresh stock solution at 10 mM in neat DMSO. Twofold serially dilute substrate in neat DMSO to a range of working concentrations at 50× the final substrate concentrations. Dilute each of the serial dilutions another 25-fold in reaction buffer to prepare a new range of working concentrations at 2× the final substrate concentrations. In general, the final substrate concentrations should be from ~0.5 to ~4× $K_m$ to allow accurate determination of $V_{max}$ and $K_m$. When this range is not possible, simply use a practical range of substrate concentrations either far less than $K_m$ to estimate the $V_{max}/K_m$ ratio for a high $K_m$ substrate or far more than $K_m$ to estimate $V_{max}$ for a low $K_m$ substrate.
4. Mix the 2× enzyme working solutions with the 2× substrate working solutions (1:1) in a black, opaque (for fluorescent substrate) or clear (for chromogenic substrate) flat bottom 384-well microtiter plate to initiate the reaction. Mix well. Include 1× substrate only controls by mixing the 2× substrate working solutions with reaction buffer (1:1). Optionally include 1× enzyme only and buffer only controls. Immediately monitor the reactions and controls on a detector. For fluorescent substrate Ac-Cys(Eu)-Asp-Asp-Met-Glu-Glu-Abu-[COO]-Ala-Ser-Ala-Lys(QSY7)-amide, detect time resolved fluorescence every 2 min for at least 2 h at 620 nm with excitation at 337 nm. For chromogenic substrate Ac-Asp-Thr-Glu-Asp-Val-Val-Pro-Nva-O-4-phenylazophenyl ester, detect absorbance at 370 nm every 2 min for at least 2 h.
5. Export the raw data and plot the time-dependence of product fluorescence or absorbance in GraphPad Prism 5. Fit the initial linear portion of the data to determine the initial velocities at

each enzyme and substrate concentration combination. The $Y$-axis intercepts should be close to the readings of substrate only controls. Not all enzyme and substrate concentrations will give usable initial portion of the data. Reactions that are too fast or too slow are not usable, only those that result in an accurate determination of initial velocities will be usable. Re-plot the usable initial velocities against substrate concentrations at each enzyme concentration. Fit the data to the Michaelis-Menten equation $Y = V_{max}/(1 + K_m/X)$, where $X$ is substrate concentration and $Y$ is the observed fluorescence or absorbance. If individual $V_{max}$ and $K_m$ values are not reliably determined, their ratio will still be reliable. Accurately determined $V_{max}$ should linearly increase with the enzyme concentrations while $K_m$ should be constant within experimental errors.

6. Decide on the optimal enzyme and substrate concentrations (*see* **Note 2**) to use in subsequent assays (*see* Subheadings 3.2 and 3.3). The optimal enzyme and substrate combination should result in nearly linear product formation within the time window of observation (usually ~2 h), with a signal-to-noise ratio preferably >10.

### 3.2 Estimation of Apparent Inhibition Potency

Another range-finding experiment is performed under the established steady state conditions to estimate the apparent inhibition potency by widely varying inhibitor concentrations with half log serial dilutions.

1. Thaw the frozen enzyme stock and warm up the frozen substrate powder and refrigerated reaction buffer as described in Subheading 3.1, **step 1**. Take out compound stock solutions.
2. Serially dilute compounds in neat DMSO in a 384-well polypropylene microtiter plate, with a range of working concentrations at 100× the final (usually 10 pM to 100 μM) compound concentrations. Thus, the highest working concentration is 10 mM and the lowest working concentration is 1 nM. This wide ranging, seven orders of magnitude can be accomplished in 15-point serial dilutions by half log (~3.16×, e.g., 10 μL added to 21.6 μL). Include no compound (i.e., DMSO only) controls at the 16th wells across the plate.
3. Dissolve substrate powder to prepare a fresh stock solution at 100× final substrate concentration as determined in Subheading 3.1. The volume for 100× substrate solution is determined by the number of assay points. For every full 384-well plate assay with final 50 μL per well, 250–300 μL of 100× substrate solution should be prepared.
4. Dilute the thawed enzyme stock in reaction buffer at working stock concentration of 2× the final enzyme concentration as determined in Subheading 3.1.

5. Add 0.5 μL each of serially diluted 100× compound solutions and 0.5 μL of 100× substrate solution to 24 μL reaction buffer in a black, opaque (for fluorescent substrate) or clear (for chromogenic substrate) flat bottom 384-well microtiter plate. Mix well. Do replicates (duplicate or triplicate).
6. Add 25 μL of 2× enzyme solution to initiate the reaction. Mix well. Immediately monitor the reactions and controls on a detector. For fluorescent substrate Ac-Cys(Eu)-Asp-Asp-Met-Glu-Glu-Abu-[COO]-Ala-Ser-Ala-Lys(QSY7)-amide, detect time resolved fluorescence every 2 min for at least 2 h at 620 nm with excitation at 337 nm. For chromogenic substrate Ac-Asp-Thr-Glu-Asp-Val-Val-Pro-Nva-*O*-4-phenylazophenyl ester, detect absorbance at 370 nm every 2 min for at least 2 h.
7. Export the raw data and plot the time-dependence of product fluorescence or absorbance in GraphPad Prism 5. Inspect the full range of inhibitions both in the initial period and near the end of observation. Confirm that final steady state is established for at least some of the compound concentrations. Select the highest compound concentration causing up to ~95 % *initial* inhibition (for 2-step slow binding inhibitors) or *final* inhibition (for 1-step slow binding inhibitors) and the lowest compound concentration giving down to ~5 % *final* inhibition for progress curve analysis next. Fit the data to get a preliminary estimate of the apparent inhibition potency and applicable kinetic constants (*see* **Note 4**).

### 3.3 Progress Curve Analysis

Based on the above range-finding experiments, 1.5- or 2-fold serially diluted inhibitors in replicates are mixed with substrate before enzyme is added to initiate the reactions. The reactions are then monitored for an extended period of time (*see* **Note 3**) under steady state conditions. Initial estimates of equilibrium and kinetic parameters are obtained from progress curve analysis. The experiments are usually repeated, or redesigned and repeated. Preincubation/dilution experiment (*see* **Note 1**) may also be needed.

1. Thaw the frozen enzyme stock and warm up the frozen substrate powder and refrigerated reaction buffer as described in Subheading 3.1, **step 1**.
2. Serially dilute compounds in neat DMSO in a 384-well polypropylene microtiter plate, with a range of working concentrations at 100× the final compound concentrations as determined in Subheading 3.2, **step 7**. The recommended fold of dilution is 1.5 or 2. Include no compound (i.e., DMSO only) controls across the plate.
3. Prepare a solution of 100× final substrate concentration, a solution of 2× final enzyme concentration, and mixtures of

compound and substrate as described in Subheading 3.2, **steps 3–5**.

4. Initiate and monitor the reaction as described in Subheading 3.2, **step 6**.
5. Export the raw data and plot the time-dependence of product fluorescence or absorbance in GraphPad Prism 5. Inspect the quality of the data. Fit the data to derive the apparent inhibition potency and applicable kinetic constants (*see* **Note 4**). An example is given to illustrate the kinetic profiling of Boceprevir and Telaprevir (*see* **Note 5**).
6. Repeat the experiment, with or without refinement for compound concentration range. Perform preincubation/dilution experiment if very slow binding inhibition is suspected (*see* **Notes 1**, **3**, and **4**).

## 4 Notes

The following guidelines should be carefully followed in experimental design and data analysis to generate trustworthy kinetic profiling data.

1. With or without preincubation: progress curve analysis can be started with or without preincubation between enzyme and inhibitor. The comparison and contrast for both modes of experiment are outlined in Table 2. In the "no preincubation" mode, the reaction is initiated by adding enzyme to mixtures of substrate and dose-varied inhibitor often arrayed in microtiter wells. The reaction is monitored for a sufficient length of time with proper sampling rate to allow the observation of any time-dependent inhibition that is not due to significant substrate depletion, product inhibition and/or enzyme instability. In the "preincubation/dilution" mode, the reaction is often initiated by diluting preincubated mixtures of enzyme and dose-varied inhibitor into a solution containing substrate. Usually, only competitive or noncompetitive inhibitor is preincubated with the enzyme, since an uncompetitive inhibitor will not bind well to the enzyme in the absence of substrate (or the uncompetitive co-substrate for a bi-substrate system). Substrate depletion and/or product formation is then monitored to observe time-dependent enzyme activity recovery from inhibition. A small dilution (~10-fold or less) is occasionally used to monitor the reestablishment of steady state for the enzyme, inhibitor, and their complex(es), allowing comparable kinetic profiling which is more simply achieved without preincubation/dilution. A large (>100 fold) dilution (called a jump dilution) to a detectable concentration of active enzyme (i.e., above the

**Table 2**
**Progress curve analysis with or without preincubation**

| | No preincubation | Preincubation/dilution |
|---|---|---|
| Reaction initiation | Adding enzyme to premixed inhibitor (dose-varied) and substrate (usually fixed near or above $K_m$, with exceptions) | Diluting preincubated enzyme and inhibitor (dose-varied or fixed) into a solution usually with substrate (occasionally without substrate for very slow activity recovery) |
| Reaction monitoring | Observing both dose- and time-dependent inhibitions<br>Final steady state velocities less than initial steady state velocities | Observing both dose- and (preincubation and reaction) time-dependent recovery from inhibitions<br>Final steady state velocities more than initial steady state velocities |
| Major advantages | Simple progress curve analysis; full kinetic profiling possible | Reversibility check; more direct measurement of (overall) dissociation rate constant |
| Major disadvantages | Long reaction time for very slow inhibition may incur undesirably substantial substrate depletion, product inhibition and/or enzyme activity loss over time | Enzyme must be stable during long preincubation<br>Difficult to achieve enzyme binding with uncompetitive inhibitors in the absence of allosteric substrate |

lower limit of the assay) and final inhibitor concentrations well below the apparent inhibition constant is often used to assess the practical reversibility of inhibition and to more directly measure the dissociation rate constant. The following discussion assumes large dilutions are used after preincubation for achieving these objectives, unless otherwise noted for head-to-head comparison with and without preincubation. The preincubation time should be well within the timeframe of enzyme stability and long enough to achieve equilibrium. Sometimes the preincubation time is varied before large dilutions to assess an irreversible inactivator's inactivation constant, but this can often be more simply done without preincubation. For very slow binding inhibitors, the preincubated enzyme/inhibitor may also be massively diluted into a solution without substrate. Aliquots taken at various time points during an extended period commensurate with the overall slow dissociation are then spiked with a substrate solution for brief monitoring (usually <10-min) of enzyme activity recovery.

2. Concentrations of reagents: the progress curve analysis calls for judicial choice of reagents' concentrations. Table 3 summarizes the key guideline for concentrations. The final enzyme concentration for either the "no preincubation" or "preincubation/dilution" mode will have to be (1) well above the lower limit

**Table 3**
**Guideline for choice of reagents' concentrations**

| | No preincubation | Preincubation/dilution |
|---|---|---|
| Enzyme concentration | Well above lower limit of detection, below upper limit for steady state velocity and preferably, to avoid tight binding condition, also below the inhibitor's $K_i^{app}$ or $K_i^{*app}$ in the presence of substrate | Up to ~1,000× $K_i$ during preincubation, usually dilute a large fold so that the enzyme concentration is well below the inhibitor's $K_i^{app}$ in the presence of substrate and below the upper limit for steady state velocity, but well above the lower limit for detecting the recovery of activity |
| Substrate concentration | Near or preferably well above $K_m$ (especially for highly potent, competitive inhibitors) or well below $K_m$ (especially for highly potent, uncompetitive inhibitors), unless limited by solubility, availability or upper limit of detectability | Saturating ($\gg K_m$) substrate concentration in assay buffer to minimize *competitive* inhibitor rebinding, unless limited by solubility, availability or upper limit of detectability |
| Inhibitor concentration | Range of dose that produces ~5–95 % initial and final steady state inhibitions | Nearly stoichiometric to enzyme during preincubation (but doses are varied widely for head to head comparison with "no preincubation" experiment) |

of the assay so that sufficient signal for enzyme activity is detectable, (2) well below the upper limit for steady state enzyme kinetics so that substrate depletion and potential product inhibition are insignificant throughout the course of the assay, and (3) preferably also below the inhibitor's apparent inhibition constant in the presence of substrate to avoid tight binding condition. When working with very diluted enzyme concentrations (relative to compound's apparent $K_i$ but usually low to sub-nM), proper care should be taken to minimize enzyme loss due to sticking to plastic wells by including detergents below critical micelle concentrations (e.g., 0.01 % Triton X-100 or 0.005 % Tween-20) or ~0.1 % bovine serum albumin in the assay buffer. For "preincubation/dilution" experiment, the concentration of enzyme (and of nearly stoichiometric compound, unless for head-to-head comparison with the "no preincubation" experiment) should be high enough during preincubation to allow rapid and near saturation binding between enzyme and inhibitor. The preincubation should also be long enough to allow isomerization and/or modification to completion. This is usually accomplished in ~30 min with concentration up to 1,000 times the intrinsic inhibition constant ($K_i$) of the enzyme–inhibitor complex, provided that the dissociation rate constant is faster than $10^{-6}$/s and the maximal isomerization or modification rate is faster than $10^{-3}$/s. Longer time and/or high concentration will be needed for rare cases outside the above rate limits.

For both modes of experiment, the final substrate concentration should be near, or preferably well above the Michaelis-Menten constant ($K_m$) in most cases, in conjunction with low enough enzyme concentration, to minimize substrate depletion. Substrate in great excess over its $K_m$ also has the advantage of increasing the apparent inhibition constant of a potent, competitive inhibitor to avoid or minimize tight binding condition. It should be noted that, however, for substrate-*uncompetitive* inhibitors, substrate concentrations well *below* the $K_m$ will be helpful to avoid or minimize tight binding condition by increasing the apparent inhibition constant of a highly potent, *uncompetitive* inhibitor (*see* Table 1 legend for this reversal effect). In "preincubation/dilution" experiments, substrate in great excess over the $K_m$ also minimizes *competitive* inhibitor rebinding after dissociation, allowing dissociation rate constant to be more accurately determined. Notable exceptions to this general rule of thumb occur when the substrate has limited solubility above its $K_m$ or limited or costly supply, or when the high substrate concentration causes the signal to exceed the upper limit of the dynamic range of detection. In these cases, one will have to use substrate concentration below the $K_m$ in conjunction with a low enough enzyme concentration to create a near steady state condition with no more than ~15 % but still detectable substrate depletion during the course of the assay. If this is not possible, a more efficient substrate and/or more sensitive detection method will have to be employed.

Inhibitor concentrations should be chosen to provide a wide range (~5–95 %) of both initial and final steady state inhibition for progress curve analysis in the "no preincubation" mode. Designing the experiment to have a wide range of inhibitor concentrations (from well above the larger to well below the smaller of the apparent inhibition constants) will yield a set of progress curves that leave very little unoccupied "white space" below the uninhibited linear line, creating a nicely spaced shell-like pattern (e.g., *see* Fig. 1). Sometimes one may perform finer (~1.5-fold) dilutions around the critical concentrations near the apparent initial and overall inhibition constants ($K_i^{app}$ and $K_i^{*app}$) to better define these parameters [22]. In "preincubation/dilution" mode to measure the "curving up" transition's observed rate constant ($k_{obs}$) approximating either $k_{off}$ (for 1-step SBI) or $k_6$ (backward isomerization/modification rate constant for 2-step SBI), one usually preincubates compound and enzyme with near stoichiometric binding, followed by a jump dilution into a substrate-containing assay buffer to yield

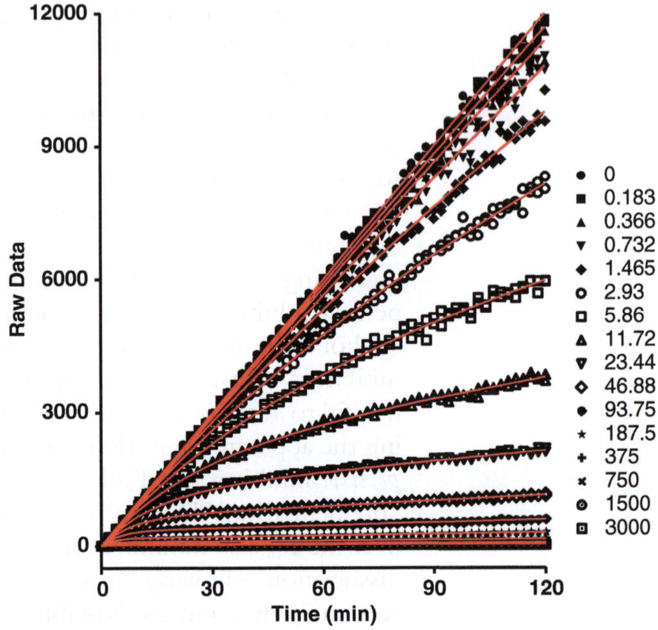


**Fig. 1** Global progress curve analysis (GPCA) of simulated data. Simulated data at the indicated compound doses (0–3,000 nM) were generated in GraphPad Prism 5 using equations for 2-step, competitive, slow but not tight binding model (Table 5) with the following parameters: $K_i = 100$ nM, $K_i^* = 1$ nM, $k_6 = 0.006$/min, [S] = $K_m = 5{,}000$ nM, $v_0 = 100$ arbitrary unit/min. A 2 % relative noise was randomly introduced to the simulated data. The data were then fit to the same equations to derive the globally floated parameters: $K_i = 102.9$ nM, $K_i^* = 1.024$ nM, $k_6 = 0.006192$/min, which are close to the expected values. Data are in *dark symbols* and fittings in *red lines*

a final concentration well below the apparent inhibition constant (for 1-step) or the apparent overall inhibition constant (for 2-step). Maintaining stoichiometric concentrations of inhibitor to enzyme is important for preincubation/dilution experiments (unless in head to head dose-dependent comparison with the "no preincubation" experiment), since significantly below or above stoichiometric binding will shrink the window between initial and final steady state velocities, making the observation of curvature difficult [23].

3. Time window of observation: the time window of observation is critical for measuring accurately the kinetic and equilibrium constants. Lack of informed attention to the time window of observation can result in misleading parameters and missed opportunity to identify potent, slow binding inhibitors. In general, a progress curve analysis will fail to identify slow binding inhibitors that have much longer residence time than the window of observation. For example, a short, 30-min observation

**Table 4**
**Guideline for time window of observation**

| Observation | Shape of 2-h progress curves without preincubation | Shape of 2-h progress curves after overnight preincubation | Interpretation of the inhibitor type | Need for redesign? |
|---|---|---|---|---|
| 1. | Linear | Linear, equivalent slopes | Fast binding inhibitor (FBI) | No need to redesign, unless it is FTBI |
| 2. | Curving down to new steady state velocities | Curving up to similar steady state velocities | Slow binding inhibitor (SBI) with moderate residence time | No need to redesign, unless it is STBI |
| 3. | Linear | Linear with equivalent slope for uninhibited control, but significantly lower slopes or slower recovery with compound | Slow binding inhibitor (SBI) with very long residence time | Need to lengthen preincubation time and the time window of observation |

will under-report the true potency for, and fail to identify, slow binding inhibitors that dissociate (for 1-step) or isomerize/modify (for 2-step) much slower than 30-min. In the drug industry's drive to discover efficacious anti-infective agents that have long residence time lasting many hours, it is imperative that the time window of observation, with adequate sampling rate to define the curvature well for at least some of the compound doses, be adequately designed in every progress curve analysis. Table 4 outlines the guideline for time window of observation.

If 2-h long progress curves are *linear* across all compound doses with comparable slopes for a given dose between data sets in both the "no preincubation" and the "preincubation/dilution" experiment (the latter with long preincubation, typically up to overnight, followed by a large dilution), then for all practical purposes the inhibitor is deemed *fast* binding with residence time too short to define without a faster sampling method (such as a stopped flow device), see Observation 1 (Table 4).

If the curvatures approaching similar new steady state velocities at a given compound dose are observed with adequate sampling rate in 2-h progress curves with and without overnight preincubation, then the inhibitor must be slow binding with moderate residence time around the time window of 2-h observation and substantially less than the overnight preincubation time, see Observation 2 (Table 4).

If, however, after overnight preincubation and dilution, the observed linear or slightly curvilinear progress curves have significantly lower slopes or depressed recovery rate at a give compound dose while the uninhibited control after preincubation/dilution gives comparable slope as in the "no preincubation" experiment (indicating no time-dependent loss of enzyme activity during preincubation), then the inhibitor must be slow binding with residence time much longer than the 2-h time window of observation and possibly even longer than the overnight preincubation time, see Observation 3 (Table 4). One may have to redesign the experiment with longer preincubation and longer time window of observation, as long as the enzyme is stable and comparable near steady state kinetics is maintained for uninhibited reaction controls with and without preincubation.

For practical convenience, we prefer to perform the progress curve analysis without preincubation first. If dose-dependent curvatures with a new steady state established for some of the compound doses are observed, one may proceed directly to data analysis without running long preincubation experiment. Only if linear progress curves are observed across all doses would one consider verifying if there is time-dependent, very slow inhibition or time-independent, very fast inhibition. Redesigning the assay may also be necessary if tight binding is suspected.

4. Data analysis: Table 5 lists the key equations that have been worked out for progress curve analysis (often without preincubation) according to the inhibitor types [17–21]. Readers should not be confused by the variant notations for otherwise equivalent formulas used by different authors and be aware of occasional typographical errors in some of the publications. In the "preincubation/dilution" mode, the initial steady state velocities will necessarily be smaller than the final steady state velocities for SBI (*see* Eq. 12 in [18]). Attention should be paid to the inherent assumptions with each model. For 2-step SBI a fast first step followed by a slower second step is often assumed. This rapid equilibrium approximation is inadequate when the two steps are of comparable rates for dissociation and isomerization/modification, which may be better described by a more general steady state approximation [19, 24].

   For linear or curvilinear progress curves across all compound doses, the analysis is usually straightforward. One can perform a global progress curve analysis (GPCA) fitting all the curves at once by globally floating shared parameters (*see* Fig. 1). Alternatively, one can replot the dose-dependent observables or their ratios ($v_i/v_0$, $v_s/v_0$, $k_{obs}$) obtained from individual or local progress curve analysis before fitting to the appropriate equations

**Table 5**
**Key equations for progress curve analysis (no preincubation)**

| Type of *competitive* inhibition | Equations for GraphPad Prism 5 |
|---|---|
| 1-Step fast binding (tight or not) | $v_0 = k_{cat}$*Enzyme/(1 + $K_m$/Substrate)<br>$K_i^{app} = K_i$*(1 + Substrate/$K_m$)<br>$Q$ = Enzyme + Compound + $K_i^{app}$<br>$v_i = v_0$*(1 − ($Q$ − SQRT($Q$*$Q$ − 4*Enzyme*Compound))/2/Enzyme)<br>$Y$ = background + $v_i$*$X$ |
| 1-Step slow (but not tight) binding | $K_i^{app} = K_i$*(1 + Substrate/$K_m$)<br>$v_s = v_0$/(1 + Compound/$K_i^{app}$)<br>$k_{obs} = k_{off}$*(1 + Compound/$K_i^{app}$)<br>$Y$ = background + $v_s$*$X$ + ($v_0 - v_s$)/$k_{obs}$*(1 − exp(−$k_{obs}$*$X$)) |
| 1-Step slow and tight binding | $K_i^{app} = K_i$*(1 + Substrate/$K_m$)<br>$Q$ = SQRT[($K_i^{app}$ − Enzyme + Compound)$^2$ + 4*$K_i^{app}$*Enzyme]<br>$v_s = v_0$*(Enzyme − Compound − $K_i^{app}$ + $Q$)/2/Enzyme<br>$k_{obs} = k_{off}$*$Q$/$K_i^{app}$<br>$g$ = Enzyme/Compound*(1 − $v_s/v_0$)$^2$<br>$Y$ = back-ground + $v_s$*$X$ + ($v_0 - v_s$)*(1 − $g$)/$k_{obs}$/$g$*LN((1 − $g$*exp(−$k_{obs}$*$X$))/(1 − $g$)) |
| 2-Step slow (but not tight) binding (rapid equilibrium approximation) | $K_i^{app} = K_i$*(1 + Substrate/$K_m$)<br>$K_i^{starapp} = K_i^{star}$*(1 + Substrate/$K_m$)<br>$v_s = v_0$/(1 + Compound/$K_i^{starapp}$)<br>$v_i = v_0$/(1 + Compound/$K_i^{app}$)<br>$k_{obs} = k_6$*$v_i/v_s$<br>$Y$ = background + $v_s$*$X$ + ($v_i - v_s$)/$k_{obs}$*(1 − exp(−$k_{obs}$*$X$)) |
| 2-Step slow and tight binding | Use numerical integration method, or redesign assays using lower enzyme and/or higher ($\gg K_m$, for competitors) or lower ($\ll K_m$, for uncompetitors) substrate to create slow but not tight binding condition |

$X$ is time, $Y$ is the observed signal of product formation or substrate depletion. Constant parameters are those whose values are fixed in the constraint tab of GraphPad Prism 5 during the progress curve analysis, such as concentrations (in the same unit) of compound (equal to values in each column header), substrate, and enzyme (floated only for enzyme active site titration), uninhibited velocity ($v_0$), and Michealis-Menten constant ($K_m$ in the same unit as the above concentrations). Narrowly constrained variables include background and sometimes uninhibited velocity ($v_0$) to allow small but notable variations in assay setup. Globally shared and floated parameters include inhibition constant ($K_i$) and turnover rate ($k_{cat}$) for 1-step FBI, and also enzyme concentration for 1-step FTBI, $K_i$ and $k_{off}$ for 1-step SBI and also enzyme concentration for 1-step STBI, and $K_i$, $K_i^*$ and $k_6$ for 2-step SBI. For STBI, one should exclude the no compound control during the global fitting (after estimating $v_0$ value from the control), due to division by zero error. For uncompetitive and noncompetitive inhibitors or bi-substrate enzymes, please see the legend to Table 1 for formulas relating apparent $K_i$ or $K_i^*$ with $K_m$ or apparent $K_m$. $K_i^*$ is rewritten as Ki$^{star}$ in the equations to avoid confusion with multiplication symbol *

to derive apparent initial and overall inhibition constants [12] (Fig. 2). The results between the two methods are comparable to each other and, with the notable exception of very small $k_6$, reasonably close to the preset parameters used in generating the simulated data. This correspondence of preset versus fit parameters provides adequate validation for the progress curve

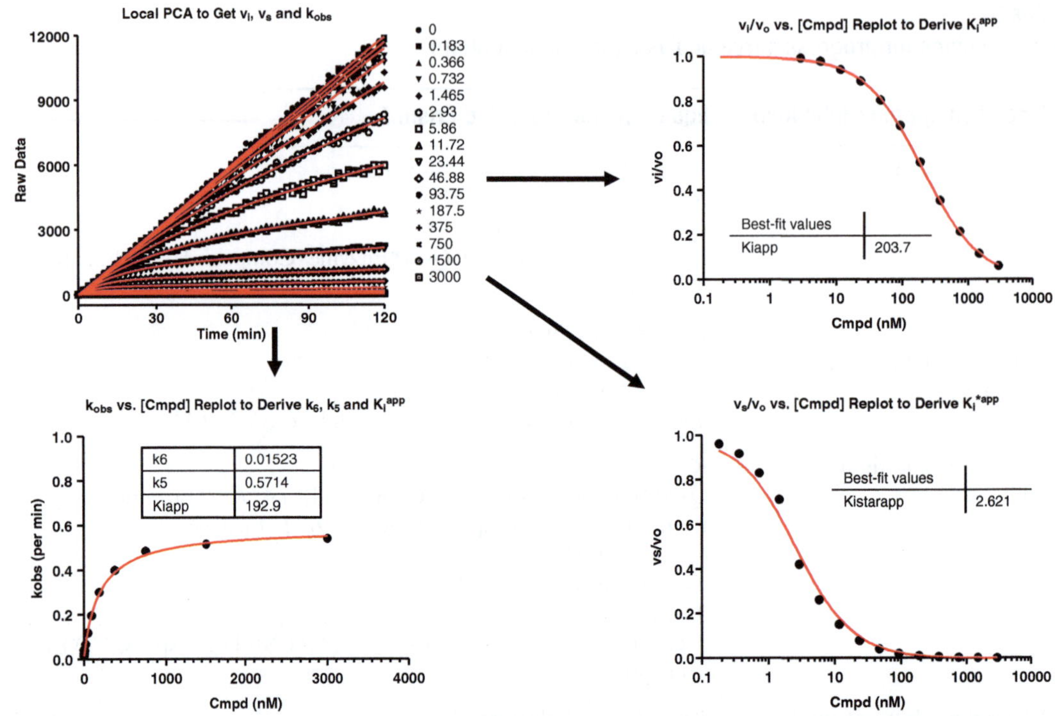


**Fig. 2** Local progress curve analysis (LPCA) plus replot method. Data were simulated as in Fig. 1. The data were then fit to $Y = \text{background} + v_s{}^*X + (v_i - v_s)/k_{obs}{}^*(1 - \exp(-k_{obs}{}^*X))$ to derive $v_i$, $v_s$ and $k_{obs}$ at each compound dose, and $v_0$ ($= v_i = v_s$) for no compound control. Then $v_i/v_0$ and $v_s/v_0$ ratios were replotted (*right side*) against compound concentrations and fit to $Y = 1/(1 + X/K_i^{app})$ and $Y = 1/(1 + X/K_i^{starapp})$ respectively, to derive apparent $K_i$ ($K_i^{app}$) and apparent $K_i^*$ ($K_i^{starapp}$) values which are divided by $1 + [S]/K_m$ (which is 2 in this example) to calculate intrinsic $K_i$ and $K_i^*$. The $k_{obs}$ versus [compound] replot (*lower left*) was fit to $Y = k_6 + k_5/(1 + K_i^{app}/X)$ to derive $k_6$, $k_5$ and apparent $K_i$. Notice that while calculated $K_i$ and $K_i^*$ values are reasonably close to the expected values, $k_6$ is overestimated significantly (0.015 per min versus the expected 0.006 per min) by this LPCA plus replot method, compared to the more accurate $k_6$ value from GPCA method (Fig. 1). In general, when $k_6$ is small and close to zero, LPCA plus replot method does not give reliable result. Data are in *dark symbols* and fittings in *red lines*

analysis, especially the more reliable GPCA method (Fig. 1) which we recommend.

For curvilinear progress curves of SBI, the analysis is dependent on the correct model used. One can always fit the same data to either 1-step or 2-step SBI model separately and compare the goodness of fit. Alternatively, one can take advantage of GraphPad's feature of head-to-head comparison between two different models (Fig. 3). Evidence in favor of 1-step or 2-step may also be identified from local progress curve analysis where all curves are individually fit to derive key observable parameters (as in Fig. 2). For example, the initial velocities ($v_i$) for 1-step SBI are usually unaffected by the inhibitor concentrations, whereas progressively slower initial velocities are observed for 2-step SBI (this can be difficult to ascertain if the isomerization/modification is fast).

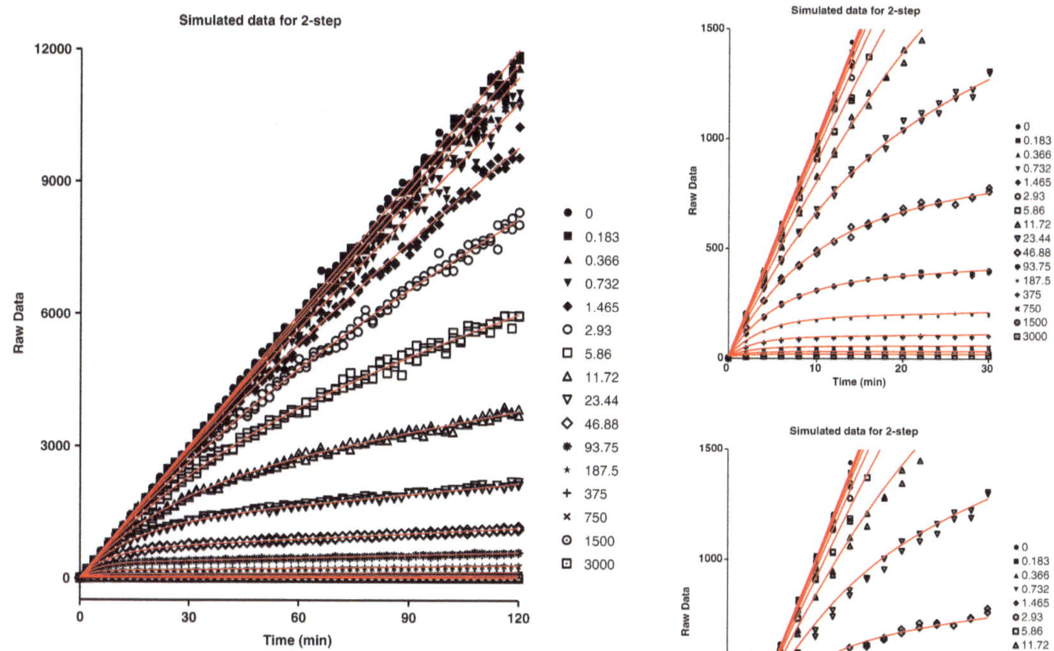


| Simpler model | Global Progress Curve Analysis for 1-Step SBI |
|---|---|
| Probability it is correct | 0.01% |
| Alternative model | Global Progress Curve Analysis for 2-Step SBI |
| Probability it is correct | 99.99% |
| Ratio of probabilities | 6716.81 |
| Preferred model | Global Progress Curve Analysis for 2-Step SBI |
| Difference in AICc | 17.62 |

**Fig. 3** Discriminating between 1-step versus 2-step model by comparative GPCA. Data for 2-step SBI were simulated as in Fig. 1. The data were then fit to a simpler model (1-step, slow binding inhibition) and an alternative model (2-step, slow binding model) using the comparison feature of Prism 5. Shown on the *right side* are curve fitting at the early time points to highlight the goodness of fitting between the most probable 2-step model (*upper right*) and the unlikely 1-step model (*lower right*). Data are in *dark symbols* and fittings in *red lines*

The rates of curvature (as measured by $k_{obs}$) are linearly dependent on the inhibitor concentrations for 1-step SBI but curved to a horizontal asymptote for 2-step SBI. Sometimes it is not easy to distinguish between 1-step versus 2-step SBI when the first step is rather weak in affinity or when the second step is relatively fast. In that case, the 2-step model degenerates into 1-step.

Depending on the dissociation rate (for 1-step) or isomerization/modification rate and relative potency for initial versus overall inhibition (for 2-step), all or only some of the parameters may be reasonably determined from a typical 2-h progress curve analysis (Table 6). The number of globally shared and floated parameters should be kept to a minimum (up to 3 in most cases). No kinetic parameters can be derived for FBI and FTBI, unless with a faster sampling method (such as a stopped flow device). Recognizing the limitations of progress curve

**Table 6**
**Parameter determinability from a 2-h global progress curve analysis**

| $k_{off}$ or $k_6$ ($s^{-1}$) | 1-step binding | 2-step binding |
|---|---|---|
| $\geq 10^{-2}$ | $K_i$ only, unless with faster sampling | $K_i^*$ only, unless with faster sampling |
| $\sim 10^{-3}$ | $K_i$, $k_{off}$ | $K_i$, $K_i^*$, $k_6$, $k_5$<br>If $K_i/K_i^* < \sim 30$ |
| $\sim 10^{-4}$ | $K_i$, $k_{off}$ | $K_i$, $K_i^*$, $k_6$, $k_5$<br>If $\sim 10 < K_i/K_i^* < \sim 300$ |
| $\sim 10^{-5}$ | $k_{on}$ ($=k_{off}/K_i$) only, unless with longer sampling | $K_i$, and $K_i^*$ ($k_6$ and $k_5$ also if $\sim 100 < K_i/K_i^* < \sim 3{,}000$) |
| $\leq 10^{-6}$ | $k_{on}$ ($=k_{off}/K_i$) only, unless with far longer sampling | $K_i$ only (unless $K_i/K_i^* > \sim 1{,}000$) |

Active enzyme concentration may also be determined under tight binding conditions. See the last paragraph about tight binding inhibitors in the *Data Analysis* section

analysis is important in designing experiment, analyzing and reporting data.

Data analysis for tight binding inhibitors warrants special considerations. Tight binding often occurs when lead optimization improves the apparent inhibition potency ($IC_{50}$) near and beyond the assay limit imposed by the active enzyme concentration. Tight binding can also be identified by a plot of $v_0/v_i$ or $v_0/v_s$ ratios versus inhibitor concentrations. A concave-up curve in the ratio range under ~100 (and asymptotic beyond ~100) will be observed for tight binding inhibitors, while such plot is linear for non-tight binding inhibitors (see relevant equations in Table 5 for this behavior). If possible, it is best to convert a tight binding inhibitor to a classical non-tight binding inhibitor by using lower concentration of enzyme (well below the apparent inhibition constant) and/or far higher ($\gg K_m$, for competitive inhibitor) or far lower ($\ll K_m$, for uncompetitive inhibitor) concentration of substrate so that active enzyme concentration is less than 10 % of $K_i^{app}$ and $K_i^{*app}$. If this is not possible, or when the need for active site titration arises, one may resort to the following strategies. For 1-step STBI, it is relatively straightforward to determine both the active site concentration of enzyme and possibly other parameters such as $k_{off}$ and $K_i$ (or at least their ratio which is $k_{on}$, *see* Tables 5 and 6). The active enzyme concentration, but not the apparent inhibition constant ($K_i^{app}$), is best determined when the active enzyme concentration is far greater than $K_i^{app}$ (super tight binding condition). Both the active enzyme concentration and $K_i^{app}$ can be simultaneously determined under moderate tight binding condition when they are within tenfold of each other [25]. $K_i^{app}$, but not the active site enzyme concentration, can be well determined when $K_i^{app}$ is

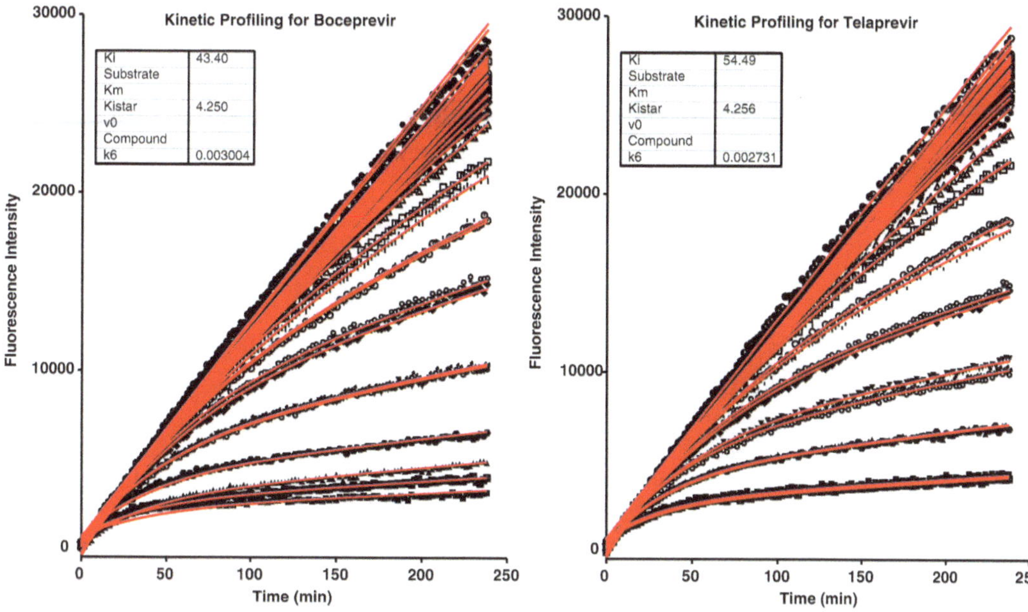


**Fig. 4** Progress curve analysis for Boceprevir and Telaprevir. Reactions were initiated by adding NS4A-tethered single chain NS3 protease of genotype 1b to a premixed solution of Boceprevir (*left*) or Telaprevir (*right*) and custom substrate peptide containing Europium fluorophore and QSY7 quencher, Ac-Cys(Eu)-Asp-Asp-Met-Glu-Glu-Abu-[COO]-Ala-Ser-Ala-Lys(QSY7)-amide. The final enzyme, dose-varied compound and substrate concentrations were 0.1 nM, 0–300 nM, and 25 nM, respectively, in the reaction buffer. The reaction was run in duplicate and monitored at 25 °C every 2 min for 4 h. Four hours were used to allow more reliable determination of the slow deacylation rate ($k_6$). Time resolved fluorescence was detected at 620 nm with excitation at 337 nm on a PHERAstar FS plate reader. The data were fit to a 2-step, slow binding model (Table 5) to derive the globally shared parameters including $K_i$, $K_i^*$ and $k_6$ (Table 7). Data are in *dark symbols* and fittings in *red lines*. Other parameters in Table 7 are calculated according to the proper formulas under Table 1

far greater than enzyme active site concentration (the preferred SAR-generating assay condition). Under all these conditions, it is preferable to float the active enzyme concentration during data fitting, unless it is predetermined accurately from an independent experiment or a two-staged fitting [22, 25]. For 2-step STBI, the data analysis is more complicated, since explicit integrated equation describing this type of inhibition is not available. It is best to use numerical integration method for data fitting. Numerical integration can also be useful for 1-step STBI when $k_{off}$ is very slow [26].

5. An example: progress curve analysis for Boceprevir and Telaprevir.

To illustrate the utility of progress curve analysis, we show in Fig. 4 such an analysis for Boceprevir and Telaprevir against genotype 1b HCV NS3 protease. The curvilinear progress curves were evident during the 4 h of monitoring inhibitions, suggesting that they are slow binding inhibitors. Mechanistic and structural studies have already established that Boceprevir and Telaprevir are slowly reversible, mechanism-based acylating

**Table 7**
**Kinetic profiling for Boceprevir and Telaprevir with HCV-1b NS3/4A protease**

| Parameter | Boceprevir | Telaprevir |
|---|---|---|
| $K_i$ (nM) | 43 ± 3.2 | 54 ± 4.3 |
| $K_i^*$ (nM) | 4.3 ± 0.21 | 4.3 ± 0.22 |
| $k_6$ (min$^{-1}$) | 0.0030 ± 0.00017 | 0.0027 ± 0.00016 |
| $k_5$ (min$^{-1}$) | 0.027 ± 0.0049 | 0.031 ± 0.0059 |
| $k_5/K_i$ (M$^{-1}$ s$^{-1}$) | 10,500 ± 2,700 | 9,600 ± 2,600 |
| $\tau$ (h) | >5.6 | >6.2 |

Parameters are defined in Table 1, derived from Fig. 4 and calculated with formulas from Table 1. Standard errors are also indicated

agents [13, 14]. Global fitting of the progress curves revealed that both drugs are comparably active against this enzyme, with low nM initial and overall affinities (Table 7). The lower estimate of the drug–target residence time as approximated by the reciprocal of $k_6$ (per legend to Table 1) is about 6 h. The in vivo residence time is unknown as it could be shortened due to higher body temperature (37 °C compared to 25 °C at which our experiment was conducted) and lengthened due to the effect of potential rebinding in the subcellular microenvironment [5]. The three times a day dosage for both Boceprevir and Telaprevir suggests that the in vivo residence time is unlikely more than 8 h. The acylation efficiency as approximated by $k_5/K_i$ (usually equivalent to $k_6/K_i^*$ if $K_i \gg K_i^*$) is on the order of ~10,000 M$^{-1}$ s$^{-1}$. The maximal acylation rate as approximated by $k_5$ ($=k_6(K_i/K_i^* - 1)$) is ~0.03/min. These levels of acylation efficiency and acylation rate assure adequate and relatively rapid target acylation during the initial loading phase of drug administration.

## Acknowledgment

The authors gratefully acknowledge the technical assistance of Edward DiNunzio, the careful reading of the manuscript by Michael Kavana, and the enthusiastic managerial support from Christine Brideau.

## References


1. Zhang R, Monsma F (2010) Binding kinetics and mechanism of action: toward the discovery and development of better and best in class drugs. Expert Opin Drug Discov 5:1023–1029
2. Copeland RA (2010) The dynamics of drug-target interactions: drug-target residence time and its impact on efficacy and safety. Expert Opin Drug Discov 5:305–310
3. Lu H, Tonge PJ (2010) Drug-target residence time: critical information for lead optimization. Curr Opin Chem Biol 14:467–474
4. Swinney DC (2009) The role of binding kinetics in therapeutically useful drug action. Curr Opin Drug Discov Dev 12:31–39
5. Vauquelin G (2010) Rebinding: or why drugs may act longer in vivo than expected from their in vitro target residence time. Expert Opin Drug Discov 5:927–941
6. Copeland RA, Pompliano DL, Meek TD (2006) Drug-target residence time and its implications for lead optimization. Nat Rev Drug Discov 5:730–739
7. Tummino PJ, Copeland RA (2008) Residence time of receptor-ligand complexes and its effect on biological function. Biochemistry 47:5481–5492
8. Zhang R, Monsma F (2009) The importance of drug-target residence time. Curr Opin Drug Discov Dev 12:488–496
9. Fang Y (2012) Ligand-receptor interaction platforms and their applications for drug discovery. Expert Opin Drug Discov 7:969–988
10. Vauquelin G (2012) Determination of drug-receptor residence time by radioligand binding and functional assays: experimental strategies and physiological relevance. Med Chem Commun 3:645–651
11. Andersson K, Karlsson R, Lofas S et al (2006) Label-free kinetic binding data as a decisive element in drug discovery. Expert Opin Drug Discov 1:439–446
12. Copeland RA (2005) Evaluation of enzyme inhibitors in drug discovery: a guide for medicinal chemists and pharmacologists. Wiley., Hoboken, NJ, pp 141–213
13. Kwong AD, Kauffman RS, Hurter P, Mueller P (2011) Discovery and development of telaprevir: an NS3-4A protease inhibitor for treating genotype 1 chronic hepatitis C virus. Nat Biotechnol 29:993–1003
14. Njoroge FG, Chen KX, Shih NY, Piwinski JJ (2008) Challenges in modern drug discovery: a case study of boceprevir, an HCV protease inhibitor for the treatment of hepatitis C virus infection. Acc Chem Res 41:50–59
15. Taremi SS, Beyer B, Maher M et al (1998) Construction, expression, and characterization of a novel fully activated recombinant single-chain hepatitis C virus protease. Protein Sci 7:2143–2149
16. Zhang R, Beyer BM, Durkin J et al (1999) A continuous spectrophotometric assay for the hepatitis C virus serine protease. Anal Biochem 270:268–275
17. Cha S (1976) Tight-binding inhibitors-III. A new approach for the determination of competition between tight-binding inhibitors and substrates-inhibition of adenosine deaminase by coformycin. Biochem Pharmacol 25: 2695–2702
18. Morrison JF, Walsh CT (1988) The behavior and significance of slow-binding enzyme inhibitors. Adv Enzymol Relat Areas Mol Biol 61:201–301
19. Sculley MJ, Morrison JF, Cleland WW (1996) Slow-binding inhibition: the general case. Biochim Biophys Acta 1298:78–86
20. Szedlacsek SE, Duggleby RG (1995) Kinetics of slow and tight-binding inhibitors. Methods Enzymol 249:144–180
21. Williams JW, Morrison JF, Duggleby RG (1979) Methotrexate, a high-affinity pseudosubstrate of dihydrofolate reductase. Biochemistry 18:2567–2573
22. Murphy DJ (2004) Determination of accurate KI values for tight-binding enzyme inhibitors: an in silico study of experimental error and assay design. Anal Biochem 327:61–67
23. Copeland RA, Basavapathruni A, Moyer M, Scott MP (2011) Impact of enzyme concentration and residence time on apparent activity recovery in jump dilution analysis. Anal Biochem 416:206–210
24. Kuzmic P (2008) A steady state mathematical model for stepwise “slow-binding” reversible enzyme inhibition. Anal Biochem 380:5–12
25. Kuzmic P, Elrod KC, Cregar LM et al (2000) High-throughput screening of enzyme inhibitors: simultaneous determination of tight-binding inhibition constants and enzyme concentration. Anal Biochem 286:45–50
26. Plesner IW, Bulow A, Bols M (2001) Accurate determination of rate constants of very slow, tight-binding competitive inhibitors by numerical solution of differential equations, independently of precise knowledge of the enzyme concentration. Anal Biochem 295:186–193
27. Cheng Y, Prusoff WH (1973) Relationship between the inhibition constant (KI) and the concentration of inhibitor which causes 50 per cent inhibition (I50) of an enzymatic reaction. Biochem Pharmacol 22:3099–3108
28. Yang J, Copeland RA, Lai Z (2009) Defining balanced conditions for inhibitor screening assays that target bisubstrate enzymes. J Biomol Screen 14:111–120

# Chapter 7

# A Novel Hepatitis C Virus NS5B Polymerase Assay of De Novo Initiated RNA Synthesis Directed from a Heteropolymeric RNA Template

**Eric Ferrari and Hsueh-Cheng Huang**

## Abstract

The hepatitis C virus (HCV) NS5B RNA-dependent RNA polymerase is essential for viral replication and a clinically validated antiviral target. Discovery of HCV polymerase inhibitors is greatly facilitated by the availability of a suitable biochemical assay using purified NS5B. We describe here a novel NS5B polymerase assay of de novo initiated RNA synthesis directed from a synthetic heteropolymeric RNA template modified with dideoxycytidine at its 3′-end. This assay has been successfully used for screening and characterization of both initiation and elongation inhibitors of the HCV NS5B polymerase.

**Key words** Hepatitis C virus, NS5B, RNA-dependent RNA polymerase, De novo RNA synthesis, Heteropolymer RNA template, Nucleotides, Initiation, Elongation, Polymerase inhibitors

## 1 Introduction

HCV, a positive-strand RNA virus of the *Flaviviridae* family, encodes a nonstructural protein 5B (NS5B) which possesses RNA-dependent RNA polymerase (RdRp) activity essential for replication of the viral RNA genome. The HCV NS5B polymerase has thus become an intensely pursued target for antiviral therapy, and a number of nucleoside and non-nucleoside inhibitors of NS5B have demonstrated clinical efficacy [1, 2]. The search for novel NS5B inhibitors generally begins with high-throughput screening of a chemical compound library in a biochemical assay using purified HCV NS5B. Towards this end we have developed a novel NS5B RdRp assay of de novo synthesis from a chemically modified heteropolymeric RNA template that has proved suitable for inhibitor screening and characterization [3–6].

Purified HCV NS5B polymerase has been shown to utilize a variety of homo- and hetero-polymeric RNA templates [7].

Edwin Yunhao Gong (ed.), *Antiviral Methods and Protocols*, Methods in Molecular Biology, vol. 1030,
DOI 10.1007/978-1-62703-484-5_7, © Springer Science+Business Media, LLC 2013

While assays employing homopolymer templates (e.g., poly-C primed with oligo-G, or poly-A primed with oligo-U) have been characterized and successfully used to identify NS5B inhibitors [8, 9], RNA synthesis directed from heteropolymer templates more closely mimics viral replication. A heteropolymer template also allows identification of nucleoside inhibitors without base specificity, which would not be possible under a homopolymer format.

Multiple modes of RNA synthesis have been described in vitro for the HCV polymerase: primer extension, either through self-priming (copy-back) or exogenously added primer; de novo initiation without a protein or RNA primer; a non-templated terminal nucleotide transferase (TNTase) activity; and end-to-end template switching [7, 10–14]. The presumed in vivo mechanism of NS5B-catalyzed viral replication is through de novo initiation, not primer-dependent elongation [7, 12, 13]. There is also evidence that TNTase activity may not be an inherent property of the viral polymerase but a purification artifact [11]. However, most of the NS5B enzyme assays that have been reported for inhibitor screening and characterization permit multiple routes of RNA synthesis [15–18]. To develop an RdRp assay selective for de novo initiated RNA replication only, we designed synthetic heteropolymer RNA templates (DCoH75 and HCV(−)UTR67 respectively) modified with dideoxycytidine (ddC) at 3′-end that eliminate the possibility of copy-back and TNTase activities, thus ensuring de novo initiation [3].

The formation of the initiation complex comprising the NS5B polymerase, RNA template, and incoming nucleotides is inefficient under a variety of in vitro conditions. Two distinct steps have been shown to occur prior to the transition to an efficient, processive elongation mode: first, a dinucleotide is formed through the first two bound nucleotides, accompanied by frequent release of abortive dinucleotide products; this is followed by a second, rate-limiting, step extending the dinucleotide, leading to the formation of a transient trinucleotide primer [19, 20]. We have found that high concentrations of NTP corresponding to the first three incorporated nucleotides are required to facilitate the initiation steps, in contrast to the much lower concentrations of NTP necessary for elongation. Thus, the apparent NTP *Km* using the DCoH75ddC RNA template ranges from 100 to 400 μM for initiation nucleotides, and drops dramatically to 0.03–0.09 μM for subsequent processive elongation [3]. The provision of an initiating trinucleotide, however, can overcome the high requirement for initiating nucleotides, a feature that we have exploited to reduce the overall substrate concentrations in the NS5B enzyme activity assay, thereby increasing the sensitivity for identifying compounds that may be competitive with nucleotides.

Using the protocol described here, we demonstrated that the assay readout of radiolabeled nucleotide incorporation captures multiple rounds of de novo initiation and elongation cycles [3].

It is amenable to high-throughput inhibitor screening, and can be modified for detailed studies of compound mechanism of action. This novel and versatile assay has been successfully used in the discovery and characterization of both non-nucleoside initiation inhibitors and nucleoside analog chain terminators of the HCV NS5B polymerase.

## 2 Materials

Use sterile, nuclease-free, molecular grade reagents. Prepare all solutions using molecular biology grade water and store all reagents at room temperature unless specified otherwise. RNA trinucleotide GAU was obtained from Oligos, Etc. All other RNA oligonucleotides and templates were synthesized, desalted, deprotected, PAGE purified, and purchased from Dharmacon/Thermo Scientific. Follow all hazardous and radioactive waste disposal guidelines for disposal of waste materials.

### 2.1 HCV NS5B Enzyme

1. HCV NS5B (genotype 1b-Con1)-Δ21-$His_6$ enzyme: Clone, express, and purify according to Ferrari et al. [21] (*see* **Note 1**). Store at −80 °C.
2. pET-21b vector (EMD Millipore).
3. Esch*erichia coli* JM109 (DE3) cells.

### 2.2 RNA Templates and RNA Trinucleotides

Store all at −20 °C (*see* **Note 2**).

1. DCoH75ddC (5′–3′): UGUGCCGGUC UUUCUGAACG GGAUAUAAAC CUGGCCAGCU UCAUCGAACA AGU UGCCGUG UCUAUGACAU AGAUddC.
2. HCV(−)UTR67ddC (5′–3′): GUGAAGACAG UAGUUCC UCA CAGGGGAGUG AUCUAUGGUG GAGUGUCGCC CCCAAUCGGG GGCUGGddC.
3. GAU.
4. GCC.

### 2.3 Reagents, Consumables, and Equipment

1. [α-$^{33}$P]CTP (10 mCi/mL, 3,000 Ci/mmol, PerkinElmer Life Sciences) for assays using DCoH75ddC template. [α-$^{33}$P]UTP (10 mCi/mL, 3,000 Ci/mmol, PerkinElmer Life Sciences) for assays using HCV(−)UTR67ddC template. Store at 4 °C (*see* **Note 3**).
2. rNTP set: ATP, CTP, GTP, UTP; 100 mM solution (GE Healthcare). Store at −20 °C (*see* **Note 4**).
3. Assay buffer: 20 mM HEPES, pH 7.3, 60 mM NaCl, 10 mM $MgCl_2$, and 100 μg/mL bovine serum albumin (BSA). Store at 4 °C (*see* **Note 5**).

4. RNasin: Recombinant ribonuclease inhibitor (20–40 units/mL, Promega).
5. 0.5 M EDTA, pH 8.0 (*see* **Note 6**).
6. 96-well sterile, conical bottom, natural, polypropylene, 450 μL well volume microplates (Thermo Scientific Nunc) (*see* **Note 7**).
7. 96-well, DE MultiScreen$_{HTS}$ filter plates (Millipore) (*see* **Note 8**).
8. MultiScreen Vacuum Manifold (Millipore) (*see* **Note 9**).
9. Mini-Orbital shaker (Bellco).
10. 0.5 M sodium phosphate buffer ($NaPO_4$), pH 7.0 (Cellgro).
11. Isotemp oven (Fisher Scientific).
12. 96-well multiscreen harvester opaque tape sealer (Millipore).
13. 96-well TopSeal microplate press-on adhesive sealing film (PerkinElmer Life Sciences).
14. Ready Safe Liquid Scintillation Cocktail (Beckman Coulter) (*see* **Note 10**).
15. TopCount NXT microplate scintillation and luminescence counter (Packard) (*see* **Note 11**).
16. DMSO.
17. 96-well sterile, polystyrene, universal lid cover (Costar) (*see* **Note 12**).
18. 1.5 mL RNase-free microcentrifuge tubes (*see* **Note 13**).
19. Molecular biology grade water (HyClone).

## 3 Methods

If assay buffer is premade, remove from 4 °C and warm to room temperature before use. Thaw all frozen reagents at room temperature except enzyme stock (*see* **Note 1**) and store on ice immediately afterward.

### 3.1 DCoH75ddC/GAU 50 μL Reaction Assembly

1. To all sample wells of a 96-well microplate, add 29 μL of Mix A which contains 1 unit RNasin, 10.3 mM DTT, 0.172 μM each of ATP, GTP, and UTP, 0.138 μM CTP, 0.431 mM GAU, 0.172 μM DCoH75ddC, and 3 μCi (0.020 μM) [$\alpha$-$^{33}$P]CTP brought up to 29 μL with assay buffer (*see* **Note 14**).
2. Add 1 μL of 100 % DMSO or water to all sample wells. DMSO concentration is 3.3 %.
3. Add 50 μL of 0.5 M EDTA to background control wells which contain 30 μL of Mix A and DMSO and shake the plate for 1 min on an orbital shaker. EDTA concentration is 0.313 M.
4. To all sample wells of the 96-well microplate, add 20 μL of Mix B which contains 3.75 mM DTT and 0.075 μM HCV NS5B

(1b-Con1)-Δ21-$His_6$ enzyme in assay buffer (*see* **Note 15**). Cover reaction plates with plate lid to minimize evaporation during assay incubation.

5. Final component concentrations in 50 μL reaction:

   1 unit RNasin, 7.5 mM DTT, 0.1 μM each of ATP, GTP, and UTP, 0.08 μM CTP, 0.25 mM GAU, 0.1 μM DCoH75ddC, 3 μCi (0.020 μM) [α-$^{33}$P]CTP, 2 % DMSO, and 0.03 μM HCV NS5B (1b-Con1)-Δ21-$His_6$ enzyme.
6. Shake plate for 1 min and incubate at room temperature for 2 h (*see* **Note 16**).
7. After incubation, add 50 μL of 0.5 M EDTA to all sample wells except background control wells to terminate assay reaction and shake plate for 1 min. Final concentration of EDTA in all sample wells is 0.25 M.

### 3.2 Reaction Product Capture

1. Remove 96-well DE filter plate lid and pre-wet plate by adding 50 μL of 0.5 M $NaPO_4$ buffer per well. Apply vacuum until buffer draws through the filters then shut off. Add back 50 μL of 0.5 M $NaPO_4$ buffer per well (*see* **Note 17**).
2. Transfer the entire volume of samples from assay plate to DE filter plate, pipette up and down to rinse tips, apply vacuum and shut off after sample draws through filter.
3. Wash the plate with 200 μL of 0.5 M $NaPO_4$ buffer per well, apply vacuum and shut off after buffer draws through filter. Repeat wash five more times (*see* **Note 18**).
4. Remove plate from manifold and blot bottom of plate on paper towels. Remove the plastic underdrain from the bottom of the plate and blot bottom again on paper towels (*see* **Note 19**).
5. Place the plate in a 37 °C oven for 30 min to dry the filters (*see* **Note 20**).
6. Remove the plate from oven and seal bottom with tape sealer. Add 40 μL per well of scintillation cocktail, seal top of plate sealing film, and shake plate for 1 min (*see* **Note 21**).
7. Count the plate on a TopCount NXT microplate scintillation counter (*see* **Note 22**).

### 3.3 Time Dependence Analysis of NS5B Activity

NS5B(1b-Con1)-Δ21-$His_6$ activity on the DCoH75ddC template with GAU trinucleotide initiator is measured in a time-course assay using the above described conditions (*see* Subheading 3.1, **steps 1–5** and 7, and Subheading 3.2, **steps 1–7**). Reactions are terminated at 20, 40, 60, 90, 120, and 150 min. As shown in Fig. 1, enzyme activity increases in a time dependent manner and is linear up to 2.5 h at room temperature. Approximately 8,000 counts per minute (CPM) (background subtracted) is observed at the 2 h time point (*see* **Note 23**).

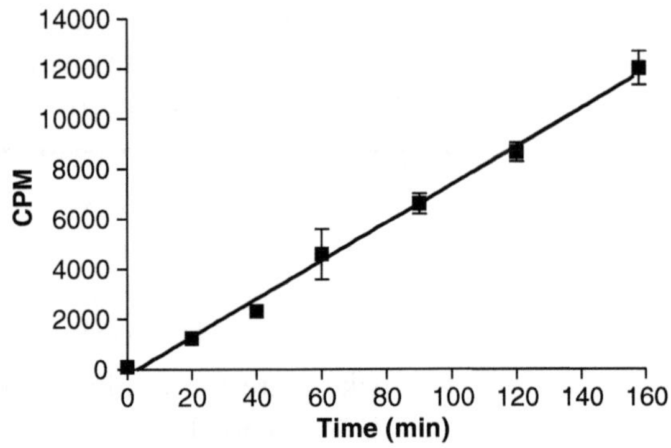


**Fig. 1** Time dependence analysis of NS5B polymerase activity. HCV NS5B (1b-Con1)-Δ21-$His_6$ polymerase activity on DCoH75ddC RNA template with GAU trinucleotide initiator as measured by [$^{33}$P]CTP incorporation. Each point represents triplicate samples

### *3.4 DCoH75ddC/GAU 96-Well Plate Reaction Assembly*

The description in Subheadings 3.1 and 3.2 is for a single 50 μL reaction. Table 1 summarizes the reagent requirements for a 96-well plate reaction assembly useful for testing compounds. To account for dead volume, sufficient reagents are included for 120 reactions, resulting in a 6 mL total volume (*see* **Note 24**). Figure 2 illustrates a typical 96-well plate assay assembly configuration for testing four compounds in duplicate.

### *3.5 Dose–Response Analysis of NS5B Inhibitors*

As described in Table 1 and Fig. 2, the assay provides a ready means for screening and evaluation of compound potency against the NS5B polymerase. Typically, a 2-h assay is performed at room temperature in the presence of 10-dose serial dilution of compounds. Percent inhibition at each compound concentration is calculated from the ratio of background-subtracted CPM to vehicle control. Dose–response curves and $IC_{50}$ values are then generated using a nonlinear three parameter fitting program in GraphPad Prism software. Figure 3 displays the dose–response of a non-nucleoside initiation inhibitor (compound A, $IC_{50} = 0.06$ μM) and a nucleoside triphosphate elongation inhibitor (compound B, $IC_{50} = 0.2$ μM) of the HCV NS5B polymerase (*see* **Note 25**).

## 4 Notes

1. HCV NS5B (1b-Con1)-Δ21-$His_6$ is a soluble 570 amino acid residue protein that results from a truncation of 21 amino acids at the C-terminus containing the membrane anchor domain of the full-length protein. NS5B(1b-Con1)-Δ21 cDNA with a polyhistidine tag ($His_6$) engineered at the C-terminus for ease of purification is cloned into the pET21b vector and transformed into JM109(DE3) cells. Details of

**Table 1**
**96-well plate assay assembly with Mix A and Mix B reagent stock concentrations, reagent volumes, and final reaction concentrations**

| **Mix A** | **3.48 mL** | **[Final]** |
|---|---|---|
| Assay Buffer | 3.291 mL | |
| RNasin 40 units/μL | 3.0 μL | 20 units/mL |
| 1 M DTT | 36 μL | 6 mM |
| 250 μM ATP | 2.4 μL | 0.1 μM |
| 250 μM GTP | 2.4 μL | 0.1 μM |
| 250 μM UTP | 2.4 μL | 0.1 μM |
| 250 μM CTP | 1.9 μL | 0.08 μM |
| 20 mM GAU | 75 μL | 0.25 mM |
| 20 μM DCoH75ddC RNA Template | 30 μL | 0.1 μM |
| 3.3 μM, 10 μCi/μL [$^{33}$P]CTP | 36 μL | 0.02 μM, 60 μCi/mL |
| **Mix B** | **2.40 mL** | **[Final]** |
| Assay Buffer | 2.390 mL | |
| 1 M DTT | 9.0 μL | 1.5 mM |
| NS5B-Δ21 Enzyme (150 μM, 9.7 mg/mL) | 1.2 μL | 30 nM |

cloning, expression and purification can be found in ref. 21. Store enzyme in small aliquots to avoid repeated freeze–thaw cycles and in a buffer containing 20 % glycerol and at least 300 mM NaCl as the protein will precipitate at low salt concentrations. Remove from freezer just before use and place immediately on ice to thaw.

2. Confirm amount of all RNAs received by resuspending in nuclease-free water then measure $OD_{260nm}$ and dilute to working solution. Avoid multiple freeze–thaws by storing in small aliquots.
3. Note that the half-life of radioactive $^{33}$P is 25 days. Fresh isotope should be used at all times to achieve maximum CPM readout. Variability among different lots of radiolabeled nucleotide is not common but has been observed. Test each new lot for proper signal over background before use. The signal-to-noise ratio should be at least five to tenfold using described conditions.
4. These are received in individual tubes. It is permissible to prepare a mixture of four NTPs and store in small aliquots to avoid multiple freeze–thaws.

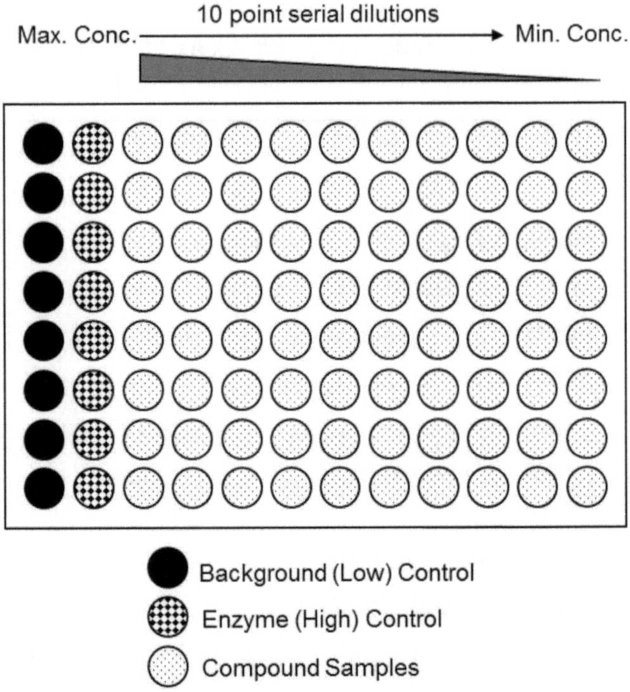


**Fig. 2** A typical 96-well plate configuration for the assay. 96-well plate assay setup showing the configuration of background and enzyme controls with 10 point serial dilutions of typically four compounds per plate tested in duplicate

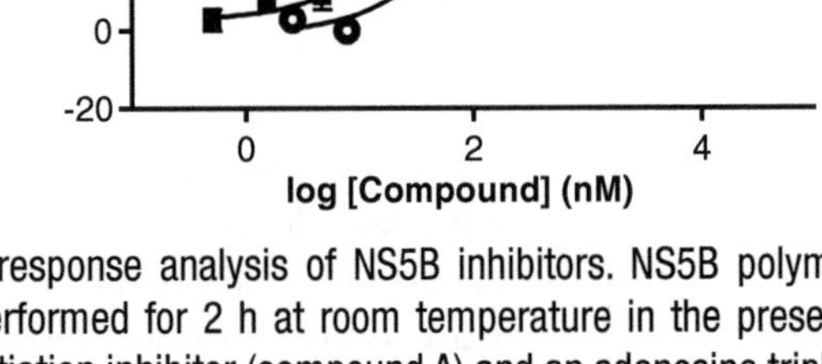


**Fig. 3** Dose–response analysis of NS5B inhibitors. NS5B polymerase activity assay was performed for 2 h at room temperature in the presence of a non-nucleoside initiation inhibitor (compound A) and an adenosine triphosphate analog chain terminator (compound B). Percent inhibition at each compound concentration was derived from the ratio of background-subtracted CPM in duplicate to vehicle control. Dose–response curves and $IC_{50}$ values were generated by a nonlinear three parameter fit program in GraphPad Prism

5. Assay buffer components are combined (except for DTT and RNasin which are added separately) and filtered through a 0.22 μm sterile filter unit. The buffer is stable for at least 2 months when stored at 4 °C. Use Fraction V graded BSA. DTT is made fresh and both DTT and RNasin are added to the buffer on the day of the assay. For 1 L, add 20 mL of 1 M HEPES, pH 7.3, 10 mL of 1 M $MgCl_2$, 12 mL of 5 M NaCl, 10 mL of 10 mg/mL BSA, bring up to 1 L with sterile, nuclease-free water and filter. Buffer components are purchased solutions except for DTT and BSA. DTT is made fresh as a 1 M solution and BSA as a 10 mg/mL solution in sterile nuclease-free water and both solutions are filtered through 0.22 μm sterile filter units.
6. EDTA is used to stop the polymerase reaction and since enzyme is added to the background controls wells (after the addition of EDTA) it is important to use fresh EDTA so there is no breakthrough enzyme activity. Background control CPM values should be no more than ~500.
7. 96-well, sterile, flat bottom microplates have been tested and found not to work well. It is difficult to transfer the entire volume of the stopped reaction samples from the flat bottom microplate to the DE filter plate, as residual sample is often found left behind.
8. DE $MultiScreen_{HTS}$ filter plates (Millipore) contain positively charged DEAE membranes, which permit capture of the negatively charged RNA reaction products. Variability among DE plates from the same lot is not common but has been observed. Problematic filter plates can sometimes be identified by the DE filters having a yellowish color which has been associated with high background CPM, whereas a good performing plate is generally bright white.
9. Residual $NaPO_4$ buffer can dry and crystallize on the manifold and in the tubing which can cause the vacuum pressure to change. It is important to maintain the maximum recommended vacuum of 4–8 Hg to efficiently capture and wash samples, so rinse manifold and tubing with nuclease-free water before and after use.
10. As of this writing Beckman Coulter is no longer in the liquid scintillation market. National Diagnostics has been recommended as a source for replacement cocktails. Ecoscint A would be the equivalent to Ready Safe but an activity comparison has not been performed using this assay.
11. Calibrate and normalize the instrument's detectors according to the manufacturer's instructions before any samples are counted.
12. Keep plates covered between reaction component additions and during incubation period. These lids are not the best for stacking plates as plates tend to slide easily.

13. Use these microcentrifuge tubes for making and storing reagent stocks and dilutions.
14. These reagent concentrations are not the final concentrations in the reaction. The final concentrations are listed in Subheading 3.1, **step 5**. Add reagents to the assay buffer in order listed to allow efficient mixing and to generate the least amount of radioactive waste. For the HCV(−)UTR67ddC/GCC 50 μL reaction assembly add 0.172 μM each of ATP, GTP, and CTP, 0.138 μM UTP, 0.431 mM GCC, 0.172 μM HCV(−)UTR67ddC, and 3 μCi (0.020 μM) [α-$^{33}$P]UTP brought up to 29 μL with assay buffer.
15. These reagent concentrations are not the final concentrations in the reaction. The final concentrations are listed in Subheading 3.1, **step 5**. Add Mix B to plate immediately after adding enzyme. Enzyme activity decreases over time, which is especially pronounced when it is the only reagent in the assay buffer.
16. Incubating for longer periods of time or increasing the concentrations of NTP in the reaction will increase the signal but make sure the reaction remains within the linear range of the assay. Refer to Fig. 1 and ref. 3.
17. Pre-wetting is required before sample capture. Plate should attach to manifold as the vacuum establishes a seal. If there is not a seal, help to create one by pressing down on the center edges of the plate. The DE filter plate lid can be used to calibrate the manifold vacuum by centering it over the manifold support grid, applying vacuum until a seal is formed and adjusting the vacuum control gauge. The lid can also be used to cover the plate between wash steps to avoid sample contamination. Adding back buffer allows for efficient transfer and mixing of sample in the well.
18. Washing the plate six times with $NaPO_4$ buffer efficiently removes any unincorporated $^{33}$P-CTP from the wells. Fewer washes will increase background CPM.
19. Blotting the plate captures any residual buffer left after filtration. Removing the plastic underdrain can be somewhat laborious. Use thumb and fore finger to grasp one corner and slowly peel away. Needle nosed pliers can also be used. Note that the underdrain and paper towels will be wet with radioactive waste so treat these as potential sources of radioactive contamination. Do not flip plate upside down while removing the underdrain to ensure no accidental detachment of filters from the bottom of the wells.
20. Dry time can be shortened to 20 min and extended to 1 h without loss of signal but extending dry time increases the risk that the filters will detach from the bottom of the wells (*see* **Note 21**).
21. Be careful handling the plate after drying. The dried filters can easily detach from the bottom of the wells and become airborne

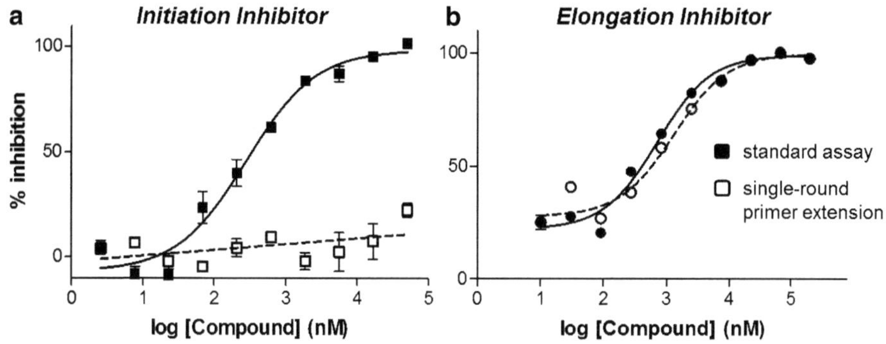


**Fig. 4** Mechanism of action evaluation of NS5B inhibitors. Mechanism of action studies were performed by comparing inhibitor dose–response in a typical NS5B polymerase assay as described in Subheading 3.5 and in a single-round primer extension assay (15 min reaction after preincubation of enzyme–primer–template for 22 h at room temperature). An initiation inhibitor (**a**) is active in an assay with multiple rounds of initiation and elongation cycles, but inactive when the assay captures only elongation. An elongation inhibitor (**b**), on the other hand, shows similar potency in both assays

if plate is tilted or turned upside down. 40 μL per well of scintillation cocktail is sufficient to cover the filter. Adding more volume does not significantly increase the signal.

22. TopCount detectors should be normalized according to manufacturer's instructions prior to using the assay protocol to count samples for the first time. Assay protocol is set up to count each well for 15 s. Raw data files are exported to Excel where background is subtracted from all samples. Data is plotted using GraphPad Prism software.
23. Similar time-course results are observed using the HCV(−) UTR67ddC template and GCC trinucleotide initiator.
24. Prepare compounds in a 96-well sterile, conical bottom, natural, polypropylene plate with compound serial dilutions in 100 % DMSO or water at 50× the final concentration according to plate map shown in Fig. 2. Transfer 1 μL per well from the compound serial dilution plate to the assay plate.
25. Mechanism of action of novel polymerase inhibitors can be evaluated by a modified form of this assay that permits single-round elongation only. Briefly, the initiating trinucleotide is replaced by 100 nM of an 11-mer RNA primer complementary to the 3′-end of the RNA template (5′-GAUCUAUGUCA-3′ in the case of DCoH75ddC template). After preincubation of enzyme-primer-template for 16–24 h at room temperature, RNA synthesis is initiated by the addition of nucleotides in the presence or absence of an inhibitor. The burst-phase activity in a reaction carried out for 15 min at room temperature represents single-round elongation of preformed initiation complexes (for details *see* ref. 3). Whereas a chain terminator will inhibit single-round primer extension, an initiation inhibitor loses its potency in such a setting (Fig. 4).

## References


1. Powdrill MH, Bernatchez JA, Götte M (2010) Inhibitors of the hepatitis C virus RNA-dependent RNA polymerase NS5B. Viruses 2:2169–2195
2. Beaulieu PL (2009) Recent advances in the development of NS5B polymerase inhibitors for the treatment of hepatitis C virus infection. Expert Opin Ther Pat 19:145–164
3. Ferrari E, He Z, Palermo RE, Huang H-C (2008) Hepatitis C virus NS5B polymerase exhibits distinct nucleotide requirements for initiation and elongation. J Biol Chem 283:33893–33901
4. Cheng CC, Huang X, Shipps GW Jr, Wang YS, Wyss DF, Soucy KA, Jiang CK, Agrawal S, Ferrari E, He Z, Huang H-C (2010) Pyridine carboxamides: potent palm site inhibitors of HCV NS5B polymerase. ACS Med Chem Lett 1:466–471
5. Anilkumar GN, Lesburg CA, Selyutin O et al (2011) I. Novel HCV NS5B polymerase inhibitors: discovery of indole 2-carboxylic acids with C3-heterocycles. Bioorg Med Chem Lett 21:5336–5341
6. Chen KX, Lesburg CA, Vibulbhan B et al (2012) A novel class of highly potent irreversible hepatitis C virus NS5B polymerase inhibitors. J Med Chem 55:2089–2101
7. Oh JW, Ito T, Lai MM (1999) A recombinant hepatitis C virus RNA-dependent RNA polymerase capable of copying the full-length viral RNA. J Virol 73:7694–7702
8. Dhanak D, Duffy KJ, Johnston VK et al (2002) Identification and biological characterization of heterocyclic inhibitors of the hepatitis C virus RNA-dependent RNA polymerase. J Biol Chem 277:38322–38327
9. Beaulieu PL, Bös M, Bousquet Y et al (2004) Non-nucleoside inhibitors of the hepatitis C virus NS5B polymerase: discovery and preliminary SAR of benzimidazole derivatives. Bioorg Med Chem Lett 14:119–124
10. Behren SE, Tomei L, De Francesco R (1996) Identification and properties of the RNA-dependent RNA polymerase of hepatitis C virus. EMBO J 15:12–22
11. Lohmann V, Körner F, Herian U, Bartenschlager R (1997) Biochemical properties of hepatitis C virus NS5B RNA-dependent RNA polymerase and identification of amino acid sequence motifs essential for enzymatic activity. J Virol 71:8416–8428
12. Luo G, Hamatake RK, Mathis DM et al (2000) *De novo* initiation of RNA synthesis by the RNA-dependent RNA polymerase (NS5B) of hepatitis C virus. J Virol 74:851–863
13. Zhong W, Uss AS, Ferrari E et al (2000) De novo initiation of RNA synthesis by hepatitis C virus nonstructural protein 5B polymerase. J Virol 74:2017–2022
14. Carroll SS, Sardana V, Yang Z et al (2000) Only a small fraction of purified hepatitis C RNA-dependent RNA polymerase is catalytically competent: implications for viral replication and in vitro assays. Biochemistry 39:8243–8249
15. Tomei L, Altamura S, Bartholomew L et al (2003) Mechanism of action and antiviral activity of benzimidazole-based allosteric inhibitors of the hepatitis C virus RNA-dependent RNA polymerase. J Virol 77: 13225–13231
16. Howe AY, Bloom J, Baldick CJ et al (2004) Novel nonnucleoside inhibitor of hepatitis C virus RNA-dependent RNA polymerase. Antimicrob Agents Chemother 48: 4813–4821
17. Carroll SS, Tomassini JE, Bosserman M et al (2003) Inhibition of hepatitis C virus RNA replication by 2′-modified nucleoside analogs. J Biol Chem 278:11979–11984
18. Lam AM, Murakami E, Espiritu C et al (2010) PSI-7851, a pronucleotide of beta-D-2′-deoxy-2′-fluoro-2′-C-methyluridine monophosphate, is a potent and pan-genotype inhibitor of hepatitis C virus replication. Antimicrob Agents Chemother 54:3187–3196
19. Liu Y, Jiang WW, Pratt J et al (2006) Mechanistic study of HCV polymerase inhibitors at individual steps of the polymerization reaction. Biochemistry 45:11312–11323
20. Harrus D, Ahmed-El-Sayed N, Simister PC et al (2010) Further insights into the roles of GTP and the C terminus of the hepatitis C virus polymerase in the initiation of RNA synthesis. J Biol Chem 285:32906–32918
21. Ferrari E, Wright-Minogue J, Fang JW et al (1999) Characterization of soluble hepatitis C virus RNA-dependent RNA polymerase expressed in *Escherichia coli*. J Virol 73: 1649–1654

# Chapter 8

# Selecting and Characterizing Drug-Resistant Hepatitis C Virus Replicon

**Inge Vliegen, Leen Delang, and Johan Neyts**


## Abstract

The current standard of care (SOC) for hepatitis C virus (HCV) genotype 1-infected patients consists of telaprevir or boceprevir in addition to pegylated interferon and ribavirin treatment. Other selective inhibitors of HCV replication are being developed. Drug pressure may, depending on the class of inhibitors, (rapidly) select for drug-resistant variants. Here we describe four different approaches to select in vitro for drug-resistant HCV subgenomic replicons.


**Key words** Drug resistance, HCV, Replicon, Genotyping, Phenotyping

## 1 Introduction

Both the lack of proofreading capacity of the HCV RNA-dependent RNA polymerase and the high turnover rate of the viral RNA are responsible for the high sequence diversity of HCV [1]. Consequently, the virus population exists in a patient as a group of genetically distinct but closely related variants, termed "quasispecies". It is therefore not surprising that variants with drug-resistance mutations occur naturally in treatment-naïve patients [2, 3]. Although these drug-resistant variants represent only minor percentages of the total virus population (frequencies <1 %), they can become the dominant viral species under antiviral pressure. In analogy, long-term in vitro cultures of HCV-infected or replicon-containing cells are also characterized by the presence of numerous mutations across the replicon genome. Subgenomic HCV replicons provide a valuable tool to identify HCV resistance to (novel) anti-HCV inhibitors. The presence of a selection marker (such as the neomycin gene) in the HCV subgenomic replicon provides a convenient tool to help studying the appearance of drug resistance. Selection of drug-resistant variants can be achieved by culturing these neomycin-replicon-containing cells under geneticin (G418) and antiviral pressure. We here provide a detailed description of the

Edwin Yunhao Gong (ed.), *Antiviral Methods and Protocols*, Methods in Molecular Biology, vol. 1030,
DOI 10.1007/978-1-62703-484-5_8, © Springer Science+Business Media, LLC 2013

different protocols that can be used to select for drug-resistant replicons. Protocols may vary depending on the results that are aimed for.

1. When culturing cells that carry HCV replicons (with a neomycin selection marker) in the presence of a single high antiviral drug concentration and G418 (Subheading 3.1) dominant resistance mutations can be selected. Cells can only survive high antiviral and G418 pressure if they carry HCV replicons with drug-resistance mutations.
2. When replicon-containing cells are passaged for several weeks in the presence of a gradually increasing concentration of antiviral drug, the evolution of (multiple) in vitro HCV replicon variants can be traced (Subheading 3.2). Thus, the emergence and disappearance of replicon variants in response to the changing selection pressure can be monitored [4].
3. In such case that the barrier to resistance of a number of molecules needs to be compared, resistance selection protocols need to be developed that are similar for each molecule (Subheading 3.3) [5].
4. "Clearance-rebound assays" can be performed to select for resistance mutations. Clearance-rebound assays are built up of two phases called the "clearance" and the "rebound" phase. During the "clearance phase", cells are generally passaged six consecutive times in the presence of antiviral drug and in the absence of G418. Following every passage in the presence of antiviral drug pressure, cells of each "clearance" passage are further cultured for three consecutive times in the absence of the antiviral drugs but in the presence of G418 ("rebound phase"). In the case that the antiviral drug is highly effective and replicon RNA is completely cleared from the cells during the "clearance phase", the cells will not survive G418-induced killing ("rebound phase"). However, in such case that the antiviral treatment is not able to completely cure the cells from their replicon or that drug-resistant variants emerge during the "clearance phase", the hepatoma cells will survive in the presence of G418 in the "rebound phase".

## 2 Materials

1. Huh 9-13 cells: Huh7 cells containing subgenomic HCV genotype 1b replicon I377/NS3-3′/wt [6].
2. Complete DMEM: Dulbecco's modified Eagle's medium (DMEM, High Glucose; Life Technologies) supplemented with 10 % fetal bovine serum (FCS) (Integro), 1× MEM nonessential amino acids (Life Technologies), 100 IU/mL

penicillin (Life Technologies), 100 μg/mL streptomycin (Life Technologies).

3. Geneticin® Selective Antibiotic (G418 Sulfate), 50 mg/mL solution (Life Technologies).
4. Culture plates (6-, 12-, 24-, 96-well plates).
5. Phosphate-buffered saline (PBS) (DPBS Powder w/o Ca, Mg, 10 L = 95.44 g, Lonza), sterilize by autoclaving.
6. 0.25 % trypsin–EDTA (1×), phenol red (Life Technologies).
7. RLT lysis buffer (included in the RNeasy Mini kit, Qiagen)
8. 1.5 mL sterile centrifugation tubes (Eppendorf).
9. TC-flasks (25 or 75 $cm^2$, TPP).
10. RNeasy Mini kit (Qiagen)-protocol animal cells I.
11. Recovery™ cell culture freezing medium, liquid (Life Technologies).
12. Cells-to-cDNA II lysis buffer (Life Technologies).
13. 96-well or 48-well PCR plates and cap strips (Bioplastics).
14. Thermal Cycler (Biometra).
15. RNase- and DNase-free $H_2O$ (Promega).
16. RT-QPRT-O32XLR One-step qRT-PCR (Eurogentec, Belgium).
17. neo-forward primer [5′-CCG GCT ACC TGC CCA TTC-3′]
    neo-reverse primer [5′-CCA GAT CAT CCT GAT CGA CAA G-3′]
    neo-probe [5′-FAM-ACA TCG CAT CGA GCG AGC ACG TAC-TAMRA-3′]
18. Fast-PCR plate 96-well, low profile (Sarstedt).
19. Optical adhesive cover for 96-well plates (Sarstedt).
20. (3-(4,5-dimethylthiazol-2-yl)-5-(3-carboxymethoxyphenyl)-2-(4-sulfophenyl)-2H-tetrazolium)/phenazinemethosulfate (MTS/PMS) (Promega).
21. Verso 1-step RT-PCR Hot-Start kit (Thermo Scientific).
22. Oligonucleotide primers for the amplification of specific HCV sequences (*see* Table 1).
23. Wizard® SV Gel and PCR Clean-Up system (Promega).

## 3 Methods

### 3.1 Cell Picking and Expansion

1. Seed Huh 9-13 cells at a density of $7.2 \times 10^4$ cells in a 12-well plate in complete DMEM containing 1 mg/mL G418 and test compound (*see* **Note 1**). Include one control well with Huh 9-13 cells grown in complete DMEM with G418 without test compound.

**Table 1**
**Primers used for HCV genotyping**

| Primer | Forward (F) or reverse (R) | Gene | Position (AA) | Product size (bp) | Sequence (5′–3′) |
|---|---|---|---|---|---|
| 1 | F | ECMV IRES | | 858 | TTGTATGGGATCTGATCTGGGGCC |
| 2 | R | NS3 | 243 | | TTAGACATATACGCCCCGAAACC |
| 3 | F | NS3 | 15 | 978 | TTGGCTGCATCATCACTAGC |
| 4 | R | NS3 | 345 | | TTCTCCAGTGCTGGACAGAG |
| 5 | F | NS3 | 205 | 901 | CCTACTGGTAGCGGCAAGAG |
| 6 | R | NS3 | 504 | | TGAGCTCGTACCAAGCACAG |
| 7 | F | NS3 | 331 | 834 | TCACCGTGCCACATCCAAACAT |
| 8 | R | NS3 | 607 | | TTTGAACGGCTCCCAGCCTATACA |
| 9 | F | NS3 | 532 | 882 | TGGGAGAGCGTCTTTACAGG |
| 10 | R | NS4B | 140 | | AATGTCCACAAGCACCTTCC |
| 11 | F | NS4B | 65 | 794 | TAGCAGGCTTGTCCACTCT |
| 12 | R | NS5A | 67 | | GACATGTCCGGTGATCTGTG |
| 13 | F | NS5A | 11 | 901 | TGGATATGCACGGTGTTGAC |
| 14 | R | NS5A | 295 | | GAGGGAATTTCCTGGACCTC |
| 15 | F | NS5A | 182 | 974 | ATACCTGGTTGGGTCACAGC |
| 16 | R | NS5B | 58 | | CCTGCAGTCTGTCAAAGGTG |
| 17 | F | NS5A | 290 | 887 | GGAGGAGGATGAGAGGGAAG |
| 18 | R | NS5B | 137 | | TGGTGGTGTCAATTGGTGTC |
| 19 | F | NS5B | 74 | 543 | GAAGGCGTCCACAGTTAAGG |
| 20 | R | NS5B | 253 | | CTTATGGCCTGTCTGGCTTC |
| 21 | F | NS5B | 223 | 1,000 | TGCTTTGACTCAACGGTCAC |
| 22 | R | NS5B | 555 | | TGTAACCAGCAACGAACCAG |
| 23[a] | F (for RT-PCR) | NS5B | 367 | 848 | CTCCTCCAATGTGTCAGTCG |
| 24[a] | R (for RT-PCR) | 3′ of poly-U tail | | | ACAGCTAGCCGTGACTAGGG |
| 25[a] | F (for sequencing) | NS5B | 367 | 729 | CTCCTCCAATGTGTCAGTCG |
| 26[a] | R (for sequencing) | In poly- U tail | | | AAAGGGAAAAAAACAGGATGGC |

[a]Different oligonucleotides are used in the RT-PCR and the sequencing PCR. Before amplification, reverse primer and template need to be mixed and denatured (2 min at 96 °C)

2. Remove medium and wash the cells once with PBS.
3. Trypsinize the cells when 90 % confluency is reached (*see* **Note 2**) or when sufficiently large colonies have developed. This is done by adding trypsin to detach the cells, resuspending the cells in complete DMEM, placing them in a sterile centrifuge tube and pelleting the cells at 600 × *g*, 6 min at room temperature.
4. Remove the medium, taking care not to disturb the pellet. Resuspend all collected cells in complete DMEM.

5. Seed $7.2 \times 10^4$ cells in a new 12-well plate in complete DMEM with G418 with the same concentration of the test compound.
6. The experiment is stopped at ≈3 weeks of culturing the cells (*see* **Note 3**).
7. Collect the cells by trypsinization, seed the collected cells from each well in a new well or culture flask (*see* **Note 4**), and let the cells expand so as to obtain sufficient cells (*see* **Note 5**) for subsequent phenotype and genotype characterization.
8. Thereafter, pipet $5 \times 10^5$ cells in a sterile centrifugation tube. Pellet the cells at 600 × *g*, 6 min at room temperature. Remove the medium and resuspend the cell pellet in 350 μL RLT lysis buffer for genotyping. The remaining of the cells from the 25 cm² TC-flask can be used for phenotyping (*see* **Note 6**).

### 3.2 Resistance Selection by Serial Passaging

1. Seed Huh 9-13 cells at a density of $3 \times 10^5$ cells in complete DMEM with G418 and 1× the $EC_{50}$ concentration of the test compound in a 25 cm² TC-flask. A 25 cm² TC-flask containing complete DMEM without compound is cocultured as a control.
2. Trypsinize and collect the cells (after washing once with PBS) when 90 % confluency is reached.
3. Seed $3 \times 10^5$ cells in a new 25 cm² TC-flask in complete DMEM with G418 and 2× $EC_{50}$ concentration of the test compound. Collect $5 \times 10^5$ cells of the cells in 350 μL RLT buffer and freeze at −80 °C for genotyping. Collect the rest of the cells in freezing medium for phenotyping.

When 90 % cell confluency is achieved, perform the same manipulations described in **steps 2–3** up to a concentration of 500× the $EC_{50}$ concentration (*see* **Note 7**).

### 3.3 Synchronic Resistance Culture

1. Seed Huh 9-13 cells at a density of $2 \times 10^5$ cells in complete DMEM with G418 containing 1×, 2×, 5×, 25×, and 125× the $EC_{50}$ concentration (*see* **Note 8**) of the test compound in a 6-well plate. An additional well of Huh 9-13 cells, containing complete DMEM and G418 without compound, is cocultured as a control.
2. Trypsinize and collect the cells (after washing once with PBS) when 90 % confluency is reached.
3. Seed $2 \times 10^5$ cells in a new 6-well plate in complete DMEM with G418 and the same concentration of the test compound. Culture the Huh 9-13 replicon-containing cells for 11 passages (5.5 weeks) under constant antiviral pressure (1×, 2×, 5×, 25×, or 125× $EC_{50}$) (*see* **Note 9**).
4. When cells survive 5.5 weeks of culture under constant antiviral pressure and G418 pressure, cell numbers are expanded to achieve a 90 % confluent 25 cm² TC-flask.

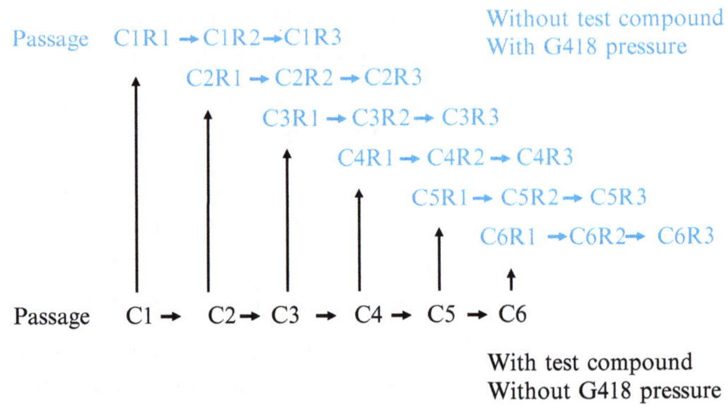


**Fig. 1** Schematic representation of the clearance-rebound assay. Six consecutive clearance passages [C1 → C6] (in the presence of a fixed concentration of test compound) are each followed by three consecutive rebound passages [R1 → R3] (with G418 pressure and without test compound)

5. Consequently, $5 \times 10^5$ cells are collected in 350 μL RLT lysis buffer and frozen at −80 °C for genotyping. The rest of the cells from the TC-flask are collected in freezing medium and stored at −80 °C for phenotyping.

### 3.4 Clearance-Rebound

1. Seed Huh 9-13 cells at a density of $3 \times 10^5$ cells in complete DMEM without G418 and containing test compound (*see* **Note 10**) in 25 cm² TC-flasks. A culture flask containing complete DMEM without compound is cocultured as a control.
2. Trypsinize and collect the cells (after washing once with PBS) when 90 % confluency is reached.
3. Seed $3 \times 10^5$ cells in a new 25 cm² TC-flask in complete DMEM without G418 with the same concentration of the test compound (clearance phase) and in a second 25 cm² TC-flask in complete DMEM with G418 and no test compound (rebound phase).
4. Cells are in total passaged six consecutive times in the presence or absence of test compound and in the absence of G418 (clearance phase). Each flask is generally passaged for three consecutive times in the presence of 1 mg/mL G418 (rebound phase) (*see* Fig. 1).
5. When cells survive passaging for three consecutive times in complete DMEM with G418 and without compound pressure, $5 \times 10^5$ cells are collected in 350 μL RLT lysis buffer for genotyping. For this, $5 \times 10^5$ cells are pipetted in a sterile centrifugation tube. Pelleting of the cells is performed at 600 × *g*, 6 min at room temperature. Medium is removed and the dried pellet is resuspended in 350 μL RLT lysis buffer and frozen at −80 °C. The rest of the cells from the 25 cm² TC-flask is collected in freezing medium and stored at −80 °C for phenotyping later.

### 3.5 Phenotyping

#### 3.5.1 Antiviral Assay

1. Seed Huh 9-13 cells at a density of $5 \times 10^3$ cells per well in a 96-well cell culture plate in complete DMEM without G418.
2. Incubate the cells for 24 h at 37 °C and 5 % $CO_2$.
3. Remove culture medium from the cells and add threefold serial dilutions of the test compound (*see* **Note 11**) in 100 μL complete DMEM without G418 per well. Use five wells as a cell control and add 100 μL complete DMEM without G418 and test compound per well.
4. Culture the cells for an additional 3 days at 37 °C and 5 % $CO_2$.
5. Remove cell culture fluid and wash the cell monolayers once with PBS.
6. Lyse the cells in 100 μL Cells-to-cDNA II lysis buffer according to the manufacturer's instructions (*see* **Note 12**).
7. Transfer lysates to a 96- or 48-well PCR plate, seal the plates, and transfer the lysate-containing PCR plate(s) to a Thermal Cycler.
8. Incubate the lysates for 15 min at 75 °C.
9. Add 100 μL of RNase- and DNase-free $H_2O$ to the lysates.
10. Lysates can now be stored at −20 °C or be used immediately for RT-qPCR without further storage.

#### 3.5.2 Quantification of HCV RNA by Real-Time RT-PCR

1. For absolute quantification of HCV replicon RNA, a standard should be generated for the target sequence that will be quantitated (in this case the neomycin gene). Prepare therefore a plasmid DNA containing the HCV Neomycin sequence.
2. Determine DNA concentrations by measuring optical density at $OD_{260nm}$ and $OD_{280nm}$.
3. Dilute HCV DNA standard in RNase- and DNase-free $H_2O$ and prepare a tenfold serial dilution starting at a concentration of 10 ng/μL.
4. Prepare the reaction mixture as in Table 2.
5. Add 20 μL of the reaction mixture to each well of the qPCR plate
6. Add 5 μL of each RNA sample (cell lysates) and standard to the appropriate wells. Use nuclease-free water as a negative control. Seal the plate with an optical adhesive cover.
7. Set the PCR plate in a 7500 Real-Time PCR system or an equivalent system after short spinning of the plate.
8. Set the reaction conditions and incubate at 48 °C for 30 min, 95 °C for 15 min, followed by 40 cycles at 95 °C for 15 s and 60 °C for 1 min.
9. Confirm the absence of amplification in the negative control and determine the $EC_{50}$ and the fold drug resistance (*see* **Note 13**).

**Table 2**
**Reaction components for one-step RT-qPCR**

| Reagent | Added volume/reaction (μL) |
|---|---|
| RNase, DNase-free $H_2O$ | 6.825 |
| Reaction buffer (2×) | 12.5 |
| neo-forward primer (60 μM) | 0.125 |
| neo-forward primer (60 μM) | 0.125 |
| neo-probe (5 μM) | 0.3 |
| RT-enzyme | 0.125 |

#### 3.5.3 Cytostatic Assay

1. Seed Huh 9-13 cells at a density of $5 \times 10^3$ cells per well in a 96-well cell culture plate in complete DMEM without G418. Culture the plate at 37 °C and 5 % $CO_2$ for 24 h.
2. Remove culture medium from the cells and add threefold serial dilutions of the test compound in 100 μL complete DMEM without G418 per well. Use five wells as a cell control and add 100 μL complete DMEM without G418 and test compound per well.
3. Culture the cells for an additional 3 days at 37 °C and 5 % $CO_2$.
4. Remove cell culture fluid and wash the cell monolayers once with PBS.
5. Add 100 μL of MTS/PMS solution to the cells.
6. Incubate the cells for 2 h at 37 °C and 5 % $CO_2$.
7. Measure the absorbance at 490 nm with a spectrophotometer (*see* **Note 14**).
8. Determine the $CC_{50}$ and compare with the $CC_{50}$ on wild-type HCV replicon (*see* **Note 15**).

### 3.6 Genotyping

#### 3.6.1 One-Step RT-PCR of HCV Specific Sequences

1. Extract total RNA using the RNeasy mini kit, according to the manufacturer's instructions and elute the RNA in 50 μL RNase- and DNase-free water (*see* **Note 16**).
2. Set up the RT-PCR reactions. The choice of primers is dictated by the region of the HCV polyprotein from which sequence information is required. Prepare the PCR amplification mixture contents as in Table 3.
3. Perform PCR amplification using the following conditions:

   50 °C for 15 min, 95 °C for 15 min, followed by 40 cycles at 95 °C for 20 s, 60 °C for 30 s, and 72 °C for 1 min, finally followed by 72 °C for 5 min.

**Table 3**
**Reaction components for one-step RT-PCR**

| Reagent | Added volume/reaction (μL) |
|---|---|
| RT enhancer | 2.5 |
| 1-step PCR hot-start master mix (2×) | 25 |
| RNase and DNase free water | 12.5 |
| Verso enzyme mix | 1 |
| Forward primer (5 μM) | 2 |
| Reverse primer (5 μM) | 2 |

4. Purify the PCR products to remove primers and other reagents from the amplification reaction. Check the length of PCR products by gel electrophoresis.
5. At this point, the end user must decide what information is desired from the sequence analysis. For population sequencing [a determination of the dominant RNA sequence(s) in the cell] the PCR products can at this stage be directly sequenced. For an estimation of the diversity of sequence variants in the original pool of HCV RNA, PCR products can be cloned into a PCR cloning vector, and individual clones can be isolated for sequencing (*see* **Note 17**). The number of clones to be sequenced is again to be decided by the user, but we typically sequence 96 clones to provide a picture of the original RNA population.
6. The HCV-specific PCR products (or clones) are now ready for DNA sequence analysis according to the method of choice for DNA sequence analysis in the end user's laboratory. The primers used to generate the PCR products (*see* Table 1 and note under Table 1) are suitable primers for sequence analysis (*see* **Note 18**).

## 4 Notes

1. The concentration of the test compound is selected following antiviral assaying; the concentration should be well below the $CC_{50}$ and be preferentially higher or at least equal to 25× the $EC_{50}$ concentration.
2. When cells suffer from compound pressure and massive cell death is observed (because of replicon disappearance), G418 pressure has to be removed till cells recover. Thereafter, cells are re-cultured with test compound and G418 pressure. Otherwise refresh the culture medium minimally once a week.

3. After ≈3 weeks of selection, colonies resistant to the test compound may develop, or, in case no resistance occurs, no colonies will be observed.
4. Depending on the amount of resistant replicon-containing cells collected, cells should be seeded after trypsinization in a new 12-well, 24-well, 6-well plate or 25 $cm^2$ TC-flask.
5. In general, one 95 % confluent 25 $cm^2$ TC-flask is sufficient for phenotyping and genotyping.
6. By freezing the cells in RLT and storing them at −80 °C, genotyping can be performed at a later time point. By collecting the rest of the cells in freezing medium and storing them at −80 °C, phenotyping can be performed later after re-culturing the cells.
7. Since certain compounds are toxic for the host cell at 500× their $EC_{50}$ concentrations (as determined by antiviral assays), drug pressure should only be increased till just below toxic concentrations of these particular compounds.
8. Since certain compounds are toxic for the host cell at the indicated concentrations (determined by antiviral assays), drug selection protocols cannot be carried out with such concentrations for these compounds.
9. When replicon-containing cells suffer (massive cell death) from compound pressure (because of replicon disappearance), G418 and compound have to be removed until cell recover. Thereafter, cells are re-cultured in the presence of the antiviral drug and G418 pressure until the predetermined 5.5 culture weeks are completed. Otherwise refresh the culture medium minimally once a week.
10. Concentrations of a particular compound should be selected such that the concentration is not cytostatic but is still able to result in at least 2-$log_{10}$ reduction in replicon replication following treatment, as determined by performing an antiviral assay in Huh 9-13 cells.
11. Start concentration of test compound and dilution steps should be based on the antiviral activity of the compound in wild-type cells; concentrations lower than the $EC_{50}$ of the test compound in wild-type cells are useless to test in (possibly) resistant replicon-containing cells.
12. Check microscopically whether cells are lysed before continuing the protocol. If not, lyse longer.
13. The 50 % effective concentration ($EC_{50}$) is defined as the concentration of compound that reduces the amount of HCV neo-RNA from the cell control by 50 %. Fold drug resistance is calculated as the ratio of the mean $EC_{50}$ in resistant replicon over the mean $EC_{50}$ in wild-type replicon.

14. MTS is bioreduced by cells into a formazan product that is soluble in tissue culture medium. The absorbance of the formazan at 490 nm can be measured directly from 96-well assay plates without additional processing. The conversion of MTS into aqueous, soluble formazan is accomplished by dehydrogenase enzymes found in metabolically active cells. The quantity of formazan product as measured by the amount of 490 nm absorbance is directly proportional to the number of living cells in culture.
15. The 50 % cytotoxic concentration ($CC_{50}$) is defined as the concentration that inhibits the proliferation of exponentially growing (control) cells by 50 %.
16. RNA can be stored at −80 °C till further usage.
17. For clonal sequencing it might be better to choose for separate RT and PCR reactions performed with high-fidelity enzymes to avoid bias introduction of viral variants by the enzymes used.
18. Mutations that were observed in both wild-type and resistant Huh 9-13 replicon-containing cells are not included in the mutational analysis as they are expected to be "adaptive mutations".

## Acknowledgments

The original work of the authors was supported by a fellowship to Leen Delang from the Research Foundation Flanders (FWO), grant G.0728.09N from the FWO and GOA/10/014 from the KU Leuven.

### References

1. Simmonds P (2004) Genetic diversity and evolution of hepatitis C virus-15 years on. J Gen Virol 85(Pt 11):3173–3188
2. Le Pogam S, Seshaadri A, Kosaka A et al (2008) Existence of hepatitis C virus NS5B variants naturally resistant to non-nucleoside, but not to nucleoside, polymerase inhibitors among untreated patients. J Antimicrob Chemother 61(6):1205–1216
3. Kuntzen T, Timm J, Berical A et al (2008) Naturally occurring dominant resistance mutations to hepatitis C virus protease and polymerase inhibitors in treatment-naive patients. Hepatology 48(6):1769–1778
4. Delang L, Vliegen I, Leyssen P, Neyts J (2012) In vitro selection and characterization of HCV replicons resistant to multiple non-nucleoside polymerase inhibitors. J Hepatol 56(1): 41–48
5. Delang L, Vliegen I, Froeyen M, Neyts J (2011) Comparative study of the genetic barriers and pathways towards resistance of selective inhibitors of hepatitis C virus replication. Antimicrob Agents Chemother 55(9):4103–4113
6. Lohmann V, Korner F, Koch J et al (1999) Replication of subgenomic hepatitis C virus RNAs in a hepatoma cell line. Science 285(5424):110–113

# Chapter 9

# A Cellular Replicon-Based Phenotyping Assay to Determine Susceptibility of Hepatitis C Virus Clinical Isolates to NS3/4A Protease Inhibitors

**Leen Vijgen, Jannick Verbeeck, Barbara Van Kerckhove, Jan Martin Berke, Diana Koletzki, Gregory Fanning, and Oliver Lenz**

## Abstract

A hepatitis C virus (HCV) replicon-based protease phenotyping assay has been developed that allows determining the susceptibility of a patient's HCV protease sequence to HCV protease inhibitors. In brief, HCV protease sequences amplified from clinical samples are cloned in a transient HCV genotype 1b replicon backbone, containing a luciferase reporter gene. These protease chimeric replicons are replication-competent when electroporated into susceptible Huh7-Lunet cells. Replication can be quantified by measuring the enzymatic activity of the luciferase protein. This assay is reproducible and robust, and has a high overall success rate for determining the phenotypic susceptibility of HCV genotype 1a and 1b patient-derived protease domains to HCV protease inhibitors. In addition, the HCV genotype 1b protease shuttle backbone also supports efficient replication of HCV genotype 4 protease sequences.

**Key words** Hepatitis C virus, HCV, NS3/4A, Protease inhibitor, Replicon, Phenotyping assay, Transient assay

## 1 Introduction

The recent approvals of the HCV protease inhibitors (PIs) Telaprevir (Incivek®/Incivo®) and Boceprevir (Victrelis®) by US Food & Drug Administration (FDA) and European Medicines Agency (EMA) have led to a significant paradigm shift in the treatment of hepatitis C. Combination of these PIs with pegylated interferon and ribavirin resulted in significantly improved sustained virologic response (SVR) rates in HCV genotype 1 infected patients, compared with pegylated interferon and ribavirin alone [1]. Nevertheless, multiple direct acting antiviral (DAA) agents with complementary modes of action will be needed to minimize viral resistance, reduce treatment duration time, increase cure rates in hard-to-treat patient populations, such as non-responders to an interferon-based regimen, and to establish all-oral interferon-free

Edwin Yunhao Gong (ed.), *Antiviral Methods and Protocols*, Methods in Molecular Biology, vol. 1030,
DOI 10.1007/978-1-62703-484-5_9, © Springer Science+Business Media, LLC 2013

treatment regimens with fewer side effects, thereby providing therapeutic options for interferon-intolerant patients.

In the development of HCV DAAs, broad-spectrum genotype coverage and the natural variability of the HCV quasispecies remain a challenge. Also, mutations conferring resistance to HCV PIs, NS5A inhibitors and NS5B polymerase inhibitors have emerged during treatment with these drug classes. In order to better understand the impact of such mutations in the background of a patient-derived protease sequence on the in vitro susceptibility to a PI, we developed a cellular replicon-based NS3/4A protease phenotyping assay [2]. In this assay, we make use of an HCV genotype 1b subgenomic replicon containing a luciferase reporter gene (clone ET) [3], that allows a transient autonomous replication when introduced into permissive Huh7-Lunet cells. In this protocol, patient-derived protease sequences are introduced into the replicon backbone using a fast and simple In-Fusion® cloning technique. This assay allows a reproducible assessment of PI susceptibility of HCV genotypes 1 and 4 protease sequences.

## 2 Materials

### 2.1 Plasmid Constructs, Cell Lines and Primers

1. Plasmid pFK_i341_PILuc-NS3-3′: Licensed from Dr. R. Bartenschlager, University of Heidelberg, Germany. This construct encodes a genotype 1b subgenomic bicistronic replicon with the poliovirus internal ribosome entry site-driven luciferase reporter and with adaptive mutations E1202G and T1280I in NS3, and K1846T in NS4B of the HCV genome, and contains two *Sac*II restriction enzyme recognition sites introduced by silent mutations at the nucleotides coding for NS3/4A protease amino acids 5–7 and 190–192.
2. Huh7-Lunet cells: Licensed from Dr. R. Bartenschlager.
3. In-Fusion primers, PCR primers and sequencing primers (*see* Table 1).

### 2.2 Restriction Enzyme Digestion and Dephosphorylation

1. *Sac*II restriction enzyme (New England Biolabs).
2. *Sca*I restriction enzyme (New England Biolabs).
3. 10× Buffer 4 solution (New England Biolabs).
4. Calf intestinal phosphatase (CIP) and corresponding 10× Buffer 3 (New England Biolabs).

### 2.3 Ligation and In-Fusion Cloning

1. T4 DNA ligase (New England Biolabs).
2. In-Fusion HD Cloning kit (Clontech Laboratories, Cat. No. 639649).
3. 1× TE buffer.
4. XL-10 Gold Competent Cells (Stratagene).
5. One Shot Top 10 Competent Cells (Invitrogen).

**Table 1**
**Primers for PCR, In-Fusion, and sequencing**

| Primer name | Application | Sequence (5'–3') |
|---|---|---|
| HCV-1a-F-out | Outer RT-PCR (HCV subtype 1a) | ctcaccgctcatgatcttgaatgc |
| HCV-1a-R-out | Outer RT-PCR (HCV subtype 1a) | gggacctcaccgctcatgat |
| HCV-1b-F-out | Outer RT-PCR (HCV subtype 1b) | atgcatgctggtgcggaa |
| HCV-1b-R-out | Outer RT-PCR (HCV subtype 1b) | gagcgccttctgtttgaattg |
| HCV-4-F-out | Outer RT-PCR (HCV genotype 4) | ctgacaccgctgcrtgygg |
| HCV-4-R-out | Outer RT-PCR (HCV genotype 4) | gatggcatctcgccrcwcatga |
| HCV-INF-FW1a | In-Fusion PCR | atggcgcctattaccgcgtacgcccagcagacaag |
| HCV-INF-REV1a | In-Fusion PCR | aatgtctgcggtaccgctggtggagaggagttgtc |
| HCV-INF-FW1b | In-Fusion PCR | atggcgcctattaccgcctactcccaacagacg |
| HCV-INF-REV1b | In-Fusion PCR | aatgtctgcggtaccgccggaggggacgagttgtc |
| HCV-INF-FW4a | In-Fusion PCR | atggcgcctattaccgcgtacgcgcagcagacccgcg |
| HCV-INF-REV4a | In-Fusion PCR | aatgtctgcggtaccgcaggggggtgttgagttgtc |
| HCV-INF-FW4d | In-Fusion PCR | atggcgcctattaccgcatacgcgcagcagacccgag |
| HCV-INF-REV4d | In-Fusion PCR | aatgtctgcggtaccgcaggaggagtggagttgtc |
| HCV-COL-FW | Colony PCR | ggctgaaggatgcccagaagg |
| HCV-COL-REV | Colony PCR | agccacctcctcgatgtttg |
| HCV-SEQ-FW | Sequencing | ggctgaaggatgcccagaagg |
| HCV-SEQ-REV | Sequencing | cgatgtaagggaggtgtgagg |

6. SOC medium (Invitrogen).
7. Ampicillin (Sigma-Aldrich).
8. Luria-Broth (LB) medium capsules (MP Biomedicals).
9. Luria-Broth (LB) agar capsules (MP Biomedicals).

#### 2.4 PCR, Gel Electrophoresis, PCR Purification and Sequencing

1. NucliSens®easyMAG™ disposables and RNA extraction reagents (BioMérieux).
2. 10 ng/mL stock solution of yeast tRNA (Ambion).
3. SuperScript III™ One-Step RT-PCR System with Platinum® Taq DNA Polymerase High Fidelity (Invitrogen, Cat. No. 12574-030).
4. Phusion High Fidelity PCR Master Mix with HF Buffer (Finnzymes, Cat. No. F-531S).
5. Expand High Fidelity PCR System (Roche Applied Science, Cat. No. 11732650001).
6. 100-bp DNA ladder.

7. GeneRuler™ 1 kb Plus DNA Ladder.
8. UltraPure™ 10× TAE buffer (Invitrogen). Make a working solution of 1× TAE buffer with DNase-free $H_2O$.
9. Blue Juice™ Gel Loading Buffer (Invitrogen).
10. Certified molecular biology agarose (Bio-Rad Laboratories NV). To make a 1 % agarose gel, add 1 g of agarose powder to 100 mL of 1× TAE working solution. Heat in microwave until the solution is homogeneous. After cooling down at room temperature, poor the gel solution into a gel tray and let stand until the gel is solidified.
11. 10,000× Sybr® Safe DNA gel stain (Invitrogen). Add 10 μL of gel stain to the 1 % agarose gel solution while the gel solution is cooling down.
12. Flash Gel™ DNA System, 1.2 % agarose (Lonza), and corresponding loading dye.
13. Qiaquick Gel Extraction Kit (Qiagen, Cat. No. 28704).
14. 96–100 % ethanol.
15. 3 M NaAc (pH 5.5).
16. QIAquick PCR Purification Kit (Qiagen, Cat. No. 28104).
17. Big Dye Terminator v.3.1 Cycle Sequencing Kit (Applied Biosystems, Cat. No. 4337455).
18. NanoDrop spectrophotometer (Thermo Fisher Scientific).

### 2.5 Plasmid Purification and In Vitro Transcription

1. PureLink HiPure Plasmid DNA Purification kit for Maxi Preparation of Plasmid DNA (Invitrogen, Cat. No. K2100-06).
2. PureLink HiPure Plasmid DNA Purification kit for Midi Preparation of Plasmid DNA (Invitrogen, Cat. No. K2100-04).
3. Isopropanol.
4. 70 % ethanol.
5. T4 MEGAscript High Yield Transcription kit (Ambion, Cat. No. AM1334M).
6. RNeasy Mini kit (Qiagen, Cat. No. 74104).

### 2.6 Transient Replicon Assay

1. Cytomix: 120 mM KCl, 0.15 mM $CaCl_2$, 10 mM $K_2HPO_4$/$KH_2PO_4$ (pH 7.6), 25 mM HEPES (1 M stock solution, Gibco, cell culture grade), 2 mM EGTA, 5 mM $MgCl_2$. Adjust to pH 7.6 with KOH, sterilize by filtration, and store at room temperature.
2. 2 mM ATP.
3. 5 mM glutathione.
4. Dulbecco's modified Eagle's medium (DMEM) (Sigma-Aldrich).
5. 10 % fetal calf serum (FCS).
6. 1 % L-glutamine.

7. 0.04 % gentamicin.
8. SteadyLite Plus solution (PerkinElmer).
9. GenePulser electroporation system (Bio-Rad Laboratories NV).
10. 384-well plates (PerkinElmer).
11. ViewLux (PerkinElmer).

## 3 Methods

The procedures described in this section are outlined in Fig. 1.

### 3.1 Preparation of Replicon Shuttle Backbone

1. Perform a digestion reaction with *Sac*II in order to excise the protease domain. Mix 10 μg of plasmid pFK_i341_PILuc-NS3-3′ with 5 μL *Sac*II (20 U/μL) and 5 μL 10× Buffer 4. Add $H_2O$ to a total volume of 50 μL. Incubate the reaction mixture at 37 °C for 4 h. The digested linear plasmid now lacks HCV NS3/4A protease ("shuttle replicon backbone lacking NS3/4A protease").

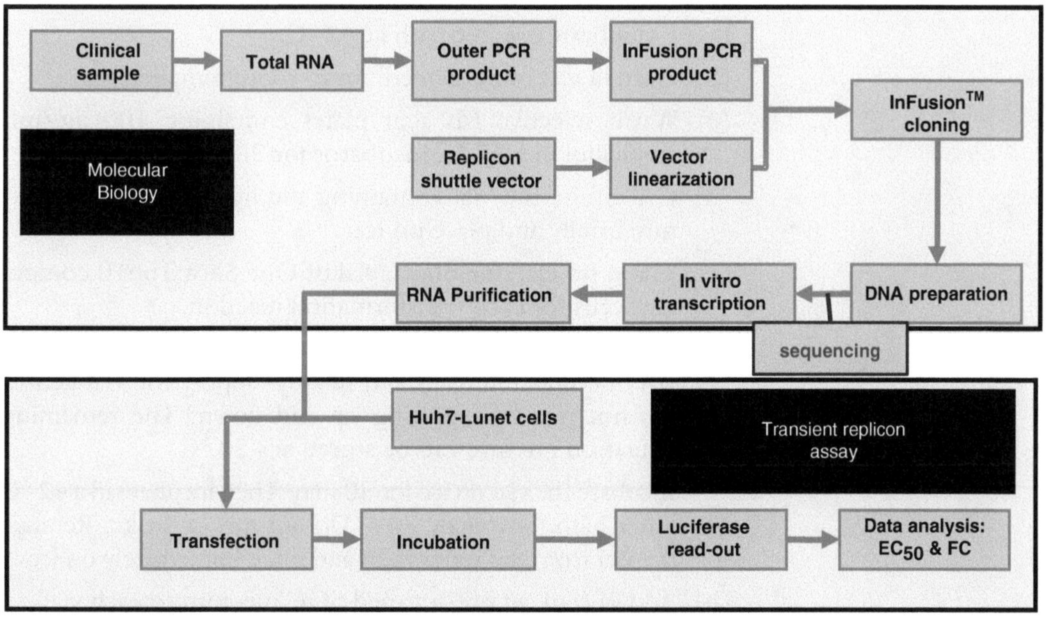


**Fig. 1** A flow diagram of the HCV transient replicon assay. HCV RNA is extracted from clinical plasma samples from HCV genotype 1 or genotype 4 infected patients. The NS3/4A protease region is amplified with sequence specific primers (outer PCR). The outer PCR product is subjected to an In-Fusion PCR with specific In-Fusion primers. The In-Fusion PCR product is cloned into a linearized replicon shuttle backbone lacking NS3/4A protease, which contains a luciferase reporter gene. DNA of this backbone is purified and sequenced to make sure that the cloning reaction is successful. The replicon shuttle backbone now contains the patient-derived NS3/4A protease sequence. After in vitro transcription of this backbone DNA, RNA is purified and used to transfect Huh7-Lunet cells. After incubation of the cells with a compound, the luciferase signal is measured. The $EC_{50}$ and FC (fold changes) values are calculated using appropriate computer software

2. Load 50 µL of the reaction mixture with 10 µL of Blue Juice™ gel loading buffer on a 1 % agarose gel containing 1× Sybr Safe DNA gel stain and run the gel for 1 h (100 V, 100 mA).
3. Perform a gel extraction of the linear shuttle replicon backbone lacking HCV NS3/4A protease (approximately 12 kb fragment) using the QIAquick Gel Extraction kit according to the manufacturer's instructions. Elute the DNA in 50 µL of Buffer EB.
4. Take 2 µL of eluted DNA and determine the concentration of the purified linear shuttle replicon backbone lacking NS3/4A protease using a NanoDrop spectrophotometer.
5. Religate the shuttle replicon backbone lacking NS3/4A protease. Mix 2 µL of 10× Buffer T4 DNA ligase with 1 µL T4 DNA ligase and add 200 ng (=50 ng per 3 kb, which equals 200 ng for 12 kb) of shuttle replicon backbone lacking NS3/4A protease to the reaction mixture. Add $H_2O$ to a total volume of 20 µL. Gently mix the reaction by pipetting up and down and microcentrifuge briefly. Incubate the reaction mixture at room temperature for 1 h. Chill on ice.
6. Transform One Shot Top10 competent cells with 3 µL of religated shuttle replicon backbone lacking HCV NS3/4A protease using the following procedure:
   (a) Equilibrate a water-bath at 42 °C.
   (b) Warm a vial of SOC medium to room temperature.
   (c) Warm selective LB agar plates containing 100 µg/mL ampicillin in a 37 °C incubator for 30 min.
   (d) Centrifuge the vial containing the ligation reaction mixture briefly and place on ice.
   (e) Thaw, on ice, one 50 µL vial of One Shot Top10 competent cells for each transformation reaction.
   (f) Pipette 3 µL of ligation reaction mixture directly to the vial of competent cells and mix by tapping the vial gently. Do not mix by pipetting up and down. The remaining religation mixture can be stored at −20 °C.
   (g) Incubate the vial on ice for 30 min. Then incubate in a 42 °C water bath for exactly 30 s. Do not mix or shake. Remove the vial from the water bath and place immediately on ice.
   (h) Add 250 µL of pre-warmed SOC medium to each vial.
   (i) Shake the vial at 37 °C for 1 h at 225 rpm in a shaking incubator.
   (j) Spread 150 µL of the transformation mixture on a selective LB agar plates containing 100 µg/mL ampicillin. The remaining transformation mixture can be stored at 4 °C and plated out the next day if necessary.
   (k) Place the agar plates upside down and incubate at 37 °C overnight (*see* **Note 1**).

7. Inoculate a single colony from the overnight cultured agar plate in 250 mL LB medium containing 100 μg/mL ampicillin and grow overnight at 37 °C.
8. Perform a maxiprep purification using the PureLink HiPure Plasmid DNA purification kit for maxi preparation of plasmid DNA according to the manufacturer's instructions. After air-dried the pellet, resuspend the DNA pellet in 500 μL TE buffer (*see* **Note 2**). Store the purified DNA at −20 °C (*see* **Note 3**).

### 3.2 Linearization and Dephosphorylation of the Replicon Backbone

1. Perform a digestion with *Sac*II in order to linearize the maxi-prepared replicon shuttle backbone lacking NS3/4A protease. Mix 20 μg of replicon shuttle backbone lacking NS3/4A protease with 30 μL of *Sac*II (20 U/μL) and 30 μL of 10× Buffer 4, and add $H_2O$ to a total volume of 300 μL. Incubate the reaction mixture at 37 °C for 2 h. Perform ethanol precipitation according to the following instructions: add 1/10 volume 3 M NaAc (pH 5.5) and at least 2 volumes (600 μL) isopropanol. Mix gently. Centrifuge at maximum speed (≥15,294 × *g*) for 30 min. Discard the supernatant carefully to obtain an intact pellet. Wash the pellet with 200 μL of 70 % ethanol. Centrifuge the reaction tube at maximum speed (≥15,294 × *g*) for 10 min. Discard the supernatant carefully and air-dry the pellet. Resuspend the pellet in 240 μL of $H_2O$.
2. To increase the success rate of the digestion reaction, repeat **step 1**, using 240 μL of purified restriction digested product from **step 1**, 30 μL of *Sac*II (20 U/μL), and 30 μL 10× Buffer 4.
3. To increase the yield of digested shuttle replicon backbone lacking NS3/4A protease, repeat **step 1**, using 240 μL of purified restriction digest product from **step 2**, 30 μL of *Sac*II (20 U/μL) and 30 μL 10× Buffer 4. Before ethanol precipitation, perform a dephosphorylation reaction using 300 μL of the restriction digest mixture, 10 μL CIP (10 U/μL), 40 μL 10× Buffer 3, and 50 μL of $H_2O$. Incubate the reaction mixture at 37 °C for 1 h. Perform ethanol precipitation (*see* **step 1**). After air-drying the pellet, dissolve the DNA in 50 μL of $H_2O$.
4. Take 2 μL of the purified, linearized replicon shuttle backbone lacking NS3/4A protease to determine the DNA concentration.

### 3.3 PCR Amplification of Patient-Derived HCV NS3/4A Protease Sequences (Amino Acids 5–191)

1. Perform the RNA extraction of an HCV-positive clinical sample. Take 256 μL of HCV-positive plasma and add this volume to a NucliSens® easyMAG™ disposal vessel and proceed immediately with the RNA extraction by following the manufacturer's instructions. Elute the RNA in 60 μL of elution buffer (*see* **Note 4**).
2. Perform an outer RT-PCR for amplification of the NS3/4A sequence from a clinical sample:

(a) Prepare the following master mix on ice: 12.5 μL 2× Superscript III reaction mix, 0.63 μL of yeast tRNA (4,800 ng/mL), 0.25 μL of both 20 μM forward primer and 20 μM reverse primer (Table 1), 0.5 μL Superscript III RT/Platinum Taq polymerase, and 5.87 μL RNase/DNase-free $H_2O$. Add 5 μL of extracted RNA and mix gently by pipetting up and down.

(b) Perform the PCR on a thermocycler using the following conditions: cDNA synthesis at 47 °C for 30 min, denaturation at 94 °C for 2 min, 50 cycles of a denaturation step at 94 °C for 15 s, annealing step at 57 °C for 30 s, and elongation step at 68 °C for 3 min. Perform a final extension at 68 °C for 5 min.

3. Perform a PCR reaction with HCV subtype-specific In-Fusion primers (Table 1) using the Phusion High Fidelity PCR Master Mix according to the following procedure:

   (a) For each reaction, mix 25 μL of 2× Phusion Master Mix, 0.50 μM of forward primer, 0.50 μM of reverse primer, and add water to a total volume of 49 μL. Add 1 μL of the outer RT-PCR product (=template DNA).

   (b) Perform the PCR on a thermocycler using the following conditions: denaturation at 98 °C for 3 min, followed by 35 cycles of a denaturation step at 98 °C for 10 s, annealing step at 60 °C for 30 s, and elongation step at 72 °C for 30 s. Perform a final extension at 72 °C for 7 min.

4. Load 5 μL of PCR product on a 1.2 % Flash Gel DNA cassette and 2.5 μL of a 100-bp DNA ladder. Run the gel for 30 min. Visualize the band pattern with a UV light transilluminator.

5. Purify the positive amplicons with an approximate length of 593 bp using the QIAquick PCR purification kit according to the manufacturer's instructions, and elute the DNA in 30 μL of buffer EB.

6. Take 2 μL of the purified DNA and determine the DNA concentration.

#### 3.4 In-Fusion Cloning of Patient-Derived HCV NS3/4A Protease Sequences into the Replicon Shuttle Backbone Lacking NS3/4A protease

1. Ligate the NS3/4A protease amplicon derived from the clinical sample in the linearized replicon shuttle backbone lacking NS3/4A protease (*see* Subheading 3.2) by using the In-Fusion HD Cloning Kit. Use an insert–vector ratio of 2:1. Calculate the amount of ng vector and insert needed with the online molar ratio calculator of Clontech (Available at http://bioinfo.clontech.com/infusion/molarRatio.do).

   (a) Prepare the In-Fusion reaction mix: add 1 μL of diluted insert and 2 μL of In-Fusion premix to the needed amount of vector. Add $H_2O$ to obtain a total volume of 10 μL (*see* **Note 5**).

(b) Incubate the reactions at 50 °C for 15 min. Add 40 μL of TE buffer (=1/5 dilution) to the reaction mixture. The replicon shuttle backbone now contains the patient-derived HCV NS3/4A protease.

2. Transform XL-10 Gold competent cells with the replicon shuttle backbone containing patient-derived HCV NS3/4A protease.

   (a) Add 5 μL of beta-mercaptoethanol to the cells and incubate at room temperature for 10 min.

   (b) Transfer 100 μL of cells per sample to a 5 mL tube.

   (c) Add 2.5 μL of diluted In-Fusion reaction mixture (*see* **step 1b**) to the appropriate 5 mL tube.

   (d) Incubate the cells on ice for 30 min.

   (e) Heat shock the cells at 42 °C in a water-bath for 30 s, and incubate the cells on ice for 2 min.

   (f) Add 900 μL SOC medium to the cells and incubate at 37 °C for 1 h.

   (g) To concentrate the cell mixture, transfer the cells to a 1.5 mL tube, and centrifuge for 30 s at 6,797 × *g*. Discard the majority of the supernatant, leaving a volume of approximately 150 μL in the tube.

   (h) Carefully resuspend the pellet of bacterial cells in the supernatant and plate the total volume onto ampicillin-containing (100 μg/mL) LB agar plates.

   (i) Place the agar plates upside down and incubate at 37 °C overnight.

3. Check the quality of the transformation reaction with colony PCR using the Expand High Fidelity PCR system according to the following instructions:

   (a) For each sample, pick ten colonies randomly.

   (b) Perform a PCR reaction with primers adjacent to the insert locus: forward primer located in the EMCV IRES and reverse primer located in the NS3/4A helicase region (Table 1).

   (c) Prepare the following PCR mix: mix 2 μL of 10× PCR buffer with $MgCl_2$, 0.7 μL of 10 mM dNTPs, 0.3 μL of Expand HF Enzyme mix (5 U/μL), and 0.06 μL of forward and 0.06 μL of reverse primer (100 μM stock solution). Add water to a total volume of 20 μL. Add a single colony to each reaction tube (*see* **Note 6**).

   (d) Use the following PCR cycling program: initial denaturation at 95 °C for 10 min, 35 cycles of denaturation at 95 °C for 30 s, primer annealing at 55 °C for 30 s, and elongation at 72 °C for 1 min 30 s, and perform a final extension step at 72 °C for 7 min.

(e) Load 5 μL of PCR product on a 1.2 % Flash Gel DNA cassette and 2.5 μL of a 100-bp DNA ladder. Run the gel for 30 min. Visualize the band pattern with a UV light transilluminator (*see* **Note 7**).

4. After successful transformation, culture the remaining colonies present on the agar dishes by adding 2 mL of LB medium. Wash off the colonies by scraping them with a spatula.
5. Transfer the suspension to 23 mL of LB medium.
6. Incubate the cultures at 37 °C for at least 7 h while shaking at 225 rpm.
7. Purify the plasmids using the PureLink™ HiPure Plasmid DNA purification kit for midi preparation of plasmid DNA to obtain high-concentration plasmid preparations according to the manufacturer's instructions, and resuspend the pellet in 100 μL TE buffer after air-drying. Store the purified DNA at −20 °C (*see* **Note 3**).
8. Take 2 μL of the purified replicon shuttle backbone containing patient-derived HCV NS3/4A protease, and determine the DNA concentration.

### 3.5 *In Vitro* Transcription of the Replicon Shuttle Backbone Containing Patient-Derived HCV NS3/4A protease

1. Linearize the purified replicon shuttle backbone(s) containing patient-derived HCV NS3/4A protease (from **step** 7 under Subheading 3.4) according to the following instructions:
   (a) Take 10 μg of each plasmid sample and add $H_2O$ to a total volume of 42 μL.
   (b) Add 3 μL (=40 U) of *Sca*I restriction enzyme, and 5 μL of corresponding buffer (10×) to obtain a total volume of 50 μL.
   (c) Incubate the reaction mixture at 37 °C for at least 4 h.
2. Purify the linearized plasmids using the QIAquick PCR Purification kit (*see* Subheading 3.3, **step 5**). Elute the digested DNA in 30 μL of buffer EB.
3. Take 2 μL of each obtained plasmid DNA to determine the DNA concentration.
4. Perform the in vitro transcription by using the T7 MEGAscript High Yield Transcription kit.
   (a) Mix 2 μL of ATP solution, 2 μL of CTP solution, 2 μL of GTP solution, 2 μL of UTP solution, 2 μL of 10× reaction buffer, 1 μg of linearized (digested) template DNA (from **step 2** under Subheading 3.5) (*see* **Note 8**), and 2 μL of RNA polymerase enzyme mix. Add water to a total volume of 20 μL.
   (b) Incubate the reaction mixture at 37 °C for 4.5 h.
   (c) Add 1 μL of TurboDNase to each reaction tube to hydrolyze the remaining template DNA.
   (d) Incubate at 37 °C for another 15 min.

5. Purify the in vitro transcribed RNA by using the RNeasy Mini Kit according to manufacturer's instruction and elute the RNA in 30 μL of RNase-free $H_2O$.
6. Take 2 μL of the RNA samples and quantify the RNA spectrophotometrically.
7. Make RNA aliquots of 10 μg and store at −80 °C until further use.
8. Analyze the RNA on a 1.2 % Flash Gel RNA cassette according to the following instructions:
   (a) Add 2.5 μL of a 1/10 dilution of the RNA to 2.5 μL of loading dye.
   (b) Incubate at 65 °C for 15 min.
   (c) Load the gel and run the gel at room temperature for 8 min (220 V).
   (d) Incubate the gel at room temperature for at least 30 min before taking a picture. Clear bands of approximate 9,000 bp should be visible.

### *3.6 Transient Replicon Assay*

1. Freshly add 2 mM ATP (pH 7.6, adjusted with KOH) and 5 mM glutathione to the cytomix.
2. Resuspend $1 \times 10^7$ Huh7-Lunet cells in 1 mL of cytomix.
3. Transfer 400 μL of this cell suspension to an electroporation cuvette (gap width 0.4 cm; Bio-Rad) containing 10 μg of in vitro transcribed replicon RNA (*see* Subheading 3.5, **step 5**).
4. Place the cuvette in the electroporation chamber (Gene-pulser BioRad) and pulse once (settings: 960 μF; 270 V; expected time constant: ~20 ms).
5. Transfer the cells to DMEM cell culture medium supplemented with 10 % FCS, 1 % L-glutamine and 0.04 % gentamicin.
6. Seed 30 μL of the cell suspension at 4,000 cells/well in a 384-well plate containing 10 μL of compounds in a 9-point dilution series and quadruplicate wells for each compound dilution. Incubate the plates at 37 °C in a humidified $CO_2$ atmosphere for 48 h.
7. Add SteadyLite Plus in a 1:1 ratio, incubate for 10–20 min in the dark, and measure luminescence using the ViewLux apparatus.
8. Calculate the 50 % effective concentration ($EC_{50}$) values based on the luminescence from compound-treated wells and cell controls. Use the median luminescence signal of medium control wells as 100 % inhibition, and use the median luminescence signal of cell control wells as 0 % inhibition. For each concentration of compound, the median luminescence signal of the quadruplicate wells is used for generating the inhibition curve. The $EC_{50}$ is calculated using computer software such as the GraphPad Prism.

## 4 Notes

1. To confirm the correct sequence of the shuttle replicon backbone without the NS3/4A protease fragment, perform miniprep purification using the QIAprep Miniprep according to manufacturer's instructions, followed by standard Sanger sequencing using the following protocol:
   (a) Mix 1 μL of eluted DNA per sequencing reaction to 1 μL of Big Dye Terminator Mix, 2.5 μL of 2.5× Dilution Buffer and 3.5 μL of $H_2O$ to obtain a final volume of 8 μL.
   (b) Add 2 μL of a 5 μM sequencing primer solution and mix gently by pipetting up and down.
   (c) Put the tubes in a thermocycler and subject them to the following sequencing protocol: 25 cycles of denaturation at 96 °C for 10 s, annealing at 50 °C for 5 s and elongation at 60 °C for 4 min.
   (d) Purify the samples by using the QIAquick PCR purification kit (*see* Subheading 3.3, **step 5**).
   (e) Load the samples on an ABI3730XL Analyzer (Applied Biosystems) or other sequencing analyzer.
   (f) Perform sequence analysis using SeqScape software (Applied Biosystems) or other sequencing software.
2. Insoluble particles may be present in the elution tube with the purified DNA. These particles do not influence the quality of the DNA and can be easily removed by centrifuging the DNA solution at high speed at room temperature for 1 min. Transfer the supernatant (=DNA sample) into a fresh tube.
3. To avoid repeated freezing and thawing of DNA, store the purified DNA at 4 °C for immediate use or aliquot the DNA and store at −20 °C for long-term use.
4. If HCV viral load is <1,000 IU/mL, load 600 μL of sample on the easyMAG, and elute in 25 μL elution buffer.
5. As a negative control, prepare the reaction mixture without insert. In this way, it is possible to evaluate the presence of vector background.
6. Do not overload the PCR reaction mix with template DNA, since this will cause inhibition of the PCR reaction.
7. The difference in amplicon length will allow distinction between colonies containing the insert (± 1,200 bp) and colonies containing uncut backbone (± 600 bp). The percentage of colonies containing the insert should be ≥50 %.
8. If the concentration of the linearized DNA is too low, use 8 μL of DNA in the reaction mixture or double the mix volume.

## References


1. Pawlotsky JM (2011) The results of phase III clinical trials with telaprevir and boceprevir presented at the Liver Meeting 2010: a new standard of care for hepatitis C virus genotype 1 infection, but with issues still pending. Gastroenterology 140:746–754
2. Vijgen L, Lenz O, Berke JM et al (2010) A cellular replicon-based phenotyping assay to determine susceptibility of clinical isolates to NS3/4A protease inhibitors. Antivir Ther 15(Suppl 2):A116
3. Lohmann V, Körner F, Koch U et al (1999) Replication of subgenomic hepatitis C virus RNA in a hepatoma cell line. Science 285: 110–113

# Chapter 10

# Expression and Purification of Hepatitis C Virus Protease from Clinical Samples

Liesbet van der Helm


## Abstract

This chapter describes the procedures for production of recombinant hepatitis C virus (HCV) NS3 protease from clinical samples, which can be used in the biochemical assays to assess the impact of different drug-resistant mutations in the NS3 protein from patients under therapy of protease inhibitors. It gives the details of expression and purification of NS3 protease using the pCold vectors that contains a promoter derived from the cold-shock genes to drive the expression of NS3 protein. This robust protocol enables a medium-throughput production of HCV NS3 proteins of all genotypes and sequences derived from patient specimen.


**Key words** Hepatitis C Virus, NS3 protease, Protein expression, Protein purification

## 1 Introduction

For years, the standard of care for patients with chronic HCV infection consisted of a combination of weekly injections of pegylated interferon-α (PegIFN) and twice-daily oral administrations of ribavirin (RBV). Treatment duration was generally long (up to 48 weeks) and the rate of sustained viral response (SVR) was low in case of (depending on the geographical locations) genotype 1 patients (40–50 %). The therapy is also associated with a range of side effects including flu-like symptoms, anemia, and depression [1]. In 2011, PegIFN and RBV as the backbone regimen in combination with one of two protease inhibitors, telaprevir and boceprevir, were approved for treatment of genotype 1 patients. These two new combination regimens can significantly shorten the treatment duration and exhibit improved SVR rates up to 79 % (for genotype 1).

HCV belongs to the *Flaviviridae* family and can be divided into six major genotypes and multiple subtypes. It contains a positive-sense, single-stranded RNA genome with size of approximately 9.6 kb. The genome is translated into a large polyprotein of

Edwin Yunhao Gong (ed.), *Antiviral Methods and Protocols*, Methods in Molecular Biology, vol. 1030,
DOI 10.1007/978-1-62703-484-5_10, © Springer Science+Business Media, LLC 2013

approximately 3,100 amino acids that is subsequently cleaved co-translationally and post-translationally into structural (C, E1, E2, p7) and nonstructural proteins (NS2, NS3, NS4A, NS4B, NS5A, and NS5B). NS3 is a multifunctional protein and consists of a serine protease domain and a helicase domain. Together with its cofactor NS4A, NS3 cleaves the nonstructural genes of the polyprotein. The ability to reconstitute its enzymatic activity in vitro allowed for in-depth understanding of protease-substrate interactions and made the NS3/4A serine protease a clinically approved antiviral target [2].

Drug resistant mutations or variants may arise in the NS3 protein when patients are treated with the NS3/4A protease inhibitors. Subgenomic replicon assays as well as biochemical assays are used in parallel to evaluate the impact of drug resistant mutations in vitro to support the clinical studies. This implies the need to generate sufficient quantities of purified recombinant NS3/4A protease for the enzymatic assay. When expressing multiple genotypes and/or mutants of the NS3/4A protease, low solubility and large variations in expression levels of the recombinant proteins can become an issue. Two strategies are typically used to overcome these problems: the first approach is to change the expression strain or modify the expression conditions (i.e., culture medium, induction temperature, addition of highly soluble fusion partners like Maltose Binding Protein (MBP) or NusA) [3]; the second approach, when the first approach resulted in no or minimal improvement, is to modify the expression construct by optimizing the codons for bacterial expression or, as described below, by using a different promoter.

Cold-shock expression vectors, pCold™ DNA, are designed for efficient protein expression utilizing a promoter derived from the *cspA* gene, one of the cold-shock genes. Cold shock induction of gene expression is carried out by addition of isopropyl-β-D-1-thiogalactopyranoside (IPTG) following temperature downshift. Downstream of the *cspA* promoter, a *lac* operator is inserted. Constitutive expression of the *lacI* gene prevents leaky expression of the genes at 37 °C. A translation-enhancing element (TEE) and a 6xHis tag are also inserted at downstream of the *cspA* promoter. The TEE, a five-codon sequence that encodes MNHKV, has been shown to enhance the translation initiation. As this product utilizes a promoter derived from *E. coli*, most *E. coli* strains can be utilized as expression host [4, 5].

Here, a detailed protocol using the pCold vectors to allow high level expression of soluble NS3 proteases of multiple mutants and various clinical isolates is provided, which can subsequently be used in the biochemical assays in combination with a synthetic NS4A cofactor (cleavage of a specific substrate, RetS1).

## 2 Materials

### 2.1 Bacterial Transformation and Protein Expression

1. pColdII vector (Takara).
2. Rosetta 2 (DE3) competent cells (Merck Millipore).
3. 34 mg/mL chloramphenicol stock solution in 100 % ethanol, aliquot into 20 mL and store at −20 °C (*see* **Note 1**).
4. 50 mg/mL carbenicillin stock solution in MilliQ $H_2O$, aliquot and store at −20 °C (*see* **Note 1**).
5. Luria–Bertani (LB) agar plates: 10 g/L tryptone, 5 g/L yeast extract, 10 g/L NaCl, 15 g/L agar. Add required amount of capsules (MP Biomedicals) to 250 mL of deionized water. Autoclave for 20 min at 121 °C. Cool at room temperature until the solution is ~50 °C. Add appropriate antibiotics for selection (50 μg/mL carbenicillin and 34 μg/mL chloramphenicol). Pipette 25 mL to each petri dish. Leave the plates uncovered until the agar is solidified but no longer than 30 min to avoid shrinking of the agar.
6. LB medium: 10 g/L tryptone, 5 g/L yeast extract, 10 g/L NaCl. Add required amount of capsules (MP Biomedicals) to 1 L of deionized water. Autoclave the media for 20 min at 121 °C. Add the antibiotics required for selection (50 μg/mL carbenicillin and 34 μg/mL chloramphenicol) when the solution cools down to room temperature.
7. SOC (Super Optimal Broth with catabolite-repression) medium.
8. 100 mM zinc chloride stock solution in MilliQ $H_2O$. Stored at 4 °C.
9. 100 mM IPTG stock solution in MilliQ $H_2O$. Stored at −20 °C.
10. Complete, EDTA-free protease inhibitor (Roche).
11. Petri dishes (100 mm diameter, sterile).
12. Sonifier (e.g., Branson Digital Sonifier 250).
13. Ultra Yield Flask (Thomson).

### 2.2 Protein Purification

1. Imidazole ULTROL grade (Calbiochem) (*see* **Note 2**).
2. High-salt lysis buffer: 25 mM Tris–HCl, pH 7.5, 10 % glycerol, 1 mM tris(2-carboxyethyl)phosphine (TCEP), 750 mM NaCl, 2 % Foscholine-12 (*see* **Note 3**).
3. Washing buffer: high-salt lysis buffer with 40 mM imidazole. Add 0.54 g imidazole to 200 mL of high-salt wash buffer and adjust the pH to 7.5 (*see* **Note 4**).
4. Elution buffer: 25 mM Tris–HCl, pH 7.5, 10 % glycerol, 1 mM TCEP, 750 mM NaCl, 2 % Foscholine-12, 500 mM imidazole.

5. His MultiTrap™ FF plates (GE Healthcare).
6. 96 deep-well plates.
7. Aluminum adhesive sealing foil for 96-well plates.
8. Swinging bucket centrifuge with holders for 96 deep-well plates (max. speed needed is 500 × *g*).
9. Plate shaker (e.g., MTS4, IKA-schutter).

### 2.3 Protein Analysis

1. NuPage® Novex 4–12 % Bis–Tris Midi gel 1.0 mm, 26 wells (Life Technologies).
2. NuPage® LDS sample buffer (4×) (Life Technologies).
3. NuPage® Reducing agent (10×) (Life Technologies).
4. Sample buffer/reducing agent mix: add 1 mL of NuPage® LDS sample buffer and 400 μL of NuPage® reducing agent to an Eppendorf tube and mix.
5. NuPage® Antioxidant (Life Technologies).
6. NuPage® MES SDS running buffer (20×) (Life Technologies): 1× running buffer is prepared in MilliQ $H_2O$.
7. BenchMark® His-tagged protein standard (Life Technologies).
8. Instant Blue (Expedeon).
9. Coomassie Plus-The Better Bradford™ Assay kit (Thermo Scientific) (*see* **Note 5**).
10. 2 mg/mL of albumin standard.
11. Spectrophotometer with 595 nm filter.

## 3 Methods

### 3.1 Cloning of NS3 Protease

The NS3 protease domain (amino acids 1–181) of HCV was inserted into a plasmid vector pColdII. The constructs contain an N-terminal 6xHis tag followed by a TEE and the NS3 coding sequence (Fig. 1). For clinical samples of HCV, the NS3 region was sequenced, and cloned into the same vector. All plasmid constructs are synthesized and sequenced by GeneArt (Life Technologies).

### 3.2 Bacterial Transformation and Protein Expression

1. Thaw 20 μL of Rosetta 2 (DE3) competent cells on ice and mix gently to ensure that the cells are evenly suspended.
2. Transform Rosetta 2 (DE3) with the pCold vector containing the gene to be expressed. Add 1 μL of plasmid DNA (50 ng/μL) to 20 μL of thawed Rosetta 2 (DE3) cells. Stir gently to mix.
3. Place the tubes on ice for 5 min.
4. Heat the tubes for 30 s in a 42 °C water-bath, do not shake. Then place the tubes on ice for 2 min.

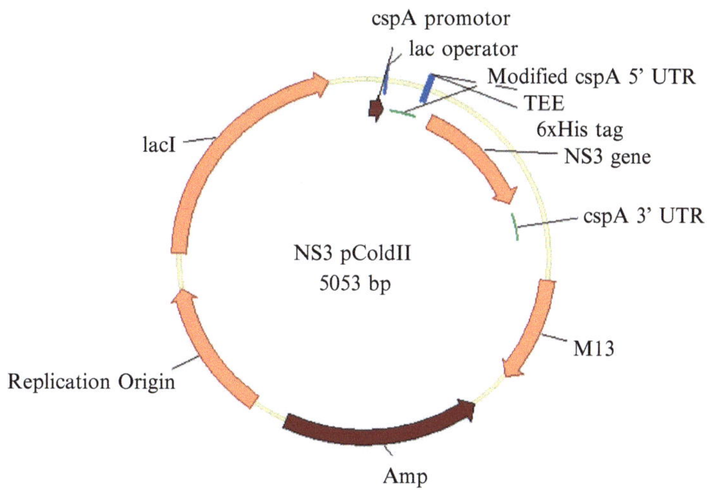


**Fig. 1** The plasmid map of the NS3 expression construct generated using nucleic acid analysis software VectorNTI. The NS3 gene is inserted into the multiple cloning sites of the pColdII vector. The protein expression is under the control of the *cspA* promoter. A TEE sequence and a 6xHis tag are inserted downstream of the promoter at the N-terminal of the NS3

5. Add 250 μL of SOC medium (equilibrated to room temperature). Transfer the tube to a shaking incubator and incubate for 60 min at 37 °C while shaking at 200 rpm.
6. Plate 50 μL on a LB-agar plate containing carbenicillin (50 μg/mL) and chloramphenicol (34 μg/mL). Place the plate upside down and incubate at 37 °C for 16 h.
7. To set up a starter culture, inoculate 3–5 colonies from the freshly transformed plate to 5 mL of LB medium containing 50 μg/mL carbenicillin and 34 μg/mL chloramphenicol.
8. Incubate at 37 °C while shaking at 200 rpm, until the $OD_{600nm}$ reaches 1.0.
9. Prepare the expression cultures, one for each expression construct, by adding 4 mL of the starter culture to 500 mL LB medium containing 50 μg/mL carbenicillin and 34 μg/mL chloramphenicol in a 2 L Ultra Yield Flask.
10. Incubate the bacterial culture at 37 °C while shaking at 200 rpm until $OD_{600nm}$ = 0.4–0.6. At this point, take a 1 mL sample to be evaluated on a NuPage® Novex 4–12 % Bis-Tris Midi gel (*see* **Note 6**).
11. Cool the cultures to 15–20 °C in ice water (*see* **Note 7**). Meanwhile let the incubator cool to 15 °C.
12. Induce the cultures by adding IPTG to a final concentration of 0.4 mM and $ZnCl_2$ to a final concentration of 50 μM (*see* **Note 8**). Place the induced cultures in a 15 °C incubator and incubate for 16 h while shaking at 200 rpm.

13. After incubation, transfer the bacterial cultures to a centrifuge tube. Take a 1 mL sample to evaluate the expression on a NuPage® Novex 4–12 % Bis–Tris Midi gel (*see* **Note 6**).
14. Harvest the bacterial cells by centrifugation at 6,000 × *g* for 10 min. Discard the supernatant.
15. Resuspend the pellet in 13 mL of high-salt lysis buffer containing complete EDTA-free protease inhibitors. Lyse the bacteria by three cycles of sonication on ice, 1 min each (10 s on and 10 s off).
16. Centrifuge for 45 min at 20,000 × *g* to pellet the insoluble fraction. Transfer the supernatant (lysates) to a fresh tube and continue with the purification as described in Subheading 3.3.

### 3.3 Protein Purification

The His MultiTrap™ FF plates are 96-well filter plates pre-packed with pre-charged Ni Sepharose™ 6 fast flow resin. The top and bottom of the plate are covered with seals to prevent leakage of the storage solution. Three different proteins can be purified on one His MultiTrap™ FF plate (four columns for each protein). Use a swinging bucket centrifuge for all centrifugation steps. The following procedures are adapted from the manufacturer's instructions:

1. Remove the bottom and top seals. Place the His MultiTrap™ FF plate on the top of a 96 deep-well plate. Centrifuge the plates for 2 min at 500 × *g* to remove all storage solution from the $Ni^{2+}$ resin.
2. Equilibrate the $Ni^{2+}$ resin with high-salt lysis buffer (binding buffer). First, add 500 μL of MilliQ water per well. Centrifuge at 500 × *g* for 2 min. Next, add 500 μL of high-salt lysis buffer per well. Centrifuge at 500 × *g* for 2 min. Repeat the last step once (*see* **Note 9**).
3. Cover the bottom of the HisMultiTrap plate with an alu-foil sticker to avoid leakage of the clarified cell lysates during the binding. Pipette 400 μL of the clarified cell lysates to each well using a multichannel pipette. Incubate at room temperature for 60 min while shaking on a shaker. Check if the resin is resuspended to a homogeneous solution and no resin sits on the bottom of the well.
4. Remove the alu-foil sticker from the bottom. Position the plate on the top of a collection plate and remove the unbound fraction by centrifugation at 100 × *g* for 4 min (or until the wells are empty).
5. Place the His MultiTrap™ FF plate on top of a fresh collection plate. Then apply 500 μL of washing buffer using a multichannel pipette. Mix by gently shaking the plate horizontally for 1 min. Remove the washing buffer by centrifugation at 500 × *g* for 2 min. Repeat the washing step once.

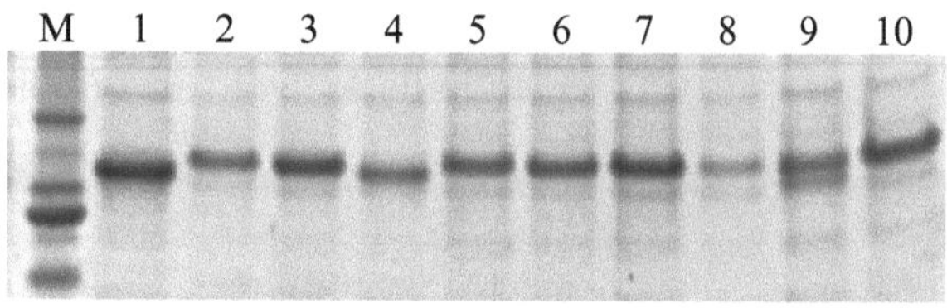


**Fig. 2** A stained protein gel of different HCV NS3 proteases expressed from clinical samples. The NS3 proteins were expressed in the pColdII vector and purified using the His MultiTrap™ FF plates (*lanes 1–10*). *Lane M* is the BenchMark protein ladder

6. Position the His MultiTrap™ FF plate on a fresh collection plate. Add 40 μL of the elution buffer to each well. Mix by gently shaking the plate horizontally for 1 min. Collect the elution buffer by centrifugation at 500×*g* for 2 min. Repeat the elution step once. Pool the elution fractions and proceed with the analysis on gel and the quantification of the yield using the Bradford assay.

### 3.4 Analysis of Expressed Protein

#### 3.4.1 Protein Analysis Using a NuPage Protein Gel

1. Prepare samples by adding 14 μL of sample buffer/reducing agent mix either to 26 μL of purified protein; or to 12 μL of clarified cell lysates (or expression sample) supplemented with 14 μL of MilliQ water.
2. Heat the samples at 90 °C for 10 min on a heating block.
3. Place the NuPage® Novex 4–12 % Bis–Tris Midi gels in a midi gel system and fill the buffer tank with 1× MES buffer.
4. Add 435 μL of antioxidant to the four upper chambers.
5. Load 20 μL of protein samples onto the gel. Load 10 μL of BenchMark protein marker on each gel to determine the size of the proteins in the samples.
6. Run at 200 V for 40 min.
7. When the run is finished, remove the gel from the cassette and stain the gel with the Instant Blue solution for 10 min up to 1 h.
8. Destain the gel with MilliQ water. Refresh the MilliQ water until the background of the gel is clear. A stained protein gel of different HCV NS3 proteases expressed from clinical samples is shown in Fig. 2.

#### 3.4.2 Determination of the Protein Concentration

1. Dilute 2 mg/mL albumin standard to 1.6 mg/mL by adding 720 μL of 2 mg/mL to 180 μL elution buffer. Then prepare a serial dilution of albumin standard: 1,600, 800, 400, 200, 100, and 50 μg/mL.
2. Make twofold serial dilutions of the protein samples.
3. Pipette 10 μL of each albumin standard or protein samples to the appropriate wells of a 96-well plate with clear flat bottom.

4. Add 300 μL of the Coomassie Plus reagent to each well and mix gently to avoid air bubbles.
5. Incubate the plate at room temperature for 10 min.
6. Measure the absorbance at a wavelength of 595 nm with a spectrophotometer.
7. Prepare a standard curve using the $OD_{595nm}$ values of albumin. Calculate the average and correct with the blank (elution buffer only). Plot the corrected standards vs. its concentration in μg/mL. Use the standard curve to determine the protein concentration of each unknown sample. Take the dilution factor of the samples into consideration.

## 4 Notes

1. Carbenicillin and chloramphenicol are added for the selection of the pCold plasmid.
2. High purity of imidazole should be used since it absorbs UV radiation at 280 nm. The absorbance of imidazole can vary depending on its source and purity, but elution buffer containing 250 mM imidazole usually has an $A_{280nm}$ of 0.2–0.4.
3. A buffer optimization was performed and we concluded that a high salt concentration and Foscholine-12 are essential for the solubility of the expressed protein.
4. Addition of imidazole to the washing and elution buffers raises the pH. Adjust the pH to 7.5 by adding 10 M HCl dropwise.
5. To quantify proteins in the elution buffer containing imidazole, the Bradford protein assay is recommended. This assay is more tolerant for higher concentrations of imidazole than the Lowry and the biuret assays. Always use elution buffer without protein as a blank.
6. A 1-mL culture sample was taken for evaluation on a protein gel. Measure the $OD_{600nm}$ of the bacterial sample. Centrifuge for 5 min and remove the supernatant. Resuspend the pellet in water (100 μL of MilliQ water for 1 $OD_{600nm}$, for example, if an $OD_{600nm}$ is 0.8, add 80 μL of water). Take 6 μL of resuspended pellet to prepare the samples for gel electrophoresis.
7. The cultures are cooled to 15 °C prior to induction with IPTG to enhance the expression of soluble protein. Place the cultures in ice water and monitor the temperature.
8. Zinc is a cofactor of the HCV NS3 protease. The zinc ion is bound far away from the catalytic site and is essential for the structural integrity of the protein [6].

9. Reducing agents may be used in the sample and buffers. Do not leave His MultiTrap™ FF plates with buffers including reducing agents when not in use. When using dithiothreitol (DTT) instead of TCEP as a reducing agent, a brown color of the resin can occur due to the reduction of the $Ni^{2+}$.

## References


1. Chatel-Chaix L, Baril M, Lamarre D (2010) Hepatitis C virus NS3/4a protease inhibitors: a light at the end of the tunnel. Viruses 2: 1752–1765
2. Lamarre D, Thibault D, Bousquet C (2004) Sensitivity of NS3 serine proteases from hepatitis C virus genotypes 2 and 3 to the inhibitor BILN 2061. J Virol 78(14):7352–7359
3. Sørensen H, Mortensen K (2005) Soluble expression of recombinant proteins in the cytoplasm of *Escherichia coli.* Microb Cell Fact 4:1
4. Qing G, Ma L, Khorchid A et al (2004) Cold-shock induced high-yield protein production in *Escherichia coli.* Nat Biotechnol 22(7): 877–882
5. Baneyx F, Mujacic M (2003) Cold-inducible promotors for heterologous expression. Methods Mol Biol 205:1–18
6. Stempniak M, Hostomska Z (1997) The NS3 proteinase domain of hepatitis C virus is a zinc-containing enzyme. J Virol 71(4): 2881–2886

# Chapter 11

# Surface Plasmon Resonance as a Tool to Select Potent Drug Candidates for Hepatitis C Virus NS5B Polymerase

Erna Cleiren, Benoit Devogelaere, and Katleen Fierens

## Abstract

Surface plasmon resonance (SPR)-based optical biosensors have been widely used to study biomolecular interactions, and applied to many areas of drug discovery including target identification, fragment screening, lead compound selection, early ADME (absorption, distribution, metabolism and excretion), and quality control. These biosensors allow the following of a biomolecular interaction in real time to monitor kinetics and determine affinity. In this chapter, we describe an SPR-based assay to measure the interaction between hepatitis C virus NS5B polymerase (wild type and/or mutants) and a small-molecule inhibitor. Viral polymerase proteins are captured on a $Ni^{2+}$-nitrilotriacetic acid sensor surface while the small-molecule inhibitors are passed over the surface. In this way kinetics and affinity of the enzyme–inhibitor interactions can be measured, making it possible to select potent and promising lead candidates.

**Key words** Surface plasmon resonance, Real-time kinetics, HCV NS5B polymerase, Interaction studies, Small-molecule inhibitors

## 1 Introduction

The use of surface plasmon resonance (SPR) technology has steadily increased since its discovery [1]. It is a leading tool in biomedical research to monitor biomolecular interactions in real time. The main components of an SPR-based optical biosensor instrument are (a) an exchangeable sensor chip coated with a gold layer on which a ligand can be immobilized, (b) a microfluidic/liquid-handling system which controls a constant flow rate while the binding partner (analyte) is injected in aqueous solution over the gold surface, and (c) an optical detector system that monitors the changes in SPR signal caused by binding events occurring on the sensor chip surface in real time [2].

Real-time complex formation and dissociation are represented in the typical plot of resonance signal (response units, RU) as a function of time, also known as the sensorgram (Fig. 1). Sensorgram analysis provides information about affinity as well

Edwin Yunhao Gong (ed.), *Antiviral Methods and Protocols*, Methods in Molecular Biology, vol. 1030,
DOI 10.1007/978-1-62703-484-5_11, © Springer Science+Business Media, LLC 2013

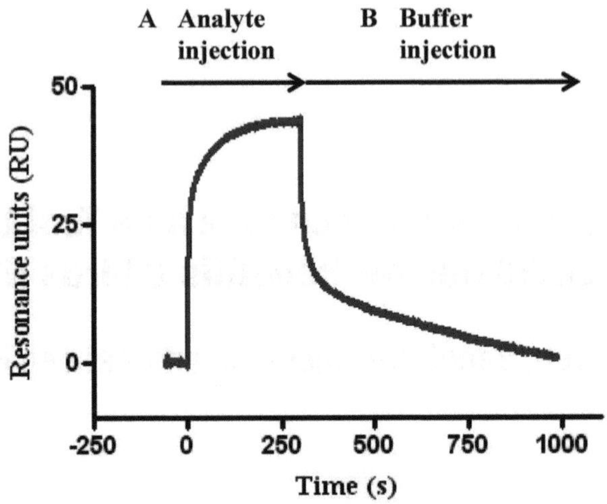


**Fig. 1** Typical binding cycle of a biomolecular interaction. (**a**) When the analyte is injected over the sensor surface, the sensorgram shows an increasing response due to molecular interactions. The response remains constant after the interaction has reached equilibrium. (**b**) The analyte is gradually replaced by buffer and the response decreases as the complex dissociates

as kinetics of the biomolecular interactions including distinct association and dissociation rate constants.

In this way many basic questions in drug discovery can be answered, for example: Does the analyte/inhibitor bind to a specific target? How strong is the interaction between analyte/inhibitor and the target? What is the residence time of this inhibitor? Does a specific mutation in a binding pocket of a target protein affect the kinetics or affinity (and hence inhibitory activity) of the inhibitor? This information is very useful to rank potential drug candidates.

To evaluate the interaction between NS5B polymerase and a small-molecule inhibitor, we have developed an SPR assay based on the combination of single-cycle kinetics and a universal target protein capturing approach (based on the affinity of the $HIS_6$-tagged protein for $Ni^{2+}$-loaded nitrilotriacetic acid (NTA)). Using a general capture immobilization method the protein can remain flexible and all protein molecules are similarly oriented, leaving the binding pocket intact and accessible.

The SPR protocol consists of three main steps: sensor chip preparation, target protein capturing, and small-molecule inhibitor injection (Fig. 2). $Ni^{2+}$ is injected over the NTA sensor chip followed by the injection of the target protein. Several target proteins can be immobilized on one sensor chip, allowing simultaneous analysis of multiple proteins, which, in our case, were hepatitis C virus (HCV) NS5B polymerase wild type and several allosteric site mutants.

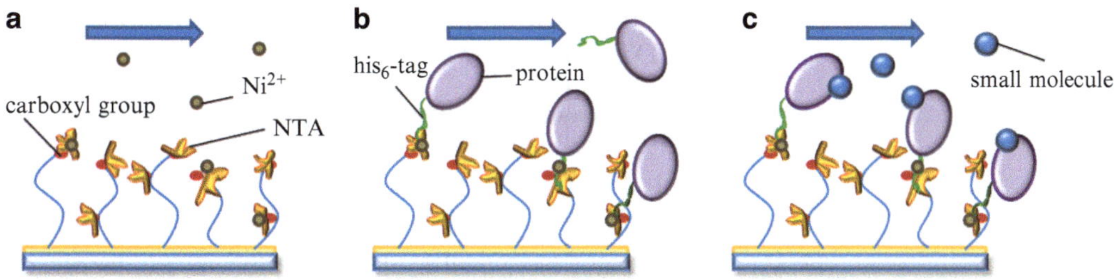


**Fig. 2** Overview of the main steps in the Biacore assay. Carboxymethylated dextran is pre-immobilized with nitrilotriacetic acid (NTA) and $Ni^{2+}$ is injected over the sensor surface (**a**). The $HIS_6$-tagged protein is captured via $Ni^{2+}$/NTA chelation (**b**). Small molecules like the HCV NS5B inhibitors are subsequently injected to measure their interaction with this enzyme (**c**)

The interaction of small-molecule inhibitors targeting HCV NS5B polymerase was measured by subsequently injecting increasing compound concentrations (five in all) in one binding cycle. This single-cycle kinetics method was found to be much more time efficient compared with the normal multi-cycle kinetics (one compound concentration per one binding cycle). Moreover, this method strongly reduced the consumption of inhibitor and reagents and allowed us to accurately determine the dissociation rate constants ($k_{off}$) of tight-binding compounds expeditiously. Furthermore, the described method enables to determine whether the mutations in different binding pockets of the enzyme resulted in a change of affinity of the small-molecule inhibitors for the mutant enzyme compared with the wild-type HCV NS5B polymerase. Based on kinetics and affinity values, promising chemical series can be selected and qualitatively ranked for further lead compound optimization.

## 2 Materials

All solutions are prepared using Water GPR RECTAPUR (BDH Prolabo).

### *2.1 Proteins*

$HIS_6$-tagged HCV NS5B polymerase wild-type (WT) NS5BΔC21 and mutants were produced and purified as reported previously: the coding sequences for the different NS5B isolates lacking 21 C-terminal residues (NS5BΔC21) were amplified and subcloned into the pET21b plasmid. The NS5BΔC21 expression constructs were then transformed into *E. coli* Rosetta 2 (DE3) for NS5B production, and protein was purified as reported before [3].

### *2.2 Immobilization of Enzymes*

1. Immobilization buffer: 20 mM 3-(*N*-morpholino)propanesulfonic acid buffer (MOPS), pH 7.4, 500 mM NaCl, 1 mM dithiothreitol (DTT), 50 μM EDTA, 0.005 % Tween-20 (*see* **Note 1**).
2. 500 μM $NiCl_2$.

***2.3 Ligand–Analyte Interaction***

1. Running buffer: 20 mM Tris–HCl (pH 7.4), 150 mM NaCl, 50 μM EDTA, 1 mM DTT, 0.005 % Tween-20, 5 % DMSO (*see* **Notes 1** and **2**).
2. Regeneration solutions:

   Solution 1: 350 mM EDTA (pH 8), 10 % DMSO, 500 mM NaCl, 0.1 % CM5, 0.2 % CHAPS. Solution 2: 1× PBS, 0.05 % SDS.
3. Analyte preparation: Small-molecule inhibitors are dissolved in 100 % DMSO at a concentration of 5 mM.

***2.4 Consumables***

1. Series S Sensor chips NTA (GE Healthcare).
2. 1 L filter units MF75 PES (Nalgene).
3. Microplate foil (96-well) (GE Healthcare).
4. Microplate 96-well (GE Healthcare).
5. Rubber caps type 2 (GE Healthcare).
6. Rubber caps type 5 (GE Healthcare).
7. Plastic vials, 7 mm (GE Healthcare).
8. Plastic vials, 15 mm (GE Healthcare).

***2.5 Equipment***

1. Biacore T100 (GE Healthcare).
2. Sample and reagent rack, type 1 (GE Healthcare).
3. Single-channel pipettes (10, 100, 200, 1,000 μL).

## 3 Methods

***3.1 Immobilization of Target Protein on Chip Surface***

1. Dock a new NTA sensor chip in the instrument.
2. Set analysis and rack temperature to 25 °C.
3. Prime the chip surface three times with running buffer. This is recommended after insertion of the sensor chip in the Biacore T100 to wet the surface and to remove particles (*see* **Notes 3** and **4**).
4. Inject regeneration solution 1 and subsequently regeneration solution 2 for 3 min each at a flow rate of 10 μL/min, to remove metal ions that may still be present after previous runs. Afterwards, perform an extra wash and stabilization step for 60 s with the running buffer.
5. Inject the nickel solution (500 μM $NiCl_2$) for 1 min at a flow rate of 10 μL/min. Nickel ions will bind to the NTA surface and the baseline will increase with approximately 40 RU. Afterwards, wash the system with the running buffer.
6. Dilute the purified $HIS_6$-tagged polymerases in immobilization buffer to a final concentration of 100 nM and inject the diluted enzyme over the Ni-NTA sensor chip for 2–3 min at a flow rate

**Table 1**
**Preparation of analyte solutions**

| | Step 1 | Step 2 | Step 3 | Step 4 | Step 5 | Step 6 |
|---|---|---|---|---|---|---|
| Analytes | 5 μL of 5 mM stock (100 % DMSO) | 100 μL of solution step 1 | 100 μL of solution step 2 | 100 μL of solution step 3 | 100 μL of solution step 4 | 100 μL of solution step 5 |
| Added dilution buffer | 95 μL of 0 % DMSO buffer | 900 μL of 5 % DMSO buffer | 900 μL of 5 % DMSO buffer | 900 μL of 5 % DMSO buffer | 900 μL of 5 % DMSO buffer | 900 μL of 5 % DMSO buffer |
| Final analyte concentration | 250 μM | 25 μM | 2.5 μM | 250 nM | 25 nM | 2.5 nM |

of 10 μL/min. The baseline should increase around 3,000 RU (*see* **Notes 5** and **6**). Spot 1 on the sensor surface is usually left empty and serves as a suitable control surface to detect analyte nonspecific binding. Spot 2 is used to capture HCV NS5B polymerase wild type. Spot 3 and 4 can be used to immobilize different polymerase mutants.

### 3.2 Interactions of Non-nucleoside Polymerase Inhibitors with Wild-Type Polymerase and Polymerase Mutants

1. Dilute the test compounds (concentration ranges between 2.5 and 25 μM) according to the dilution scheme described in Table 1 (*see* **Note 7**).
2. Inject (flow rate 30 μL/min at 25 °C) five increasing concentrations of compound or five blank samples (no compounds) for a period of 300 s each in one single cycle. To monitor dissociation, inject running buffer over the sensor surface for a period of 1,200 s.
3. Sequentially inject regeneration buffer solution 1 for 3 min and regeneration buffer solution 2 for 2 min at a flow rate of 10 μL/min so that nickel is stripped (and thus the proteins) resulting in a complete removal of proteins and analytes from the sensor surface. Regeneration is followed by a wash with running buffer and a stabilization period of 300 s (*see* **Note 8**).
4. Repeat the procedure of target protein capturing and analyte injection for each new analyte and blank sample. An overview of the stepwise protocol is presented in Fig. 3.

### 3.3 Analysis of Binding Kinetics and Affinity

1. To obtain an actual sensorgram, subtract the response of the blank reference spot and the zero concentration injections from the analyte binding cycles, also called double referencing [4].
2. Analyze data using simultaneous nonlinear regression analysis (global fitting) adapted for single-cycle kinetics with Biacore

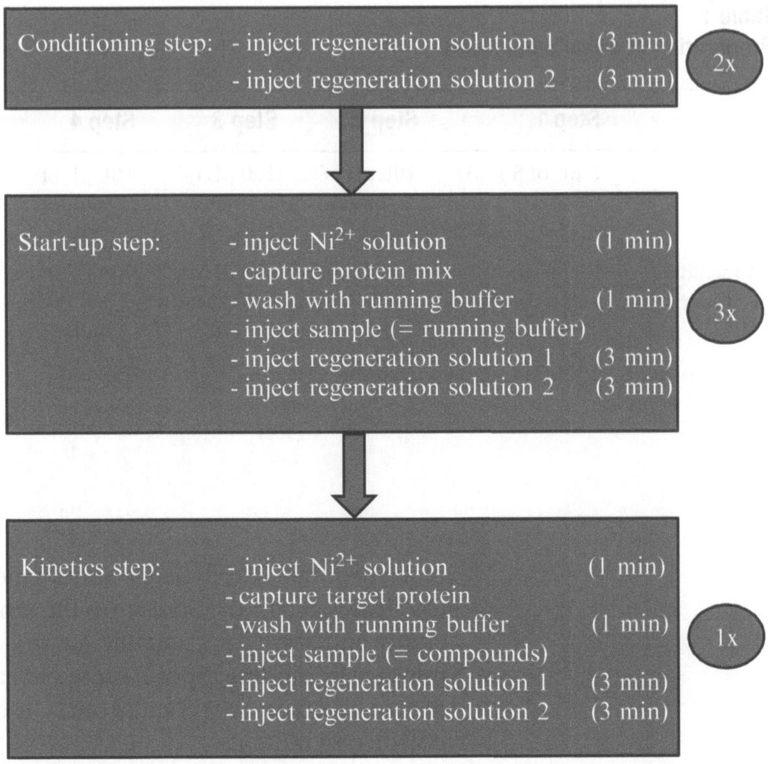


**Fig. 3** Overview of the different steps in the assay protocol. The Biacore T100 Control Software includes a Method Builder tool which guides the user through several assay steps. The conditioning step prepares the system before starting with the selected samples. In the start-up cycle, a mixture of used proteins is immobilized on the chip and running buffer is used as a sample. In the third step, the samples of which the interactions should be analyzed are injected

T100 BiaEval evaluation software 2.0 (GE Healthcare). More detailed information about the software can be found in Biacore T100 Software Handbook (GE Healthcare).

3. Perform every analysis in at least two independent experiments.
4. Verify that the fitted values for association rate constant $k_{on}$, dissociation rate constant $k_{off}$, or affinity constant $K_D = k_{off}/k_{on}$ are reproduced within 1 order of magnitude. Typical sensorgrams are displayed in Fig. 4.

## 4 Notes

1. All buffers are degassed and filtered. The latter is needed to remove particles (0.22 μm filters, Millipore) in order to prevent clogging of the Biacore system. Add Tween-20 after filtration (final concentration is 0.005 %) to prevent excess foaming. Tap on the bottle to remove any remaining air bubbles.

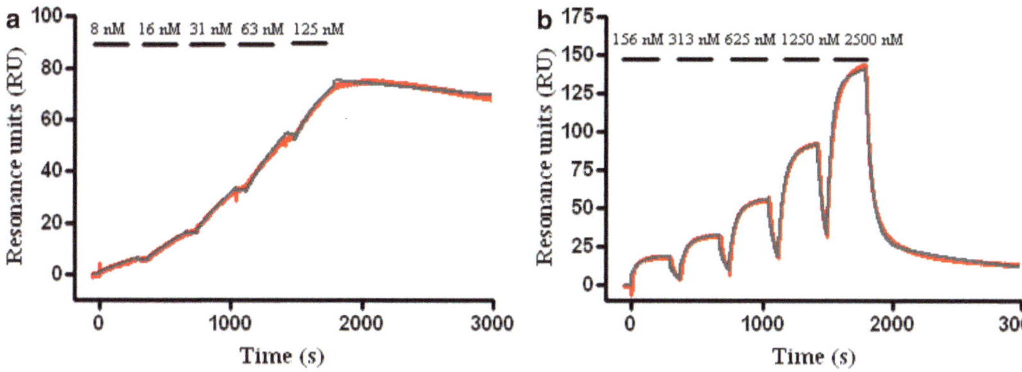


**Fig. 4** Examples of single-cycle sensorgrams representing the interaction of an HCV NS5B inhibitor to HCV NS5B polymerase wild type (**a**), and mutant (**b**). *Red* traces represent the reference-subtracted data, while the *black* traces represent the fitted binding curves

2. One liter of running buffer is prepared without DMSO. Remove 50 mL of running buffer and replace with 50 mL of 100 % DMSO. The final DMSO concentration in the running buffer is 5 %. Keep the 50 mL running buffer without DMSO for further sample preparation (0 % DMSO buffer).
3. Before inserting a sensor chip into the instrument, check both sides of the chip for particles. Although it is recommended to insert the sensor chip immediately in the instrument to avoid contamination, little dust particles may still be attached to the surface. Therefore, it is recommended to clean the sensor surface with Rectapur water and dry both sides with compressed air. Avoid touching the sensor surface and ensure that no liquid or particles are attached before assembling the sensor chip back into the sheath. Any contamination could lead to false measurements.
4. It is recommended to prime the system a couple of times with running buffer to ensure a stable baseline. Alternatively, one can also choose to perform several buffer injections over the sensor surface before injecting the analyte.
5. The advantage of using $HIS_6$-tagged proteins is the homogeneous orientation of the ligand molecules on the sensor chip surface. As this immobilization is only a capturing method and not covalent immobilization, ligand leakage may occur during the run. Serious ligand leakage can be solved by covalent binding of the ligand to the surface using amine coupling chemistry.
6. To avoid mass transport limitations or target protein aggregation it is better to use low immobilization levels in the kinetic experiments. A good rule of thumb is to apply the following formula:

$$R_{max} = R_{ligand} \times MW_{analyte} / MW_{ligand} \times n$$

$R$ = response units, MW = molecular weight, $n$ = stoichiometry (analyte/ligand).

$R_{ligand}$ is the immobilized level of the ligand expressed in RU. The surface binding capacity is usually expressed as $R_{max}$ which is the maximum analyte binding response. It is recommended to aim for a low $R_{max}$ of typically 50–100 RU.

7. When the affinity range is unknown a priori, it is advisable to test a broad range of analyte concentrations (nM to μM). Five to eight analyte concentrations (including a zero concentration) should be tested.
8. It is recommended to include a number of start-up cycles in each run. Start-up cycles are similar to analysis cycles except that a dummy sample (buffer) replaces the analyte sample. A newly docked chip often shows some instability during the first cycles. Such instability can be observed and effectively neutralized in the start-up cycles [5].

## References

1. Jönsson U, Malmqvist M (1992) In: Turner APF (ed) Advances in biosensors 2, JAI, London, pp. 291–336
2. Cooper MA (2003) Label-free screening of bio-molecular interactions. Anal Bioanal Chem 377:834–842
3. Nyanguile O, Devogelaere B et al (2010) 1a/1b Subtype profiling of non-nucleoside polymerase inhibitors of hepatitis C virus. J Virol 84: 2923–2934
4. Myszka DG (1999) Improving biosensor analysis. J Mol Recognit 12:279–284
5. GE Healthcare (2007) Biacore T100 software handbook BR-1006-48 Edition AE. Uppsala, Sweden

# Chapter 12

# Amplification and Sequencing of the Hepatitis C Virus NS3/4A Protease and the NS5B Polymerase Regions for Genotypic Resistance Detection of Clinical Isolates of Subtypes 1a and 1b

**Diana Koletzki, Theresa Pattery, Bart Fevery, Leen Vanhooren, and Lieven J. Stuyver**

## Abstract

Genotypic testing based on subtype-specific amplification and population Sanger sequencing for two nonstructural (NS) protein-coding regions, the NS3/4A protease and the NS5B polymerase, of the hepatitis C virus (HCV) genome is described here. The protocols include the molecular steps for RNA extraction, one-step RT-PCR followed by inner PCR and population Sanger sequencing, to obtain the sequence information of the target regions from the clinical isolates of HCV subtypes 1a and 1b, which can be used to detect any sequence change in the viral genome as for example caused by the development of drug resistance in these two common viral targets.

**Key words** Hepatitis C virus (HCV), Genotyping, NS3/4A protease, NS3 protease, NS5B polymerase, Amplification, Sanger sequencing

## 1 Introduction

Chronic hepatitis C virus (HCV) infection affects approximately 180 million people worldwide, with 3 million new infections each year [1], and is one of the leading causes of cirrhosis and liver failure. Until recently, the standard therapy of choice (SOC) for the treatment of chronic hepatitis C consisted of pegylated interferon alpha (pegIFN) given by weekly injections in combination with twice-daily oral ribavirin (RBV). However, due to only moderate treatment success measured as sustained virologic response in combination with significant adverse events, there is a medical need to improve efficacy, safety, and tolerability for treatment of patients chronically infected with HCV. A major milestone towards efficacy was reached in 2011 with the approval of the very first direct acting

Edwin Yunhao Gong (ed.), *Antiviral Methods and Protocols*, Methods in Molecular Biology, vol. 1030,
DOI 10.1007/978-1-62703-484-5_12, © Springer Science+Business Media, LLC 2013

antiviral (DAA) drugs against HCV, the protease inhibitors boceprevir (Victrelis™) and telaprevir (Incivek™, Incivo™), which need to be taken as a combination therapy together with pegIFN and RBV. To further improve therapy, there are currently numerous other DAA candidate drugs in clinical development targeting various regions of the viral genome, as for instance the viral NS5B polymerase. Drugs acting against targets of infectious agents usually raise the question of selection pressure and development of resistance against these molecular entities. With lessons learned from other viruses, in particular from antiretroviral therapy against the human immunodeficiency virus (HIV-1), any possible drug resistance development needs to be closely monitored during clinical development. As such, there was a need to develop robust and standardized protocols for the assessment of nucleotide and amino acid variation within certain HCV drug targets during the course of clinical trials. Since most clinical trials were initially focused on the two major drug targets of HCV, the NS3/4A protease and the NS5B polymerase, we provide the genotyping protocols to assess their respective sequence information.

HCV is a small enveloped, single-stranded, positive-sense RNA virus with a genome of approximately 9.6 kb. Discovered in 1989 by Choo et al. [2], it soon became clear that HCV shows an amazingly high nucleotide sequence diversity, caused by its error-prone RNA-dependent RNA polymerase. Sequence analysis of the viral genomes led to the classification of six major genotypes, which can be further subdivided into several subtypes per genotype [3, 4]. HCV genotypes and subtypes exploit a nucleotide sequence variability of typically about 30–35 % and 20–25 %, respectively [3]. This sequence variability poses a challenge for the development of robust amplification and sequencing protocols of certain viral genomic regions, including the viral NS3/4A protease and NS5B polymerase domains. As a consequence, we decided to develop subtype-specific amplification and sequencing protocols, leading to the design of subtype-specific primers. In this chapter, we provide the experimental details for the two major subtypes 1a and 1b within genotype 1. The HCV subtypes and genotypes can be determined using commercially available tests, e.g. VERSANT® HCV Genotype 2.0 Assay (LiPa; Siemens), TRUGENE® HCV Genotyping Assay (Siemens), and RealTi*me* HCV Genotype II Assay (Abbott). The subtypes 1a and 1b were selected for this protocol because initially most clinical trials were and still are being performed on patients infected with genotype 1 and subtypes 1a and 1b are the most prominent strains found among patients infected with genotype 1 in Western countries [5]. For the assay design we looked for conserved areas outside the drug target region to design the amplification primers. Sequencing primers were placed if possible in conserved regions and with a frequency

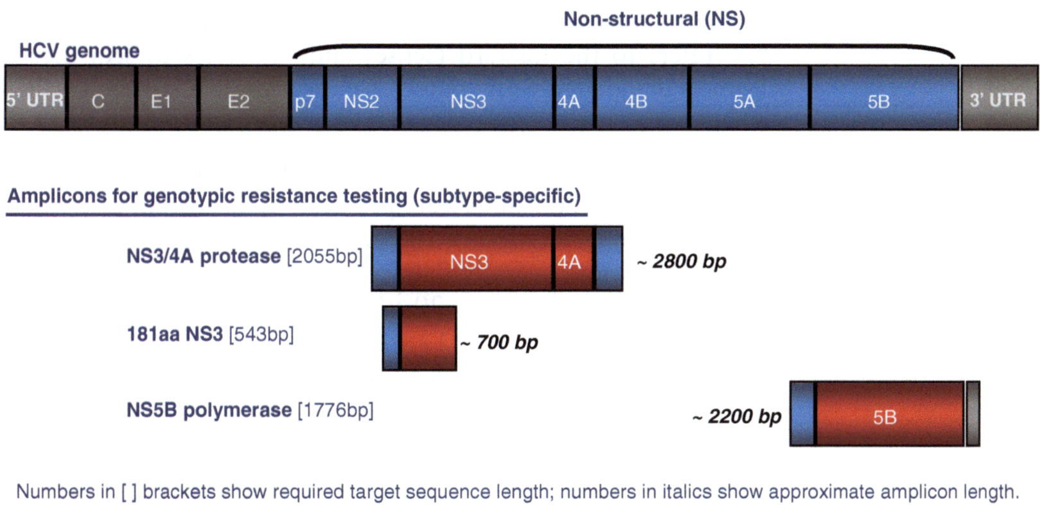


**Fig. 1** Overview of the amplification products

to cover the target region of interest in both sequencing directions. Cycling conditions were kept identical per target. Using this approach, the subtype information of the clinical isolates is all that required to select the appropriate primers before proceeding with the protocols.

We provide methods to assess sequence information of the complete NS3/4A protease, the NS3 protease domain (covering the N-terminal 181 amino acids of the NS3 gene), and the complete NS5B polymerase gene. A schematic overview of the amplicons is given in Fig. 1. The two protocols for the viral protease can be used either independently or in combination. In the latter case, we recommend starting with the protocol for the complete NS3/4A protease region, and in case amplification fails to apply the NS3 protocol on the extracted RNA. In most cases, this will improve the clinical sensitivity of the assay.

## 2 Materials

All solutions and reagents are prepared using Ultrapure™ DNase/RNase-free water (Invitrogen) and disposable DNase/RNase-free filter tips.

### *2.1 Replicon Cell Lines*

HCV subtypes 1a and 1b replicon cell lines: Hepatoma cell lines (provided by Apath, USA, and REBLIKON GmbH, Germany) that were stably transfected with either the subgenomic or the genomic HCV replicons.

### 2.2 Primers

In the following nomenclature, "R" refers to reverse (antisense), "F" refers to the forward (sense) and "s" refers to sequencing. Primers designated with "outer" are used in the one-step RT-PCR whereas the ones designated with "inner" are used in the inner PCR.

1. All oligonucleotides are ordered desalted (not HPLC purified) and lyophilized. Stocks are prepared by dissolving in water to a final concentration of 100 μM and stored at −20 °C. Ready-to-use primer solutions are prepared by diluting the 100 μM stock with water to a 20 μM working solution that can be stored for a week at 4 °C.
2. HCV-1a NS3/4A protease primer sequences for amplification:

   HCV1a-NS3/4A_R_outer, 5′-GGGACCTCACCGCTCATGAT-3′.

   HCV1a-NS3/4A_R_inner, 5′-CTCACCGCTCATGATCTTGAATGC-3′.

   HCV1a-NS3/4A_F_outer, 5′-CTCACCGCTCATGATCTTGAATGC-3′.

   HCV1a-NS3/4A_F_inner, 5′-CGTGCAAATGGCCATCATCAAG-3′.
3. HCV-1a NS3/4A protease primer sequences for sequencing:

   1a-NS2-F1s, 5′-GCGCTTACTGGCACCTATG-3′.

   1a-NS3-F1s, 5′-AGGCACGCCGATGTCAT-3′.

   1a-NS3-R2s, 5′-CGGGACCTTGGTGCTCTT-3′.

   1a-NS3-F2s, 5′-CGGCACTGTCCTTGACCA-3′.

   1a-NS3-R3s, 5′-GAGTCGAAGTCGCCGGTA-3′.

   1a-NS3-F3s, 5′-CCGAGACTACAGTTAGGCTACG-3′.

   1a-NS3-R4s, 5′-GCATGTCATGATGTATTTGGTG-3′.

   1a-NS4b-R1s, 5′-ACGAGGACCTTCCCCAGT-3′.
4. HCV-1b NS3/4A protease primer sequences for amplification:

   HCV1b-NS3/4A_R_outer, 5′-GAGCGCCTTCTGTTTGAATTG-3′.

   HCV1b-NS3/4A_R_inner, 5′-CTGTTTGAATTGCTCGGCGAG-3′.

   HCV1b-NS3/4A_F_outer, 5′-ATGCATGCTGGTGCGGAA-3′.

   HCV1b-NS3/4A_F_inner, 5′-TGGTGCGGAAAGTCGCTGG-3′.
5. HCV-1b NS3/4A protease primer sequences for sequencing:

   1b-NS2-F1s, 5′-GGTCATTATGTCCAAATGGC-3′.

   1b-NS3-F1s, 5′-CGGCAGCTCGGACCTTTA-3′.

   1b-NS3-R2s, 5′-CACTTGGAATGTCTGCGGTAC-3′.

   1b-NS3-F2s, 5′-GATGAGTGCCACTCAACTGACT-3′.

1b-NS3-R3s, 5′-CGTCTGTTGCCACGACAA-3′.

1b-NS3-F3s, 5′-CTATGACGCGGGCTGTG-3′.

1b-NS3-R4s, 5′-AGCCGTATGAGACACTTCCAC-3′.

HCV1b-NS3/4A_R_inner, 5′-CTGTTTGAATTGCTCGGC GAG-3′.

6. HCV-1a NS3 protease primer sequences for amplification:

   1a-180aa-R-outer, 5′-GCTGCCGGTGGGAGCATG-3′.

   1a-180aa-R-inner, 5′-GAGCATGCAGGTGGGCCAC-3′.

   1a-180aa-F-outer, 5′-GCGGCGACATCATCAACGG-3′.

   1a-180aa-F-inner, 5′-CATCAACGGCTTGCCCGTCTC-3′.

7. HCV-1a NS3 protease primer sequences for sequencing:

   1a-180aa-R-inner, 5′-GAGCATGCAGGTGGGCCAC-3′.

   1a-180aa-F-inner, 5′-CATCAACGGCTTGCCCGTCTC-3′.

8. HCV-1b NS3 protease primer sequences for amplification:

   1b-180aa-R-outer, 5′-GCTGCCAGTGGGAGCGTG-3′

   1b-180aa-R-inner, 5′-GGGAGCGTGTAGATGGGCCAC-3′.

   1b-180aa-F-out,5′-GCGTGTGGGGACATCATCTTAGG-3′.

   1b-180aa-F-inner, 5′-TCATCTTAGGCCTGCCCGTCTC-3′.

9. HCV-1b NS3 protease primer sequences for sequencing:

   1b-180aa-R-inner, 5′-GGGAGCGTGTAGATGGGCCAC-3′.

   1b-180aa-F-inner, 5′-TCATCTTAGGCCTGCCCGTCTC-3′.

10. HCV-1a NS5B polymerase primer sequences for amplification:

    HCV1a-NS5B_R_outer, 5′-AGGCCGGAGTGTTTACCCC AAC-3′.

    HCV1a-NS5B_R-inner, 5′-GGAGTGTTTACCCCAACCTT CA-3′.

    HCV1a-NS5B_F_outer, 5′-TGACTATGAACCACCTGTGG TCC-3′.

    HCV1a-NS5B_F_inner, 5′-CACCTGTGGTCCATGGCTG-3′.

11. HCV-1a NS5B polymerase primer sequences for sequencing:

    HCV1a-NS5B_F_inner, 5′-CACCTGTGGTCCATGGCTG-3′.

    1a-NS5a-F2s, 5′-CGGTGGTCCTCACCGAA-3′.

    1a-NS5b-F1s, 5′-CATCAACTCCGTGTGGAAAG-3′.

    1a-NS5b-R1s, 5′-CAGCGGGTATCATACGAGAA-3′.

    1a-NS5b-F2s, 5′-GCACCATGCTCGTGTGTG-3′.

    1a-NS5b-R2s, 5′-GTCATCAGTATCATCCTCGCC-3′.

    1a-NS5b-F3s, 5′-CGACTCCATGGTCTTAGCG-3′.

    HCV1a-NS5B_R-inner, 5′-GGAGTGTTTACCCCAACCTT CA-3′.

12. HCV-1b NS5B polymerase primer sequences for amplification:
    HCV1b-NS5B_R_outer, 5′-GGCCTGGAGTGGTTAGCTC CCC-3′.
    HCV1b-NS5B_R_inner, 5′-GAGTGGTTAGCTCCCCGTT CA-3′.
    HCV1b-NS5B_F_outer, 5′-TAGAGTCCTGGAAGGACCC GG-3′.
    HCV1b-NS5B_F_inner,5′-TGGAAGGACCCGGACTACG-3′.
13. HCV-1b NS5B polymerase primer sequences for sequencing:
    HCV1b-NS5B_F_inner,5′-TGGAAGGACCCGGACTACG-3′.
    1b-NS5a-F2s, 5′-CACGGGTGCCCATTGC-3′.
    1b-NS5b-F1s, 5′-AAGGAGATGAAGGCGAAGG-3′.
    1b-NS5b-R1s, 5′-CATCACGGCCTGAGGAAG-3′.
    1b-NS5b-F2s, 5′-TCGCTCACAGAGCGGCT-3′.
    1b-NS5b-R2s, 5′-TGGAGGAGCATGATGTTATCA-3′.
    1b-NS5b-F3s, 5′-CGACTCCATGGTCTTAGCG-3′.
    HCV1b-NS5B_R-inner, 5′-GAGTGGTTAGCTCCCCGTT CA-3′.

### 2.3 RNA Extraction, One-Step RT-PCR, and Inner PCR

1. EasyMAG RNA extraction reagents (BioMérieux, The Netherlands).
2. NucliSens®easyMAG™ disposables (BioMérieux, The Netherlands).
3. 10 ng/mL stock solution of yeast tRNA (Ambion).
4. SuperScript III™ One-step RT-PCR System with Platinum® Taq DNA Polymerase High Fidelity (Invitrogen).
5. Expand™ High Fidelity DNA Polymerase Kit (Roche).
6. 25 mM dNTPs (Applied Biosystems).

### 2.4 Agarose Gel Electrophoresis

1. MB agarose.
2. 10× TBE buffer (Invitrogen).
3. 1 % ethidium bromide solution.
4. TrackIt™ 1 kb Plus DNA ladder.
5. 6× DNA loading dye.
6. Power supply (Bio-Rad).

### 2.5 PCR Purification and Sanger Sequencing

1. QIAquick 96 PCR purification kit (Qiagen).
2. Big Dye Terminator kit (Applied Biosystems).
3. DyeEx 96-well plate kit (Qiagen).
4. ABI3730XL Analyzer (Applied Biosystems) or any other sequencing analyzer.

5. SeqScape sequence analysis software (Applied Biosystems).
6. BioEdit sequence alignment software for Win95/98/NT/2000/XP/7 (free use).

## 3 Methods

### 3.1 RNA Extraction

#### 3.1.1 Clinical Samples

1. Take the samples out of the −80 °C freezer and place them immediately on ice (*see* **Note 1**).
2. Pipette 600 μL of the sample (plasma, serum) into the NucliSens®easyMAG™ disposal vessels and proceed immediately with the RNA extraction (option: lysis on board) by following the protocol of the manufacturer. Elute RNA in 25 μL elution buffer. Alternatively, use 256 μL sample input and elute in 60 μL elution buffer.
3. Transfer the eluted RNA into new tubes and proceed immediately with the PCR amplification. RNA samples may be stored in cryotubes (single use only) at −80 °C until further use.

#### 3.1.2 Replicon RNA Control

To prepare replicon RNA, resuspend 10,000 replicon cells in 1 mL of the respective culture medium and treat like plasma samples for RNA extraction. RNA is eluted in 25 μL elution buffer. The resulting replicon RNA is diluted 1:10 in DNase/RNase-free water and stored at −80 °C as 6 μL aliquots until further use.

### 3.2 One-Step RT-PCR (NS3/4A Protease, NS3 Protease, and NS5B Polymerase Regions)

1. Take the RNA samples out of the −80 °C freezer and place them immediately on ice.
2. Thaw the samples slowly by placing them in a refrigerator (4 °C).
3. In the "pre-PCR lab" (*see* **Note 2**), place a new PCR plate in a precooled (4 °C) PCR plate-cooling chamber. Mark the plate and the sample numbers.
4. Also in the "pre-PCR lab", prepare a fresh mix containing DNase/RNase-free water, 2× Superscript III reaction mix, yeast tRNA, forward and reverse primers, and SuperScript III RT/Platinum Taq polymerase. Mix according to Table 1. Prepare the mix on ice (or a cooling block). Pipette 20 μL/well of the mix to the required number of wells of a new PCR plate. With exception of the primers, the one-step RT-PCR mix is identical for all target regions. Seal the PCR plate and transfer to a P3 lab where the viral RNAs are extracted.
5. Take the RNA samples out of the refrigerator and place them in a plate holder on the bench.

**Table 1**
**Composition of one-step RT-PCR mix**

| Reagents (*to be pipeted in this order*) | Stock concentration | Final concentration | Volume per sample (μL) |
|---|---|---|---|
| DNase/RNase-free water | | | 5.88 |
| 2× Superscript III reaction mix | 2× | 1× | 12.50 |
| Yeast tRNA | 4,800 ng/mL | 120 ng/mL | 0.63 |
| Forward primer | 20 μM | 0.2 μM | 0.25 |
| Reverse primer | 20 μM | 0.2 μM | 0.25 |
| Superscript III RT/Platinum Taq polymerase | | | 0.50 |
| Total volume/sample | | | 20.00 |
| Sample RNA volume (or controls) | | | 5.00 |

6. In case of several RNA samples, ensure to transfer the samples in the correct order into the one-step RT-PCR plate. For replicon RNA control extracted using easyMAG (*see* Subheading 3.1.2), a 1:10 dilution should be used.
7. Transfer the appropriate volume of sample RNA (5 μL) into the corresponding well of the prepared PCR plate and mix gently by pipetting up and down. Repeat for the rest of the RNA samples.
8. Seal the PCR plate and centrifuge briefly to collect the reaction mixture at the bottom of the wells. Transfer the PCR plate as quickly as possible into the "nested-PCR laboratory". Place the prepared PCR plate in the thermocycler sample block and close the lid. Choose the appropriate program for the PCR reaction (*see* Table 2) and run the program.
9. Remove the plate from the thermocycler at any time during the final hold at 12 °C.
10. Proceed with the inner PCR step or store at −20 °C.

#### 3.3 Inner PCR (NS3/4A Protease, NS3 Protease, and NS5B Polymerase Regions)

1. In the "pre-PCR lab", place a new PCR plate in a precooled (4 °C) PCR plate-cooling chamber. Mark the plate and the sample numbers.
2. Also in the "pre-PCR lab", prepare a fresh mix containing DNase/RNase-free water, the respective buffer, dNTPs, forward and reverse inner primers, and the expand high-fidelity enzyme, and pipette the required volume (19 μL) in the required number of wells of the PCR plate (for volumes refer to Table 3).

**Table 2**
**Cycling conditions for one-step RT-PCR**

| | NS3/4A protease | | | NS3 protease | | | NS5B polymerase | | |
|---|---|---|---|---|---|---|---|---|---|
| **Program step** | **Temp. (°C)** | **Time** | **Cycles** | **Temp. (°C)** | **Time** | **Cycles** | **Temp. (°C)** | **Time** | **Cycles** |
| cDNA synthesis | 47 | 30 min | 1 | 52 | 30 min | 1 | 47 | 30 min | 1 |
| Denaturation | 94 | 2 min | 1 | 94 | 2 min | 1 | 94 | 2 min | 1 |
| Amplification | 94<br>57<br>68 | 15 s<br>30 s<br>3 min | 50 | 94<br>58<br>72 | 15 s<br>30 s<br>1 min | 50 | 94<br>59<br>68 | 15 s<br>30 s<br>2.5 min | 50 |
| Final extension | 68 | 5 min | 1 | 72 | 5 min | 1 | 68 | 5 min | 1 |
| Hold | 12 | ∞ | Hold | 12 | ∞ | Hold | 12 | ∞ | Hold |

**Table 3**
**Composition of inner PCR mix**

| Reagents (*to be pipeted in this order*) | Stock concentration | Final concentration | Volume per sample (μL) |
|---|---|---|---|
| DNase/RNase-free water | | | 15.62 |
| Buffer 1 or 2 (Roche)[a] | 10× | 1× | 2.00 |
| dNTPs | 25 mM | 0.35 mM | 0.28 |
| Forward primer | 20 μM | 0.40 μM | 0.40 |
| Reverse primer | 20 μM | 0.40 μM | 0.40 |
| Expand HF polymerase | 5 U/μL | 0.075 U/μL | 0.30 |
| Total volume/sample | | | 19.00 |
| One-step RT-PCR product (or controls) | | | 1.00 |

[a]Use buffer 2 for the NS3/4A and the NS3 protease; use buffer 1 for the NS5B polymerase

3. Transfer the plate to the "nested-PCR lab" in order to add the outer PCR product (*see* **Note 3**).
4. Take the one-step RT-PCR plate out of the thermocycler or from the refrigerator and place it in a plate holder on the bench in the "nested-PCR lab".
5. Transfer the appropriate volume of outer PCR product (1 μL) into the corresponding well of the prepared inner PCR plate and mix gently by pipetting up and down. Repeat for the rest of the samples. After use, store the one-step RT-PCR plate with the remaining outer PCR product at −20 °C.
6. Close all wells with micro caps, and subsequently transfer to the "post-PCR lab".

**Table 4**
**Cycling conditions for inner PCR**

| | NS3/4A protease | | | NS3 protease | | | NS5B polymerase | | |
|---|---|---|---|---|---|---|---|---|---|
| **Program step** | **Temp. (°C)** | **Time** | **Cycles** | **Temp. (°C)** | **Time** | **Cycles** | **Temp. (°C)** | **Time** | **Cycles** |
| Denaturation | 94 | 2 min | 1 | 94 | 2 min | 1 | 94 | 2 min | 1 |
| Amplification | 94<br>57<br>68 | 15 s<br>30 s<br>3 min | 30 | 94<br>58<br>72 | 15 s<br>30 s<br>1 min | 30 | 94<br>59<br>68 | 15 s<br>30 s<br>2.5 min | 30 |
| Final extension | 68 | 5 min | 1 | 72 | 5 min | 1 | 68 | 5 min | 1 |
| Hold | 12 | ∞ | Hold | 12 | ∞ | Hold | 12 | ∞ | Hold |

7. Spin the liquid quickly down with a centrifuge to collect the reaction mixture at the bottom of the wells, place the plate on the thermocycler sample block, and close the lid. Choose the appropriate program for the inner PCR reaction (*see* Table 4).
8. Remove the plate from the thermocycler at any time during the final hold at 12 °C.
9. Proceed with the PCR-purification or store the reaction products at 4 °C until further processing. For long-term storage, store the inner PCR products at −20 °C.

### 3.4 Purification of PCR Products

Purify the inner PCR products with the QIAquick 96 PCR purification kit according to the manufacturer's instruction and elute in 100 μL of EB buffer (provided with the kit).

### 3.5 Agarose Gel Electrophoresis

1. Prepare a 1 % agarose gel by dissolving 3 g of agarose in 300 mL of 1× TBE buffer. Heat for 2–3 min in the microwave until a clear solution is obtained. Cool down the agarose solution to approximately 60 °C (*see* **Note 4**).
2. Add 15 μL of the ethidium bromide stock solution and mix by careful swaying (*see* **Note 5**). Prepare the gel. Let it solidify for approximately 30 min. Pour an appropriate volume of 1× TBE buffer onto the gel and remove the comb. Make sure that every slot is filled with buffer.
3. Prepare a 96-well plate by adding 2 μL of loading buffer into each well.
4. Take 4 μL of the inner PCR product and add it into the appropriate well containing the loading buffer.
5. Pipette 7 μL of the TrackIt™ 1 kb Plus DNA ladder into the first and the last slot of each row of the gel.
6. Transfer the prepared samples into the remaining slots of the gel and put the gel into the electrophoresis chamber.

**Table 5**
**Cycling conditions for sequencing reaction**

| | Sequencing reaction | | |
|---|---|---|---|
| **Program step** | **Temp. (°C)** | **Time** | **Cycles** |
| Denaturation | 96 | 10 s | 25 |
| Annealing | 50 | 5 s | |
| Elongation | 60 | 4 min | |
| Hold | 4 | ∞ | Hold |

7. Run the gel at a constant voltage of 150 V (400 mA) for approximately 45 min.
8. Take the gel out of the chamber and visualize the DNA samples under UV light. Store the file electronically as applicable.
9. Check the size of the PCR products and compare with the expected amplicon length (*see* Fig. 1). If the PCR amplification is successful, a major amplified band of the expected size should be observed.
10. Store the purified PCR products at 4 °C until further processing. For long-term storage, store the purified inner PCR products at −20 °C.

### 3.6 Sanger Sequencing

1. Prepare a sequencing mix by adding 1 μL of the purified inner PCR product per sequencing primer reaction to 1 μL of Big Dye terminator mix, 2.5 μL of 2.5× dilution buffer, and 3.5 μL of water to give a final volume of 8 μL. If using more than one sequencing primer per sample, multiply the volumes accordingly.
2. Mark a fresh PCR plate.
3. Dispense 8 μL of the sequencing mix vertically per well into the 96-well PCR plate.
4. For each single sequencing reaction, add 2 μL of the 5 μM sequencing primer solution into the corresponding well and mix gently by pipetting up and down. Repeat this step for each sequencing reaction of every sample.
5. Close all wells with micro caps.
6. Place the prepared sequencing plate in the thermocycler sample block and close the lid. Choose the appropriate program for the sequencing reaction (*see* Table 5).
7. Remove the plate from the thermocycler at any time during the final hold at 4 °C.
8. Proceed with the purification of the sequencing reaction using DyeEx 96 well plate kit from Qiagen according to the manufacturer's instruction.

9. Load the plates on an ABI3730XL Analyzer or other sequencing analyzer according to the manufacturer's instructions.

### 3.7 Sequence Analysis

1. Perform the sequence analysis using SeqScape software (Applied Biosystems) or alternative suitable software according to the supplier's instructions.
2. For the assembly of the sequence reads into a contig, it is necessary to provide a reference sequence. Reference sequences can be downloaded from public databases such as the European HCV Database (http://euhcvdb.ibcp.fr/euHCVdb/) or the Los Alamos HCV database (http://hcv.lanl.gov/content/sequence/HCV/ToolsOutline.html).
3. Currently, the following reference sequences are recommended for alignment of clinical isolates: for HCV subtype 1a isolates (HCV-1a strain H77 with accession number NC004102) and for HCV subtype 1b isolates (HCV-1b strain con1 with accession number AJ238799). SeqScape allows the generation of a mutation report on nucleotide and amino acid levels of viral sample sequences based on alignment against the chosen reference sequence.
4. For viral sequence analysis with focus on drug resistance monitoring, it is important to apply the concept of reference comparison. This means that the viral sequence of each patient prior to antiviral treatment serves as a unique reference to monitor potential changes on nucleotide or amino acid level over time (*see* **Note 6**). This implies that viral sequences of a certain patient obtained from different time points in the course of treatment may be analyzed in a single alignment (*see* **Note 7**). The alignment and analysis can be performed using alignment software as for instance BioEdit.

## 4 Notes

1. RNA extraction is recommended to be performed in a laboratory of a required biosafety level as applicable to laboratory guidelines. Keep in mind that clinical samples like plasma are infectious and coinfection with other viruses or pathogens may be a possibility. We perform the RNA extraction in a dedicated nucleic acid extraction area in a P3 laboratory. Wear personal protection clothing and personal protective equipment (e.g., safety goggles) as required.
2. Make the one-step RT-PCR mix in the "pre-PCR lab" in which all RT and PCR reagents, such as primers, probes, reverse transcriptase, Taq polymerase, buffers, probe master, dNTPs, and $H_2O$, are stored and the master mixes for RT and PCR are prepared. Seal the plate and transfer to the P3 lab to

add the RNA. The one-step RT-PCR is recommended to be performed in the "nested-PCR lab". Wear personal protection clothing and personal protective equipment as required.

3. Make the inner PCR mix in the "pre-PCR lab". Close the wells with micro caps and transfer to the "post-PCR lab". The inner PCR itself and all succeeding steps can be performed in a laboratory where PCR and DNA amplification are handled ("post-PCR lab").
4. The 1 % agarose will remain as liquid if stored until further use in a water bath at the appropriate temperature.
5. As ethidium bromide is mutagenic, handle with caution and wear protective gloves.
6. A detailed guide of genotypic and phenotypic analysis for HCV drug resistance monitoring can be found in a consensus document prepared by the HCV DRAG [6]. It provides recommendations with a focus on submission of data to regulatory authorities.
7. Recently, clinically relevant HCV drug resistance mutations in the NS3 protease and the NS5B polymerase region have been published in a consensus table and figure from the HCV DRAG. It provides an overview about drug resistance mutations of several HCV DAA which are either already approved or in late-stage clinical development [7].

## Acknowledgement

The authors would like to thank Stéphanie Dumont for her invaluable contribution to the assay development.

### References

1. Lauer GM, Walker BD (2001) Hepatitis C virus infection. N Engl J Med 345(1):41–52
2. Choo QL et al (1989) Isolation of a cDNA clone derived from a blood-borne non-A, non-B viral hepatitis genome. Science 244(4902):359–362
3. Simmonds P et al (2005) Consensus proposals for a unified system of nomenclature of hepatitis C virus genotypes. Hepatology 42(4): 962–973
4. Kuiken C, Simmonds P (2009) Nomenclature and numbering of the hepatitis C virus. Methods Mol Biol 510:33–53
5. Simmonds P (2001) The origin and evolution of hepatitis viruses in humans. J Gen Virol 82(Pt 4):693–712
6. Kwong AD et al (2011) Sequence and phenotypic analysis for resistance monitoring in hepatitis C virus drug development: recommendations from the HCV DRAG. Gastroenterology 140: 755–760
7. HCV Phenotype Working Group, HCV Drug Development Advisory Group (2012) Clinically relevant HCV drug resistance mutations figures and tables. Ann Forum Collab HIV Res 14(2): 1–10

# Chapter 13

# A Southern Blot Assay for Detection of Hepatitis B Virus Covalently Closed Circular DNA from Cell Cultures

**Dawei Cai, Hui Nie, Ran Yan, Ju-Tao Guo, Timothy M. Block, and Haitao Guo**

## Abstract

Chronic hepatitis B remains a substantial public health burden affecting approximately 350 million people worldwide, causing cirrhosis and liver cancer, and about 1 million people die each year from hepatitis B and its complications. Hepatitis B is caused by hepatitis B virus (HBV) infection. As an essential component of the viral life cycle, HBV covalently closed circular DNA (cccDNA) is synthesized and maintained at low copy numbers in the nucleus of infected hepatocytes, and serves as the transcription template for all viral RNAs. Therefore, cccDNA is responsible for the establishment of viral infection and persistence. The presence and longevity of cccDNA may also explain the limitations of current antiviral therapy for hepatitis B. Thus, understanding the mechanisms underlying cccDNA formation and regulation is critical in understanding the HBV pathogenesis and finding a cure for hepatitis B. Here we describe a protocol for HBV cccDNA extraction and detection in detail. The procedure includes two major steps: (1) HBV cccDNA extraction by Hirt protein-free DNA extraction method and (2) HBV cccDNA detection by Southern blot analysis. The method is straightforward and reliable for cccDNA assay with cell culture samples, and it is useful for both HBV molecular biology and antiviral research.

**Key words** HBV, cccDNA, Hirt extraction, Southern blot

## 1 Introduction

Hepatitis B virus (HBV) is the causative agent for a chronic viral infection of the liver called "chronic hepatitis B," and its most severe consequence, hepatocellular carcinoma (HCC) [1]. HBV is a partially double-stranded DNA virus that belongs to the *Hepadnaviridae* family. The virion has an outer lipid envelope, inside of which there is an icosahedral nucleocapsid composed of 180 or 240 copies of viral core proteins. The nucleocapsid encloses the genomic relaxed circular DNA (rcDNA) and a viral DNA polymerase which is covalently attached to 5′ end of one of the DNA strands [2, 3].

Edwin Yunhao Gong (ed.), *Antiviral Methods and Protocols*, Methods in Molecular Biology, vol. 1030,
DOI 10.1007/978-1-62703-484-5_13, 

HBV is a naturally noncytopathic virus that mainly infects hepatocytes, although it has been detected in extrahepatic sites [4, 5]. Although tissue culture models of HBV infection are limited [6, 7], studies of viral entry indicate that infection of the hepatocyte is thought to be through endocytosis upon the engagement to a yet-to-be identified plasma membrane receptor [8]. The viral genome is then transported into the nucleus, where it converts to covalently closed circular DNA (cccDNA), which is the transcription template for all viral RNAs [9, 10]. A longer than full genome-length viral RNA transcript, called pregenomic RNA (pgRNA), is exported into the cytoplasm and packaged into nucleocapsid, inside of which it undergoes reverse transcription to generate viral minus-strand DNA, and then plus-strand DNA synthesis to yield rcDNA [2, 3]. The mature nucleocapsids are either packaged by viral envelope proteins and egress as virion particles to further spread the infection or shuttled back into the nucleus to amplify the cccDNA reservoir through the so-called intracellular amplification pathway, resulting in a steady-state level of cccDNA ranging from 1 to 50 copies per infected cell [2]. Therefore, the sustained level of cccDNA in the infected cells is critical for establishment and persistence of chronic HBV infection.

Currently available HBV therapeutics, specifically interferon-α and nucleos(t)ide analogues, fail to eradicate the cccDNA reservoir from infected hepatocytes although the viral DNA replication can be substantially suppressed [11, 12]. Consequently, failure to repress or eliminate the cccDNA results in viral rebound after the cessation of therapy [12, 13]. Moreover, the cccDNA could "inherit" and amplify drug-resistant mutations and evade nucleos(t)ide treatment [11].

Therefore, there is an urgent need for the development of novel therapeutic agents that directly target the cccDNA formation and maintenance. To serve such a purpose, a reliable cccDNA measurement method is required. Here we describe a protocol being used for HBV cccDNA extraction and detection from cell cultures [14–16], which is modified from previously reported methods for avihepadnavirus cccDNA assay, i.e., duck hepatitis B virus (DHBV) [17, 18].

For cccDNA extraction from HBV-replicating cells, a Hirt protein-free DNA extraction procedure is employed [19]. In principle, sodium dodecyl sulfate (SDS) is first applied to break down all the lipid membranes and viral capsids to release all the viral nucleic acids. A high concentration of salt (NaCl) is then added to precipitate down high-molecular-weight cellular chromatin and protein covalently bonded DNA in SDS–protein complexes, including HBV DNA replicative intermediates with viral polymerase covalently attached. Protein-free DNA stays in the supernatant and can be further purified by organic phenol. HBV cccDNA, a protein-free DNA, is one of the major viral products of this process. Another protein-free viral DNA, specifically deproteinized rcDNA (DP-rcDNA), which is a putative precursor of cccDNA, can be extracted simultaneously as well [14, 20].

Currently, there are two major means of detecting HBV cccDNA. One of those is a polymerase chain reaction (PCR)-based method, such as real-time PCR and rolling circle PCR [21, 22]. Both PCR methods amplify cccDNA by specifically designed primers. Although the sensitivity of PCR detection is satisfactory, the major drawback is that the PCR amplification signals not only come from the cccDNA but also can come from other co-purified viral DNA, such as replicative intermediates in total DNA preparation, or DP-rcDNA from Hirt extraction, especially during the later PCR cycles.

Alternatively, the cccDNA can be directly detected by Southern blot analysis through DNA hybridization, which not only detects the cccDNA but also distinguishes the cccDNA from other viral DNA species by differential migration during agarose gel electrophoresis. In this chapter, we describe in detail the Southern blot procedure for HBV cccDNA detection and quantification. Although this protocol is not amenable to high-throughput screening for antiviral compounds, it could serve as a suitable and reliable method for hit validation in secondary antiviral assays [16].

## 2 Materials

### 2.1 Extraction of cccDNA from Cell Culture

1. 1 M Tris–HCl, pH 7.5.
2. 0.5 M ethylenediamine tetraacetic acid (EDTA), pH 8.0.
3. TE buffer (10:10): 10 mM Tris–HCl, pH 7.5 and 10 mM EDTA.
4. TE buffer (10:1): 10 mM Tris–HCl, pH 7.5 and 1 mM EDTA.
5. Phenol: Equilibrated with 10 mM Tris–HCl (pH 8.0) and 1 mM EDTA.
6. Phenol/chloroform: Phenol:chloroform:isoamyl alcohol (25:24:1), saturated with 10 mM Tris–HCl (pH 8.0) and 1 mM EDTA.
7. 10 % SDS.
8. 5 M NaCl.
9. 100 % ethanol, and 70 % ethanol.

### 2.2 Agarose Gel Electrophoresis

1. 10× DNA gel loading buffer: 10 mM EDTA (pH 8.0), 50 % (V/V) glycerol, 0.25 % (W/V) bromophenol blue.
2. 1× TAE buffer: 0.04 M Tris base, 0.04 M glacial acetic acid, 1 mM EDTA, pH 8.2–8.4. Prepare 30× stock solution, and store at room temperature.
3. Agarose (molecular biology grade).

#### 2.3 Probe Preparation for HBV DNA Detection

1. pGEM-3Z vector (Promega) with an insertion of genome-length HBV fragment (i.e., EcoRI linearized), in which the transcription of sense RNA is under the control of the prokaryotic SP6 promoter.
2. 800 Ci/mmol [$\alpha$-$^{32}$P] UTP (PerkinElmer).
3. Riboprobe® in vitro transcription system-SP6 (Promega).
4. 5 M $NH_4OAc$.
5. 10 mg/mL yeast RNA (Ambion).
6. Isopropanol.
7. Formamide.

#### 2.4 Southern Blotting and DNA Hybridization

1. Depurination buffer: 0.2 M HCl (*see* **Note 1**).
2. Denaturing buffer: 0.5 M NaOH, 1.5 M NaCl.
3. Neutralization buffer: 1.5 M NaCl, 1 M Tris–HCl, pH 7.4.
4. 20× SSC: 3 M NaCl, 0.3 M sodium citrate.
5. Amersham Hybond—XL membrane (GE Healthcare).
6. Whatman 3 MM chromatography paper.
7. UV crosslinker.
8. EKONO™ hybridization buffer (G-Biosciences).
9. Wash buffer: 0.1 % SDS, 0.1× SSC.
10. Phosphorimager exposure and scan system.

## 3 Methods

#### 3.1 Cell Culture and Compound Treatment

1. Seed HBV stable cell lines, such as HepG2.2.15 [23], or tetracycline-inducible (tet-off) HepAD38 [24], or tetracycline-inducible (tet-off) HepDE19/DES19 cells [14], in collagen-coated 6-well plates at approximately 100 % confluence with appropriate culture medium (*see* **Notes 2** and **3**).
2. When the cell monolayer reaches complete confluence (normally 24–48 h post seeding), change the culture medium with fresh medium (*see* **Note 4**). For tetracycline-inducible cell lines, remove the tetracycline from culture medium to induce HBV replication.
3. For antiviral assays, add testing compounds at serial non-cytotoxic doses in the culture medium, and set solvent (i.e., dimethyl sulfoxide) as treatment control (*see* **Note 5**).
4. Repeat compound treatment with medium change every other day; harvest the cells at day 10 post treatment.

### 3.2 cccDNA Extraction from Cell Culture

1. Lyse cells by adding 1.5 mL of TE buffer (10:10) and 0.1 mL of 10 % SDS into each well of a 6-well cell culture plate. Gently mix and incubate the plate for 30 min at room temperature.
2. Transfer the viscous cell lysate to a 15 mL centrifuge tube (*see* **Note 6**). Add 0.4 mL of 5 M NaCl and gently invert the tube ten times. Let the tube sit at 4 °C for at least 16 h to efficiently precipitate proteins and protein-associated DNAs.
3. Centrifuge at 14,500 × *g* for 30 min at 4 °C. Transfer the supernatant to a fresh 15 mL centrifuge tube. To remove the residual protein from the supernatant, add an equal volume of phenol (approx. 2 mL) to the supernatant and mix thoroughly by hand shaking for 10 s (*see* **Note 7**). Centrifuge at 3,500 × *g* for 10 min at 4 °C, and transfer the aqueous phase to a fresh 15 mL tube. Repeat phenol extraction step. Add equal volume of phenol/chloroform (approx. 2 mL), mix thoroughly by hand shaking, centrifuge at 3,500 × *g* for 10 min at 4 °C, and transfer the aqueous phase to a fresh 15 mL tube.
4. Add two volumes of 100 % ethanol (approx. 4 mL) and mix thoroughly by inverting the tube several times. Incubate at room temperature overnight to precipitate DNA (*see* **Note 8**).
5. On the second day, centrifuge the tube at 3,500 × *g* for 30 min at 4 °C, and discard the supernatant. Add an equal volume of 75 % ethanol (approx. 2 mL) and gently flick the tube to wash the DNA pellet. Centrifuge at 3,500 × *g* for 15 min at 4 °C.
6. Discard the supernatant. Allow the pellet to air-dry for about 10 min at room temperature. Dissolve the DNA pellet in 25 μL TE buffer (10:1).

### 3.3 Agarose Gel Electrophoresis

1. Prepare 150 mL of 1.2 % gel by mixing 1.8 g agarose, 5 mL of 30× TAE buffer, and 145 mL of distilled water in a microwavable flask. Microwave the flask until the agarose is thoroughly dissolved. Let it cool down to about 50 °C at room temperature. Pour the liquid gel into the gel tray (12 cm × 10 cm) with an appropriate comb inserted. Let the gel solidify at room temperature before removal of the comb.
2. Prepare 600 mL of gel running buffer by diluting 20 mL of 30× TAE buffer with deionized water.
3. Prepare samples by adding approximately 3 μL of 10× loading buffer to 25 μL of each sample. Pipette up and down to mix well.
4. Assemble submarine gel-running unit by correctly positioning the gel box, adding running buffer into the chamber, and loading 28 μL of each sample into a separate well. Connect the gel-running unit to a power supply and run gel at 25 V overnight.

### 3.4 Southern Blot

1. Disconnect the gel-running unit and take out the gel carefully.
2. Submerge the gel in a tray containing freshly prepared 0.2 M HCl. Agitate gently for 10 min at room temperature to depurinate the DNA samples from the gel. The bromophenol blue dye in the gel should gradually turn grey during this process.
3. Rinse the gel three times with distilled water.
4. Submerge the gel in a tray of denaturing buffer, and agitate gently for 1 h at room temperature. The loading dye in the gel should turn back to blue during this process.
5. Rinse the gel three times with distilled water.
6. Submerge the gel in a tray of neutralization buffer, and agitate gently for at least 1 h at room temperature.
7. Soak the Hybond-XL membrane (12 × 10 cm) in a tray with deionized water for 5 min at room temperature. Replace the water with 20× SSC and continue to agitate for 15 min.
8. To transfer DNA from the gel to the Hybond-XL membrane, assemble the transfer apparatus as follows: Layer two sheets of Whatman 3 MM filter paper in the gel transfer tray and pour 20× SSC transfer buffer on top to completely wet the Whatman papers. Smooth out the bubbles in between the Whatman papers and the tray gently by a plastic roller, and get rid of the excess amount of the transfer buffer. Place the gel on top of the Whatman papers with the side containing the loading wells facing down; make sure that there is no air bubble between the Whatman papers and the gel. Place the presoaked Hybond-XL membrane on top of the gel and use a plastic roller to get rid of the air bubbles between the gel and the membrane. Layer two sheets of 20× SSC pre-wet Whatman papers (12 × 10 cm) on top of the membrane; use plastic roller to remove air bubbles in between. Place a stack of dry paper towels (about 4 in.) precut to the gel size on top of the Whatman papers. Seal the transfer tray with plastic wrap to prevent buffer evaporation during transfer. Put some weight (approx. 500 g such as metal plate, casserole dish) on top of the transfer apparatus (Fig. 1). Let the assembled transfer apparatus sit still on a flat surface and transfer for 24–48 h (*see* **Note 9**).
9. After gel transfer, disassemble the blot apparatus and flip over the membrane with the gel still attached, and mark the position of each loading well on the membrane by a pencil (*see* **Note 10**). Wrap up the gel and paper towels by plastic wrap and discard.
10. Cross-link the Hybond-XL membrane in a UV crosslinker chamber with UV energy dosage at 120 mJ/cm$^2$. The membrane can be directly subject to hybridization, or sandwiched between two sheets of Whatman filters and stored at −20 °C.

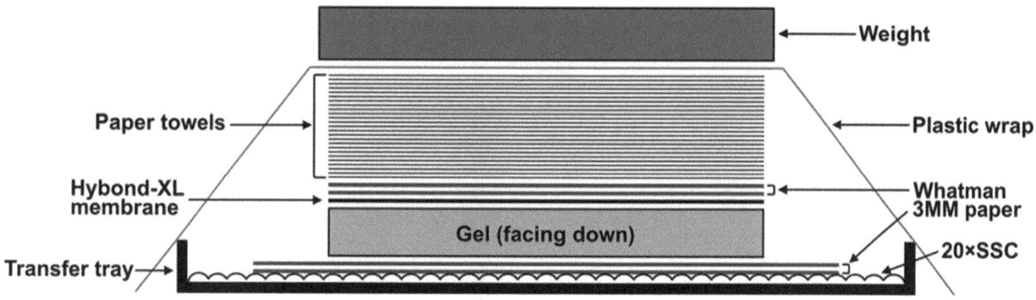


**Fig. 1** Assembly of HBV DNA Southern blot transfer apparatus. See text for details

### *3.5 Probe Preparation for HBV DNA Detection (See Note 11)*

1. Prepare rNTP (supplied with Riboprobe® system) solution by mixing 50 μL of each 10 mM rATP, 10 mM rGTP, 10 mM rCTP, and 0.2 mM UTP.
2. Add 4 μL of 5× reaction buffer, 4 μL of rNTP solution prepared from the previous step, 2 μL of 100 mM DTT, 0.25 μg of SalI-linearized pGEM-HBV DNA template (*see* **Note 12**), 1 μL of RNasin, and 1 μL of SP6 polymerase into a nuclease-free tube. Adjust the volume to 14 μL with nuclease-free water before adding 6 μL of [$\alpha$-$^{32}$P] UTP into the mixture. Incubate the reaction at 37 °C for 1 h.
3. To remove the DNA template, add 1 μL of RNase-free RQ1 DNase to the reaction and incubate at 37 °C for 15 min.
4. Add 15 μL of 5 M $NH_4OAc$ to stop the reaction. To precipitate the RNA probe, add 113 μL of nuclease-free water, 2 μL of yeast RNA, and 150 μL of isopropanol; gently mix; and incubate at room temperature for 10 min.
5. Centrifuge the mixture at 12,000 × *g* for 30 min at 4 °C, and discard the supernatant carefully without disturbing the pellet. Dissolve the probes in 400 μL of deionized formamide, followed by measurement of the counts per minute (CPM) of acid-insoluble $^{32}$P with scintillation counter (PerkinElmer). Store the probes at −20 °C (*see* **Note 13**).

### *3.6 Hybridization (See Note 11)*

1. Place the cross-linked membrane in a hybridization tube with the DNA-binding side facing the center of the tube. Add 5 mL of EKONO™ hybridization buffer, and pre-hybridize the membrane by rotating the hybridization tube at 65 °C for 1 h in a hybridization oven (*see* **Note 14**).
2. Replace the pre-hybridization buffer with 5 mL of fresh EKONO™ hybridization buffer, and add HBV riboprobes with $1 \times 10^7$ CPM. Rotate the hybridization tube at 65 °C overnight.
3. Discard the hybridization solution on the following day, and wash the hybridization membrane with approximately half a tube of wash buffer. Rotate at 65 °C for 30 min.

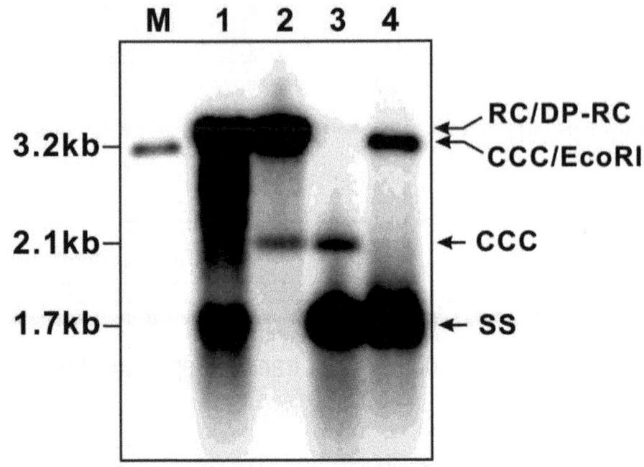


**Fig. 2** Detection of HBV DNA by a Southern blot assay. A typical pattern of HBV cytoplasmic core DNA (*lane 1*) and Hirt DNA (*lane 2*) upon Southern blot hybridization is shown. *RC* relaxed circular DNA, *DP-RC* deproteinized RC DNA, *CCC* covalently closed circular DNA, *SS* single-stranded (-) DNA, *M* HBV genomic length DNA marker. The approximate size of each HBV DNA species is labeled on the *left*. To further validate the authenticity of HBV cccDNA, the Hirt DNA sample in *lane 2* has been heated to 85 °C for 5 min before gel loading, a condition that denatures DP-rcDNA into SS DNA, while the cccDNA stays undenatured and its electrophoretic mobility remains unchanged (*lane 3*). The DNA sample from *lane 3* is further digested with EcoRI, in which condition the cccDNA is linearized to a genome-length double-stranded DNA (*lane 4*). Reproduced from [15] with permission from Elsevier

4. Discard the wash buffer, and replace with half a tube of fresh wash buffer. Continue to wash at 65 °C for 1 h.
5. After the second wash, take out the membrane and dry it with paper towels, making sure that the membrane is not cracked or wrinkled during this process. Seal the membrane with plastic wrap. Place the membrane in a phosphorimager cassette with the DNA-binding side facing up. Layer the phosphor imaging screen on top facing the DNA-binding side of the membrane. Close the cassette tightly and expose for 1–2 h in the dark.
6. Scan the phosphor imaging screen with phosphorimager system. Store the membrane properly and erase the signal in the phosphor imaging screen with intense light for reuse.
7. A typical phosphor image of HBV DNA Southern blot is shown in Fig. 2 (also *see* refs. 14, 15). Quantify the signal intensity of the cccDNA bands with software provided by the phosphorimager system. When testing antiviral compounds, compare the signals from treatment wells with solvent control well and determine the compound's effective concentration inhibiting cccDNA production by 50 % ($EC_{50}$ values) (*see* **Note 15**).

## 4 Notes

1. 0.2 M HCl needs to be freshly prepared, as the solution will gradually lose the acid concentration over time.
2. Collagen-coated plate is ideal for cells to form confluent monolayers, especially for HepG2-based cells.
3. HepG2.2.15 cell line is cultured with DMEM/F12 medium supplemented with 10 % fetal bovine serum, 100 U/mL penicillin, and 100 μg/mL streptomycin, plus 400 μg/mL G418. Tetracycline-inducible (tet-off) cell lines are maintained in the same way as the HepG2.2.15, but with the addition of 1 μg/mL tetracycline.
4. To obtain a high level of HBV replication in tetracycline-inducible cell lines, it is critical to withdraw the tetracycline when the cells reach confluence, as HBV replication will arrest cell growth [25, 26].
5. Addition of the compounds at the start of induction of HBV replication, before the cccDNA has formed, will expose the establishment phase of the cccDNA metabolism to possible inhibition, but will still allow detecting agents that are able to remove the cccDNA itself after it becomes established.
6. Use wide-mouth plastic transfer pipette to transfer the lysate into centrifuge tube.
7. Use leakproof tube with screw cap. Do not shake the tube too vigorously or too long, which may physically nick the supercoiled cccDNA.
8. Because high concentration of salt is used in Hirt extraction, incubation of DNA solution at low temperature will result in a lot of salt co-precipitating with DNA.
9. When assembling the Southern blot transfer apparatus, avoid possible short circuits of capillary liquid flow. The DNA will not be efficiently transferred if the blot short-circuits. Use parafilm strips to seal the edges of the gel.
10. After transfer, the gel becomes thinner. Use the pencil tip to penetrate the loading wells and mark their positions on the membrane. This step is for labeling the DNA-binding side of the membrane and tracking the running lanes with orientation.
11. All radioactive materials and procedures should be handled strictly by following the institutional laboratory safety regulation rules.
12. Linearization of the plasmid pGEM-HBV by SalI is an optimal step to produce "run-off" transcripts derived from the HBV sequence only. Other restriction enzyme, which singly cuts the vector sequences downstream of SP6-HBV transcription cassette but does not leave a 3′ overhang, can also be used.

13. $^{32}$P-radiolabeled HBV DNA probes prepared by random priming or end labeling can also be used in the hybridization.
14. If using other commercial hybridization buffers, follow the pre-hybridization and hybridization conditions recommended by the manufacturers.
15. In HBV stable cell lines, the cccDNA is exclusively synthesized from the rcDNA through the "intracellular amplification pathway". If cccDNA reduction is observed under compound treatment, the effect of test compound on viral core DNA replication and cccDNA stability should be analyzed as previously described [16].

## Acknowledgements

We thank Dr. Andrea Cuconati (Institute for Hepatitis and Virus Research) for critical reading of the manuscript. This work was supported by the NIH grants (R01AI094474, R21AI088424) and Hepatitis B Foundation through an appropriation of the Commonwealth of Pennsylvania. Haitao Guo is the Bruce Witte fellow of the Hepatitis B Foundation.

### References

1. Lok AS, McMahon BJ (2009) Chronic hepatitis B: update 2009. Hepatology 50:661–662
2. Seeger C, Mason WS (2000) Hepatitis B virus biology. Microbiol Mol Biol Rev 64:51–68
3. Block TM, Guo H, Guo JT (2007) Molecular virology of hepatitis B virus for clinicians. Clin Liver Dis 11(4):685–706, vii
4. Dejean A, Lugassy C, Zafrani S et al (1984) Detection of hepatitis B virus DNA in pancreas, kidney and skin of two human carriers of the virus. J Gen Virol 65:651–655
5. Naumova AK, Favorov MO, Keteladze ES et al (1985) Nucleotide sequences in human chromosomal DNA from nonhepatic tissues homologous to the hepatitis B virus genome. Gene 35:19–25
6. Gripon P, Rumin S, Urban S et al (2002) Infection of a human hepatoma cell line by hepatitis B virus. Proc Natl Acad Sci U S A 99:15655–15660
7. Schulze-Bergkamen H, Untergasser A, Dax A et al (2003) Primary human hepatocytes–a valuable tool for investigation of apoptosis and hepatitis B virus infection. J Hepatol 38: 736–744
8. Huang HC, Chen CC, Chang WC et al (2012) Entry of hepatitis B virus into immortalized human primary hepatocytes by clathrin-dependent endocytosis. J Virol 86:9443–9453
9. Guo H, Mao R, Block TM, Guo JT (2010) Production and function of the cytoplasmic deproteinized relaxed circular DNA of hepadnaviruses. J Virol 84:387–396
10. Rabe B, Vlachou A, Pante N et al (2003) Nuclear import of hepatitis B virus capsids and release of the viral genome. Proc Natl Acad Sci U S A 100:9849–9854
11. Zoulim F (2005) New insight on hepatitis B virus persistence from the study of intrahepatic viral cccDNA. J Hepatol 42:302–308
12. Zoulim F, Mason WS (2012) Reasons to consider earlier treatment of chronic HBV infections. Gut 61:333–336
13. Werle-Lapostolle B, Bowden S, Locarnini S et al (2004) Persistence of cccDNA during the natural history of chronic hepatitis B and decline during adefovir dipivoxil therapy. Gastroenterology 126:1750–1758
14. Guo H, Jiang D, Zhou T, Cuconati A, Block TM, Guo JT (2007) Characterization of the intracellular deproteinized relaxed circular DNA of hepatitis B virus: an intermediate of covalently closed circular DNA formation. J Virol 81:12472–12484
15. Zhou T, Guo H, Guo JT, Cuconati A, Mehta A, Block TM (2006) Hepatitis B virus e antigen production is dependent upon covalently closed circular (ccc) DNA in HepAD38 cell

cultures and may serve as a cccDNA surrogate in antiviral screening assays. Antiviral Res 72:116–124
16. Cai D, Mills C, Yu W, Yan R, Aldrich CE, Saputelli JR, Mason WS, Xu X, Guo JT, Block TM, Cuconati A, Guo H (2012) Identification of disubstituted sulfonamide compounds as specific inhibitors of hepatitis B virus covalently closed circular DNA formation. Antimicrob Agents Chemother 56:4277–4288
17. Tuttleman JS, Pourcel C, Summers J (1986) Formation of the pool of covalently closed circular viral DNA in hepadnavirus-infected cells. Cell 47:451–460
18. Guo H, Mason WS, Aldrich CE et al (2005) Identification and characterization of avihepadnaviruses isolated from exotic anseriformes maintained in captivity. J Virol 79:2729–2742
19. Hirt B (1967) Selective extraction of polyoma DNA from infected mouse cell cultures. J Mol Biol 26:365–369
20. Gao W, Hu J (2007) Formation of hepatitis B virus covalently closed circular DNA: removal of genome-linked protein. J Virol 81:6164–6174
21. Margeridon S, Carrouee-Durantel S, Chemin I et al (2008) Rolling circle amplification, a powerful tool for genetic and functional studies of complete hepatitis B virus genomes from low-level infections and for directly probing covalently closed circular DNA. Antimicrob Agents Chemother 52:3068–3073
22. Mazet-Wagner AA, Baclet MC, Loustaud-Ratti V et al (2006) Real-time PCR quantitation of hepatitis B virus total DNA and covalently closed circular DNA in peripheral blood mononuclear cells from hepatitis B virus-infected patients. J Virol Methods 138:70–79
23. Sells MA, Chen M, Acs G (1987) Production of hepatitis B virus particles in hepG2 cells transfected with cloned hepatitis B virus DNA. Proc Natl Acad Sci U S A 84:1005–1009
24. Ladner SK, Otto MJ, Barker CS et al (1997) Inducible expression of human hepatitis B virus (HBV) in stably transfected hepatoblastoma cells: a novel system for screening potential inhibitors of HBV replication. Antimicrob Agents Chemother 41:1715–1720
25. Wang T, Zhao R, Wu Y et al (2011) Hepatitis B virus induces G1 phase arrest by regulating cell cycle genes in HepG2.2.15 cells. Virol J 8:231
26. Gearhart TL, Bouchard MJ (2011) The hepatitis B virus HBx protein modulates cell cycle regulatory proteins in cultured primary human hepatocytes. Virus Res 155:363–367

# Chapter 14

# In Vitro Phenotyping of Recombinant Hepatitis B Virus Containing the Polymerase/Reverse Transcriptase Gene from Clinical Isolates

**Yang Liu and Kathryn M. Kitrinos**


## Abstract

This chapter describes the procedures for the generation of recombinant hepatitis B virus (HBV) that contains the polymerase/reverse transcriptase (pol/RT) gene from clinical isolates. It also contains procedures for phenotypic testing of the recombinant HBV in HepG2 cells for susceptibility to nucleos(t)ide analogs. Amplification and cloning of the pol/RT gene as opposed to the full-length HBV genome from a clinical isolate allows for clinical samples with lower viral loads to be successfully analyzed. Furthermore, it enables a direct comparison of sequence changes in the pol/RT region between isolates.


**Key words** Hepatitis B virus, HBV, Drug resistance, PCR, DNA cloning, Transient transfection, Phenotype assay, Cell culture, Real-time PCR

## 1 Introduction

Hepatitis B virus (HBV) is the causative agent of chronic hepatitis B (CHB) infection in about 350 million people worldwide, leading to approximately 1 million annual deaths by resultant illness such as cirrhosis and hepatocellular carcinoma [1]. HBV is a double-stranded DNA virus that utilizes reverse transcription for virus replication, which is driven by the viral enzyme polymerase/reverse transcriptase (pol/RT) [2]. HBV pol/RT lacks a proofreading function, which results in a high mutation rate during virus replication [3]; therefore, the virus population in an infected patient exists as a quasispecies of closely related isolates. HBV pol/RT is the viral target of all the nucleos(t)ide analogs currently marketed for the treatment of HBV, including lamivudine (LAM), adefovir dipivoxil (ADV), telbivudine (LdT), entecavir (ETV), clevudine (L-FMAU), and tenofovir (TFV) [4]. All known drug resistance mutations reside within the pol/RT region [5, 6].

Edwin Yunhao Gong (ed.), *Antiviral Methods and Protocols*, Methods in Molecular Biology, vol. 1030,
DOI 10.1007/978-1-62703-484-5_14, 

Due to the negative effects that can be caused by the emergence of drug resistance mutations in CHB-treated patients, such as the loss of virus suppression and resumption of liver disease progression, resistance surveillance is a key component for the treatment management of CHB patients. Resistance surveillance comprises two components, genotypic evaluation of the pol/RT for potential resistance-associated substitutions and phenotypic evaluation of the resistance-associated mutations by in vitro drug susceptibility testing in cell culture. Previously, Tenney et al. [7] cloned the majority of the HBV pol/RT region (amino acids 17–311) from clinical isolates into an HBV laboratory strain in order to evaluate in vitro drug susceptibility; however, this method does not allow for the evaluation of the full pol/RT gene (amino acids 1–344). Zhu et al. [8] developed a procedure for cloning full-length HBV quasispecies populations from clinical isolates in order to evaluate in vitro drug susceptibility. However, amplifying and cloning the full-length HBV (3.2 kb) is a technically challenging task and it is impractical for use in large-scale clinical trials. The current method describes an HBV phenotyping assay based on cloning the entire HBV pol/RT region (amino acids 1–344) from clinical samples into a plasmid vector containing a full-length HBV laboratory strain lacking the pol/RT gene under the direction of a CMV promoter. This method allows for the evaluation of the entire pol/RT region while enabling analysis of clinical samples with lower viral loads (PCR amplification cutoff=400 copies/mL). HBV DNA replication of the recombinant virus and its susceptibility to nucleos(t)ide analogs are assessed by transient transfection of the viral DNA construct into HepG2 cells followed by intracellular core DNA extraction and quantitation by quantitative PCR (qPCR).

## 2 Materials

### 2.1 HBV DNA Extraction, PCR Amplification, and DNA Cloning

1. 50× TAE buffer: 2 M Tris–acetate, 100 mM EDTA (Teknova).
2. 1× TE buffer: 10 mM Tris–HCl, pH 8.0, 1 mM EDTA.
3. HotStarTaq Master Mix kit (Qiagen).
4. High Fidelity PCR Master Mix (Roche).
5. Ultrapure agarose.
6. 1.2 % E-gel (Invitrogen) (*see* **Note 1**).
7. Gel loading dye, blue (6×).
8. 1 kb DNA ladder.
9. Ethidium bromide.
10. Ethanol.
11. Isopropanol.
12. QIAamp DNA Blood kit (Qiagen).

13. QIAquick PCR Purification kit (Qiagen).
14. QIAquick Gel Extraction kit (Qiagen).
15. HiSpeed Plasmid Maxi kit (Qiagen).
16. EcoRI-HF enzyme (New England Biolabs).
17. SphI-HF enzyme (New England Biolabs).
18. Antarctic phosphatase enzyme (New England Biolabs).
19. T4 DNA ligase enzyme (New England Biolabs).
20. Max Efficiency® DH5α competent cells (Invitrogen).
21. LB Broth with ampicillin-100: 1 % tryptone, 0.5 % yeast extract, 1.0 % NaCl, 100 μg/mL ampicillin.
22. LB Ampicillin-100 agar plate, 100 mm plate.
23. Rattler Plating Beads (Zymo Research).
24. SOC: 2 % tryptone, 0.5 % yeast extract, 10 mM NaCl, 2.5 mM KCl, 10 mM $MgCl_2$, 10 mM $MgSO_4$, 20 mM glucose (Invitrogen).
25. Primers for first-round PCR (*see* **Note 2**):

    5 μM Primer OS1: 5′-GCCTCATTTTGTGGGTCACCA TA-3′.

    5 μM Primer P1R: 5′-TCACGGTGGTTTCCATGCGA-3′.
26. Primers for second-round PCR:

    10 μM Primer HBV RT-F: 5′-G<u>GAATTC</u>CACTGCCTTCC ACCAAGCTC-3′ (*see* **Notes 2** and **3**).

    10 μM Primer HBV RT-R: 5′-GGTTCCAC<u>GCATGC</u>GCTG ATGG-3′ (*see* **Notes 2** and **4**).
27. Primers for sequencing (*see* **Note 2**):

    10 μM Primer CHBV3: 5′-CCTGCTGGTGGCTCCAG TTC-3′.

    10 μM Primer BB: 5′-CTGGATGTGTCTGCGGCG-3′.

    10 μM Primer HBV 786R: 5′-GGACTCAAGATGTTGTAC AGACTTGGCC-3′.

    10 μM Primer EE: 5′-GTTCCGCAGTATGGATCG-3′.
28. Plasmid DNA pHY92 and pRTAN (*see* **Note 5**).
29. 0.2 mL PCR tubes (Phenix Research Products).
30. E-gel electrophoresis system (Invitrogen) (*see* **Note 1**).
31. Nanodrop 1000 spectrophotometer (Thermo Scientific).
32. Thermomixer (Eppendorf).
33. C1000 thermal cycler (Bio-Rad).

### 2.2 Transfection and Cell Culture

1. Tenofovir (TFV), adefovir (AFV), emtricibine (FTC), lamivudine (LAM), telbivudine (LdT), and entecavir (ETV) (Moravek Biochemicals) (*see* **Note 6**).
2. 1× DPBS.
3. DMEM-F12.
4. Fetal bovine serum.
5. 0.5 % Trypsin–EDTA.
6. Penicillin/streptomycin, 100×.
7. Fugene 6 (Roche).
8. NP-40 (*see* **Note 7**).
9. Turbo DNA-free kit (Applied Biosystems).
10. QuickExtract DNA Extraction Solution (Epicentre Biotechnologies).
11. TaqMan Universal Master Mix (Applied Biosystems).
12. Vector pEGFP-C2 (Clontech).
13. Primers for qPCR (*see* **Note 2**):

    20 μM Primer HBV F: 5′-CCGTCTGTGCCTTCTCATCTG-3′.

    20 μM Primer HBV R: 5′-AGTCCAAGAGTYCTCTTATGYAAGACCTT-3′.

    20 μM Primer AMP F: 5′-GTATGCGGCGACCGAGTT-3′.

    20 μM Primer AMP R: 5′-TGCTATGTGGCGCGGTATT-3′.
14. Probes for qPCR (*see* **Notes 2** and **8**):

    100 μM HBV Probe: FAM-CCGTGTGCACTTCGCTTCACCTCTGC-BHQ.

    100 μM AMP MGB Probe: VIC-TCTTGCCCGGCGTCA-MGBNFQ.
15. 100 mL reagent reservoir.
16. 96-well cell culture plate.
17. V-bottom 96-well plate (Greiner Bio One).
18. 96-well PCR plate (Phenix Research Products).
19. 96-well white real-time PCR plate (Roche).
20. Aluminum PCR plate sealer (Sorenson).
21. Countess® Automated cell counter (Invitrogen).
22. Titer plate shaker (Thermo Scientific).
23. Eppendorf centrifuge model 5804 (Eppendorf).
24. Tissue culture incubator (Thermo Scientific).
25. MJ Research PTC-100 thermal cycler (Thermo Scientific).
26. Roche 480 real-time PCR machine (Roche).
27. TableCurve 2D program (Systat).

## 3 Methods

Carry out all procedures at room temperature unless otherwise specified (*see* **Notes 9** and **10**).

### 3.1 Isolation of HBV from Patient Serum

1. Thaw all patient serum and/or plasma samples at room temperature.
2. Use the QIAamp DNA Blood kit to extract DNA by following the manufacturer's instructions and elute DNA in 200 μL buffer AE. If samples are for immediate use, proceed to PCR amplification (*see* Subheading 3.2). Samples that are not immediately used should be stored at −20 °C.

### 3.2 PCR Amplification of pol/RT Sequences for Cloning

This round of PCR is required for serum/plasma samples with lower HBV DNA levels (<$10^6$ copies/mL) (*see* **Note 11**). For serum/plasma samples with higher HBV DNA levels (>$10^6$ copies/mL), skip to second-round PCR steps (*see* Subheading 3.2.2).

#### 3.2.1 First-Round PCR

1. Set up PCR reaction(s) in 0.2 mL PCR tubes. The reaction mixture per serum/plasma sample is serum/plasma HBV DNA 20 μL, primer OS1 (5 μM) 4 μL, primer P1R (5 μM) 4 μL, DNase/RNase-free water 22 μL, and 2× HotStarTaq Master Mix 50 μL. The final reaction volume is 100 μL.
2. Close the tubes, briefly spin the tubes to capture the entire PCR mixture at the bottom of the tube, and amplify the HBV DNA in a C1000 thermal cycler with the following conditions: 95 °C for 15 min; 94 °C for 1 min, 58 °C for 30 s, and 72 °C for 1.5 min, 35 cycles; 72 °C for 7 min; and hold at 4 °C. Upon completion of amplification, store the amplicon at 2–8 °C until needed.

#### 3.2.2 Second-Round PCR

This round of PCR amplification will add the EcoRI/SphI restriction enzyme sites to the flanking ends of the HBV pol/RT gene for subsequent cloning.

1. Set up PCR reaction(s) in 0.2 mL PCR tubes. The reaction mixture per sample is first-round PCR product from Subheading 3.2.1, **step 2**, or serum/plasma HBV DNA 2.5 μL, primer HBV RT-F (10 μM) 1.5 μL, primer HBV RT-R (10 μM) 1.5 μL, DNase/RNase-free water 19.5 μL, and 2× High Fidelity Master Mix 25 μL. The final reaction volume is 50 μL.
2. Close the tubes, briefly spin down the tubes to capture the entire PCR mixture at the bottom of the tube, and amplify DNA in a C1000 thermal cycler with the following conditions: 94 °C for 2 min; 95 °C for 30 s, 58.5 °C for 30 s, and 72 °C for 1.5 min, 35 cycles; 72 °C for 7 min; and hold at 4 °C. Upon completion of amplification, store the amplicon at 2–8 °C until needed.

### 3.3 PCR Product Purification and EcoRI/SphI Digestion (See Note 12)

1. Mix 5 μL of PCR sample(s) from Subheading 3.2.2, **step 2**, with 1 μL 6× gel loading dye and run on a precast 1.2 % E-gel for 30 min using the E-gel electrophoresis system along with 0.25 μg of 1 kb DNA ladder (5 μL) plus 1 μL 6× gel loading dye. The PCR product(s) should then be visualized using a Molecular Imager Gel Doc XR system or an equivalent system to see if a product of ~1.2 kb is detectable. Furthermore, a rough estimation of the amount of PCR product(s) should be determined by comparing the 1.2 kb band of each PCR product with the 1.5 kb band from the 1 kb DNA ladder. The approximate DNA mass of the 1.5 kb band from a 0.25 μg 1 kb DNA ladder load is 18 ng. If the intensity of the 5 μL PCR product is the same or stronger than the 1.5 kb band from the 1 kb DNA ladder, there is enough PCR product to continue with **step 2** (*see* **Note 13**).
2. Use the QIAquick PCR Purification kit to purify PCR product(s) by following the manufacturer's instructions. Elute PCR product(s) in 30 μL buffer EB.
3. Quantify purified PCR product(s) by running 2 μL mixed with 1 μL 6× gel loading dye and 3 μL distilled water on a precast 1.2 % E-gel along with 0.25 μg of 1 kb DNA ladder (5 μL) mixed with 1 μL 6× gel loading dye for 30 min using the E-gel electrophoresis system. The PCR product(s) should then be visualized using a Molecular Imager Gel Doc XR system or an equivalent system to see if a product of ~1.2 kb is detectable, and to have a rough estimation of the amount of each PCR product (*see* **step 1**). If the intensity of the 2 μL PCR product is the same or stronger than the 1.5 kb band from the 1 kb DNA ladder, there is enough PCR product to continue with **step 4**.
4. Digest 0.5–1 μg PCR product(s) in a total volume of 34 μL in a 1.5 mL Eppendorf tube. Incubate the reaction at 37 °C for 2 h using a thermomixer (Eppendorf) or other incubation equipment. The reaction mixture per PCR product is 0.5–1 μg purified PCR product from **step 3**, 3.4 μL NEB 10× Buffer 4, 1.0 μL EcoRI-HF, 1.5 μL SphI-HF, and DNase/RNase-free water up to 34 μL (*see* **Note 14**).
5. Purification of digested PCR product: If only a 1.2 kb band is observed on the E-gel when the PCR product(s) is analyzed in **step 3**, use the Qiagen QIAquick PCR purification kit to purify PCR product(s) as per the manufacturer's instructions. Add 30 μL Buffer EB to elute the purified PCR product(s) and continue to **step 6**. If there are nonspecific PCR products observed on the E-gel when the PCR product(s) is analyzed in **step 3**, the entire digestion needs to be separated on a 1 % agarose gel containing 0.5 μg/mL ethiduim bromide in 1× TAE buffer (not E-gel) (*see* **Note 15**). The digested PCR product(s) should be mixed with 7 μL of 6× loading dye and

then run along with 0.25 μg of 1 kb DNA ladder mixed with 1 μL 6× loading dye at 100 V for 1 h. The digested PCR product(s) should be visualized using a Molecular Imager Gel Doc XR system or an equivalent system and the 1.2 kb band should be removed from the gel and placed into a 1.5 mL Eppendorf tube (*see* **Note 16**). The DNA should then be extracted from the gel piece using QIAquick Gel Purification kit as per the manufacturer's instructions. Elute DNA in 30 μL buffer EB and continue to **step 6**.

6. The purified EcoRI- and SphI-digested PCR product(s) along with the purified EcoRI- and SphI-digested pHY92 fragment (*see* **Note 14**) are quantified by running 2 μL mixed with 1 μL of 6× gel loading dye and 3 μL water along with 0.25 μg 1 kb DNA ladder mixed with 1 μL 6× gel loading dye on a precast 1.2 % E-gel for 30 min using the E-gel electrophoresis system. Then visualize the gel as in **step 1**. If the intensity of the 2 μL digested PCR product is the same or stronger than the 1.5 kb band from the 1 kb DNA ladder, there is enough digested PCR product to continue to Subheading 3.5, **step 1**.

### 3.4 Vector Preparation (See Note 17)

1. In a fresh 1.5 mL Eppendorf tube, mix 3 μg pRTAN plasmid (*see* **Note 5**), 3 μL 10× NEB buffer 4, 1 μL EcoRI-HF, 1.5 μL SphI-HF, and add DNase/RNase-free water to a final volume of 30 μL. Incubate at 37 °C for 2 h using a thermomixer or other incubation equipment.
2. Prepare 100 mL of a 1 % agarose gel containing 0.5 μg/mL ethiduim bromide in 1× TAE buffer (do not use an E-gel). Load 1 μL digested pRTAN along with 100 ng undigested pRTAN and 0.25 μg 1 kb DNA ladder (each sample mixed with 1 μL 6× gel loading dye) and run the gel at 100 V for 1 h. A complete digest of pRTAN should run as a single band at around 6 kb. If digestion is complete, purify digested pRTAN using the QIAquick PCR Purification kit as per manufacturer's instructions. Add 30 μL Buffer EB to elute the purified DNA.
3. In order to prevent the pRTAN plasmid from religating to itself during ligation, mix 30 μL purified digested pRTAN vector with 3.5 μL of 10× Antarctic phosphatase buffer and 2 μL Antarctic phosphatase. Incubate the mixture at 37 °C for 1 h using a thermomixer or other incubation equipment. Inactivate the Antarctic phosphatase by incubating the mixture at 65 °C for 10 min.
4. Purify the digested, phosphatased pRTAN vector using the QIAquick PCR purification kit by following the manufacturer's instructions. Add 30 μL Buffer EB to elute the purified DNA.
5. Run 1 μL digested pRTAN vector mixed with 1 μL of 6× gel loading dye alongside 0.5 μg 1 kb DNA ladder mixed with 1 μL of 6× gel loading dye on 1.2 % E-gel for 30 min.

The digested vector should be visualized using a Molecular Imager Gel Doc XR system or an equivalent system to see if a band of ~6 kb is detectable, and to have a rough estimation of the amount of each PCR product (*see* Subheading 3.3, **step 1**, for details). The approximate DNA mass of the 6 kb band from a 0.5 μg 1 kb DNA ladder load is 50 ng. If the intensity of the 1 μL digested vector is the same or stronger than the 6 kb band from the 1 kb DNA ladder, there is enough digested vector to continue to Subheading 3.5, **step 1**. To confirm the concentration, also perform $OD_{260nm}$ reading using Nanodrop 1000 spectrophotometer or an equivalent system to measure the concentration of pRTAN as per manufacturer's instructions. Adjust concentration to 50 ng/μL. The prepared vector should be stored at −20 °C until ready for use.

### 3.5 Ligation, Transformation, Plasmid Maxiprep, and Sequence Analysis

1. Prepare ligation reaction(s) in 0.2 mL PCR tubes and incubate in a C1000 thermal cycler or other incubation equipment at 12 °C overnight (*see* **Note 18**). The ligation mix per sample is 250–500 ng EcoRI- and SphI-digested, purified PCR product (from Subheading 3.3, up to 16 μL), 2 μL of 10× T4 ligase buffer, 1 μL EcoRI/SphI/Antarctic phosphatase-treated pRTAN vector (from Subheading 3.4, 50 ng/μL), 1 μL T4 ligase, and DNase/RNase-free water up to 20 μL. After incubation, continue to **step 3** or store at −20 °C until ready for use.
2. For ligation assay controls, add 16 μL water or 250 ng of the 1.2 kb EcoRI–SphI-digested fragment of pHY92 (*see* **Note 14**) with up to 16 μL water into two separate 0.2 mL PCR tubes for a negative or a positive control, respectively. The other ligation reagents will be used at the same volumes as noted in **step 1**. The control ligations are incubated with the same conditions as noted in **step 1**.
3. Make a 1/5 dilution of each ligation mixture (including the positive/negative controls from **step 2**) in 1× TE buffer and use 1 μL of the diluted ligation for transformation. Transform all diluted ligation(s) using Max Efficiency DH5α Competent Cells along with the pUC19 plasmid control provided with the competent cells. Follow the manufacturer's instructions with these modifications: After the 2-min incubation on ice following the heat shock at 42 °C, add 900 μL SOC media to each transformation. Immediately pipette 100 μL of each transformation/SOC media mixture onto an LB/Amp agar plate and spread the mixture on the plate using rattler plating beads. Incubate the plates in a 37 °C incubator overnight. Pipette the remaining 900 μL transformation/SOC media mixture into a 500 mL Erlenmeyer flask containing 100 mL LB media and 100 μg/mL ampicillin. Incubate in a 37 °C incubator shaking at 250 rpm overnight.

4. Inspect the LB/Amp agar plates and determine if it is reasonable to proceed with the transformations (*see* **Note 19**). Use the HiSpeed Plasmid Maxi kit to extract plasmid DNA from the overnight culture(s) in the Erlenmeyer flasks. Follow the manufacturer's instructions and elute the plasmid DNA in 500 μL buffer TE. Plasmid DNA can be stored at 4 °C until ready for use.
5. Quantify concentration of maxiprep using Nanodrop 1000 spectrophotometer or an equivalent system as per the manufacturer's instructions.
6. Sequence the HBV pol/RT from the plasmid DNA samples using standard population-based sequence technology. The following four primers are used for sequencing the pol/RT gene: CHBV3, BB, HBV 786R, and EE (*see* **Note 2**).

### 3.6 Cell Culture and Transfection (See Note 20)

1. HepG2 cells should be maintained in DMEM-F12 culture media containing 10 % fetal bovine serum and 50 μg/mL penicillin/streptomycin by splitting twice every week. Low-passage (<20) cells should be used for transfection. Approximately one T175 flask of confluent HepG2 cells is enough for three 96-well plate assays.
2. Using freshly confluent (exponential growth) HepG2 cells, aspirate the media, and wash the cells once with 25 mL 1× DPBS (for a T175 flask). Trypsinize the cells by adding enough 0.5 % trypsin–EDTA to cover the bottom of the flask (7.5 mL for a T175 flask) and incubate at 37 °C for 5 min. Add an equivalent amount of DMEM-F12 to the flask to dilute the trypsin (7.5 mL for a T175 flask), tap the flask to loosen cells, and pipette vigorously along the bottom of the plate to resuspend the HepG2 cells. Count the resuspended HepG2 cells using an automated cell counter or other cell-counting equipment and dilute cells to a final concentration of $3.6 \times 10^5$ cells/mL in DMEM-F12 culture media containing 10 % fetal bovine serum and 50 μg/mL penicillin/streptomycin.
3. To prepare the transfection mixture(s) (*see* **Note 21**), add the appropriate volume of plasmid DNA (100 ng/well) to the appropriate volume of serum-free DMEM-F12 media (*see* Table 1). Then add the appropriate amount of Fugene 6 reagent (0.3 μL/well). Be careful not to allow Fugene 6 reagent to come into direct contact with the side of the tube. A replication-incompetent plasmid pRTAN used for a negative assay control and a GFP-expressing plasmid pEGFP-C2 used for a transfection efficiency control will also be transfected following the same methodology. Both controls are used in two wells of each 96-well assay plate (*see* Fig. 1 for example of 96-well plate layout). Incubate all transfection mixtures at room temperature for 15 min.

**Table 1**
**Example calculations for transfection reactions**

| Sample | Plasmid DNA concentration (ng/μL) | Number of wells | Input plasmid DNA needed (ng) | Input plasmid DNA needed (μL) | Fugene 6 (μL) | Serum-free media (μL) | HepG2 cells (μL, at $3.6 \times 10^5$ cells/mL) |
|---|---|---|---|---|---|---|---|
| 1 | 100.0 | 1 | 100.0 | 1.0 | 0.3 | 8.7 | 140 |
| 1 | 100.0 | 60 | 6,000.0 | 60.0 | 18.0 | 522.0 | 8,400 |
| pRTAN | 98.0 | 4 | 400.0 | 4.1 | 1.2 | 34.7 | 560 |
| pEGFP-C2 | 912.0 | 4 | 400.0 | 0.4 | 1.2 | 38.4 | 560 |

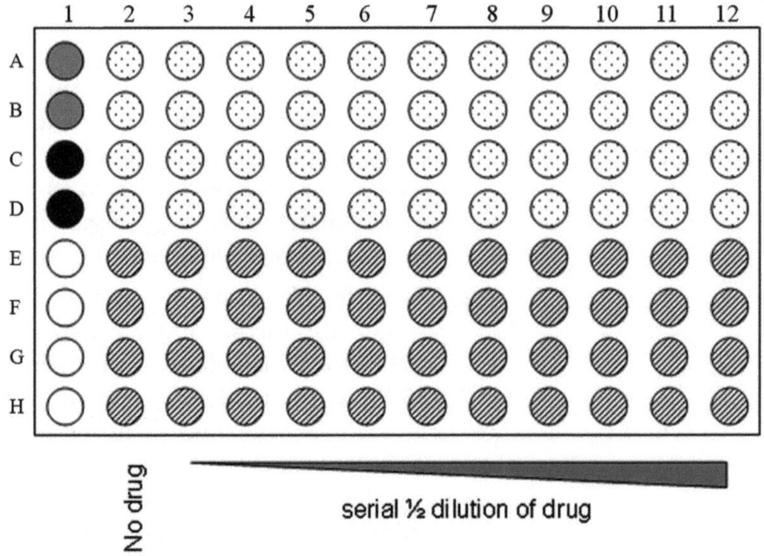


**Fig. 1** Phenotype assay plate layout. The *dot-filled circles* (rows A–D, columns 2–12) indicate cells transfected with one construct, and the *striped circles* (rows E–H, columns 2–12) indicate cells transfected with a different construct. The *grey-filled circles* (A1/B1) are cells transfected with the replication-incompetent plasmid pRTAN. The *black circles* (C1/D1) are cells transfected with the GFP-expressing plasmid pEGFP-C2. *Open circles* (E1, F1, G1, and H1) are untransfected HepG2 cells

4. Add each DNA–Fugene transfection mixture to the appropriate amount of resuspended cells ($5.4 \times 10^4$ cells/well) with DMEM-F12 media containing 10 % fetal bovine serum and 50 μg/mL penicillin/streptomycin. Aliquot 150 μL transfection mixture/well to a 96-well cell culture plate (*see* Fig. 1 for example of 96-well plate layout). Use a reservoir and a multichannel pipette for plating a large amount of cells. Two recombinant viruses can be tested per 96-well assay plate and each assay plate also includes two wells with pRTAN-transfected

cells (wells A1 and B1), two wells with GFP-expressing plasmid-transfected cells (wells C1 and D1), and four wells with untransfected HepG2 cells (wells E1, F1, G1, and H1).

5. Add 50 μL of twofold serial dilutions of drug or media alone (*see* **Note 22**) in each well containing 150 μL transfected cells. The wells containing the assay controls (pRTAN, GFP-expressing plasmid, and untransfected cells) receive 50 μL of media alone. Incubate the cells at 37 °C, 5 % $CO_2$, overnight.
6. Check transfection efficiency by observing the wells transfected with GFP plasmid using a fluorescent microscope 24 h after transfection. Approximately 10–15 % of HepG2 cells should be expressing GFP in a typical transfection (*see* **Note 23**).
7. Aspirate media from the 96-well plate and wash once with 150 μL 1× DPBS per well. Be careful not to suck up the cell monolayer during aspiration (*see* **Note 24**). Aspirate the DPBS from the 96-well plate. Add 150 μL/well of DMEM-F12 culture media containing 10 % fetal bovine serum and 50 μg/mL penicillin/streptomycin, and then add 50 μL of twofold serial dilutions of drug in each well or 50 μL media for the no-drug wells. Incubate the cells at 37 °C, 5 % $CO_2$, for 3 days.
8. After 3 days of incubation, add fresh media/drug to each 96-well plate by repeating **step 7**. Incubate the cells at 37 °C, 5 % $CO_2$, for another 3 days.
9. After 3 more days of incubation (7 days in total from transfection), aspirate cell culture media from the 96-well plate and wash the cells once with 150 μL 1× DPBS per well. Aspirate the DPBS from the 96-well plate. The plates are then ready for DNA extraction or can be stored at −20 °C until ready to use.

### 3.7 Extraction of Intracellular Replicated HBV DNA (See Note 25)

1. Add 100 μL of 0.33 % NP-40 to each well of the harvested cell culture plate (make 10 mL 0.33 % NP-40/96-well plate). Cover the plate with a plate sealer and shake slowly at speed 2 on a titer plate shaker for 5 min at room temperature.
2. Break up cell monolayer in NP-40 solution by pipetting the solution up and down several times until the solution appears homogenous. Transfer the solution to a new v-bottom 96-well plate (*see* **Note 26**). Cover the plate with a plate sealer and spin plate for 5 min at 370×*g*, 4 °C, in an Eppendorf centrifuge model 5804.
3. While the v-bottom plates are spinning, make up DNase mix using components from the Turbo DNA-free kit (DNase mix per 96-well plate is 650 μL $dH_2O$, 650 μL 10× Turbo DNase Buffer, and 200 μL TurboDNase enzyme). Add 15 μL of DNase mix to each well of a fresh v-bottom plate.
4. Remove v-bottom plates containing NP-40/cell lysates from the centrifuge and remove plate sealers. Transfer 50 μL of

lysate supernatant from the old v-bottom plate to a new v-bottom plate containing the DNase mix, being careful not to come into contact with the cell lysate pellet at the bottom of the old v-bottom plate (*see* **Note 27**). Mix upon addition and cover the plate with a plate sealer. Incubate the plate for 60 min at 37 °C in a tissue culture incubator (*see* **Note 28**).

5. After the 60-min incubation, add 12 μL of inactivation beads from the Turbo DNA-free kit (*see* **Note 29**) to each well of the plate, cover the plate with a plate sealer, shake the plate at speed 2 on a titer plate shaker at room temperature for 1 min, and then spin the plate for 5 min at 1,479×*g* at room temperature in an Eppendorf centrifuge model 5804.
6. While the v-bottom plates are spinning, aliquot 65 μL/well of Epicentre QuickExtract DNA Extraction Solution to a fresh 96-well PCR plate. Remove v-bottom plates containing NP-40/cell lysates from the centrifuge, remove the plate sealer, remove 35 μL of DNase-treated supernatant from the pelleted inactivation beads, and add it to the PCR plate containing the Epicentre QuickExtract DNA Extraction Solution, mixing upon addition. Cover the plate with an aluminum plate sealer, briefly spin down the plate at 1,479×*g* for 5 s, and incubate the PCR plates at the following conditions in an MJ Research PTC-100 thermal cycler: 65 °C for 6 min; 98 °C for 2 min; and hold at 4 °C. After incubations, plates can be stored at −20 °C or real-time PCR can be run immediately.

### 3.8 Quantitative Multiplex PCR to Quantify HBV DNA

1. Prepare reaction mix for the appropriate number of assay plates (*see* Table 2). For each 96-well plate to be tested, set up reaction mix for 100 wells.
2. Add 40 μL of reaction mix to each well of a 96-well white real-time PCR plate.
3. Add 5 μL of extracted DNA from mock-transfected HepG2 cells (wells E1, F1, G1, H1, *see* Fig. 1) to the wells on the real-time PCR plate that will contain the plasmid DNA standard curve (wells A1, B1, C1, and D1, *see* Fig. 2 for qPCR setup map). Add 5 μL of the standards to real-time PCR plate wells A1, B1, C1, and D1 (*see* **Note 30**). Additionally, add the following to the wells in column 1 on the real-time PCR plate: 10 μL extracted DNA from pRTAN-transfected cells (wells A1, B1, *see* Fig. 1) to wells E1 and F1, 10 μL extracted DNA from untransfected HepG2 cells (wells E1, F1, G1, H1, *see* Fig. 1) to well G1, and 10 μL water to well H1.
4. For columns 2 through 12 on the real-time PCR plate, add 10 μL from the extracted DNA plate wells to the corresponding wells of the qPCR plate.
5. Seal the qPCR plate with an optical adhesive cover from Applied Biosystems and spin down for 5 s at 1,479×*g*.

**Table 2**
**Example calculations for qPCR reaction mixture**

| Reaction mix reagents | 1rxn (μL) | Per 96-well plate (μL) |
|---|---|---|
| TaqMan Universal PCR Master Mix, no UNG | 25 | 2,500 |
| HBV F (900 nM final from 20 μM stock) | 2.25 | 225 |
| HBV R (900 nM final from 20 μM stock) | 2.25 | 225 |
| HBV probe (100 nM final from 10 μM stock) | 0.5 | 50 |
| AMP F (900 nM final from 20 μM stock) | 2.25 | 225 |
| AMP R (900 nM final from 20 μM stock) | 2.25 | 225 |
| AMP MGB probe (100 nM final from 10 μM stock) | 0.5 | 50 |
| $H_2O$ | 5 | 500 |
| Sample | 10 | |
| Total | 50 | |

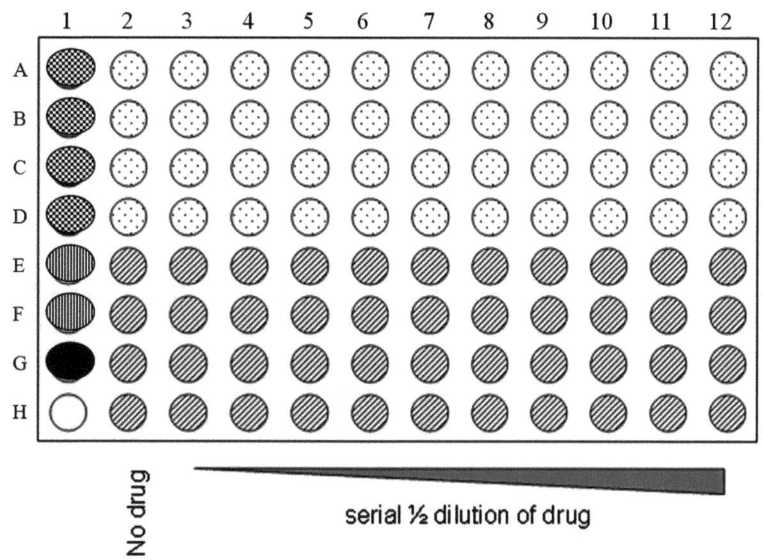


**Fig. 2** qPCR plate layout. The *dot-filled circles* (rows A–D, columns 2–12) indicate cells transfected with one construct, and the *striped circles* (rows E–H, columns 2–12) indicate cells transfected with a different construct. The *checkerboard circles* (column 1, A–D) include 5 μL extracted DNA from untransfected HepG2 cells and $1.0 \times 10^6$ (A1), $1.0 \times 10^5$ (B1), $1.0 \times 10^4$ (C1), and $1.0 \times 10^3$ (D1) copies of standard HBV DNA each. The *vertical striped circles* (E1 and F1) include 10 μL extracted DNA from replication-incompetent plasmid pRTAN-transfected cells. The *black-filled circle* (G1) includes 10 μL extracted DNA from untransfected HepG2 cells. The *open circle* (H1) includes 10 μL of water

Harvest Date:
qPCR Date:

| | pHY92 | | | | | | |
|---|---|---|---|---|---|---|---|
| TFV (μM) | Row 1 | Row 2 | Row 3 | Row 4 | Average | STDEV | CV |
| 0 | 1.23E+05 | 1.22E+05 | 1.13E+05 | 1.33E+05 | 1.23E+05 | 8.31E+03 | 6.76 |
| 0.39 | 1.06E+05 | 1.09E+05 | 1.08E+05 | 1.13E+05 | 1.09E+05 | 2.77E+03 | 2.55 |
| 0.78 | 9.89E+04 | 9.85E+04 | 9.67E+04 | 9.02E+04 | 9.61E+04 | 4.04E+03 | 4.20 |
| 1.56 | 8.82E+04 | 8.56E+04 | 8.99E+04 | 9.15E+04 | 8.88E+04 | 2.52E+03 | 2.84 |
| 3.13 | 7.40E+04 | 7.81E+04 | 8.51E+04 | 8.01E+04 | 7.93E+04 | 4.60E+03 | 5.79 |
| 6.25 | 6.57E+04 | 5.73E+04 | 6.56E+04 | 7.49E+04 | 6.59E+04 | 7.19E+03 | 10.91 |
| 12.50 | 3.37E+04 | 3.66E+04 | 3.26E+04 | 3.67E+04 | 3.49E+04 | 2.07E+03 | 5.93 |
| 25.00 | 2.31E+04 | 2.60E+04 | 2.56E+04 | 3.04E+04 | 2.63E+04 | 3.03E+03 | 11.55 |
| 50.00 | 2.62E+04 | 2.37E+04 | 2.26E+04 | 2.25E+04 | 2.38E+04 | 1.72E+03 | 7.25 |
| 100.00 | 1.69E+04 | 1.58E+04 | 1.38E+04 | 1.68E+04 | 1.58E+04 | 1.46E+03 | 9.25 |
| 200.00 | 7.96E+03 | 7.08E+03 | 6.83E+03 | 5.67E+03 | 6.88E+03 | 9.44E+02 | 13.71 |

| TFV (μM) | | % | |
|---|---|---|---|
| 0 | | 100.00 | |
| 0.39 | | 88.58 | |
| 0.78 | | 78.23 | |
| 1.56 | | 72.30 | |
| 3.13 | | 64.58 | |
| 6.25 | | 53.63 | |
| 12.50 | | 28.41 | |
| 25.00 | | 21.39 | |
| 50.00 | | 19.34 | |
| 100.00 | | 12.87 | |
| 200.00 | | 5.60 | |

**Fig. 3** Example file for calculating $EC_{50}$ values. Quadruplicate HBV DNA copy numbers at each drug concentration are averaged. Furthermore, for each drug concentration, the percent copy number relative to the no-drug control copy number is determined. STDEV = standard deviation, CV (coefficient variation) = STDEV/Average × 100. The CVs of less than 20 are generally acceptable

Place the plate into the Roche 480 real-time PCR machine, select the Multi Color Hydrolysis Probe detection format, and run the program according to the manufacturer's specifications using the following parameters: 50 °C for 2 min; 95 °C for 10 min; and 95 °C for 15 s, 60 °C for 1 min, 40 cycles.

6. When the qPCR is completed, export and save the real-time PCR file in appropriate folder according to the manufacturer's software, and calculate the HBV/AMP copy number based on the standard curves (*see* **Notes 30** and **31**).
7. Since each clinical sample is tested in quadruplicate across the dose range, the quadruplicate HBV copy numbers at each drug concentration should be averaged. Furthermore, for each drug concentration the percent copy number relative to the no-drug copy number should be determined (*see* Fig. 3 for template).
8. Use TableCurve 2D software or an equivalent linear regression analysis program to analyze the data and generate an $EC_{50}$ value. Figure 4 is an example of $EC_{50}$ of pHY92 to TFV generated by TableCurve 2D (*see* **Notes 32** and **33**).

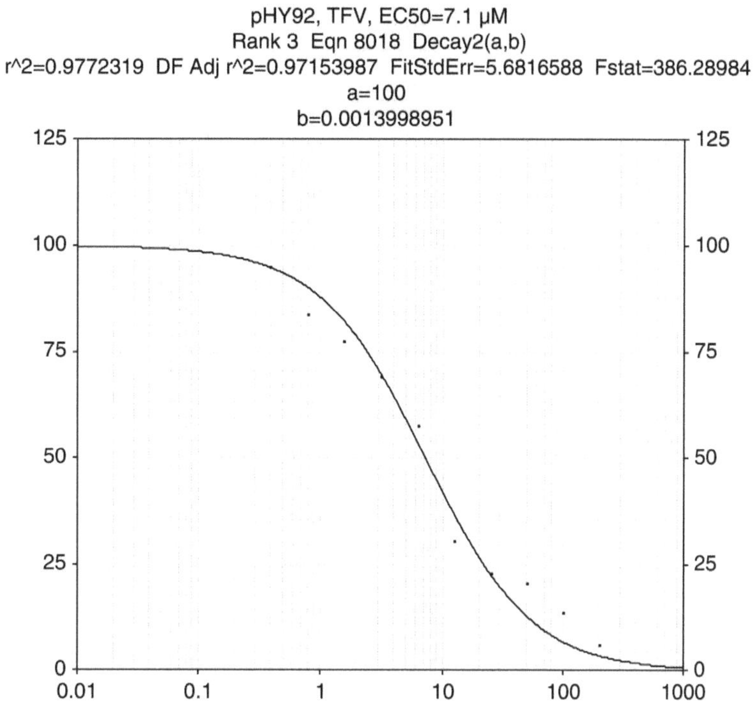


**Fig. 4** Example of tenofovir $EC_{50}$ curve for the pHY92 plasmid. The $EC_{50}$ curve was generated using the TableCurve 2D software

## 4 Notes

1. Other gel systems or pouring your own gel can be substituted.
2. Primers and probes were designed by aligning HBV sequences from different genotypes and identifying conserved regions. Primers are synthesized by Integrated DNA Technologies and shipped lyophilized. Upon receipt, resuspend at 100 µM in 1× TE buffer. Store at −20 °C once resuspended in TE buffer. Dilute primers to working concentrations (*see* Subheading 2.1, **items 25–27** and Subheading 2.2, **items 13** and **14**) prior to use.
3. The EcoRI restriction site for cloning PCR products into pRTAN is included in the primer (underlined).
4. The SphI restriction site for cloning PCR products into pRTAN is included in the primer (underlined).
5. pHY92 is a plasmid DNA that expresses a wild-type, genotype A laboratory strain of HBV under the control of a CMV promoter [9]. pRTAN is a plasmid vector containing the full-length pHY92 HBV genome, minus the majority of pol/RT (amino acids 17–311) [10]. Both of the constructs contain unique EcoRI and SphI sites, which flank the 5′ and 3′ ends of the pol/RT region, respectively.

6. Make 100 mM drug stock using double-distilled $H_2O$ and store at −20 °C for future use.
7. Make a stock solution of 10 % NP-40 with double-distilled $H_2O$ and store at room temperature for later use. Dilute 10 % NP-40 with 1× DPBS to make a fresh working solution of 0.33 % NP-40 for lysing cells each time cell extractions are performed.
8. The real-time PCR probes are synthesized by Applied Biosystems and shipped at 100 μM in 1× TE buffer and should be protected from light. Dilute probes to working concentrations (*see* Subheading 2.2, **item 14**) prior to use.
9. Laboratory personnel performing methods that require experimentation to be performed in the Bio-safety Level 3 facility (BSL3) and or the Bio-safety Level 2 facility (BSL2) should have documented training in the proper use of these facilities and be familiar with the Safety and Bio-safety manuals. Proper personal protective equipment shall be worn at all times in the BSL-2 and BSL-3 facilities. PPE for the BSL-2 are protective eyewear, gloves, and lab coat. PPE for the BSL-3 are protective eyewear, booties, gown, two pairs of gloves, and a mask. All other experimentation performed in other laboratories requires personnel to wear protective eyewear, a pair of gloves, and lab coat.
10. There are numerous steps in this procedure that allow for downtime during which subsequent steps can be performed as time permits.
11. Generally serum samples from posttreatment time points require this PCR step.
12. After each step in this section, products can be stored at −20 °C until ready to use.
13. The subsequent enzymatic digestion step (*see* Subheading 3.3, **step 4**) requires at least 200 ng of PCR product. In case there is an insufficient amount of PCR product, replicate PCRs can be carried out to obtain sufficient quantities.
14. For a positive control, digest 1 μg pHY92 plasmid DNA (described in **Note 5**) in parallel which contains a full-length HBV pol/RT. The digested pHY92 should be analyzed on an agarose gel prior to proceeding to Subheading 3.3, **step 5**, and should only yield two bands (1.2 and ~6 kb). If not, the enzymatic digestion failed and Subheading 3.3, **step 4**, should be repeated. In order to analyze the pHY92 digestion, prepare a 1.0 % agarose gel containing 0.5 μg/mL ethiduim bromide in 1× TAE buffer (not E-gel) and load the entire digested pHY92 mixed with 7 μL 6× loading dye in one well along with 0.25 μg of 1 kb DNA ladder mixed with 1 μL 6× loading dye in another well and run gel at 100 V for 1 h. The pHY92 digestion should be visualized on a Molecular Imager Gel Doc XR system or an

equivalent system and the gel piece containing the 1.2 kb band should be removed from the gel and placed into a 1.5 mL Eppendorf tube. The DNA should then be extracted from the gel piece using the QIAquick Gel Purification kit by following the manufacturer's instructions. Elute DNA in 30 μL buffer EB. The purified EcoRI–SphI fragment from pHY92 should be quantified (Subheading 3.3, **step 6**) and used as a positive ligation control (Subheading 3.5, **step 2**).

15. Need to load multiple wells or tape wells together on the gel comb prior to pouring the gel to be able to run the entire PCR product on the gel.
16. Minimize UV exposure time to avoid damaging DNA when cutting out DNA bands.
17. After each step in this section, products can be stored at −20 °C until ready to use.
18. Ligations can also be carried out at room temperature for 1 h so that transformation steps can be performed on the same day.
19. If there are fewer colonies obtained from the digested, purified PCR product ligation than the vector-only negative control, this is an indication that the digested PCR product was not successfully ligated into the digested pRTAN vector and the steps in Subheading 3.5 should be repeated. If the LB/Amp agar plate containing the positive control (EcoRI/SphI-digested pHY92 ligated into pRTAN) has less than 100 colonies, this is an indication that the transformation efficiency was low and Subheading 3.5, **step 3**, should be repeated.
20. For each recombinant virus, a single dose–response curve is generated from quadruplicate 11 point drug curves (includes no drug) in a 96-well plate format in order to derive a single 50 % effective concentration value ($EC_{50}$). Run 2–3 replicates per virus. Replicate $EC_{50}$ values for each recombinant virus should be within twofold of each other; if not, this is an indication that the assay did not work and should be repeated. It is recommended to start the transfections on a Monday to avoid weekend feedings. Use multichannel pipettes whenever possible.
21. For a single dose–response curve which requires 44 wells of cells, transfect enough cells for 60 wells to enable the use of multichannel pipettes during Subheading 3.6, **step 4**.
22. Twofold serial dilutions of drug are made in DMEM-F12 media containing 10 % fetal bovine serum and 50 μg/mL penicillin/streptomycin. The dose range for TFV, adefovir, and emtricitabine is 0, 0.39, 0.78, 1.56, 3.13, 6.25, 12.5, 25, 50, 100, and 200 μM. The dose range for LAM is 0, 0.08, 0.16, 0.31, 0.63, 1.25, 2.5, 5, 10, 20, and 40 μM. The dose range for LdT is 0, 0.1, 0.2, 0.39, 0.78, 1.56, 3.13, 6.25, 12.5, 25, and 50 μM. The dose range for ETV is 0, 1.0, 2.0, 3.9, 7.8, 15.6, 31.3, 62.5, 125, 250, and 500 nM.

23. Lack of green fluorescence from GFP-transfected cells indicates a failed transfection and all transfections should be repeated.
24. Gently place the aspirator at the edge of the well bottom in order not to disturb the cells.
25. Up to four plates can reasonably be extracted at one time. Twelve-well multichannel pipettes should be used in all of the steps in Subheading 3.7.
26. This step is one of the most critical during the extraction of intracellular HBV DNA. Be sure to transfer all of the NP-40/ cell lysate solution to a new v-bottom plate.
27. Place pipette at the edge of the well bottom to avoid the cell pellet.
28. Digest for at least 1 h to make sure that all the remaining input plasmid DNA are completely digested.
29. Vortex the tube containing the inactivation beads at full speed for 30 s before loading the inactivation beads into a 96-well plate.
30. Standards are dilutions of a linearized plasmid containing one copy each of HBV DNA and the ampicillin resistance gene. The standard HBV DNA copies as well as ampicillin resistance gene copies used in this assay are $1.0 \times 10^6$, $1.0 \times 10^5$, $1.0 \times 10^4$, and $1.0 \times 10^3$ copies. The standard curve is used to calculate the HBV copy number of each well.
31. The ampicillin resistance gene copy number of testing samples can be calculated based on the amp standard curve. If the ampicillin resistance gene values in the recombinant virus wells (columns 2–12) are comparable to the HBV values in the same wells, this may indicate that the input DNA plasmids from the transfection step were not effectively digested during DNA extraction. This result can be confirmed by high HBV signals from the pRTAN wells (E1 and F1 in Fig. 2), which contains cells transfected with the replication-incompetent HBV plasmid pRTAN. If there is input DNA plasmid contamination, the assay should be repeated.
32. Typically the phenotypic testing of each recombinant RT construct is performed in three independent experiments and the average $EC_{50}$ along with standard deviation are reported. The average $EC_{50}$ of recombinant RT constructs from post-therapy points are compared to those from baseline and differences reported as fold change: $EC_{50}$ post-therapy/$EC_{50}$ baseline.
33. This assay has not been approved by the United States Food and Drug Administration and is labeled For Research Use Only. Results should not be used for diagnostic purposes.

## References


1. Lee WM (1997) Hepatitis B virus infection. N Engl J Med 337:1733–1745
2. Seeger C, Mason WS (2000) Hepatitis B virus biology. Microbiol Mol Biol Rev 64: 51–68
3. Doo E, Liang TJ (2001) Molecular anatomy and pathophysiologic implications of drug resistance in hepatitis B virus infection. Gastroenterology 120:1000–1008
4. Zoulim F, Perrillo R (2008) Hepatitis B: reflections on the current approach to antiviral therapy. J Hepatol 48(Suppl 1):S2–S19
5. Locarnini S (2003) Hepatitis B viral resistance: mechanisms and diagnosis. J Hepatol 39:S124–S132
6. Zoulim F, Buti M, Lok AS (2007) Antiviral-resistant hepatitis B virus: can we prevent this monster from growing? J Viral Hepat 14 (Suppl 1):29–36
7. Tenney DJ, Levine SM, Rose RE et al (2004) Clinical emergence of entecavir-resistant hepatitis B virus requires additional substitutions in virus already resistant to Lamivudine. Antimicrob Agents Chemother 48:3498–3507
8. Zhu Y, Curtis M, Snow-Lampart A et al (2007) In vitro drug susceptibility analysis of hepatitis B virus clinical quasispecies populations. J Clin Microbiol 45:3335–3341
9. Yang H, Westland C, Xiong S, Delaney WE (2004) In vitro antiviral susceptibility of full-length clinical hepatitis B virus isolates cloned with a novel expression vector. Antiviral Res 61:27–36
10. Zhu Y, Curtis M, Borroto-Esoda K (2011) HBV DNA replication mediated by cloned patient HBV reverse transcriptase genes from HBV genotypes A-H and its use in antiviral phenotyping assays. J Virol Methods 173:340–346

# Part III

## Flaviviruses

# Chapter 15

# Cell-Based Antiviral Assays for Screening and Profiling Inhibitors Against Dengue Virus

**Edwin Yunhao Gong, Marleen Clynhens, Tania Ivens, Pedro Lory, Kenny Simmen, and Guenter Kraus**

## Abstract

Dengue, a mosquito-borne virus of the *Flaviviridae* family, is reemerging as one of the most important human pathogens in tropical and subtropical regions of the world. It is estimated that 2.5 billion people live in areas at risk for transmission of dengue virus (DENV). Furthermore, it causes significant morbidity and mortality with 50–100 million infections per year. Currently, there are no vaccines commercially available and no effective antiviral drugs for treatment of DENV infections. In this chapter, we describe a plaque reduction assay and a cell-based high-throughput antiviral assay for identifying inhibitors against DENV. The latter is a homogeneous high-throughput assay that has been fully validated according to the Food and Drug Administration (FDA) guidelines for assay validation and can be used for screening compound libraries.

**Key words** Dengue virus, Dengue antiviral, High-throughput screening, CPE inhibition assay, Plaque assay

## 1 Introduction

The dengue virus (DENV), a member of the *Flavivirus* genus in the *Flaviviridae* family, is endemic in more than 100 countries, mainly in tropical and subtropical regions including most of Southeast Asia, Southern and Central America, and the Caribbean and South Pacific regions (for reviews *see* refs. 1–3). Four related but distinct serotypes of DENV exist in nature and are transmitted to humans through the bite of infected mosquitoes, primarily by two species, *Aedes aegypti* and *Aedes albopictus*. The virus threatens almost half of the world's population with estimated 50–100 million infections, 500,000 hospitalizations, and 24,000 deaths annually. DENV infection causes a spectrum of disease in humans, ranging from dengue fever to a severe life-threatening syndrome called dengue hemorrhagic fever/dengue shock syndrome. Currently, there is no licensed

Edwin Yunhao Gong (ed.), *Antiviral Methods and Protocols*, Methods in Molecular Biology, vol. 1030,
DOI 10.1007/978-1-62703-484-5_15, © Springer Science+Business Media, LLC 2013

vaccine available and therapeutic options are limited. The treatment of infection is entirely supportive.

DENV is an enveloped positive-sense single-stranded RNA virus with a genome of approximately 11 kb. The viral genome encodes a single polyprotein that is then cleaved co-translationally and post-translationally by host and virus-encoded proteases into three structural and seven nonstructural proteins. In theory, any of these viral proteins and host proteins can be targeted for antiviral drug development. In particular, the NS3 protease and NS5 RNA polymerase remain the most attractive viral targets based on the experience of HIV and hepatitis C virus antiviral drug development.

In this chapter, we describe two cell-based antiviral screening and profiling assays: a plaque reduction assay and a high-throughput screening assay for identifying inhibitors against DENV [4]. A protocol for growing DENV stocks is also presented.

## 2 Materials

### 2.1 Cells, Virus, and Reagents

1. Vero cells (African green monkey kidney cells): European Collection of Cell Cultures (ECACC).
2. BHK-21 cells (baby hamster kidney cells): American Type Culture Collection (ATCC).
3. DENV-2 (serotype 2) NGC (New Guinea C) strain: National Collection of Pathogenic Viruses (NCPV, UK).
4. Eagle's minimum essential medium (MEM) without L-glutamine.
5. 10× MEM without L-glutamine and sodium bicarbonate (Gibco).
6. Heat-inactivated fetal calf serum (FCS).
7. 200 mM L-glutamine (100×).
8. Gentamicin solution (50 mg/mL).
9. 100 mM sodium pyruvate (100×).
10. MEM nonessential amino acids (100×).
11. 0.05 % trypsin–EDTA.
12. Vero cell growth medium (MEM/10 % FCS): Eagle's MEM supplemented with 10 % FCS, 2 mM L-glutamine, and 0.02 mg/mL gentamicin.
13. Vero cell maintenance medium (MEM/2 % FCS): Same as the growth medium but supplemented with 2 % FCS.
14. BHK-21 cell growth medium: Eagle's MEM supplemented with 10 % FCS, 2 mM L-glutamine, 1× MEM nonessential amino acids, 1× sodium pyruvate, and 0.02 mg/mL gentamicin.

15. BHK-21 cell maintenance medium: Same as the BHK-21 cell growth medium but supplemented with 2 % FCS.
16. 2× BHK-21 cell maintenance medium: 2× MEM diluted from 10× MEM without L-glutamine and sodium bicarbonate (**item 5**) and then supplemented with 4 % FCS, 2× L-glutamine, 2×7.5 % $NaHCO_3$, 0.04 mg/mL gentamicin. Adjust to the final volume with autoclaved deionized $H_2O$.
17. Dulbecco's phosphate-buffered saline (D-PBS), no calcium chloride, no magnesium chloride.
18. 10 % formalin in D-PBS.
19. 1 % crystal violet solution (Sigma-Aldrich).
20. 1.5 % methylcellulose in water: Weigh 1.5 g of methylcellulose in an Erlenmeyer flask and add 100 mL of deionized $H_2O$ into the flask. Place the cap loosely on the bottle. Put a stirring bar into the flask and place on a magnetic stirrer. Heat the flask under stirring until boiling. Then allow the flask to cool to room temperature. Continue stirring the flask overnight.
21. Trypan blue stain (Gibco) (optional).
22. ATPLite™ 1-step kit (PerkinElmer).
23. Reference compounds:
    (a) Interferon α-2a (Roferon A, Roche).
    (b) 6-Azauridine (Sigma-Aldrich).
    (c) 2′-C-Methylcytidine (Sigma-Aldrich).
    (d) Ribavirin (Sigma-Aldrich).

### 2.2 Consumables

1. 6-well, flat-bottom plates: Corning® Costar® cell culture plates.
2. Nunc® TripleFlasks (F8542, Sigma-Aldrich).
3. Greiner Lumitrac 384-well plate (VWR).
4. White-view 384-well plates (white wall and transparent bottom) (PerkinElmer).
5. Absorbent pad.

### 2.3 Equipment

1. Centrifuge.
2. Magnetic stirrer.
3. Multidrop combi liquid dispenser (Thermo Scientific).
4. ViewLux™ (PerkinElmer) or any luminescent reader.
5. Cell counter (Coulter Electronics LTD) (optional).
6. Microplate shaker.
7. Light-box.

## 3 Methods

Carry out all procedures at a biosafety cabinet unless otherwise specified. The handling and manipulation of the DENV are restricted in our biosafety level three laboratory based on the Belgian biosafety regulation.

### 3.1 Dengue Virus Cultivation and Harvesting

1. Seed Vero cells in T175 flasks using the Vero cell growth medium (MEM/10 % FCS) and incubate at 37 °C in a humidified 5 % $CO_2$ incubator (37 °C, 5 % $CO_2$) overnight. The cell monolayer should be sub-confluent (~90 %) on the following day.
2. On the following day, prepare a viral dilution at a multiplicity of infection (MOI) of 0.1 $TCID_{50}$/cell (the 50 % tissue culture infective dose per cell) in MEM/2 % FCS. Prepare 15 mL of virus dilution for each flask to be infected.
3. Remove the old culture medium from flasks and wash the sub-confluent cells once with D-PBS.
4. Add 15 mL of the virus dilution to each flask.
5. Swirl the flask gently to spread the viral inoculum evenly to cover the whole-cell monolayer.
6. Incubate the cells at 37 °C, 5 % $CO_2$, for 2 h with occasional swirling.
7. Discard the viral inoculum.
8. Rinse the infected cells once with D-PBS.
9. Add 18 mL of fresh MEM/2 % FCS to each flask and incubate the infected cells at 37 °C, 5 % $CO_2$, for 5–6 days until the viral cytopathic effect (CPE) reaches ~90–100 %.
10. Collect the supernatant into sterile 50 mL Falcon tubes and centrifuge at 3,700 × *g* and 4 °C for 10 min to remove the cell debris.
11. After centrifugation, pool the supernatants from all tubes into a sterile bottle or a fresh T175 flask.
12. Add heat-inactivated FCS to a final concentration of 20 % to the harvested supernatant (*see* **Note 1**). Aliquot 1 mL to each cryovial and store in a −80 °C freezer.
13. Determine the titer of the virus stock by either a plaque assay to calculate the plaque-forming units per mL (PFU/mL) or the 50 % end-point (e.g., Reed–Muench method) to calculate the $TCID_{50}$/mL.

### 3.2 Antiviral Testing of Anti-dengue Compounds by a Plaque Assay

The DENV plaque assay can be performed using different cell lines such as Vero cells, Vero E6 cells, and BHK-21 cells. Because of prolonged incubation period (6–7 days) for DENV plaques to develop on Vero cells, many cells are lost at the end of the assay.

On the other hand, only a 4-day incubation period is needed when the plaque assay is performed using the BHK-21 cells. Therefore, the protocol described here is performed using the BHK-21 cells.

1. Prepare a BHK-21 cell suspension at a density of $1.0 \times 10^5$ cells/mL in BHK-21 cell growth medium.
2. Seed 6-well plates (*see* **Note 2**) with 3 mL of the BHK-21 cell suspension per well (final $3 \times 10^5$ cells/well). Prepare duplicate wells for each compound concentration.
3. Incubate the plate at 37 °C, 5 % $CO_2$, overnight.
4. On the following day, check the BHK-21 cells with a microscope. The cell monolayer should be sub-confluent (~90 %).
5. Thaw a DENV stock quickly at 37 °C in a water-bath with shaking. Make an appropriate dilution of the virus (100 PFU per mL) in the BHK-21 cell maintenance medium.
6. Remove the old culture medium from the 6-well plates by inverting the plate over an absorbent pad.
7. Rinse the cells with D-PBS once.
8. Add 1 mL of diluted virus into each well (100 PFU per well).
9. Incubate the plates at 37 °C, 5 % $CO_2$, for 1.5–2 h. Swirl the plates periodically to spread the virus evenly over the cell monolayer.
10. During the incubation period, perform tenfold (or fivefold) serial dilutions of test compounds to prepare 2× final test concentrations in 2× BHK-21 cell maintenance medium (*see* Subheading 2.1, **item 16**; also *see* **Note 3**). Then, prepare methylcellulose overlay by mixing equal volumes of each compound dilution (2× final test concentrations) with 1.5 % methylcellulose in water (in this overlay mixture, the compound concentration is 1× in 0.75 % methylcellulose with 1× maintenance medium). For the control wells containing no compound, prepare methylcellulose overlay by mixing equal volumes of 2× BHK-21 cell maintenance medium with 1.5 % methylcellulose in water. Prepare enough volumes for the duplicate wells.
11. After incubation, remove the viral inoculum from the plates by inverting the plates over an absorbent pad.
12. Add 3 mL/well of the methylcellulose overlay containing various concentrations of test compound to the corresponding wells of 6-well plates.
13. Add 3 mL of methylcellulose overlay containing no compound to the control wells.
14. Incubate the plates at 37 °C, 5 % $CO_2$, for 4 days.
15. Remove the methylcellulose overlay from plates by inverting the plate over an absorbent pad.

16. Add 2 mL of 10 % formalin to each well to fix the cells. Incubate at room temperature for 10 min.
17. Discard the formalin from the plates by inverting the plates over an absorbent pad.
18. Stain the fixed cell monolayer by adding 2–3 drops of 1.0 % crystal violet to each well. Leave at room temperature for 5–10 min.
19. Remove excess of crystal violet by rinsing the plates in a gentle stream of tap water.
20. Air-dry the plates.
21. Place the plates upside down on a light-box. Count the plaques in each well using a colony counter and average the plaque numbers in the duplicate wells for each test concentration.
22. Enter the data in a standard statistical program, such as GraphPad Prism, and calculate the $EC_{50}$ (effective concentration achieving 50 % of inhibition of virus plaques) values for test compounds.

### 3.3 A Cell-Based High-Throughput Antiviral Assay for Screening Inhibitors of DENV

The assay principle is based upon the inhibition of viral-induced CPE by a test compound. The ATP levels of metabolically active cells protected by the presence of antiviral compounds are measured by bioluminescence [5]. The antiviral efficacy of test compounds is expressed as the $EC_{50}$ value defined as the concentration of compound achieving 50 % inhibition of the virus-reduced luminescence signals as compared with the uninfected cell control. The toxicity of test compounds ($CC_{50}$: 50 % cytotoxic concentration) is measured in the absence of virus infection and in parallel with the activity testing. The hits are selected by comparing the selectivity index ($CC_{50}/EC_{50}$). Compared with the dye uptake assay, this ATP/luminescence assay gives a wider kinetic range and a better signal-to-noise ratio. Importantly, this assay is homogeneous (mix and measure) and permits a high-throughput screening mode [5]. In addition, it can pick up inhibitors of any stage of DENV replication because it is a whole infection assay. A flow diagram of this assay is illustrated in Fig. 1.

#### 3.3.1 Compound Plating

Different plate formats can be used to plate the test compounds:

For primary screening of compound libraries, depending on the size of the library, we typically use the following:

(a) 80 compounds per plate with four single test concentrations of each compound. Alternatively, 320 compounds per plate with four single test concentrations on four separated plates, one concentration on each plate.
(b) 320 compounds per plate with a single test concentration of each compound.

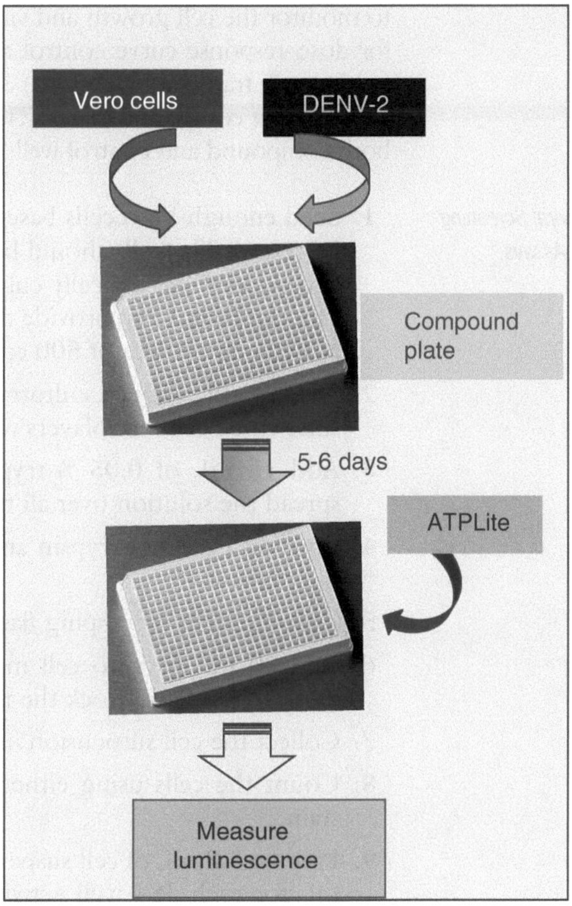


**Fig. 1** A flow diagram of the cell-based high-throughput antiviral assay for screening anti-dengue compounds

For hit confirmation and compound profiling, the plate formats are typically the following:

(a) 8 compounds per plate, nine test concentrations (four or fivefold serial dilutions) with quadruplicate wells for each test concentration. In this format, columns 19–22 are virus controls containing no compound; columns 23 and 24 are cell controls (uninfected cells containing no compound).

(b) 16 compounds per plate, 5 serially diluted (four or fivefold) test concentrations with quadruplicate wells for each test concentration. Columns 21 and 22 are virus controls and columns 23 and 24 are uninfected cell controls.

The test compounds are serially diluted from their stock solutions using the Vero cell maintenance medium (*see* **Note 3**) and plated in white opaque 384-well plates (Greiner Lumitrac 384-well plates) in a volume of 10 μL per well using a 96-channel liquid handler. In order to monitor the cell growth and viral CPE, the reference compounds for dose–response curve control are plated in a separate white-view (white wall, transparent bottom) 384-well plate and tested in parallel with the test compound plates. The final concentration of DMSO in both compound and control wells is 0.5 %.

#### 3.3.2 Antiviral Screening and Toxicity Assays

1. Seed enough Vero cells based on the number of test plates to be tested. The cells should be confluent at the day of antiviral experiment. We typically culture the Vero cells in triple flasks. One triple flask can provide three parallel growth surfaces with a total culture area of 500 $cm^2$.
2. Discard the old cell culture medium from a triple flask and rinse three cell monolayers with 60 mL D-PBS.
3. Add 15 mL of 0.05 % trypsin–EDTA to a triple flask and spread the solution over all three surfaces.
4. Pour off excess of trypsin and incubate the flask at 37 °C for 2–3 min.
5. Dislodge cells by tapping flask with the palm of hand.
6. Add 30 mL of Vero cell maintenance medium (MEM/2 % FCS) to the flask. Rock the flask to dislodge the cells.
7. Collect the cell suspension in a sterile Falcon tube.
8. Count the cells using either a cell counter or a trypan blue stain.
9. Prepare 5.8 mL of cell suspension at a density of $1 \times 10^5$ cells/mL for each 384-well activity plate (the optimal number of cells in each well is 1,500 and the volume added to each well is 15 μL).
10. Prepare 11.6 mL of cell suspension at a density of $5 \times 10^4$ cells/mL for each 384-well toxicity plate (the optimal number of cells in each well is also 1,500 and the volume added to each well is 30 μL).
11. Thaw a DENV-2 stock quickly in a 37 °C water-bath with shaking. Make an appropriate dilution of the virus in MEM/2 % FCS to result in an MOI of 0.1 $TCID_{50}$/cell. Prepare 5.3 mL for each 384-well activity plate (15 μL/well).
12. In a biosafety cabinet, use a multidrop combi liquid dispenser to dispense cells and virus into the test plates:
    For the activity test plates (containing test compounds or no compounds in 10 μL of MEM/2 % FCS in each well):
    (a) Dispense 15 μL of MEM/2 % FCS into the wells of columns 23 and 24 (cell controls).
    (b) Dispense 15 μL of cell suspension ($10^5$ cells/mL) into all wells of the test plate.
    (c) Dispense 15 μL of the virus dilution into all wells of columns 1–22.

For the toxicity test plates (also containing test compounds or no compounds in 10 μL of MEM/2 % FCS in each well):

(a) Dispense 30 μL of MEM/2 % FCS into the wells of columns 23 and 24 (medium controls).

(b) Dispense 30 μL of cell suspension ($5 \times 10^4$ cells/mL) into all wells of columns 1–22.

13. Incubate the plates at 37 °C, 5 % $CO_2$, for 5–6 days.

14. At the end of incubation, check the plates with reference compounds under an inverted microscope. The CPE in the virus-control wells should reach ~100 %. The ATP levels in compound-treated and untreated cells are measured by a luminescent assay using an ATPLite™ 1-Step kit. We use ViewLux™ to measure the luminescence. The following procedures are recommended by the manufacturer:

    (a) Warm up all reagents to room temperature before use.

    (b) Reconstitute the lyophilized substrate solution by adding the appropriate volume of buffer to the substrate bottle. Mix the contents by inversion and leave the solution to stand for 5 min (*see* **Notes 4** and **5**).

    (c) Add 40 μL of the reconstituted reagent to each well of both activity and toxicity plates using a multidrop combi liquid dispenser (*see* **Note 4**).

    (d) Shake the plates for 2 min at 700 rpm using an orbital microplate shaker.

    (e) Program Viewlux to take a 0.1-s integrated reading of each plate.

15. Export the test results to an Excel sheet. Calculate the percentage of inhibition (% inhibition) based on the reduction of the luminescent signals in the compound-treated wells relative to the cell controls. If duplicate or quadruplicate wells are used for each drug concentration, average the luminescent readings in those wells. Calculate the $EC_{50}$ and $CC_{50}$ of test compounds using computer software, such as GraphPad Prism.

## 4 Notes

1. We noticed that the titer of DENV stocks can decrease up to tenfold during storage at −80 °C freezer. Supplementing with 20 % heat-inactivated FCS can help to stabilize the virus titers for a longer storage time.

2. We found that different brands of 6-well plates can affect the quality of plaque reduction assay. For BHK-21 cells, we recommend using the 6-well, flat-bottom cell culture plates from

Corning® Costar® (*see* Subheading 2.2, **item 1**) for the dengue plaque reduction assay.

3. For antiviral testing, we use medium with reduced serum concentration of 2 % (maintenance medium, MEM/2 % FCS) to reduce protein binding of some compounds.
4. Reconstituted reagent is light sensitive. If a clear bottle is used, wrap the bottle with aluminum foil to avoid light exposure. Once the solution is added to the test plates, we place an opaque empty plate on the top of the plate stack.
5. The reconstituted reagent is stable for 1 month at −80 °C, 48 h at 2–8 °C, and 8 h at 15–25 °C.

## References


1. Halstead S (2007) Dengue. Lancet 370: 1644–1652
2. Whitehead S, Blaney J, Durbin A, Murphy B (2007) Prospects for a dengue virus vaccine. Nat Rev Microbiol 5:518–528
3. Gubler D, Kuno G, Markoff L (2007) Flaviviruses. In: Knipe D, Howley P et al (eds) Fields Virology, 5th edn. Lippincott Williams and Wilkins, New York, pp 1153–1252
4. Gong E, Ivens T, Clynhens M, Lory P, Kraus G (2010) A high throughput method for identifying inhibitors against dengue viruses. International Publication Number: WO2010/130748 Al
5. Gong E, Ivens T, Van den Eynde C et al (2008) Development of robust antiviral assays for profiling compounds against a panel of positive-strand RNA viruses using ATP/luminescence readout. J Virol Methods 151:121–125

# Chapter 16

# A Duplex Real-Time RT-PCR Assay for Profiling Inhibitors of Four Dengue Serotypes

**Edwin Yunhao Gong, Alexandra Smets, Nick Verheyen, Marleen Clynhens, Emmanuel Gustin, Pedro Lory, and Guenter Kraus**

## Abstract

We have developed a duplex real-time RT-PCR assay for profiling antiviral inhibitors of four dengue virus (DENV) serotypes. In this assay, the primers and the probe for amplifying DENV were designed in the conserved regions of the genome after aligned more than 300 nucleotide sequences of four dengue serotypes deposited in the GeneBank. To discriminate the antiviral activity from the cytotoxicity of compounds, a housekeeping gene of the Vero cells, β-actin, was used to design the primers and the probe for the second set of PCR as an internal control, which is used to normalize the RNA levels of dengue-specific PCR due to the cellular toxicity of test compounds. For compound profiling, the duplex PCR is performed using LightCycler® in a single tube to simultaneously amplify both the dengue target gene and the Vero cell housekeeping gene from the compound-treated Vero cell lysates. This assay was validated against a panel of reference compounds. The results show that the universal primers and probe in this duplex RT-PCR assay can efficiently amplify all four dengue serotypes and that the PCR efficiency for both the dengue target gene and the Vero cells β-actin gene is 100 %.

**Key words** Dengue virus, Duplex RT-PCR, Real-time PCR, Antiviral profiling

## 1 Introduction

The genome of dengue virus (DENV) is a positive-sense single-stranded RNA molecule with a size of approximately 11 kb. Based on the antigenic similarity, specifically on the basis of recognition by the plaque reduction neutralization test (PRNT), DENV are grouped into four distinct serotypes. In epidemic regions, all four serotypes circulate and cause a similar spectrum of diseases (for reviews *see* refs. 1–3). At amino acid level, the positional homology of the DENV genomes among serotypes is 63–68 %; within the serotype, the genetic similarity among the strains can exceed 90 % [3].

Edwin Yunhao Gong (ed.), *Antiviral Methods and Protocols*, Methods in Molecular Biology, vol. 1030, DOI 10.1007/978-1-62703-484-5_16, © Springer Science+Business Media, LLC 2013

To establish a compound-profiling assay for different dengue serotypes using real-time RT-PCR as an end-point, more than 300 nucleotide sequences of four dengue serotypes deposited in the GeneBank were aligned. Conserved regions in the 3′-UTR of the dengue genome were selected to design the primers and the probe. To discriminate the antiviral activity from the cytotoxicity of compounds, another set of PCR primers and probe were designed to amplify the Vero cell β-actin gene, a housekeeping gene of the Vero cells. The amplification conditions of this duplex qRT-PCR were optimized and validated using a panel of reference compounds. The results demonstrate that our universal dengue probe and primers can amplify all four dengue serotypes in 100 % efficiency; the PCR efficiency for the housekeeping gene β-actin is also 100 %.

To profile antiviral inhibitors against different dengue serotypes, the Vero cells seeded on a 96-well plate are infected with DENV and incubated with different concentrations of test compounds. After 3 days of incubation, the infected cells are lysed with a Cells-to-CT™ kit (Ambion). The cell lysates are directly used for the reverse transcription reaction. The duplex real-time PCR is then performed using a Lightcycler (Roche Applied Science). The Cp value of the dengue-specific PCR amplification for each compound concentration is normalized by that of the housekeeping gene. The $EC_{50}$ of a test compound is calculated based on the relative quantitation [4–7].

## 2 Materials

### *2.1 Virus, Cells, and Media*

1. DENV-1 TC974666 strain: National Collection of Pathogenic Viruses (NCPV, UK) (*see* **Note 1**).
2. DENV-2 NGC (New Guinea C) strain: NCPV (*see* **Note 1**).
3. DENV-3 H87 strain: NCPV (*see* **Note 1**).
4. DENV-4 H241 strain: NCPV (*see* **Note 1**).
5. Vero cells (African green monkey kidney cells): European Collection of Cell Cultures (ECACC).
6. Vero cell growth medium (MEM/10 % FCS): Eagle's minimum essential medium (MEM) supplemented with 10 % fetal calf serum (FCS), 2 mM L-glutamine, and 0.02 mg/mL gentamicin.
7. Vero cell maintenance medium (MEM/2 % FCS): Same as the growth medium but supplemented with 2 % FCS.
8. Dulbecco's phosphate-buffered saline (D-PBS), no calcium chloride, no magnesium chloride (Gibco).
9. Trypan blue stain (Gibco) (optional).

#### 2.1.1 Kit for Cell Lysis

Cells-to-CT™ bulk lysis reagents containing DNase I-cells to Ct lysis solution, cells to Ct stop solution (Ambion, Cat. No. 4391851C).

### 2.2 Reverse Transcription (RT)

1. $MgCl_2$.
2. dNTP mix.
3. Expand high fidelity buffer (10×) with $MgCl_2$ (Roche, Cat. No. 11759 167 001).
4. Expand reverse transcriptase (Roche, Cat. No.11785834001).
5. Protector RNase inhibitor (Roche, Cat. No. 03 335 402 001).
6. UltraPure™ DNase/RNase-free distilled water.

### 2.3 Lightcycler Kits

1. LightCycler® 480 probes master 2× conc. (Roche, Cat. No. 04 887 301 001).
2. LightCycler® 480 probes master $H_2O$ PCR grade (Roche).

### 2.4 Primers and Probes

1. DENV primers and probe (*see* **Note 2**):

   Forward primer ($F_{DENV}$): 5′-CGGTTAGAGGAGACCCCTC-3′.

   Reverse primer ($R_{DENV}$): 5′-GAGACAGCAGGATCTCTGG TC-3′.

   Probe ($P_{DENV}$): FAM-5′-AAGGACTAGAGGTTAGAGGAGA CCCCCC-3′-BHQ1 (amplicon: 167 bp).
2. β-Actin primers and probe (*see* **Note 2**):

   Forward primer ($F_{actin}$): 5′-GGCCAGGTCATCACCATT-3′.

   Reverse primer ($R_{actin}$): 5′-ATGTCCACGTCACACTTCATG-3′.

   Probe ($P_{actin}$): JOE-5′-TTCCGCTGCCCTGAGGCTCTC-3′-BHQ1 (amplicon: 134 bp).

### 2.5 Consumables

1. LightCycler® 480 multiwell plate 96, white (Roche, Cat. No. 04729692001).
2. 96-well transparent flat bottom plates (Corning).
3. Disposable RNase- and DNase-free filter pipette tips.
4. Reagent reservoirs.
5. 96-well PCR plates for RT (Westburg).
6. LightCycler® 480 sealing foil for PCR (Roche).
7. Microseal B film sealing tape for RT (BioRad).
8. Aluminum sealing tape (Nunc).

### 2.6 Equipment

1. Coulter cell counter (Beckman) (optional).
2. Multidrop combi liquid dispenser (Thermo Scientific).
3. LightCycler® 480 instrument (Roche Applied Science).

4. Thermocyclers PE9700 (PerkinElmer).
5. Multiple channel pipettes (10, 100, 200, 1,000 μL).
6. Centrifuge (for 96-well plates).

## 3 Methods

Follow good laboratory practice and guidelines when performing PCR (*see* **Note 3**).

### 3.1 Preparation of Compound-Treated DENV-Infected Cells

1. Prior to antiviral testing, seed enough (depending on the number of compounds or plates to be tested) Vero cells in T175 flasks or triple flasks in MEM/10 % FCS and incubate at 37 °C for 3 or 4 days in a 5 % $CO_2$ incubator. The cells should be confluent at the day of the experiment.
2. Preparation of test compound plates:
   Perform fourfold serial dilutions of test compounds with MEM/2 % FCS (*see* **Note 4**) to prepare 4× final concentrations and pipette 25 μL of each drug testing concentration into the duplicate wells of a 96-well cell culture plate. We typically test nine concentrations for each compound. In this plate format, columns 1–9 are serially diluted compounds, column 10 is the virus control (no compound), column 11 is the cell control (uninfected cells containing no compound), and column 12 is the medium control (medium only).
3. Prepare a Vero cell suspension at a density of $2.0 \times 10^5$ cells/mL using MEM/2 % FCS (the final density is $1.0 \times 10^4$ cells per well in 50 μL).
4. Thaw DENV stocks quickly in a 37 °C water-bath with shaking. Make appropriate dilutions of different serotypes in MEM/2 % FCS:

   DENV-1/TC974666 at a multiplicity of infection (MOI) of 1 $TCID_{50}$/cell.

   DENV-2/NGC at an MOI of 0.1 $TCID_{50}$/cell.

   DENV-3/H87 at an MOI of 0.05 $TCID_{50}$/cell.

   DENV-4/H241 at an MOI of 0.5 $TCID_{50}$/cell.
5. Use a multidrop combi liquid dispenser or a multiple channel pipette to dispense cells and virus into the test plates that contain 25 μL of test compounds in columns 1–9 and 25 μL of medium with DMSO (same concentration as in columns 1–9) in columns 10–12:
   (a) Dispense 75 μL of MEM/2 % FCS into all wells in column 12 (medium controls) and 25 μL to all wells in column 11 (cell controls).

(b) Dispense 50 μL of cell suspension ($2 \times 10^5$ cells/mL) into all wells of columns 1–11.

(c) Dispense 25 μL of virus suspension into all wells of columns 1–10.

6. Incubate the plates at 37 °C for 3 days in a 5 % $CO_2$ incubator. Then proceed with the cell lysis as described in Subheading 3.2.

### 3.2 Preparation of Cell Lysates

The cell lysate is prepared using the Cells-to-CT™ bulk lysis reagents containing DNase I-cells to Ct lysis solution, cells to Ct stop solution according to the manufacturer's instructions:

1. Before experiment, thaw the stop solution. Invert or flick the tube several times to mix thoroughly, and place on ice.
2. Prepare an appropriate volume (master mix) of the cells to Ct lysis buffer with DNase I (1/100 dilution: 49.5 μL lysis buffer + 0.5 μL DNase I for each well of a 96-well plate).
3. After 3-day incubation (from **step 6** under Subheading 3.1), check the infected cells with an inverted microscope.
4. Remove the culture medium from the compound-treated plates carefully using a multiple channel pipette.
5. Wash the compound-treated infected cells with 100 μL/well of ice-cold D-PBS three times.
6. Carefully remove D-PBS as much as possible from the wells. If not proceeding with the cell lysis immediately, the plates can be sealed with aluminum sealing tape and stored at −80 °C overnight (*see* **Note 5**).
7. Add 50 μL/well of lysis buffer with DNase I to each well.
8. Mix the lysis reaction by pipetting up and down five times (*see* **Note 6**).
9. Incubate the lysis reaction at least for 5 min at room temperature. Check under a microscope if all cells are lysed.
10. Add 5 μL/well of stop solution to the wells and mix by pipetting up and down five times (*see* **Note 6**).
11. Incubate the plates at room temperature for 2 min.
12. Store the plates on ice if used immediately in the RT step. If not proceeding immediately, seal the plates with aluminum sealing tape and store at −80 °C.

### 3.3 Reverse Transcription

1. In a pre-PCR laboratory, prepare master mixes A and B as in Tables 1 and 2, respectively.
2. Dispense 7.57 μL of master mix A to each well of a 96-well PCR plate. Seal the plate and transfer to a post-PCR laboratory.
3. In the post-PCR laboratory, add 5 μL of cell lysates to the corresponding wells.

**Table 1**
**Components of the RT master mix A**

| | Stock concentration (μM) | Final concentration (μM) | Volume/reaction (μL) |
|---|---|---|---|
| DNase/RNase-free $H_2O$ | | | 7.27 |
| $R_{DENV}$ | 20 | 0.27 | 0.15 |
| $R_{actin}$ | 20 | 0.27 | 0.15 |

**Table 2**
**Components of the RT master mix B**

| | Stock concentration | Final concentration | Volume/reaction (μL) |
|---|---|---|---|
| Expand high fidelity buffer | 10× | 1× | 2 |
| $MgCl_2$ | 25 mM | 3.5 mM | 2.8 |
| dNTPs | 10 mM | 1 mM | 2 |
| RNase inhibitor | 40 U/μL | 1 U/μL | 0.5 |
| Expand RT | 40 U/μL | 0.33 U/μL | 0.13 |

**Table 3**
**Reaction parameters of the reverse transcription**

| | Temperature (°C) | Time |
|---|---|---|
| Reverse transcription | 50 | 30 min |
| Denaturation | 99 | 5 min |
| Hold | 4 | Hold |

4. Centrifuge the PCR plate briefly to collect the liquid to the bottom of the wells.
5. On a thermocycler, run a denaturing program at 75 °C for 5 min, and then at 4 °C on hold.
6. At the end of denaturing program, centrifuge the plates briefly to collect the liquid to the bottom of the wells.
7. Immediately add 7.43 μL of master mix B to the corresponding wells and run an RT program on a thermocycler as in Table 3.
8. At the end of RT reaction, keep the plates on ice if proceeding immediately with real-time PCR. If not used immediately, seal the plates and store the cDNA at a −20 °C freezer.

**Table 4**
**Components of the real-time PCR master mix**

| | Stock concentration | Final concentration | Volume/reaction (μL) |
|---|---|---|---|
| DNase/RNase-free $H_2O$ | | | 7.74 |
| Roche probes master 2× | 2× | 1× | 12.5 |
| $F_{DENV}$ | 20 μM | 0.3 μM | 0.38 |
| $R_{DENV}$ | 20 μM | 0.3 μM | 0.38 |
| $P_{DENV}$ | 20 μM | 0.1 μM | 0.13 |
| $F_{actin}$ | 20 μM | 0.3 μM | 0.38 |
| $R_{actin}$ | 20 μM | 0.3 μM | 0.38 |
| $P_{actin}$ | 20 μM | 0.1 μM | 0.13 |

**Table 5**
**The cycling parameters for the real-time PCR using LightCycler®**

| Step | Temperature (°C) | Time | Ramp rate | |
|---|---|---|---|---|
| Preincubation/denaturation | 95 | 10 min | 4.4 | |
| Denaturation | 95 | 10 s | 4.4 | |
| Annealing | 58 | 1 min | 2.2 | 40 cycles |
| Elongation | 72 | 1 s | 4.4 | |
| Cooling | 40 | 10 s | 1.5 | |

### *3.4 Real-Time PCR Assay*

1. In the pre-PCR laboratory, prepare the real-time PCR master mix as in Table 4.
2. Dispense 22 μL of PCR master mix to each well of a LightCycler multiwell plate 96 using a multiple channel pipette.
3. Transfer the Lightcycler plates to the post-PCR laboratory.
4. Add 3 μL of cDNA to each well except the wells of no-template negative control.
5. Seal the plate with a LightCycler® 480 sealing foil.
6. Centrifuge the plate briefly to mix the PCR components and collect the liquid to the bottom of the wells.
7. Place the plate on a LightCycler® 480 machine and run a cycling program as in Table 5.

### *3.5 Data Analysis*

Data analysis is performed using in-house-developed software tools. Detailed descriptions of the calculation methods can be found in refs. 4–7. The calculation starts from the $C_p$ values measured for a gene of interest (DENV), and a housekeeping gene used as reference (REF = β-actin). We first calculate the $\Delta C_p$ value for every well:

$$\Delta C_p = C_p(\text{DENV}) - C_p(\text{REF}).$$

Every plate contains a number of virus control wells, in which no inhibition is expected. Using the median value of the $\Delta C_p$ values of these wells as $R_0$, we calculate

$$\Delta\Delta C_p = \Delta C_p - R_0.$$

The well inhibition value ($I$) is determined with the formula below, and in the assumption that the PCR efficiency η equals 1:

$$I = 1 - (1+\eta)^{-\Delta\Delta C_p}.$$

To determine the $IC_{50}$, dose–response curve is analyzed by interpolation. If multiple replicates are present for a compound concentration, the median of these replicates is taken. If the median inhibition values are less than 50 % at all concentration, the highest concentration is reported with a ">" censor. If all inhibitions are above 50 %, the lowest concentration is reported with a "<" censor. If the value for a concentration is exactly 50 %, this is reported. Else the algorithm looks for an interval bounded by median inhibitions $I_l < 50$ % and $I_h > 50$ % and corresponding concentrations $C_l$ and $C_h$ so that $C_l < C_h$. The $IC_{50}$ is interpolated on this interval, using linear interpolation in the log-concentration domain:

$$IC_{50} = \exp\left(\ln(C_h) - \ln\left(\frac{C_h}{C_l}\right) \times \left(\frac{I_h - 0.5}{I_h - I_l}\right)\right).$$

The $IC_{50}$ must be uniquely defined, i.e., if more than one such interval is found, the curve is rejected. All dose–response curves reported for active compounds are reviewed by the operator.

## 4 Notes

1. According to the Belgian biosafety regulation, the handling and manipulation of the DENV are restricted in a biosafety level 3 laboratory. Once the infected cells are lysed, the reverse transcription and the real-time PCR steps can be performed in a regular laboratory.
2. All primers and probes were synthesized and HPLC purified by ID&T.
3. To avoid PCR contamination, good laboratory practice and guidelines for PCR must be followed. We designated different laboratories to be used for PCR: pre-PCR laboratory and

post-PCR laboratory. All RT and PCR reagents, such as primers, probes, reverse transcriptase, Taq polymerase, buffers, probe master, dNTPs, and DNase/RNase-free $H_2O$, are stored in the pre-PCR laboratory. The master mixes for RT and PCR are also prepared in the pre-PCR laboratory. Viral RNA or DNA including cDNA and PCR products are never allowed to enter the pre-PCR laboratory. RNase- and DNase-free filter pipette tips must be used in all steps. Pipettes are regularly irradiated using UV light in a mini PCR working station. Real-time PCR machines and addition of DNA or RNA templates are only allowed in the post-PCR laboratory.

4. For antiviral testing, we use medium with reduced serum concentration at 2 % (maintenance medium, MEM/2 % FCS) to reduce the protein binding of some compounds.
5. After washing with cold D-PBS, the plates can be sealed with aluminum sealing tape and stored at a −80 °C freezer overnight. On the next day, proceed with the lysis reaction.
6. To avoid bubble formation, mix with the pipette set at 35 μL and expel the solution without emptying the pipette tip completely.

## References


1. Halstead S (2007) Dengue. Lancet 370: 1644–1652
2. Whitehead S, Blaney J, Durbin A, Murphy B (2007) Prospects for a dengue virus vaccine. Nat Rev Microbiol 5:518–528
3. Gubler D, Kuno G, Markoff L (2007) Flaviviruses. In: Knipe D, Howley P et al (eds) Fields virology, 5th edn. Lippincott Williams & Wilkins, New York, pp 1153–1252
4. Schmittgen TD, Livak KJ (2008) Analyzing real-time PCR data by the comparative $C_T$ method. Nat Protoc 3(6):1101–1108
5. Pfaffl MW (2001) A new mathematical model for relative quantification in real-time RT-PCR. Nucleic Acids Res 29:e45
6. Livak KJ, Schmittgen TD (2001) Analysis of relative gene expression data using real-time quantitative PCR and the 2(-Delta Delta C(T)) Method. Methods 25:402–408
7. Pfaffl MW (2006) Relative quantification. In: Dorak MT (ed) Real-time PCR. Taylor & Francis, New York, pp 63–80

# Chapter 17

# High-Throughput Screening Using Dengue Virus Reporter Genomes


## Wolfgang Fischl and Ralf Bartenschlager



## Abstract

Large-scale antiviral drug discovery and systematic identification of host factors promoting or restricting virus replication require robust, scalable, and cost-effective assay systems that allow for high throughput and automation. Here we describe the construction and application of full-length infectious dengue virus reporter genomes for such screening applications.



**Key words** Dengue virus, Luciferase, High-throughput screening, Reporter virus, RNA interference


## 1 Introduction

Dengue is the most prevalent mosquito-borne viral disease causing estimated 50–100 million infections annually with a rapidly growing incidence over the past decades especially in tropical countries [1]. Dengue virus (DENV), the causative agent of the disease, is transmitted by mosquitoes of the genus *Aedes* to humans and accounts for most arboviral infections worldwide. In accordance to the prevalence of the vector, DENV is endemic in tropical and subtropical areas in Southeast Asia, the Pacific, Africa, Australia, and the Americas with more than 2.5 billion individuals being at risk of infection.

DENV is a positive-sense RNA virus belonging to the genus *Flavivirus* within the family *Flaviviridae*. There are four antigenically distinct serotypes and although primary infection is often asymptomatic or manifests as self-limiting dengue fever, secondary infections with a heterologous serotype are often accompanied with more severe courses such as dengue hemorrhagic fever (DHF) or dengue shock syndrome (DSS). Despite growing effort and an urgent medical need, there is currently no approved vaccine or antiviral therapy available for the prevention or treatment of DENV infection. Antiviral therapy can also be used for prevention.


Edwin Yunhao Gong (ed.), *Antiviral Methods and Protocols*, Methods in Molecular Biology, vol. 1030, DOI 10.1007/978-1-62703-484-5_17, © Springer Science+Business Media, LLC 2013

With the advent of state-of-the-art high-throughput screening systems, the development of therapeutic strategies combating dengue will be significantly facilitated and accelerated. Standard detection methods relying on the quantification of viral genomes or antigens such as plaque assay, cytopathic effect-reduction assay, or real-time reverse transcriptase polymerase chain reaction are often time consuming, semiquantitative, and/or cost intensive and, therefore, not well applicable to high-throughput applications. An alternative to these approaches is the use of a genetically engineered reporter virus system. It is based on the insertion of an easily measurable reporter-gene, encoding for luciferase or a fluorescent protein such as green fluorescent protein, into the viral genome, thus providing a surrogate marker for viral RNA replication and allowing fast detection and quantification. Here we describe the construction of such DENV reporter genomes and their application to high-throughput screening.

## 2 Materials

Prepare all solutions using ultrapure deionized water and analytical grade reagents. When working with RNA, always wear gloves and use RNase-free $H_2O$, reagents, and consumables.

Phenol and chloroform should be handled in a fume hood and diligently disposed according to waste disposal regulations.

### 2.1 Reagents

1. *Xba*I, *Sac*I, *Ngo*MIV, *Age*I, *Asc*I, and *Eco*RI restriction enzymes (NEB).
2. NEBuffer 4 (NEB).
3. 100× BSA (NEB).
4. Calf intestine phosphatase (NEB).
5. FideliTaq and 10× PCR buffer (USB).
6. 10 mM dNTP mix (Peqlab).
7. T4 DNA ligase (Fermentas).
8. Nucleospin® Gel and PCR Clean-up kit (Macherey-Nagel).
9. Nucleospin® Plasmid kit (Macherey-Nagel).
10. Nucleobond® PC500 kit (Macherey-Nagel).
11. $m^7G(5')ppp(5')G$ RNA cap structure analog (NEB); prepare 5 mM stock in RNase-free $H_2O$ and store at −20 °C.
12. RNasin® Ribonuclease inhibitor (40 U/μL, Promega).
13. SP6 RNA polymerase (20 U/μL, NEB).
14. RQ1 RNase-free DNase (1 U/μL, Promega).
15. 2 M sodium acetate, pH 4.5.
16. Water-saturated phenol, pH < 5.0; store at 4 °C.

17. Chloroform.
18. Isopropanol.
19. RNase-free water.
20. Native coelenterazine (P.J.K.): Prepare 1 mM stock in methanol and store at −70 °C; protect from light.
21. D-Luciferin (P.J.K): Prepare 1 mM stock in 25 mM glycyl-glycin and store at −70 °C; protect from light.
22. 1 M DTT; store at −20 °C.
23. 100 mM ATP (pH 7.8, adjust with KOH).
24. 70 % ethanol.
25. Competent *E. coli* DH5α.

### 2.2 Buffers and Media

1. 5× SP6 reaction buffer: 400 mM HEPES, 80 mM $MgCl_2$, 10 mM spermidin, 200 mM DTT in RNase-free $H_2O$; store at −20 °C.
2. rNTP-mix: 25 mM rATP, 25 mM rCTP, 25 mM rUTP, 12.5 mM rGTP (Roche), in RNase-free $H_2O$; store at −20 °C.
3. Growth medium: Dulbecco's modified Eagle's medium (Life technologies) supplemented with 10 % fetal calf serum (PAA), 2 mM L-glutamine, 1× nonessential amino acids, 100 U/mL penicillin, and 100 μg/mL streptomycin; store at 4 °C.
4. Phosphate-buffered saline (PBS).
5. Cytomix: 120 mM KCl, 0.15 mM $CaCl_2$, 10 mM potassium phosphate buffer (pH 7.6), 25 mM HEPES (pH 7.6), 2 mM EGTA, 5 mM $MgCl_2$; adjust the pH to 7.6 using KOH. Freshly add 2 mM ATP (pH 7.6, adjusted with KOH) and 5 mM glutathione. Store at room temperature.
6. Luciferase lysis buffer: 1 % (v/v) Triton X-100 (*see* **Note 1**), 10 % (v/v) glycerol, 25 mM glycyl-glycine (pH 7.8), 15 mM $MgSO_4$, 4 mM EGTA; store at 4 °C. Add 1 mM DTT directly before use.
7. Luciferase assay buffer: 25 mM glycyl-glycine (pH 7.8), 15 mM potassium phosphate buffer (pH 7.8), 15 mM $MgSO_4$, 4 mM EGTA. Store at room temperature (*see* **Note 2**).

### 2.3 Cell Lines

All cell lines are maintained in standard growth medium (*see* Subheading 2.2) at 37 °C and 5 % $CO_2$ and are routinely subcultured twice a week at a split ratio of 1:5–1:10.

1. BHK-21, baby hamster kidney cells (ATCC number: CCL-10).
2. Vero E6, African green monkey kidney cells (ATCC number: CRL-1586).
3. Huh-7, human hepatoma cells (JCRB number: JCRB0403) or Huh-7.*fluc* (*see* **Note 3**).

### 2.4 Equipment

1. Routine cell culture equipment: Class II biological safety cabinets, $CO_2$ incubators, tissue culture-grade plasticware, and consumables.
2. 384-well microplates, tissue culture treated, white, μClear (Greiner).
3. GenePulser® II electroporation system (Bio-Rad).
4. Multidrop Combi Reagent Dispenser (Thermo Scientific).
5. Matrix Hydra DT Pipetting Workstation (Thermo Scientific).
6. Mithras LB940 Multimode microplate reader (Berthold).
7. Centrifuge suitable for microtiter plates.

### 2.5 Oligonucleotides

The following oligonucleotides were used for cloning of the dengue reporter virus construct (restriction sites are underlined):

1. pFK-fwd: 5′-CAG CAT CCT GCG ATG CAG ATC CGG-3′.
2. 2A.MCS_rev: 5′-GCA TGT TAG AAG ACT TCC CCT GCC CTC GGC TCT ACC GGT GGC GCG CCC AGA GAT CTG CTC TCT AAT TAA AA-3′.
3. 2A_fwd: 5′-GGC AGG GGA AGT CTT CTA ACA TGC GGG GAC GTG GAG GAA AAT CCC GGG CCC ATG AAT AAC CAA CGG AAA AAG GCG-3′.
4. Capsid_rev: 5′-GTT CTC CGT TAC GTG TGG TTA AAT-3′.
5. C103.AscI_rev: 5′-TAA TGG CGC GCC GTG AGA ATCT CTT TGT CAG CTG TTG-3′.
6. AscI.Rluc_fwd: 5′-TTA AGG CGC GCC ATG ACT TCG AAA GTT TAT GAT CCA-3′.
7. Rluc.AgeI_rev: 5′-ATA CCG GTT TGT TCA TTT TTG AGA ACT CGC TC-3′.

## 3 Methods

### 3.1 Cloning of Infectious DENV Reporter cDNA Clones

The infectious cDNA clone pFK-DVs served as parental construct for cloning of the dengue reporter virus construct DENV-R2A (Fig. 1a). Plasmid pFK-DVs (Fig. 1b) was obtained by insertion of a synthetic copy of the full-length genomic sequence of DENV-2 strain 16681 [2] (Genbank Accession NC_001474) into the low-copy plasmid vector pFK [3]. For ease of cloning, the synthetic sequence of the 16681 strain (called DENV $16681^{syn}$) was modified (*see* **Note 4**). Moreover, the parental vector pFK was modified by insertion of the SP6 promoter upstream of the DENV 5′-NTR to allow synthesis of authentic viral RNA by in vitro transcription.

The DENV reporter plasmid pFK-DVs-R2A (Fig. 1c) was constructed using the following approach: A sequence encoding the 2A peptide of *Thosea asigna* virus [4] flanked by restriction

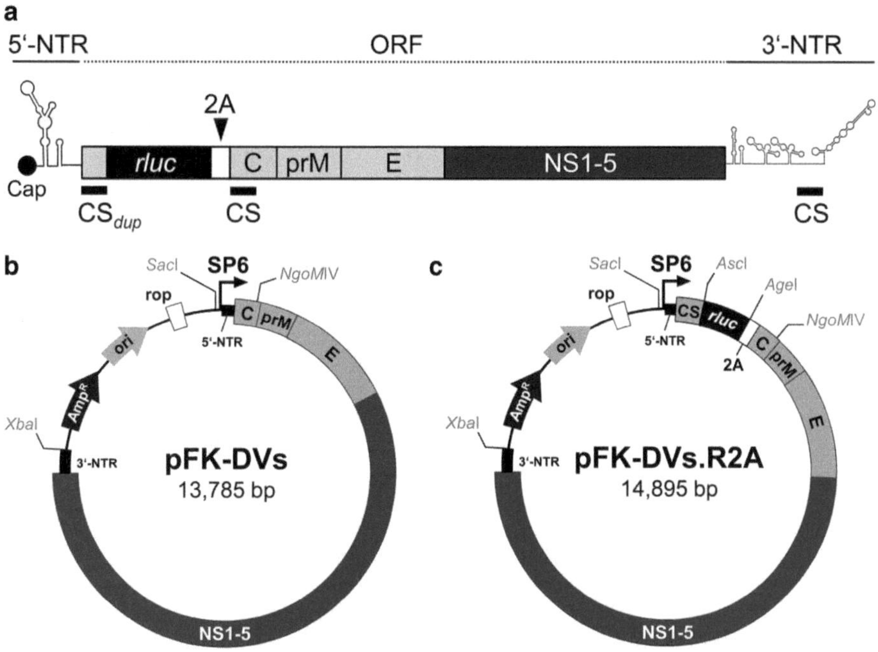


**Fig. 1** (**a**) Schematic representation of the DENV-R2A reporter virus genome organization. The first 103 nucleotides of the DENV capsid were duplicated ($CS_{dup}$) and placed upstream of the *Renilla* luciferase (*rluc*) gene to ensure proper long-range RNA–RNA interaction with the cyclization sequence (CS) at the 3′-NTR during replication. The reporter gene is followed by the 2A peptide of *Thosea asigna* virus to liberate the *Renilla* luciferase from the DENV polyprotein during/after translation. *NTR* non-translated region, *ORF* open reading frame. (**b**) Plasmid map of the parental DENV-2 strain 16681 cDNA clone pFK-DVs and (**c**) of the DENV reporter construct pFK-DVs-R2A. The approximate positions of restriction sites used in this protocol are presented. *SP6* SP6 promotor, *AmpR* β-lactamase gene (ampicillin resistance gene), *ori* origin of replication (pMB1 derived from pBR322), *rop* repressor of primer

sites for *Asc*I and *Age*I was inserted upstream of the capsid coding sequence, resulting in the intermediate plasmid pFK-DVs-2A. Next, a duplication of the first 103 nucleotides of the DENV capsid coding region was inserted upstream of the *Asc*I/*Age*I cloning site to ensure proper long-range RNA–RNA interactions during genome replication [5]. Finally, the *Renilla* luciferase gene was inserted via the *Age*I/*Asc*I restriction sites.

The complete cloning procedure is schematically outlined below and requires knowledge of standard molecular biology techniques, such as restriction digestion, ligation, and plasmid preparation, which are not given in full detail. All plasmids were propagated in the *E. coli* strain DH5α under standard growth conditions (*see* **Note 5**).

1. Prepare PCR "A": 50 ng plasmid pFK-DVs, 1× PCR buffer, 200 μM dNTPs, 0.4 μM forward and reverse primer each (pFK_fwd and 2A.MSC_rev), 0.2 U FideliTaq polymerase,

and $H_2O$ to 50 μL. Cycling conditions: 95 °C: 1.5 min → 15× [95 °C: 30 s (denaturation) → 52 °C: 30 s (annealing) → 68 °C: 30 s (elongation)] → 68 °C: 5 min → 4 °C: ∞.

2. Prepare PCR "B": 50 ng plasmid pFK-DVs, 1× PCR buffer, 200 μM dNTPs, 0.4 μM forward and reverse primer each (2A_fwd and Capsid_rev), 0.2 U FideliTaq polymerase, and $H_2O$ to 50 μL. Use same cycling conditions as described in **step 1**.
3. Separate PCR products via a 1.5 % preparative agarose gel, cut out the amplicons of correct size (PCR "A": 500 bp and PCR "B": 421 bp), and purify the DNA by using commercial spin columns. Elute DNA in 30 μL $H_2O$.
4. Prepare overlap-extension PCR "AB": 2 μL PCR "A," 2 μL PCR "B," 1× PCR buffer, 200 μM dNTPs, 0.4 μM forward and reverse primer each (pFK_fwd and Capsid_rev), 0.2 U FideliTaq polymerase, and $H_2O$ to 50 μL. Cycling conditions: 95 °C: 1.5 min → 15× [95 °C: 30 s (denaturation) → 52 °C: 30 s (annealing) → 68 °C: 60 s (elongation)] → 68 °C: 5 min → 4 °C: ∞.
5. Separate the PCR product (897 bp) on a 1.5 % preparative agarose gel and purify by using spin columns.
6. Digest 5 μg of plasmid pFK-DVs and 5–15 μL of the purified PCR product with *Sac*I and *Ngo*MIV in buffer NEB4 supplemented with 1× BSA for 1.5 h at 37 °C.
7. Add 2 μL of CIP to the restriction digest of the pFK-DV vector and incubate for 30–60 min at 37 °C.
8. Separate samples by preparative gel electrophoresis using a 1 % agarose gel. Recover and purify the pFK-DV vector (13.4 kb).
9. For insertion of the PCR fragment "AB" (499 bp) into pFK-DVs, mix vector and insert at a molar ratio of approx. 1:3 and incubate for 2 h at room temperature with T4 DNA ligase. The reaction mixture is used for transformation of *E. coli* DH5α and bacteria are streaked onto an LB-agar plate containing 100 μg/mL ampicillin.
10. Incubate plate overnight at 37 °C.
11. Inoculate ≥5 single colonies each into 5 mL LB medium containing 100 μg/mL ampicillin and incubate overnight at 37 °C.
12. Isolate plasmid from the complete medium of each bacterial culture using a DNA commercial plasmid purification system. Elute plasmid DNA in 30 μL $H_2O$.
13. Verify successful ligation and sequence fidelity by DNA sequencing.
14. We call this intermediate plasmid pFK-DVs.2A.

15. Prepare PCR "C": 50 ng plasmid pFK-DVs, 1× PCR buffer, 200 μM dNTPs, 0.4 μM forward and reverse primer each (pFK_fwd and C103.AscI_rev), 0.2 U FideliTaq polymerase, and $H_2O$ to 50 μL. Cycling conditions: 95 °C: 1.5 min → 15× [95 °C: 30 s (denaturation) → 52 °C: 30 s (annealing) → 68 °C: 30 s (elongation)] → 68 °C: 5 min → 4 °C: ∞.
16. Separate PCR product via a 1.5 % preparative agarose gel, cut out the amplicon of correct size (PCR "C": 568 bp), and purify the DNA by using commercial spin columns. Elute DNA in 30 μL $H_2O$.
17. Digest 1–5 μg (*see* **Note 6**) of plasmid pFK-DVs.2A and 5–15 μL of the PCR product with *Sac*I and *Asc*I in buffer NEB4 supplemented with 1× BSA for 1.5 h at 37 °C.
18. Add 2 μL CIP to the restriction digest of the vector pFK-DVs.2A and incubate for 30–60 min at 37 °C.
19. Separate samples by preparative gel electrophoresis (0.8 % agarose in case of the vector and 2 % agarose in case of the insert). Recover and purify 13.7 kb (vector) and 226 bp fragment (insert).
20. Ligate purified fragments using T4 DNA ligase using a molar ratio of vector:insert of approx. 1:3.
21. Proceed as described above in **steps 10–14**.
22. We call this intermediate plasmid pFK-DVs.C103.2A (*see* **Note 7**).
23. Prepare PCR "D": 50 ng plasmid pRL-SV40, 1× PCR buffer, 200 μM dNTPs, 0.4 μM forward and reverse primer each (AscI.Rluc_fwd and Rluc.AgeI_rev), 0.2 U FideliTaq polymerase, and $H_2O$ to 50 μL. Cycling conditions: 95 °C: 1.5 min → 15× [95 °C: 30 s (denaturation) → 52 °C: 30 s (annealing) → 68 °C: 60 s (elongation)] → 68 °C: 5 min → 4 °C: ∞.
24. Separate PCR product via a 1 % preparative agarose gel, cut out the amplicon of correct size (PCR "D": 953 bp), and purify the DNA by using commercial spin columns. Elute DNA in 30 μL $H_2O$.
25. Digest 1–5 μg of plasmid pFK-DVs.C103.2A and 5–15 μL of the PCR product with *Age*I and *Asc*I in buffer NEB4 for 1.5 h at 37 °C.
26. Add 2 μL of CIP to the vector restriction digest (pFK-DVs.C103.2A) and incubate for 30–60 min at 37 °C.
27. Separate vector digest by preparative gel electrophoresis (0.8 % agarose). Recover and purify the 13.7 kb fragment.
28. Purify digested PCR product by using commercial spin columns.

29. Ligate purified DNA fragment with the vector by using T4 DNA ligase and a molar ratio of vector:insert of approx. 1:3.
30. Proceed as described above in **steps 10–14**.
31. We call this plasmid pFK-DVs.R2A (Fig. 1c).
32. Retransform plasmid pFK-DVs.R2A into *E. coli* DH5α, use a single colony to inoculate 500 mL LB containing 100 μg/mL ampicillin, and grow overnight at 37 °C.
33. Isolate plasmid DNA from bacteria by using a commercial DNA plasmid purification kit (*see* **Note 8**).

### 3.2 In Vitro Transcription of RNA

1. Linearize plasmid pFK-DVs-R2A with *Xba*I (100 μL reaction): 10 μg plasmid DNA, 10 μL buffer NEB4, 1 μL 100× BSA, 5 μL *Xba*I (20 U/μL), and $H_2O$ to 100 μL. Incubate for 2 h at 37 °C.
2. Purify restriction reaction by using the NucleoSpin® Gel and PCR Clean-up kit. Elute linearized DNA in 41 μL of RNase-free $H_2O$.
3. Analyze a small aliquot of the DNA by agarose gel electrophoresis to verify complete restriction digest of the plasmid DNA.
4. Store DNA at −20 °C or proceed directly with in vitro transcription (IVT).
5. Prepare IVT reaction mix (*see* **Note 9**): 20 μL 5× SP6 reaction buffer, 12.5 μL rNTP mix (25 mM each ATP, CTP, UTP and 12.5 mM GTP), 20 μL $m^7G(5')ppp(5')G$ RNA cap structure analog (5 mM), 2.5 μL RNasin (40 U/μL), and 4 μL SP6 RNA polymerase (20 U/μL).
6. Incubate reaction for 2.5 h at 40 °C (*see* **Note 10**).
7. Add 2 μL (20 U/μL) SP6 RNA polymerase and incubate for another 2.5 h or overnight at 40 °C.
8. Add 20 μL of RNase-free DNase (1 U/μL) and incubate for 1 h at 37 °C.
9. Store RNA at −70 °C or proceed with RNA purification.
10. For RNA purification, add 60 μL of 2 M sodium acetate (pH 4.5), 440 μL of RNase-free $H_2O$, and 400 μL water-saturated phenol.
11. Mix thoroughly using a vortex mixer and incubate on ice for 10 min.
12. Centrifuge at 17,900 × *g* in a tabletop centrifuge for 10 min at 4 °C.
13. Transfer upper phase into a fresh tube and add 600 μL chloroform.
14. Vortex briefly and centrifuge at 17,900 × *g* for 3 min at room temperature.
15. Transfer upper phase into a fresh tube.

16. Add 0.7 volumes of isopropanol (equilibrated to room temperature) to precipitate the RNA and to remove unincorporated nucleotides.
17. Centrifuge at 17,900×*g* for 15 min at room temperature (*see* **Note 11**).
18. Wash pellet once with 70 % ethanol.
19. Centrifuge at 17,900×*g* for 10 min at room temperature.
20. Air-dry pellet briefly at room temperature.
21. Resuspend RNA in 100 μL RNase-free $H_2O$.
22. Incubate RNA for 5 min at 37 °C to ensure complete resuspension.
23. Quantify RNA by $OD_{260/280nm}$ measurement (*see* **Note 12**).
24. Verify integrity of RNA by agarose gel electrophoresis (*see* **Note 13**).
25. Store RNA at −70 °C.

### 3.3 Generation of Seed Virus Stocks

1. Harvest sub-confluent BHK-21 cells by trypsinization and resuspend in growth medium.
2. Count cells using a hemocytometer.
3. Pellet cells at 300×*g* for 5 min at room temperature.
4. Resuspend cells in sterile PBS (≥10 mL per 15 cm diameter dish).
5. Pellet cells at 300×*g* for 5 min at room temperature.
6. Resuspend cells in Cytomix at $1.5 \times 10^7$ cells/mL (*see* **Note 14**).
7. Combine 400 μL of cell suspension with 5–10 μg of in vitro-transcribed RNA (*see* **Note 15**).
8. Quickly transfer the mix to an electroporation cuvette (0.4 cm gap width).
9. Pulse cells once with 975 μF and 270 V using a Bio-Rad Gene Pulser II. Time constants should be between 18 and 20 ms.
10. Resuspend cells in 10 mL pre-warmed growth medium.
11. Repeat **steps 7–10** and combine cells from two electroporations. Seed the cells into a 15-cm-diameter cell culture dish and incubate for 12–16 h at 37 °C and 5 % $CO_2$.
12. Aspirate the medium to remove dead cells and debris and add 15 mL fresh growth medium containing 15 mM HEPES (*see* **Note 16**).
13. Incubate cells until a cytopathic effect (*see* **Note 17**) becomes apparent (usually between 48 and 72 h post transfection).
14. Harvest supernatant and clear from cell debris by filtration through a 0.45 μm syringe filter (*see* **Note 18**). Aliquot and store seed virus at −70 °C (*see* **Note 19**).
15. Determine virus titers by plaque assay or $TCID_{50}$ assay (*see* **Note 20**).

### 3.4 Production of Large-Scale Virus Stocks

1. Seed $6 \times 10^6$ Vero E6 cells per 15-cm-diameter dish and incubate overnight at 37 °C and 5 % $CO_2$.
2. Thaw aliquots of seed virus and adjust to $1 \times 10^5$ PFU/mL in growth medium. Prepare 10 mL of diluted seed virus per 15-cm-diameter dish of Vero E6 cells to be infected.
3. Aspirate medium from Vero E6 cells and wash once with sterile PBS.
4. Add 10 mL diluted seed virus per dish (corresponding to an MOI of approximately 0.1 PFU/cell).
5. Incubate for 2 h at 37 °C and 5 % $CO_2$ with gentle rocking.
6. Aspirate inoculum and wash cells twice with sterile PBS.
7. Add 15 mL growth medium supplemented with 15 mM HEPES.
8. Incubate at 37 °C and 5 % $CO_2$ for 4–6 days.
9. Collect supernatant and clear by filtration (0.45 μm syringe filter).
10. Aliquot virus stock and freeze at −70 °C.
11. Determine virus titer by plaque assay or $TCID_{50}$ assay (*see* **Note 20**).

### 3.5 High-Throughput Screening

This protocol has been designed for large-scale RNAi screening, but can also be easily adapted to small-molecule screening. SiRNAs are solid-phase reverse transfected [6] into Huh-7.*fluc* cells (*see* **Note 3**) and subsequently infected with DENV-R2A. Firefly luciferase (used as surrogate marker for cell number and viability) and *Renilla* luciferase (used as marker for DENV replication) are measured using a homemade dual-luciferase assay.

1. For each well of a 384-well microplate, add 1,200 Huh-7.*fluc* cells in a total volume of 50 μL growth medium by using a multidrop reagent dispenser. Use white 384-well microplates containing prespotted siRNAs in transfection reagent (*see* **Note 21**).
2. Spin plates briefly at 300 × *g* to ensure uniform distribution of cells.
3. Incubate plates for 72 h at 37 °C and 5 % $CO_2$ (*see* **Note 22**).
4. Thaw required amount of virus aliquots and adjust to $1.5 \times 10^6$ PFU/mL (*see* **Note 23**).
5. Aspirate medium from the cells using a Hydra DT pipetting workstation or a similar liquid-handling device.
6. Add 10 μL of the diluted virus inoculum to each well using a multidrop reagent dispenser or a similar liquid-handling system (*see* **Note 24**).
7. Incubate for 4 h at 37 °C and 5 % $CO_2$.

8. Add 40 μL growth medium to each well using a multidrop reagent dispenser.
9. Incubate for 44 h at 37 °C and 5 % $CO_2$ (*see* **Note 25**).
10. Aspirate medium and wash cells once with 50 μL PBS.
11. Aspirate PBS and add 20 μL luciferase lysis buffer freshly supplemented with 1 mM DTT.
12. Freeze plates once at −70 °C to ensure complete lysis. Plates can be stored frozen until further use.

### 3.6 (Dual-) Luciferase Assay

For dual-luciferase assay follow **steps 1–6**. If only *Renilla* luciferase activity is measured, directly proceed after **step 1** with **step 5** and omit the emission filter during measurement.

1. Allow plates to thaw at room temperature for 30–60 min.
2. To each well, add 30 μL luciferase assay buffer supplemented with 70 μM D-luciferin, 2 mM ATP, and 1 mM DTT using a multidrop reagent dispenser.
3. Incubate plates for 5 min in the dark (*see* **Note 26**).
4. Measure plates on a Mithras LB940 multimode reader using an integration time of 0.1 s per well.
5. To each well, add 30 μL luciferase assay buffer supplemented with 7.14 μM coelenterazine using a multidrop reagent dispenser. Proceed immediately to measurement.
6. Measure plates with a Mithras LB940 multimode reader using an integration time of 0.5 s per well and a 475 nm emission filter (*see* **Note 27**).

## 4 Notes

1. Triton X-100 significantly enhances coelenterazine autoluminescence. Due to biosafety regulations, we routinely use 1 % Triton X-100 to inactivate DENV, although the signal-to-noise ratio is inferior compared to luciferase lysis buffer containing only 0.1 % Triton X-100. If signal intensity is a concern, reduce the amount of Triton X-100 to 0.1 % if in accordance with your local biosafety regulations.
2. The luciferase assay buffer is usually stable for up to 3 weeks at room temperature. If a precipitate is visible, discard and prepare buffer freshly.
3. Huh-7.*fluc* cells were generated by lentiviral transduction of Huh-7 cells using a vector containing the firefly luciferase reporter gene. Luciferase activity can be measured as a surrogate marker for cell number and cell viability and thus allows for an additional normalization and more robust performance

of screens. Please note that a detailed protocol for the generation of this cell line is beyond the scope of this chapter.

4. To facilitate cloning the following modifications with respect to the deposited genomic sequence of the DENV-2 strain 16681 (Genbank Accession NC_001474) were introduced into our synthetic DENV $16681^{syn}$ genome: A402C (results in a unique *Ngo*MIV site in C); A1698G (generation of unique *SanD*I site in E, disruption of *Bam*HI site at nt 1,687, resulting in a unique *Bam*HI site at position 2,204); A2604G, A2605C, and A2607T (results in a unique *Mlu*I site in NS1 at position 2,603); A2679G (disruption of *Hpa*I site in NS1, resulting in a unique *Hpa*I site in NS5 at position 10,128); and C3492A (disruption of a *Sal*I site in NS2A, resulting in a unique *Sal*I site in C at position 166).

5. Full-length flavivirus cDNA clones are often considered as difficult to work with due to their instability during propagation in *E. coli* as a result of recombination events and poor yield of plasmid DNA. According to our experience the constructs described in this protocol are very stable. However, due to the low copy number of the plasmids, larger culture volumes are necessary to obtain sufficient DNA quantities for downstream applications (we recommend to perform plasmid maxipreps from 500 to 1,000 mL culture volumes). If recombination is a reoccurring issue, grow the bacterial cultures at lower temperature (30 °C).

6. As DNA yield is in general low with full-length DENV constructs, a standard plasmid mini-preparation from 1 to 10 mL bacterial culture might not yield sufficient amount of DNA for subsequent cloning steps. In this case, a larger plasmid preparation (e.g., from 100 mL culture volume) should be generated.

7. This intermediate plasmid can be used to insert various reporter genes via the *Age*I/*Asc*I cloning sites, such as luciferases (*Renilla* luciferase) or fluorescent markers (e.g., eGFP, RFP). Note that the fitness of reporter viruses can be dramatically impaired depending on the reporter gene (e.g., insertion of the 1.7 kb firefly luciferase gene is significantly less tolerated than the much smaller *Renilla* luciferase).

8. In spite of the high stability of the described plasmids, we observe sporadic recombination events during propagation of full-length infectious DENV clones in *E. coli*. Therefore, we highly recommend an analytical restriction digest of every new batch of a plasmid stock to ensure integrity of the construct. This is most easily achieved by using an enzyme with multiple restriction sites in the plasmid (e.g., *Eco*RI), thus resulting in a specific DNA fragment pattern.

9. Assemble the in vitro transcription reaction at room temperature. As an alternative to the homemade protocol, commercial transcription systems (e.g., Ambion MEGAScript® SP6) can be used. Adding the cap structure analog to the reaction is absolutely required to obtain functional viral RNA.
10. During the incubation period, the reaction often gets turbid due to the accumulation of pyrophosphate.
11. If the IVT reaction was successful, a clearly visible pellet should be observed after isopropanol precipitation.
12. A reaction with 10 μg DNA as template should typically yield 50–100 μg of in vitro-transcribed RNA.
13. Due to the high degree of secondary structure of RNA, a denaturing formaldehyde agarose gel must be used to achieve proper separation of the RNA. However, for routine verification of RNA integrity, a standard agarose gel is usually sufficient. Intact RNA should be visible as a distinct band whereas heavy smearing indicates degradation. Be sure to use a properly cleaned electrophoresis tank and freshly prepared buffer to exclude degradation that occurs during electrophoresis.
14. For each electroporation, $6 \times 10^6$ cells suspended in 400 μL Cytomix are needed. Prepare for each virus stock at least two electroporation reactions, which are then pooled and seeded into a 15-cm-diameter dish.
15. After mixing of the RNA with the cells, proceed to the electroporation immediately, since the RNA is prone to degradation by RNases present in the cell suspension.
16. Flaviviruses are highly sensitive to low pH, resulting in an irreversible conformational change of the envelope glycoprotein and thereby loss of infectivity. Always add 15 mM HEPES to the cell culture medium if viral supernatants shall be frozen, in order to avoid a pH-shift during freezing.
17. BHK-21 cells are especially sensitive to DENV-induced cytopathic effects, starting with rounding of cells 48–72 h post transfection. Cell culture supernatants should be harvested when the majority of cells start to detach.
18. As an alternative to filtration, supernatants can be cleared from cell debris by centrifugation. Large virus stocks can be easily clarified by filtration using 0.45 μm vacuum filtration systems (e.g., Millipore Steritop®). Do not use filtration devices with pore sizes below 0.45 μm, as this will result in a significant loss of infectivity.
19. Keep virus-containing supernatants on ice whenever possible. For storage longer than a few hours, DENV stocks should always be frozen at −70 °C.

20. DENV reporter viruses are slightly attenuated compared to the parental wild-type virus, resulting in reduced plaque size. Prolonged incubation of cells might therefore be necessary to obtain clearly visible plaques. Alternatively, virus titers can be determined by $TCID_{50}$ assay analogous to the protocol described by Lindenbach and coworkers [9]. Briefly, indicator cells (e.g., Vero E6 or Huh-7) are seeded into 96-well plates and subsequently infected with a tenfold serial dilution of the virus stock (typically 6–12 wells per dilution). After 3 days, cells are fixed with ice-cold methanol and immunostained for E protein. Positive wells (containing at least one cell expressing E protein) are recorded and the $TCID_{50}$ titer can be calculated according to [7, 8] or by using the Excel spreadsheet available for download from the University Heidelberg at http://www.klinikum.uni-heidelberg.de/Downloads.126386.0.html.
21. White multiwell plates give enhanced yield and at the same time minimize optical cross talk between neighboring wells. We recommend plates with a clear bottom for visual inspection of the cells, especially during assay setup.
22. The incubation time of 72 h after seeding was chosen to ensure efficient RNAi-mediated knockdown before infection and can be freely chosen according to the experimental requirements.
23. The optimal MOI should be determined empirically and is largely dependent on the cell line to be used and the desired time frame of the experiment. We usually recommend MOIs between 0.5 and 5 PFU/cell.
24. We noticed that especially in 384-well plates, infection is slightly more efficient when using small volumes of inocula. However, as this requires an additional handling step for replenishing growth medium, infection can also be carried out in larger volumes. In this case, adjust the MOI accordingly if infection rate drops significantly.
25. For a robust luciferase signal, an incubation of 48 h post infection is recommended. However, depending on the sensitivity of the cell line (BHK-21 > Vero > Huh-7: susceptibility in declining order) and the MOI used, earlier time points may be chosen, if necessary.
26. Incubation of plates with the assay buffer ensures stabilization of the firefly luciferase signal resulting in more homogenous readings.
27. By taking advantage of the different emission spectra of the two luciferases, a filter can be used instead of chemical quenching of the firefly luciferase signal. For *Renilla* luciferase measurements only, the filter should be omitted as it reduces the signal intensity.

## References


1. Guzman MG, Halstead SB, Artsob H et al (2010) Dengue: a continuing global threat. Nat Rev Microbiol 8:S7–S16
2. Kinney RM, Butrapet S, Chang GJ et al (1997) Construction of infectious cDNA clones for dengue 2 virus: strain 16681 and its attenuated vaccine derivative, strain PDK-53. Virology 230:300–308
3. Lohmann V, Korner F, Koch J, Herian U, Theilmann L, Bartenschlager R (1999) Replication of subgenomic hepatitis C virus RNAs in a hepatoma cell line. Science 285:110–113
4. Donnelly ML, Hughes LE, Luke G et al (2001) The 'cleavage' activities of foot-and-mouth disease virus 2A site-directed mutants and naturally occurring '2A-like' sequences. J Gen Virol 82:1027–1041
5. Alvarez DE, Lodeiro MF, Luduena SJ et al (2005) Long-range RNA-RNA interactions circularize the dengue virus genome. J Virol 79:6631–6643
6. Erfle H, Neumann B, Rogers P et al (2008) Work flow for multiplexing siRNA assays by solid-phase reverse transfection in multiwell plates. J Biomol Screen 13:575–580
7. Kärber G (1931) Beitrag zur kollektiven Behandlung pharmakologischer Reihenversuche. Arch Exp Pathol Pharmakol 162: 480–487
8. Spearman C (1908) The method of "right and wrong cases" ("constant stimuli") without Gauss's formulae. Br J Psychol 2:227–242
9. Lindenbach BD, Evans MJ, Syder AJ et al (2005) Complete replication of hepatitis C virus in cell culture. Science 309:623–666

# Chapter 18

# Fluorimetric and HPLC-Based Dengue Virus Protease Assays Using a FRET Substrate

**Christoph Nitsche and Christian D. Klein**


## Abstract

The number of dengue virus infections is increasing and the dengue NS2B–NS3 protease is considered a promising target for the development of antiviral therapies. Therefore, reliable and fast screening systems are needed for the discovery of new lead structures. In this chapter, we describe two dengue virus protease assays based on an internally quenched, high-affinity Förster resonance energy transfer (FRET) substrate ($K_m$ = 105 μM). A fluorimetric assay using a microtiter fluorescence plate reader can be used for high-throughput screening of a large number of compounds. Alternatively, an HPLC-based assay with fluorescence detection can be applied to confirm the compound hits and to avoid false-positive results that may arise due to the inner filter effect of some compounds.


**Key words** Dengue virus protease, NS2B–NS3, Microtiter plate assay, HPLC-based assay, Fluorimetric, FRET substrate, High-throughput screening

## 1 Introduction

Dengue fever and other diseases caused by the viruses in the *Flaviviridae* family are of special interest for ongoing research activities such as antiviral drug discovery and vaccine development. One of the pursued treatment strategies is to target the viral protease NS3 to inhibit the viral life cycle and virus replication [1]. To support this strategy, assay systems are necessary that allow a fast and reliable analysis of compounds in screening campaigns. The NS3 protein provides, apart from other functions, a serine protease functionality. This function is essential for the post- and co-translational cleavages of the viral polyprotein, which is initially formed in the host cell. For catalytic activity of NS3, the hydrophilic core sequence of the NS2B protein is required as a cofactor. Therefore, expression constructs of the protease are often hybrids of the protease domain of NS3 and the hydrophilic core sequence of NS2B, linked by a flexible sequence that usually consists of

Edwin Yunhao Gong (ed.), *Antiviral Methods and Protocols*, Methods in Molecular Biology, vol. 1030,
DOI 10.1007/978-1-62703-484-5_18, 

glycine and serine residues. The expression constructs are thus identified as " NS2B–NS3" or "NS2B–NS3pro".

The NS3 protease cleaves polypeptide sequences that bear two basic amino acids (Arg, Lys) in $P_1$ and $P_2$, with further privileged moieties found for the $P_4$–$P_1'$-sequence [2–4]. This property can be used for the development of peptide substrates, containing a fluorogenic or chromogenic group near the cleavage site, to generate a detectable signal induced by the cleavage of the substrate. Several screening systems based on peptide substrates with fluorogenic or chromogenic groups have been evaluated for dengue virus protease in recent years. These are *C*-terminal chromogenic groups that are located directly at the cleavage site such as *p*-nitroaniline (*p*NA) [2, 3] and thiobenzyl ester (SBzl) [4] or fluorogenic groups in the same position ($P_1'$) such as 7-amino-4-methylcoumarin (AMC) [2, 4–7] and 7-amino-3-carbamoylmethyl-4-methylcoumarin (ACMC) [4, 8].

The location of a bulky organic moiety in $P_1'$ directly at the cleavage site leads to the problem that the substrate is not similar to the natural or the preferred substrate sequence that bears further residues in $P_1'$–$P_4'$, which might be important for selective intermolecular recognition. To avoid this problem, various internally quenched FRET substrates using different donor–acceptor systems have been developed to assay dengue virus protease [5, 9–12]. The FRET donor–acceptor pair used for the method presented here is based on the anthranilamide donor (2-aminobenzoic acid as the unconnected fluorescent moiety) and the 3-nitrotyrosine acceptor [13], which has been successfully used in dengue virus protease substrates [9, 10, 12]. 3-Nitrotyrosine quenches most of the fluorescence of the anthranilamide donor as long as these two moieties are located within the same molecule (covalently linked by the substrate sequence) within the Förster radius [13–15]. The substrate described here comprises the sequence 2-Abz-Nle-Lys-Arg-Arg-Ser-(3-$NO_2$)-Tyr-$NH_2$ ($K_m$ = 105 μM), representing the recognition-relevant residues $P_4$–$P_1'$ with an *N*-terminal anthranilamide (2-Abz) and a *C*-terminal 3-nitrotyrosine moiety, respectively. By catalytic activity of the protease, the substrate is cleaved selectively between the arginine ($P_1$) and serine ($P_1'$) residues, resulting in two cleavage products. The fluorescence of the cleavage product 2-Abz-Nle-Lys-Arg-Arg-OH is no longer quenched by the 3-nitrotyrosine residue by FRET, resulting in increased fluorescence. If this fluorescence increase is monitored using a microtiter fluorescence plate reader, a large number of compounds can be screened towards their inhibition potential in a short time, as was shown for the dengue virus protease inhibitors described in refs. 16–19.

However, this assay system is prone to disturbance by substances whose absorption spectrum overlaps with the emission spectrum of the fluorogenic group, resulting in a lower detectable fluorescence, which is known as the "inner filter effect" [20].

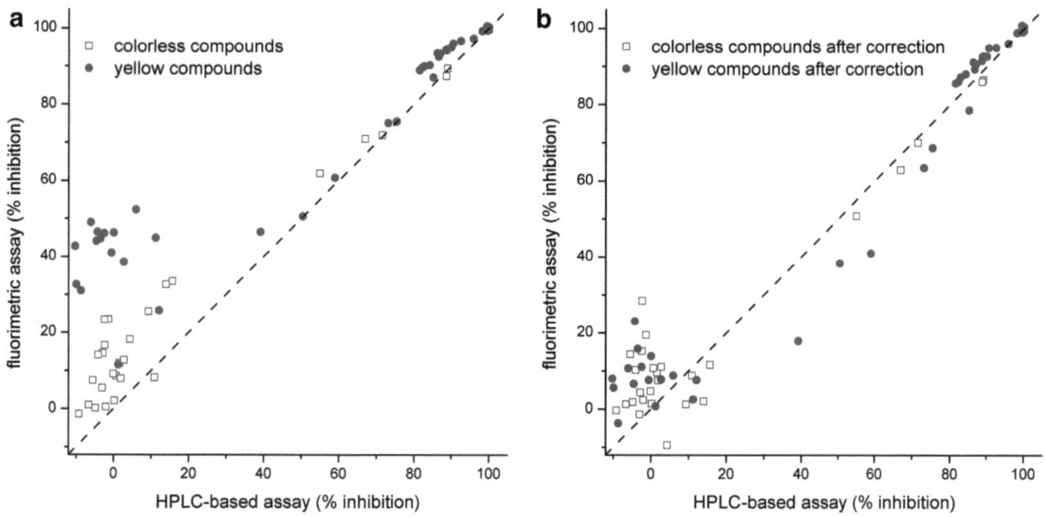


**Fig. 1** Comparison of the inhibition results of the two assay systems for selected colored and colorless compounds. The results of the fluorimetric screening assay (ordinate) are plotted against the HPLC-based results (abscissa) as %-inhibition. The dashed curves show the theoretically ideal correlation ($y = x$) between the two systems. (**a**) The results of both assay systems are shown without any correction. (**b**) The results of both assay systems are shown with consideration of the determined correction factors, due to the possible inner filter effect of the inhibitors, for the values obtained from the fluorimetric assay. In the lower inhibition range, the fluorimetric screening assay may show false-positive results, especially if the compounds are colored and thus interfere with the substrate absorption/emission process. In the higher inhibition range (>50 %) the correlation between the two assay systems without correction is as theoretically expected, independent of the color of the assayed compound

For the substrate system described here, especially yellow-colored inhibitors were found to give false-positive results due to absorbance of a fraction of the emitted light (Figs. 1 and 2). For lower inhibitory values (10–50 % at 50 μM substrate and 50 μM inhibitor), this effect has a more pronounced consequence than for higher potencies (Fig. 1). By quantifying this inner filter effect with a simple experiment using the fluorescent dye 2-aminobenzoic acid and the colored compound in the desired assay concentration, a correction factor can be calculated to reduce the risk of false-positive results (Figs. 1 and 2).

As the substrate itself is colored, due to the quenching moiety 3-nitrotyrosine, it is also able to absorb a part of the emitted light at higher concentrations in the form of an inner filter effect [20]. For a screening assay this additional effect does not need to be considered, because all results are calculated in relation to the uninhibited positive control, containing the same substrate concentration and thus resulting in an inner filter effect of the same range. However, for kinetic studies, where different substrate concentrations are used, the inner filter effect of the substrate has to be considered. Correction factors for different substrate concentrations (Table 1) can be determined using a simple protocol

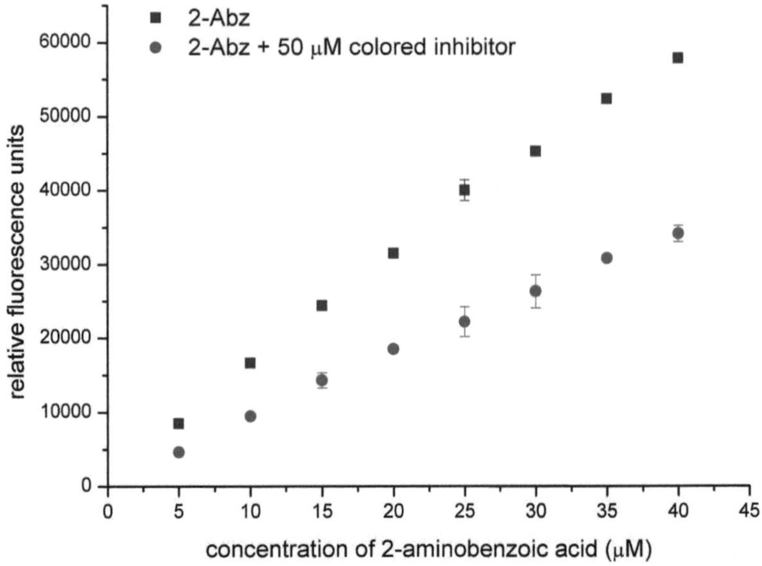


**Fig. 2** Fluorescence quenching, due to inner filter effect of a colored inhibitor, pretending a higher inhibition potential. At all analyzed concentrations of the fluorescent 2-aminobenzoic acid (2-Abz) a proportionally lower fluorescence is measured if a fluorescence-quenching inhibitor is added. In relation to the positive control (without colored inhibitor) a factor of $0.576 \pm 0.028$ for all measured values could be calculated for this compound. The reciprocal value of $1.742 \pm 0.086$ could be used as correction factor for the inner filter effect, to explain and correct possible false-positive results. Based on this linear correlation, it is adequate to use only one 2-Abz concentration to determine a correction factor for a compound, as done for all compounds shown in Fig. 1b

[20, 21]. As the inner filter effect is dependent not only on the wavelength and the concentration of the quenching compound but also on different technical parameters and instrumental configurations [20], it is recommended to recheck the given correction factors (Table 1) for the used system as described in literature [20]. The values observed with these corrections are in good correlation to the values that we found with the orthogonal HPLC-based assay so that kinetic parameters can be determined using the microtiter plate assay with inner filter correction.

However, an alternative HPLC-based assay system is recommended for all measurements beyond initial screening applications, because the HPLC-based assay is less prone to false-positive results and provides more reliable data for further applications. HPLC-based assays for dengue virus protease are used and have been described in literature [11, 22]. The assay system described here uses the same conditions and substrate as in the microtiter plate screening assay [21]. Consequently, both assay systems can be applied to a single 96-well microtiter plate to confirm the data.

**Table 1**
**Correction factors for the inner filter effect of the substrate at eight different substrate concentrations**

| Substrate concentration (μM)[a] | Correction factor[b] |
|---|---|
| 25 | 1.0492 |
| 50 | 1.0495 |
| 75 | 1.1784 |
| 100 | 1.3743 |
| 150 | 1.5357 |
| 200 | 1.6451 |
| 300 | 1.9582 |
| 400 | 2.3748 |

This correction has to be considered for every kinetic measurement, where the exact rate of turnover for a particular substrate concentration has to be determined. The apparent lower fluorescence increase has to be multiplied by the correction factor to observe the correct value. These values were calculated for this assay system according to the procedure described in the literature [20, 21]. For divergent technical equipment and parameters it is recommended to recheck the given correction factors for the used system [20]
[a]The substrate concentrations are based on the molecular weight of 985.1 g/mol without the consideration of possible TFA adducts
[b]In literature these factors are mostly found as their reciprocal values [20, 21]

For example, only the "hits" identified in the fluorimetric screening assay can be sampled for HPLC. The HPLC-based assay also employs fluorescence detection, but, in contrast to the fluorescence-based microtiter plate screening assay, spectroscopic interferences are not relevant to the inhibition results. Furthermore, inhibitory compounds with autofluorescence can be assayed, which is practically impossible with the homogeneous, fluorimetric assay system. The fluorescent substrate cleavage product 2-Abz-Nle-Lys-Arg-Arg-OH is detected as a distinct peak and can be quantified by integration of the HPLC chromatogram (Fig. 3). Disadvantages of the HPLC-based assay are the relatively low throughput, the solvent consumption, and a larger effort for data evaluation. For example, with the protocols provided here, the homogeneous fluorimetric assay can be carried out for a 96-well plate within 15 min, whereas the HPLC assay requires up to one day for the same number of samples.

The buffer used for the protease assay is based on ethylene glycol as polyol and Brij 58® as nonionic detergent at pH 9 (Tris–HCl), previously identified as optimal conditions [10]. The substrate synthesis follows the Fmoc-based protocol for solid-phase peptide synthesis [10, 23, 24].

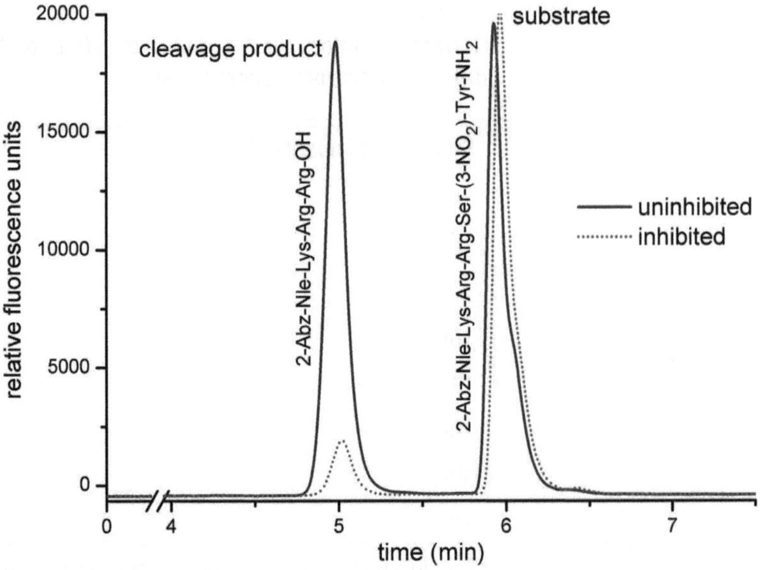


**Fig. 3** Exemplary chromatogram of the HPLC-based assay system using fluorescence detection. The first peak corresponds to the cleavage product of the enzymatic reaction. The second peak corresponds to the uncleaved substrate. For the evaluation of the inhibitory potential only the first peak area is relevant and analyzed by integration. The smaller this peak area is in comparison to the uninhibited control, the higher the inhibitory potency of the analyzed compound

## 2 Materials

### *2.1 FRET Substrate Synthesis*

1. Disposable syringe (5 mL) with bottom frit (*see* **Note 1**) and short cannula with cap.
2. Rink amide resin (loading capacity ~0.65 mmol/g).
3. Dichloromethane (DCM).
4. Dimethylformamide (DMF) (*see* **Note 2**).
5. Piperidine solution: 25 % piperidine in DMF.
6. *N*,*N*-Diisopropylethylamine (DIPEA).
7. *O*-(Benzotriazol-1-yl)-*N*,*N*,*N*′,*N*′-tetramethyluronium hexafluorophosphate (HBTU).
8. Protected amino acids: Fmoc-2-Abz-OH, Fmoc-Nle-OH, Fmoc-Lys(Boc)-OH, Fmoc-Arg(Pbf)-OH, Fmoc-Ser(*t*Bu)-OH, Fmoc-(3-$NO_2$)-Tyr-OH (*see* **Note 3**).
9. Cleavage solution: 92.5 % trifluoroacetic acid (TFA), 5.0 % triisopropylsilane (TIPS), 2.5 % deionized water. Prepare immediately before use and store at 4 °C (*see* **Note 4**).

10. Cold diethyl ether (store before use at –20 °C).
11. Degassed HPLC-grade methanol (containing 0.1 % TFA) and degassed deionized and filtered water (containing 0.1 % TFA).
12. Preparative HPLC system with preparative RP-18 column (*see* **Note 5**).

### 2.2 Fluorimetric 96-Well Microtiter Plate Assay

1. Microtiter fluorescence plate reader.
2. Black 96-well V-bottom plates.
3. Assay buffer: 50 mM Tris–HCl, pH 9, ethylene glycol (10 % v/v), Brij® 58 (0.0016 %). To prepare 200 mL assay solution, weigh 1.22 g Tris in a 250 mL glass bottle and add 180 mL deionized water, 20 mL ethylene glycol, and 3.6 mL of a solution of 0.8 mM Brij® 58 (10×CMC) (*see* **Note 6**). After a clear solution is obtained, adjust the pH with 1 M hydrochloric acid to a value of 9.0 (*see* **Note 7**). The assay buffer should be stored at 4 °C and can be used for up to 60 days.
4. Dengue virus protease (*see* **Note 8**): Cloning, expression, and purification procedures for the NS2B–NS3 protease are available in the literature [2–7, 10, 25].
5. 2 μM enzyme solution in assay buffer: The solution should be freshly prepared, mixed carefully, and stored on ice. For concentration determination 10 μL of the aliquot protease stock solution are added to 990 μL assay buffer in a standard quartz cuvette (path length = 10 mm) and the absorption at 280 nm in relation to buffer solution (without protein) is measured using a UV spectrophotometer (*see* **Note 9**).
6. 500 μM substrate solution: Weigh 1–2 mg of the freeze-dried substrate and add the related amount of assay buffer to obtain a 500 μM substrate solution, which will have an orange color (*see* **Note 10**).
7. 10 mM compound solution(s): Weigh the compound(s) in small vials to prepare 10 mM stock solutions in dimethyl sulfoxide (DMSO) or deionized water (depending on the solubility of the compound).
8. 100 μM 2-aminobenzoic acid in assay buffer, if correction factors for possible false-positive results should be determined (molecular weight: 137.1 g/mol).

### 2.3 HPLC-Based Assay

1. HPLC system with 96-well-plate auto sampler and fluorescence detector (*see* **Note 11**).
2. Degassed HPLC-grade acetonitrile (containing 0.1 % TFA).
3. 4 % TFA solution in deionized water.
4. Aluminum foil

## 3 Methods

### 3.1 FRET Substrate Synthesis

Consider that, at usual operating speed, 3 days are required for the substrate synthesis and purification. The following procedure describes the synthesis of 10–20 mg of FRET substrate. If higher amounts are desired, this protocol can easily be scaled up without any changes in the synthetic methodology.

1. Weigh five equivalents of the protected amino acid and HBTU into the same reaction vessel (2 mL capacity). Arrange the tubes in the following order: (a) 73 mg Fmoc-(3-$NO_2$)-Tyr-OH + 62 mg HBTU, (b) 63 mg Fmoc-Ser(*t*Bu)-OH + 62 mg HBTU, (c) 106 mg Fmoc-Arg(Pbf)-OH + 62 mg HBTU, (d) 106 mg Fmoc-Arg(Pbf)-OH + 62 mg HBTU, (e) 77 mg Fmoc-Lys(Boc)-OH + 62 mg HBTU, (f) 58 mg Fmoc-Nle-OH + 62 mg HBTU, and (g) 59 mg Fmoc-2-Abz-OH + 62 mg HBTU.
2. Weigh 50 mg of the Rink amide resin into the syringe with bottom frit after removing the plunger.
3. Add the plunger again and swell the resin by aspirating about 2 mL DCM through the cannula into the syringe. Add the cannula cap and shake the syringe for 15 min (*see* **Note 12**).
4. Dispense the solvent into the waste and wash the resin 3× with DMF by aspirating and dispensing the solvent into and out of the syringe.
5. To remove the Fmoc group on the resin, aspirate about 2 mL of the 25 % piperidine solution, shake for 10 min, and dispense the solution into waste. Repeat the rinsing procedure and shake for another 5 min before dispensing the solution into waste.
6. Meanwhile add 1 mL DMF to the first coupling mixture (a) and mix it till nearly all solid is dissolved.
7. Wash the resin 3× with DMF, 3× with DCM, and again 3× with DMF, by aspirating and dispensing the solvent into and out of the syringe.
8. Directly before use add around 60 μL DIPEA (ten equivalents) to the coupling solution and mix it one more time. If there is still a lot of undissolved material, add another 0.5 mL DMF.
9. Aspirate the coupling solution into the syringe and shake for 1 h.
10. Dispense the coupling solution into the waste and wash the resin 3× with DMF, 3× with DCM, and again 3× with DMF.
11. Repeat **steps 5–10** for the coupling **steps** (**b–g**).
12. Remove the last Fmoc group after the last coupling step is finished as described in **step 5**.

13. Wash the resin 3× with DMF, 3× with DCM, and 3× with diethyl ether.
14. Dry the resin in the syringe under reduced pressure for at least 1 h or overnight.
15. To cleave the substrate from the resin, aspirate 1 mL of the freshly prepared cleavage solution into the syringe with the dried resin and shake for 2–3 h (*see* **Note 13**).
16. Add 20 mL of cold diethyl ether to a 30 mL plastic tube with screw-cap and dispense the cleavage solution from the syringe into the plastic tube. The substrate will precipitate as a pale yellow solid. Store the tube at –20 °C to complete precipitation.
17. Repeat the cleavage procedure (**step 15**) with 0.5–1 mL cleavage solution for another hour to complete the process. Dispense the solution into the same plastic tube with cold diethyl ether and the substrate.
18. Centrifuge the precipitate at 4,000 × *g* for 5 min. Decant the solution from the pellet, add again 20 mL of cold diethyl ether suspense the pellet, and centrifuge/decant again. Repeat this procedure for a third time and dry the pellet in vacuum for at least 30 min or overnight.
19. For purification using preparative HPLC, dissolve the crude substrate in deionized and filtered water (*see* **Note 14**). Use methanol (0.1 % TFA) and water (0.1 % TFA) as solvents and equilibrate the RP-18 column to 10 % methanol. Inject the peptide solution and slowly increase the methanol content to 100 % (*see* **Note 15**). UV detection may be performed at 214, 254, and 280 nm. The substrate can be detected at all three wavelengths and elutes between 55 and 65 % methanol.
20. Evaporate the methanol (for example with a rotary evaporator), convert the aqueous solution into a 15 mL plastic tube with screw-cap, and freeze it in liquid nitrogen for 15 min.
21. Prick some small holes into the screw cap on the tube and freeze-dry the substrate overnight (*see* **Note 16**). Seal the tube (for example with plastic film) and store the freeze-dried substrate at –20 °C.

### 3.2 Fluorimetric 96-Well Microtiter Plate Assay

The following procedure describes the dengue virus protease assay for compound screening using 50 μM substrate, 50 μM test compound, and 100 nM protease in a total volume of 100 μL. For kinetic studies or $IC_{50}$-value determination, it is necessary to adapt the concentrations of the substrate and test compounds and to consider the correction factors for the inner filter effect of the substrate (Table 1) during data processing. Using the following procedure, 23 compounds can be screened towards their inhibition potential (%-inhibition) in triplicate in one 96-well plate (*see* **Note 17** and Table 2).

**Table 2**
**Exemplary approach for the arrangement of the test compound- and assay-solutions in a 96-well plate**

| | 1 | 2 | 3 | 4 | 5 | 6 | 7 | 8 | 9 | 10 | 11 | 12 |
|---|---|---|---|---|---|---|---|---|---|---|---|---|
| A | I-1 | I-1 | I-1 | I-9 | I-9 | I-9 | I-17 | I-17 | I-17 | I-1 | I-9 | I-17 |
| B | I-2 | I-2 | I-2 | I-10 | I-10 | I-10 | I-18 | I-18 | I-18 | I-2 | I-10 | I-18 |
| C | I-3 | I-3 | I-3 | I-11 | I-11 | I-11 | I-19 | I-19 | I-19 | I-3 | I-11 | I-19 |
| D | I-4 | I-4 | I-4 | I-12 | I-12 | I-12 | I-20 | I-20 | I-20 | I-4 | I-12 | I-20 |
| E | I-5 | I-5 | I-5 | I-13 | I-13 | I-13 | I-21 | I-21 | I-21 | I-5 | I-13 | I-21 |
| F | I-6 | I-6 | I-6 | I-14 | I-14 | I-14 | I-22 | I-22 | I-22 | I-6 | I-14 | I-22 |
| G | I-7 | I-7 | I-7 | I-15 | I-15 | I-15 | I-23 | I-23 | I-23 | I-7 | I-15 | I-23 |
| H | I-8 | I-8 | I-8 | I-16 | I-16 | I-16 | pos. | pos. | pos. | I-8 | I-16 | pos. |
| | Triplicate assay solutions (100 μL) with the final concentrations of test compound (50 μM), enzyme (100 nM), and substrate (50 μM) | | | | | | | | | 1 mM compounds in 50 μL assay buffer | | |

Columns 10, 11, and 12 are used for the preparation of 23 compound (inhibitor) solutions (I-1 … I-23) and the uninhibited positive control without test compound (pos.). Columns 1–9 are then used for pipetting the assay solutions in triplicate, including inhibitors, enzyme, and substrate at their final concentration in a final volume of 100 μL

1. Add 5 μL of the 10 mM test compound stock solutions to the corresponding columns 10–12 according to the layout in Table 2. Add 5 μL DMSO to the well of the positive control without test compound (uninhibited reference specimen).
2. Add 45 μL of the assay buffer to these wells and mix to make 1 mM compound solutions (*see* **Note 18**).
3. Pipet 3 × 5 μL of the 1 mM compound solution and the reference specimen (positive control without test compound) into the wells on the left of the plate and add 80 μL of assay buffer (Table 2).
4. Add 5 μL of the protease solution and incubate at 20 °C for 15 min.
5. Meanwhile prepare the microtiter fluorescence plate reader for the assay. The excitation wavelength for the FRET pair used here is 320 nm and the emission wavelength is 405 nm. The activity of the enzyme is recorded as relative fluorescence units per second (RFU/s) over 15 min (*see* **Note 19**).
6. After incubation, add simultaneously, and as fast as possible, 10 μL of the substrate solution to every assayed well, place the plate in the microtiter fluorescence plate reader, and start recording the cleavage reaction for 15 min.

7. For evaluation of the results calculate the average slope per second and the standard deviation for every inhibitor/enzyme combination and calculate the %-inhibition in proportion to the uninhibited control (*see* **Note 20**).
8. If correction factors, due to possible false-positive results, for compounds should be determined, use the 1 mM inhibitor solutions (Table 2) in assay buffer and pipet three times (triplicate determination) 5 μL of these solutions into other wells of a black microtiter plate (final concentration 50 μM). Add 90 μL of assay buffer and 5 μL of 2-aminobenzoic acid solution (final concentration 5 μM). For the positive control (without inner filter effect) use 95 μL of buffer, 5 μL of 2-aminobenzoic acid solution, and no inhibitor. Measure the absolute fluorescence values for all samples (same conditions as used before; **step 5**), average the triplicate values, and calculate the correction factors (deviation) in relation to the positive control (Fig. 2) [20]. If a significantly lower fluorescence is obtained for an inhibitor, this compound might be prone to false-positive results for this assay.

### 3.3 HPLC-Based Assay

This assay can be used as stand-alone method, but may also be applied to a plate that has previously been measured in the homogeneous, fluorimetric assay, in order to confirm inhibitory "hits". Although the following procedure is described for use of a 96-well plate auto sampler, alternative sampling systems can be used in an analogous way, for example, by transferring the solutions from the 96-well plate into HPLC sample vials.

1. Execute the general assay procedure as described in Subheading 3.2, **steps 1–6**.
2. Quench the enzymatic reaction after 15 min by simultaneous addition of 10 μL of 4 % TFA solution to obtain a final TFA concentration of 0.36 % (*see* **Note 21**).
3. Cover the plate with aluminum foil and store it for at least 15 min at 4 °C before starting the HPLC procedure by placing the plate into the auto sampler (*see* **Note 22**).
4. Use an HPLC gradient with acetonitrile (0.1 % TFA) and water (0.1 % TFA) as solvents, starting at 10 % and ending at 95 % of acetonitrile, before preparing for the next run with 10 % acetonitrile again. Set up the fluorescence detector to an excitation wavelength of 320 nm and an emission wavelength of 405 nm. In ideal conditions, two well-separated peaks representing the cleavage products and the uncleaved substrate should be obtained, before 55 % of acetonitrile are reached (Fig. 3). Some optimization steps may be needed for a different HPLC system (*see* **Note 23**). Measure all samples in triplicate using identical conditions.

5. Before or after the measurement of the assay solutions it is recommend to measure a "blank" sample in triplicate containing only 50 μM substrate in the assay buffer without enzyme (*see* **Note 24**).
6. For evaluation of the data, integrate all peak areas of the cleavage product in the same way, average the triplicate determinations, and calculate the standard deviations. Calculate the %-inhibition in proportion to the uninhibited control (*see* **Note 25**).

## 4 Notes

1. Solvent-resistant frits of suitable size can be purchased from different providers or trepanned from larger frits. The only important aspect is that the resin does not leak through the syringe via the outlet.
2. *N*-methyl-2-pyrrolidone (NMP) can alternatively be used as solvent for the same protocol.
3. 2-Abz-OH and (3-$NO_2$)-Tyr-OH stand for 2-aminobenzoic acid and 3-nitrotyrosine, respectively. All mentioned protected amino acids are commercially available.
4. A good mixing procedure is required for this solution. After some hours a phase separation may occur.
5. We use an ÄKTA® Purifier system (GE Healthcare, Germany) with an RP-18 pre- and main column (Reprospher, Dr. Maisch GmbH, Germany, C18-DE, 5 μm, 30 × 16 mm, and 120 × 16 mm).
6. To obtain a 0.8 mM Brij® 58 solution with a tenfold critical micelle concentration (10 × CMC) dissolve 90 mg Brij® 58 in 100 mL deionized water.
7. A calibrated pH electrode and a magnetic stirrer should be used to obtain an exact pH value.
8. We use the dengue virus serotype 2 NS3 protease domain (first 184 amino acids of NS3) connected via a flexible linker of eight glycine and one serine residues (GGGGSGGGG) to the hydrophilic core sequence (40 amino acids) of the NS2B cofactor (CF40.gly.NS3pro of Leung and coworkers) [3, 10]. Purification is performed by $Ni^{2+}$ affinity chromatography of the 6× His-tagged protein [3, 10]. 50 μL stocks of the purified protein are stored at −80 °C in 100 mM Tris–HCl, pH 7.9, 50 mM NaCl in deionized water, and glycerol (50 % v/v). Under these conditions the protein can be stored for 6 months with only a moderate loss of activity.

9. The extinction coefficient depends on the sequence of the protein. For our protease construct (*see* **Note 8**) we calculated (ExPASy ProtParam tool) an extinction coefficient of 41,940 L/(mol·cm). The extinction of the photometric measurement should ideally be between 0.1 and 0.2.
10. The molecular weight of the substrate is 985.1 g/mol. However, the substrate isolated by the described procedure may have up to four equivalents of TFA as adduct at the four basic groups. Nevertheless we are only calculating with the lower mass, because the exact number of TFA equivalents is difficult to determine and may depend on different parameters of the isolation process.
11. We use a Jasco HPLC system with an FP-2020 plus fluorescence detector, PU-980 HPLC pump, and an AS 950-10 auto sampler with an RP-18 column Phenomenex® Luna C18(2) (5 μM, 150 × 3 mm).
12. We use a laboratory shaker GFL-3005 (300 rpm, orbital motion) with a rack for plastic tubes, where the syringe(s) can be placed while shaking. To avoid losses of the solutions out of the cannula, it is necessary to aspirate some air after the solution is aspirated and attach the cannula cap carefully.
13. Make sure that the syringe is closed well. Otherwise the substrate, which is no longer attached to the resin, may be lost.
14. The amount of substrate purified by one preparative HPLC cycle depends on the column and on the sample loop one uses. Using our previously described column (*see* **Note 5**) equipped with a 1 mL sample loop, the substrate (10–20 mg) is dissolved in 1 mL water and injected twice (2 × 0.5 mL).
15. For our preparative HPLC system (*see* **Note 5**) we use the following conditions: eluent A: water (0.1 % TFA), eluent B: methanol (0.1 % TFA), flow rate: 8 mL/min, and total time: 30 min. The run is initiated with an equilibration at 10 % B (duration: 2.5 min). This is followed by a linear ramp to 100 % B within 21 min. 100 % B is maintained for 2.5 min, and followed by a switch to 10 % B, which is then maintained for re-equilibration for 4 min.
16. For the lyophilization a freeze drier is preferred. If that is not available, a high-vacuum pump with a liquid nitrogen-filled cryo trap can be used.
17. In a recommended procedure, 3 of the 12 columns of a 96-well plate are used for pipetting and diluting 23 test compounds and the uninhibited control (Table 2). The other nine columns are used for the proper assay execution with three adjacent wells bearing the same compound solution or the uninhibited control for a triplicate determination (Table 2). This approach is expedient

for the use of 8-channel pipettes. If the compounds are diluted in a separate plate, a maximum of 31 compounds can be screened in triplicate using one 96-well plate. According to this procedure, an alternative plate format for the determination of $IC_{50}$ values (e.g., three compounds per plate) is as follows: seven or eight different dilutions of one compound per column and one well for the positive control without inhibitor (dilutions in columns 10–12) are measured in triplicate (columns 1–9) as described in Subheading 3.2 (Table 2). An exemplary plate format for $K_i$ and $K_m$ determinations can use eight different substrate concentrations in rows A–H (e.g., 25, 50, 100, 150, 200, 300, 400 μM; correction factors in Table 1) and various test compound concentrations (including a blank, 0 μM) in duplicate or triplicate in the orthogonal columns (1–12 or less). If $K_i$-values should be determined, it can also be expedient to use a contrary plate format with eight different test compound concentrations (rows A–H) and a few substrate concentrations in duplicate or triplicate (columns 1–12 or less). The respective benefits of these two possible exemplary approaches depend on the kinetic data evaluation plot that is used (Lineweaver–Burk vs. Dixon).

18. Mixing of the assay solutions is achieved by aspiration and dispensing the solutions with a pipette. Alternatively mix the well contents by circular motions with the pipette tip.
19. We use a BMG Labtech Fluostar OPTIMA microtiter fluorescence plate reader.
20. An example: If the averaged value for one determined inhibitor is $4 \pm 0.5$ RFU/s and the averaged value for the uninhibited reference is found to be $10 \pm 0.5$ RFU/s, then one typically observes an inhibition potential of $60 \pm 5$ % for the compound.
21. To obtain reliable results it is important that the time interval from adding the substrate to the quenching procedure by TFA is nearly identical for the enzyme/inhibitor testing solutions and for the uninhibited enzyme reference sample (positive control). To ensure this, it is recommended to add the substrate and the TFA solutions in the same direction with the same pipetting approach over the plate (for example, with an 8-channel pipette or robot from the left to the right).
22. After the enzyme reaction is quenched, the plate can be stored for up to 48 h at 4 °C, before it is added to the HPLC system for measurements.
23. With our HPLC system (*see* **Note 11**) we use the following conditions: injection volume: 35 μL, eluent A: water (0.1 % TFA), eluent B: acetonitrile (0.1 % TFA), flow rate: 1 mL/min, and total time: 15 min. The initial conditions are as follows: 10 % B for 1 min. This is followed by a linear ramp to

55 % B (within 8.5 min), followed by a switch to 95 % B, which is maintained for 3 min. Then a switch to 10 % B is performed, which is maintained for 2.5 min.

24. Using this method one can ensure that no cleavage product peak is obtained without enzyme activity. If there is a small peak, where normally the cleavage product is located, it is important to quantify this peak area and subtract this value from all other measured cleavage product peak integrals. Without this correction, a value of 100 % inhibition cannot be obtained and all activities appear too low.
25. An example: If the averaged integral of the cleavage product peak is 692 ± 73 in the inhibitor-containing experiment and the averaged integral for the uninhibited reference is found to be 3,367 ± 149, then you have observed an inhibitory potential of 79.4 ± 2.2 % for this compound. If for the "blank" sample a small peak with an averaged integral of 112 is found, this value has to be subtracted from all peak areas, leading to the corrected inhibitory activity of 82.2 ± 2.2 %.

## Acknowledgment

The basic principles of the assay system reported here were developed by Dr. Christian Steuer [10, 21]. We thank Michael Wacker for technical assistance and solid-phase peptide synthesis. Christoph Nitsche thanks for the support by a fellowship of the Studienstiftung des deutschen Volkes.

### References

1. Lescar J, Luo D, Xu T et al (2008) Towards the design of antiviral inhibitors against flaviviruses: the case for the multifunctional NS3 protein from dengue virus as a target. Antiviral Res 80:94–101
2. Yusof R, Clum S, Wetzel M et al (2000) Purified NS2B/NS3 serine protease of dengue virus type 2 exhibits cofactor NS2B dependence for cleavage of substrates with dibasic amino acids in vitro. J Biol Chem 275:9963–9969
3. Leung D, Schroder K, White H et al (2001) Activity of recombinant dengue 2 virus NS3 protease in the presence of a truncated NS2B co-factor, small peptide substrates, and inhibitors. J Biol Chem 276:45762–45771
4. Li J, Lim SP, Beer D et al (2005) Functional profiling of recombinant NS3 proteases from all four serotypes of dengue virus using tetrapeptide and octapeptide substrate libraries. J Biol Chem 280:28766–28774
5. Gouvea IE, Izidoro MA, Judice WAS et al (2007) Substrate specificity of recombinant dengue 2 virus NS2B-NS3 protease: influence of natural and unnatural basic amino acids on hydrolysis of synthetic fluorescent substrates. Arch Biochem Biophys 457:187–196
6. Niyomrattanakit P, Winoyanuwattikun P, Chanprapaph S et al (2004) Identification of residues in the dengue virus type 2 NS2B cofactor that are critical for NS3 protease activation. J Virol 78:13708–13716
7. Shiryaev SA, Ratnikov BI, Aleshin AE et al (2007) Switching the substrate specificity of the two-component NS2B-NS3 flavivirus proteinase by structure-based mutagenesis. J Virol 81:4501–4509
8. Yin Z, Patel SJ, Wang WL et al (2006) Peptide inhibitors of dengue virus NS3 protease. Part 2: SAR study of tetrapeptide aldehyde inhibitors. Bioorg Med Chem Lett 16:40–43

9. Niyomrattanakit P, Yahorava S, Mutule I et al (2006) Probing the substrate specificity of the dengue virus type 2 NS3 serine protease by using internally quenched fluorescent peptides. Biochem J 397:203–211
10. Steuer C, Heinonen KH, Kattner L et al (2009) Optimization of assay conditions for dengue virus protease: effect of various polyols and nonionic detergents. J Biomol Screen 14:1102–1108
11. Deng J, Li N, Liu H et al (2012) Discovery of novel small molecule inhibitors of dengue viral NS2B-NS3 protease using virtual screening and scaffold hopping. J Med Chem 55: 6278–6293
12. Prusis P, Lapins M, Yahorava S et al (2008) Proteochemometrics analysis of substrate interactions with dengue virus NS3 proteases. Bioorg Med Chem 16:9369–9377
13. Meldal M, Breddam K (1991) Anthranilamide and nitrotyrosine as a donor-acceptor pair in internally quenched fluorescent substrates for endopeptidases: multicolumn peptide synthesis of enzyme substrates for subtilisin Carlsberg and pepsin. Anal Biochem 195: 141–147
14. Förster T (1948) Zwischenmolekulare Energiewanderung und Fluoreszenz. Ann Phys 437: 55–75
15. Yaron A, Carmel A, Katchalski-Katzir E (1979) Intramolecularly quenched fluorogenic substrates for hydrolytic enzymes. Anal Biochem 95:228–235
16. Steuer C, Gege C, Fischl W et al (2011) Synthesis and biological evaluation of alpha-ketoamides as inhibitors of the Dengue virus protease with antiviral activity in cell-culture. Bioorg Med Chem 19:4067–4074
17. Nitsche C, Steuer C, Klein CD (2011) Arylcyanoacrylamides as inhibitors of the dengue and West Nile virus proteases. Bioorg Med Chem 19:7318–7337
18. Nitsche C, Behnam MAM, Steuer C et al (2012) Retro peptide-hybrids as selective inhibitors of the dengue virus NS2B-NS3 protease. Antiviral Res 94:72–79
19. Mendgen T, Steuer C, Klein CD (2012) Privileged scaffolds or promiscuous binders: a comparative study on rhodanines and related heterocycles in medicinal chemistry. J Med Chem 55:743–753
20. Liu YY, Kati W, Chen CM et al (1999) Use of a fluorescence plate reader for measuring kinetic parameters with inner filter effect correction. Anal Biochem 267:331–335
21. Steuer C (2011) Medizinische Chemie der Dengue Protease und verwandter flaviviraler Proteasen. PhD thesis, University of Heidelberg
22. Khumthong R, Angsuthanasombat C, Panyim S et al (2002) In vitro determination of dengue virus type 2 NS2B-NS3 protease activity with fluorescent peptide substrates. J Biochem Mol Biol 35:206–212
23. Merrifield RB (1965) Solid-phase peptide synthesis. Endeavour 24:3–7
24. Chang C-D, Meienhofer J (1978) Solid-phase peptide synthesis using mild base cleavage of N alpha-fluorenylmethyloxycarbonylamino acids, exemplified by a synthesis of dihydrosomatostatin. Int J Pept Protein Res 11:246–249
25. Champreda V, Khumthong R, Subsin B et al (2000) The two-component protease NS2B-NS3 of dengue virus type 2: cloning, expression in Escherichia coli and purification of the NS2B, NS3(pro) and NS2B-NS3 proteins. J Biochem Mol Biol 33:294–299

# Chapter 19

# Expression and Purification of Dengue Virus NS5 Polymerase and Development of a High-Throughput Enzymatic Assay for Screening Inhibitors of Dengue Polymerase

**Edwin Yunhao Gong, Hannah Kenens, Tania Ivens, Koen Dockx, Katrien Vermeiren, Geneviève Vandercruyssen, Benoit Devogelaere, Pedro Lory, and Guenter Kraus**

## Abstract

The nonstructural protein 5 (NS5) of dengue virus (DENV) plays a central role in the virus replication. It functions as a methyltransferase and an RNA-dependent RNA polymerase. As such, it is a promising target for antiviral drug development. To develop a high-throughput biochemical assay for screening compound libraries, we expressed and purified the polymerase domain of the dengue NS5 protein in bacterial cells. The polymerase activity is measured using a scintillation proximity assay. This homogeneous and high-throughput assay enables screening of compound libraries for identifying polymerase inhibitors against DENV. In this chapter we describe the methods to express and purify the dengue NS5 polymerase from *E. coli* and a validated high-throughput enzymatic assay for screening inhibitors of NS5 polymerase.

**Key words** Dengue virus, NS5, RNA-dependent RNA polymerase, Protein expression, Protein purification, High-throughput screening assay, Scintillation proximity assay, Polymerase inhibitors

## 1 Introduction

The dengue virus (DENV), an enveloped positive-sense single-stranded RNA virus in the *Flavivirus* genus, encodes a nonstructural protein NS5 (for reviews *see* refs. 1–4). NS5 comprises two domains: an S-adenosyl methyltransferase domain and an RNA-dependent RNA polymerase (RdRp) domain. Two domains are separated by an interdomain linker region [5]. NS5 is critical for many functions, including replication, capping of the RNA genome, and possibly host cell gene regulation (for reviews *see* refs. 6, 7).

Edwin Yunhao Gong (ed.), *Antiviral Methods and Protocols*, Methods in Molecular Biology, vol. 1030,
DOI 10.1007/978-1-62703-484-5_19, © Springer Science+Business Media, LLC 2013

The replication of flaviviral RNA is thought through the formation of replicative intermediates (partially double-stranded RNA) and replicative form (double-stranded RNA) [8, 9]. The synthesis of both positive- and negative-strand RNAs during the replication is catalyzed by the RdRp, which is generally an essential component of a replication complex [9]. Due to the essential roles of the polymerase as a central player in the viral life cycles, RdRps have been the target of considerable interest for antiviral drug development based on the experiences with human hepatitis C virus [6, 7].

To develop a robust assay suitable for testing large number of compounds, we expressed the recombinant DENV-2 NS5 polymerase protein in bacterial cells and purified the enzymatically active protein using affinity chromatography. The polymerase activity is measured using a scintillation proximity assay (SPA). In this assay, a biotin-labeled primer is annealed to a poly-rC template and the primer extension is initiated in the presence of $^{3}$H-GTP and NS5 polymerase. The newly synthesized RNA with incorporation of radioactive GTP is captured through the binding of biotin to streptavidin-coupled SPA beads and the captured radioactivity is detected by a liquid scintillation counter. The assay conditions were optimized and the assay parameters were fully validated using selected reference compounds. This robust and homogeneous (mix-and-measure) assay enables a high-throughput mode for screening and profiling inhibitors against DENV NS5 polymerase. In this chapter, we describe the expression and purification of DENV NS5 polymerase and the high-throughput enzymatic assay for identifying DENV NS5 polymerase inhibitors.

## 2 Materials

### *2.1 Virus, Plasmid Vector, and Bacterial Cells*

1. DENV serotype 2 NGC (New Guinea C) strain: National Collection of Pathogenic Viruses (NCPV, UK).
2. Plasmid vector pET11a (Novagen).
3. Competent Rosetta 2 (DE3) cells (Novagen) (*see* **Note 1**).

### *2.2 Bacterial Transformation and Expression*

1. 50 mg/mL of carbenicillin stock solution: Dissolve 5 g of carbenicillin (Novagen) in 100 mL of deionized $H_2O$. Aliquot 15 mL into each Falcon™ tube and store at −20 °C.
2. 34 mg/mL of chloramphenicol stock solution: Dissolve 3.4 g of chloramphenicol (Sigma-Aldrich) in 100 mL of 100 % ethanol. Aliqout 50 mL to each Falcon™ tube and store at −20 °C.
3. Luria-Bertani (LB) agar: 10 g/L tryptone, 5 g/L yeast extract, 10 g/L NaCl, 15 g/L agar. Dissolve 10 capsules (MP Biomedicals) in 250 mL of deionized $H_2O$ and autoclave at 121 °C for 15 min. Cool the medium at room temperature until about <50 °C (hand-warm). Then add carbenicillin to a

final concentration of 100 μg/mL and chloramphenicol to a final concentration of 34 μg/mL. Pipette 30 mL of this medium to a sterile petri dish and allow the agar to be solidified at room temperature. Then store the agar plates at 4 °C.

4. LB broth: 10 g/L tryptone, 5 g/L yeast extract, 0.5 g/L NaCl, and supplemented with 100 μg/mL of carbenicillin and 34 μg/mL of chloramphenicol. Dissolve 8 capsules of LB broth (MP Biomedicals) in 500 mL of deionized $H_2O$ and autoclave for 15 min. After cooling to room temperature, add carbenicillin and chloramphenicol to final concentrations of 100 and 34 μg/mL, respectively.
5. SOC medium.
6. 100 mM Isopropyl β-D-1-thiogalactopyranoside (IPTG) stock solution: Dissolve 1.19 g of IPTG (Acros Organics) in 50 mL of sterile deionized $H_2O$.
7. 200 mM $ZnCl_2$ stock solution: Dissolve 2.72 g $ZnCl_2$ (Sigma-Aldrich) in 100 mL of sterile deionized $H_2O$. Store at 4 °C.
8. 200 mM $MgCl_2$ stock solution: Add 20 mL of 1 M $MgCl_2$ solution (Ambion) to 80 mL of sterile deionized $H_2O$. Store at 4 °C.
9. Lysis buffer: 20 mM Tris–HCl (pH 7.5), 300 mM NaCl, 10 % glycerol, 0.1 % Nonidet P-40, 4 mM $MgCl_2$, 5 mM dithiothreitol (DTT).
10. HALT protease inhibitor (Thermo Fisher Scientific).

### 2.3 SDS-PAGE

1. NuPAGE© Novex 4–12 % Bis–Tris gel 1.0 mm, 12 wells (Life Technologies) (*see* **Note 2**).
2. NuPAGE© MES SDS running buffer (20×) (Life Technologies).
3. NuPAGE© LDS sample buffer (4×) (Life Technologies).
4. NuPAGE© sample reducing agent (10×) (Life Technologies).
5. Simply Blue™ SafeStain (Life Technologies).

### 2.4 Protein Purification

1. Ni-NTA binding buffer: 100 mM Tris–HCl (pH 7.5), 5 % glycerol, 2 mM DTT, 300 mM NaCl.
2. Ni-NTA washing buffer: 40 mM imidazole, 100 mM Tris–HCl (pH 7.5), 5 % glycerol, 2 mM DTT, and 300 mM NaCl (dissolve 0.27 g imidazole in 100 mL of binding buffer and adjust pH to 7.5).
3. Ni-NTA elution buffer: 400 mM imidazole, 100 mM Tris–HCl (pH 7.5), 5 % glycerol, 2 mM DTT, 300 mM NaCl (dissolve 2.723 g imidazole in 100 mL of binding buffer and adjust pH to 7.5).
4. Dialysis buffer: 10 mM Tris–HCl (pH 7.5), 300 mM NaCl, 20 % glycerol, 5 mM $MgCl_2$.

### 2.5 Equipment for Protein Expression and Purification

1. ÄKTAxpress™ system (GE Healthcare): Used for the Ni-NTA purification and operated at 4 °C.
2. Ni-NTA column: HisTrap FF crude 5 mL (GE Healthcare).
3. Bacterial shaker.
4. Centrifuge.
5. Dialysis cassette: Slide-A-Lyzer 3.5 K (Thermo Fisher Scientific).
6. Nanodrop 1000 spectrophotometer (Isogen Lifesciences).
7. XCell SureLock™ mini-cell electrophoresis system (Life Technologies).
8. SDS-PAGE power supplier: Bio-Rad Power Pac HC (Bio-Rad).

### 2.6 NS5 Polymerase Assay

1. DNase/RNase-free distilled water, UltraPure.
2. 1 M Tris–HCl (pH 7.5), UltraPure.
3. DTT, UltraPure.
4. 2 M KCl.
5. 5 M NaCl.
6. 0.5 M EDTA (pH 8).
7. RNA secure 10 mL (Ambion).
8. 10 mM GTP (Ambion).
9. 1 M $MgCl_2$.
10. 1 M $MnCl_2$.
11. Oligo G13 RNA 13 bases, 5′ biotinylated (Proligo, Cat. No. ON19).
12. Poly-rC, 25 mg (GE Healthcare, Cat. No. 27-4220-02).
13. Streptavidin PVT SPA beads (PerkinElmer, Cat. No. RPNQ0009).
14. [8-$^3$H]guanosine 5′-triphosphate ammonium salt ($^3$H-GTP) (PerkinElmer, Cat. No. NET1201001MC, 1 mCi) (*see* **Note 3**).
15. 10× RdRp buffer: 200 mM Tris–HCl (pH 7.5), 250 mM KCl, 30 mM DTT.
16. 5× annealing buffer: 50 mM Tris–HCl (pH 7.5), 500 mM NaCl, 5 mM EDTA.

### 2.7 Consumables and Equipment for SPA

1. Greiner Lumitrac 384-well white plate (VWR).
2. TopSeal-A, clear, self-adhesive, for 384-well microplates, 100 sheets/box (PerkinElmer).
3. Multidrop combi reagent dispenser (Thermo Scientific).
4. TopCount reader (PerkinElmer) or any other liquid scintillation counter.

## 3 Methods

### 3.1 Plasmid Construct of Recombinant DENV NS5 Polymerase

The NS5 polymerase domain (amino acids 273–900) of DENV serotype 2 NGC strain was inserted into a plasmid vector pET11a. To facilitate protein purification, a 10× His tag and a Strep tag were incorporated to the N-terminal of the polymerase domain. In order to facilitate the removal of all tagged sequences, a tobacco etch virus (TEV) protease cleavage site was inserted at the immediate N-terminal of polymerase domain. The expressing cassette is 10× His-Strep tag-TEV cleavage site-polymerase domain (273–900). A codon optimization for *E. coli* was performed. The complete expression construct was synthesized and sequenced by GeneArt (Life Technologies).

### 3.2 Bacterial Transformation and Protein Expression

1. Thaw 50 μL of Rosetta 2 (DE3) competent cells in an Eppendorf tube on ice and mix gently to ensure that the cells are evenly resuspended.
2. Add 2 μL of plasmid DNA (100 ng/μL) to the competent cells on ice. Stir gently with a pipette tip to mix.
3. Place the tube on ice for 30 min.
4. Incubate the tube for 30 s in a 42 °C water bath, and then put on ice for 5 min.
5. Add 250 μL of SOC medium to the tube and incubate in a 37 °C bacterial shaker for 60 min with shaking speed at 250 rpm.
6. Pipette 100 μL of the transformation mixture onto an LB agar plate containing 100 μg/mL of carbenicillin and 34 μg/mL of choramphenicol and spread the bacterial cells using a sterilized, bent glass rod spreader.
7. Place the agar plates upside down and incubate in a 37 °C incubator overnight.
8. On the next day, pick up a single bacterial colony from the agar plate and inoculate 100 mL of LB medium containing 100 μg/mL of carbenicillin and 34 μg/mL of choramphenicol. Incubate in a 37 °C shaker for 16 h with shaking speed at 250 rpm.
9. After 16-h incubation, pipette 500 μL of the overnight bacterial culture to inoculate 500 mL of fresh LB medium (containing 100 μg/mL of carbenicillin and 34 μg/mL of choramphenicol) in a baffled-bottom flask. Depending on the amount of protein needed, more culture medium can be inoculated (*see* **Note 4**). Incubate the flask in a 37 °C shaker until the $OD_{600\ nm}$ reaches 0.6–1.0.
10. When the $OD_{600\ nm}$ reaches between 0.6 and 1.0, cool the culture on ice water to 15 °C.

11. Pipette 1 mL bacterial culture to an Eppendorf tube as a sample for evaluation of protein expression on SDS-PAGE (sample before IPTG induction) and centrifuge the sample tube at 14,000 rpm for 2 min in a microfuge. Resuspend the pellet in an appropriate amount of deionized $H_2O$: 10 μL of water for every 0.1 $OD_{600\ nm}$, i.e., use 60 μL of water for $OD_{600\ nm} = 0.6$.
12. Add IPTG (100 mM stock solution) to a final concentration of 400 μM to the remaining 500 mL of bacterial culture.
13. Add $MgCl_2$ and $ZnCl_2$ to final concentrations of 50 and 50 μM, respectively. Incubate the bacterial culture at 15 °C overnight with a shaking speed at 250 rpm.
14. On the following day, take 1 mL sample from the culture to evaluate the expression on an SDS-PAGE (sample after IPTG induction): Centrifuge the sample tube at 14,000 rpm for 2 min in a microfuge and resuspend the bacterial pellet in an appropriate amount of deionized $H_2O$ as described in **step 11**.
15. Collect all bacterial cultures into 1,000-mL centrifuge tubes.
16. Centrifuge to pellet the bacterial cells at 6,000 × *g* and 4 °C for 10 min.
17. Discard the supernatants.
18. Place the centrifuge tubes with the pellet on ice, and continue with the purification. Otherwise, store the pellets at −20 °C.
19. Weigh the pellets and resuspend the pellets in the lysis buffer: Use 10 mL of lysis buffer supplemented with 1 mL of HALT protease inhibitor for each gram of pellet.
20. Pool the bacterial lysates into a precooled bottle and keep on ice.
21. Homogenize the lysates by passing through a homogenizer, such as Emulsiflex C50, three times.
22. Take 1 mL sample from the lysates to evaluate the lysis on an SDS-PAGE and centrifuge the sample tube at 14,000 rpm for 2 min in a microfuge. Transfer the supernatant (soluble fraction) to a fresh Eppendorf tube and resuspend the pellet in 1 mL of deionized $H_2O$ (insoluble fraction).
23. Transfer the bacterial cell lysates to a precooled centrifuge tube, and centrifuge at 20,000 × *g* and 4 °C for 60 min.
24. After centrifugation, transfer the supernatant (clarified cell lysates) to a fresh tube, which can be used immediately in the protein purification steps as described in Subheading 3.4. If not used immediately, store the clarified cell lysates at −20 °C.

### 3.3 Detection of Expressed Recombinant Protein by SDS-PAGE

1. Prepare a sample mix by adding 400 μL of NuPAGE© sample-reducing agent (10×) to 1,000 μL of NuPAGE© LDS sample buffer (4×).

2. Prepare 2 Eppendorf tubes and pipette 14 µL of the sample mix and 14 µL of deionized $H_2O$ into each tube. Then add 12 µL of the samples before and after the IPTG induction to each respective tube.
3. Prepare 2 Eppendorf tubes for the samples of soluble and insoluble fractions: Pipette 14 µL of the sample mix and 20 µL of deionized $H_2O$ to each tube. Then add 6 µL of samples of soluble and insoluble fractions to the respective tube.
4. Heat all tubes on a heating block at 95 °C for 25 min and load 20 µL of each sample onto a NuPAGE© Novex 4–12 % Bis–Tris gel 1.0 mm.
5. Dilute the NuPAGE© MES SDS running buffer (20×) to prepare 1× MES SDS running buffer.
6. Run an SDS-PAGE using the XCell SureLock™ mini-cell electrophoresis system at 200 V for 35 min.
7. After electrophoresis, stain the gel with the Simply Blue™ SafeStain.

### 3.4 Purification of Enzymatically Active NS5 Polymerase

1. Connect the HisTrap FF crude 5 mL column to the ÄKTA System at 4 °C. All purification steps are performed at 4 °C.
2. Equilibrate the column with at least 5× column volume of binding buffer.
3. Load the clarified cell lysates (*see* Subheading 3.2, **step 24**) into the column with a flow rate of 0.25 mL/min and collect the flow-through (*see* **Note 5**).
4. When the sample is fully loaded, wash the column with 10× column volume of binding buffer to remove impurities and unbound materials.
5. Wash the column with 10× column volume of washing buffer to remove the contaminants. The flow rate for the washing step is 1 mL/min.
6. Elute the sample with 10× column volume of elution buffer. Maintain the flow rate at 1 mL/min.
7. Collect the eluent in 2-mL fractions. The proteins eluted are monitored and recorded at a UV wavelength of 280 nm.
8. After elution, regenerate the column by washing with 3 to 5× column volume of binding buffer.
9. Take 12 µL from each fraction and the flow-through to evaluate the purified protein on an SDS-PAGE as described in Subheading 3.3.
10. Observe the purified polymerase band on the SDS-PAGE after staining. The molecular weight of expressed DENV NS5 polymerase with tags is 77.6 kDa.

**Table 1**
**Components in the primer–template annealing mix**

| Component | Volume needed per mL (μL) |
|---|---|
| DNase/RNase-free water | 592 |
| 5× annealing buffer | 200 |
| RNA secure | 40 |
| Poly-rC (1 mg/mL or 10 μM) | 18 |
| Oligo G13 (60 ng/μL or 12 μM) | 150 |

11. Pool the fractions that contain the purified protein with the right molecular weight.
12. Dialyze the pooled fractions against 2,000 mL of dialysis buffer overnight at 4 °C using a Slide-A-Lyzer 3.5 K dialysis cassette.
13. On the next day, measure the concentration of dialyzed protein using a Nanodrop-1000 spectrophotometer by selecting program Protein A280, and entering the right extinction coefficient (180.88) and molecular weight (77.60 kDa). Use the dialysis buffer as a blank sample.
14. Aliquot 200 μL of purified DENV NS5 polymerase in each cryovial and flash-freeze in liquid nitrogen for 2 min. Then store the cryovial in a −80 °C freezer.

## 3.5 DENV NS5 Polymerase Assay and Compound Screening

### 3.5.1 Primer–Template Annealing

1. Prepare a primer–template annealing mix as in Table 1.
2. Incubate the annealing reaction at 60 °C for 10 min, then at 95 °C for 3 min, and 37 °C for 30 min.
3. At the end of incubation, dilute the reaction mixture 1 in 2 with 1× annealing buffer.
4. Store the annealed primer–template mixture at 4 °C (*see* **Note 6**).

### 3.5.2 NS5 Polymerase Screening Assay Using SPA (See **Note 7**)

1. Perform fourfold serial dilutions of test compounds with DNase/RNase-free distilled water to prepare 3× final test concentration and plate 10 μL/well of each dilution to the corresponding wells of 384-well white plates (Greiner Lumitrac 384-well plate).
2. Prepare an NTP mix as in Table 2.
3. Thaw a DENV NS5 polymerase stock from a −80 °C freezer (*see* Subheading 3.4, **step 14**).
4. Dilute the NS5 polymerase stock to 300 nM with 1× RdRp buffer.

**Table 2**
**Components in the NTP mix**

| | Final concentration | Volume needed per mL (μL) |
|---|---|---|
| DNase/RNase-free water | | 477 |
| 10× RdRp buffer | 1× | 200 |
| $MgCl_2$ | 125 mM | 48 |
| $MnCl_2$ | 125 mM | 48 |
| GTP | 30 μM | 218 |
| $^3$H-GTP | | 9 |

5. Prepare a suspension of the PVT-SPA beads at a concentration of 5 mg/mL in 0.5 M EDTA.
6. Mix equal volumes of the NS5 polymerase dilution (300 nM) and the annealed primer–template (ratio 1:1).
7. Use a multidrop combi reagent dispenser to dispense 10 μL of the NS5 polymerase/primer–template mixture to each well of a 384-well white plate that contains 10 μL of either compound dilutions or water with DMSO (as no-compound controls). Also include a negative control (serves as 100 % inhibition): The PVT-SPA beads in 0.5 M EDTA are added before the addition of NS5 polymerase, primer–template, and NTP mix (*see* **Note 8**).
8. Incubate the reaction at room temperature for 10 min.
9. Dispense 10 μL/well of the NTP mix to the NS5 polymerase/primer–template/compound mixture. The total volume of the polymerase reaction is 30 μL and the final concentrations of each component are 2 mM $MgCl_2$, 2 mM $MnCl_2$, 2.7 nM GTP-[$^3$H]GTP, 15 nM Poly-rC, 150 nM oligo G13, and 50 nM NS5 polymerase.
10. Incubate the plates at room temperature for 2 h.
11. Add 30 μL of PVT-SPA beads (5 mg/mL in 0.5 M EDTA) to each well (*see* **Note 8**).
12. Seal the plates with TopSeal-A.
13. Incubate the plates at room temperature overnight.
14. Count the radioactivity (counts per minute, CPM) using a TopCount reader for 30 s/well or any other liquid scintillation counter.
15. Calculate the percentage of inhibition of each compound dilution and the 50 % inhibitory concentration ($IC_{50}$) using computer software, such as the Graphpad Prism.

## 4 Notes

1. Rosetta™ 2 host strains are BL21 derivatives designed to enhance the expression of eukaryotic proteins that contain codons rarely used in *E. coli*. These strains supply tRNAs for 7 rare codons (AGA, AGG, AUA, CUA, GGA, CCC, and CGG) on a compatible chloramphenicol-resistant plasmid. Such strains are suitable for production of protein from target genes cloned in pET vectors by induction with IPTG.
2. Polyacrylamide and bisacrylamide are neurotoxic. Handle the gel with protective gloves.
3. Follow the local rules of good laboratory practice while working with radioactive isotopes.
4. The yield of DENV NS5 polymerase after affinity purification is approximately 10 mg per liter of bacterial culture. Based on the amount of protein needed, the bacterial culture can be scaled up if a large batch of protein is needed.
5. Load slowly, preferably overnight.
6. The annealed primer–template can be stored at 4 °C for up to 2 weeks.
7. The quenching assay:

   Some compounds can quench the signal of SPA. In this case, we perform a quenching assay as below:

   (a) Pool equal volumes of NS5-polymerase/primer–template mix with the NTP mix in a 50-mL falcon tube.

   (b) Incubate at room temperature for 2 h.

   (c) Add 1.5 mL of PVT-SPA beads (5 mg/mL in 0.5 M EDTA) per mL total mix.

   (d) Dispense 50 μL of this mixture to each well.

   (e) Seal the plates with TopSeal-A.

   (f) Incubate the plates at room temperature overnight.

   (g) Count the radioactivity as described in Subheading 3.5.2, **step 14**.
8. 0.5 M EDTA stops the polymerase reaction.

### References


1. Halstead S (2007) Dengue. Lancet 370: 1644–1652
2. Whitehead S, Blaney J, Durbin A et al (2007) Prospects for a dengue virus vaccine. Nat Rev Microbiol 5:518–528
3. Lindenbath BD, Thiel H-J, Rice CM (2007) Flaviviridae: the virus and their replication. In: Knipe D, Howley P et al (eds) Fields virology, 5th edn. Lippincott Williams & Wilkins, New York, NY, pp 1101–1152
4. Gubler D, Kuno G, Markoff L (2007) Flaviviruses. In: Knipe D, Howley P et al (eds) Fields virology, 5th edn. Lippincott Williams & Wilkins, New York, NY, pp 1153–1252

5. Yap TL, Xu T, Chen YL et al (2007) Crystal structure of the dengue virus RNA-dependent RNA polymerase catalytic domain at 1.85-angstrom resolution. J Virol 81:4753–4765
6. Malet H, Masséa N, Seliskoa B et al (2008) The flavivirus polymerase as a target for drug discovery. Antiviral Res 80:23–35
7. Rawlinson SM, Pryor MJ, Wright P et al (2006) Dengue virus RNA polymerase NS5: a potential therapeutic target? Curr Drug Targets 7: 1623–1638
8. Chu PWG, Westaway EG (1985) Replication strategy of Kunjin virus: evidence for recycling role of replicative form RNA as template in semi-conservative and asymmetric replication. Virology 140:68–79
9. Gong YE, Trowbridge R, Macnaughton TB et al (1996) Characterisation of RNA synthesis during a one-step growth curve and of the replication mechanism of bovine viral diarrhea virus. J Gen Virol 77:2729–2736

# Chapter 20

# Detection and Quantification of Flavivirus NS5 Methyl-Transferase Activities

**Siew Pheng Lim, Christophe Bodenreider, and Pei-Yong Shi**

## Abstract

Flavivirus NS5 is the most conserved protein amongst the flavivirus proteins and is an essential enzyme for viral mRNA capping and replication. It encodes a methyl-transferase (MTase) domain at its N-terminal region which carries out sequential N7 and 2′-*O* methylation, resulting in the formation of the cap1 structure on its viral RNA genome. Two key methods have been established to measure these activities in vitro: thin-layer chromatography (TLC) and scintillation proximity assays (SPA). TLC offers the advantage of direct visualization of the amounts and types of cap structures formed whilst the SPA assay is more sensitive and quantitative. It is also amenable to high-throughput compound screening. The drawback of both assays is the need for radioisotope usage. We further describe the adaptation of a nonradioactive immune-competitive fluorescence polarization assay for detection of dengue virus MTase activity.

**Key words** Flavivirus, Dengue virus, NS5, Methyl-transferase, RNA capping, Thin-layer chromatography, Fluorescence polarization assay, Scintillation proximity assay

## 1 Introduction

The *Flaviviridae* family comprises approximately 70 viruses that are primarily spread through arthropod vectors such as ticks and mosquitoes. Many members of the Flavivirus genus, such as West Nile virus (WNV), yellow fever virus, St. Louis encephalitis virus, Murray Valley encephalitis virus, and the four serotypes of dengue virus (DENV) cause a wide range of human diseases [1]. Flaviviruses are enveloped and contain a single-stranded positive-sense RNA genome of approximately 10.7 kb in length [1]. This RNA has a cap1 structure and an open reading frame that encodes a polyprotein, which is processed by both viral and cellular proteases into three structural proteins (C, M, and E) and seven nonstructural (NS) proteins (NS1, NS2A, NS2B, NS3, NS4A, NS4B, and NS5).

NS5 is the most conserved protein amongst the flavivirus proteins and has a methyl-transferase (MTase) domain and an

Edwin Yunhao Gong (ed.), *Antiviral Methods and Protocols*, Methods in Molecular Biology, vol. 1030,
DOI 10.1007/978-1-62703-484-5_20, © Springer Science+Business Media, LLC 2013

RNA-dependent RNA polymerase (RdRp) domain at the N- and C-termini, respectively [2]. It catalyzes N-7 and 2′-*O* methylations of the viral RNA cap in a sequential manner, GpppA-RNA→m$^7$GpppA-RNA→m$^7$GpppAm-RNA [3–5]. More recently, it was also reported that the dengue MTase performs 2′-*O* methylation on internal adenosines of the viral genome [6]. Both N-7 and 2′-*O* methylation events utilize *S*-adenosyl-L-methionine (AdoMet) as the methyl donor and generate *S*-adenosyl-L-homocysteine (SAH) as the by-product.

DENV NS5 MTase was initially determined to have 2′-*O* methylation activity by TLC [4]. Short capped RNA oligonucleotides are methylated by the DENV MTase with AdoMet which carries a radioactive label on its methyl group (usually tritium; [$^3$H-methyl]-AdoMet). The products are digested with enzymes (nuclease P1 and alkaline phosphatase) to release the individual nucleotides, followed by TLC separation by spotting the reaction on a silica thin-layer sheet. Individual squares containing these nucleotides are cut out and quantified using scintillation counting to determine the levels of $^3$H-methyl group transferred to the RNA substrate.

Subsequent characterization using the same assay system demonstrated that this domain possessed N7-methylase activity [3, 7–9]. Notably, polyethyleneimine cellulose plates were used for nucleotide separation and rather than using tritiated AdoMet, the GTP cap of the RNA template was radiolabeled. In addition, long in vitro-transcribed (IVT) RNA comprising the authentic viral 5′UTR sequence was requisite for induction of N7 activity. The TLC method is somewhat laborious and of low throughput. In order to screen large compound libraries for inhibitors targeting the flavivirus MTase activities, scintillation proximity assays (SPA) were developed which are more sensitive and robust [5, 10, 11]. The SPA also allowed the kinetic characterization of the flavivirus 2′-*O* and N7 MTase activities, and enabled the assays to be performed at $K_m$ concentrations of the RNA and AdoMet substrates. The optimal buffer conditions for the flavivirus N7 and 2′-*O* activities were also characterized with the TLC and SPA assays [7–11].

Graves et al. [12] described a fluorescence polarization assay to detect the catechol-*O*-methyl-transferase (COMT) activity. This method relies on a fluorescent-tagged SAH and an antibody specific for SAH, to quantify the levels of SAH produced during enzymatic reaction. It can be adapted to measure the DENV MTase activity.

### 1.1 Radioactive TLC

TLC is a separation method for analyzing the components present in the mixture. It provides information on the number of components in a mixture, their quantities, identities, and purities. By observing the appearance of a product or the disappearance of a reactant, it can also be used to monitor the progress of a reaction. Thus, for flavivirus MTase, under the correct buffer conditions with a particular RNA template and the substrate, AdoMet, one can determine the catalytic transfer of the methyl group from AdoMet

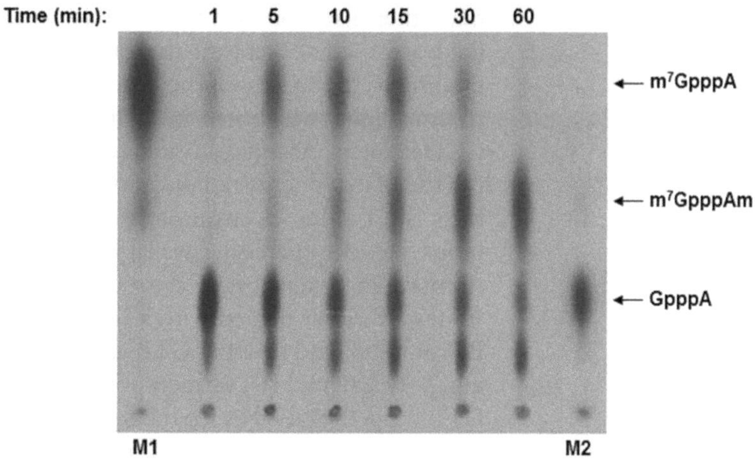


**Fig. 1** TLC time-course experiment performed with DENV MTase showing N7 methylation of the RNA substrate after reaction for 1 min, giving rise to formation of $m^7$GpppA cap structure. Progressive 2'-*O* methylation of the RNA substrate from 10 to 60 min resulted in the formation of $m^7$GpppAm. The reaction conditions are as described in Subheading 3.1. The migration rates of the products are $m^7$GpppA > $m^7$GpppAm > GpppA. The labels M1 and M2 denote the migration positions of the $m^7$GpppA and GpppA markers, respectively

to the N7 position of the GTP cap of the RNA and/or the 2′-*O* position of its penultimate adenosine. The RNA template is first capped with radioactive $^{33}$P- or $^{32}$P-GTP using the vaccinia virus N7-MTase and then incubated with the flavivirus MTase in the presence of unlabelled AdoMet [3, 7–9]. After purification and concentration, the RNA is digested with nuclease P1 and alkaline phosphatase and spotted on polyethyleneimine cellulose TLC plates. The plates are then placed in a shallow pool of $LiCl_2$ solvent, and allowed to travel up the plate by capillary action. As the solvent travels up the plate, it moves over the original spot. As the different components, $m^7$GpppAm, $m^7$GpppA, and GpppA, have different polarities, they migrate to different distances from the original spot location and show up as separate spots. When the solvent has travelled almost to the top of the plate, the plate is removed and the solvent is allowed to evaporate. The plate is exposed on an autoradiographic film overnight and the spots can be directly observed after development of the film. The positions of the spots are compared against radiolabeled $m^7$GpppAm, $m^7$GpppA, and GpppA markers. A TLC plate showing the migration positions of $m^7$GpppAm, $m^7$GpppA, and GpppA cap structures from DENV MTase N7 and 2′-*O* activities is illustrated in Fig. 1.

### 1.2 Radioactive SPA

SPA technology is widely used in assay development and biochemical screening as it allows sensitive measurement of a broad range of biological processes in a homogeneous system, without the need for wash steps. The assay can be miniaturized into 96-, 384-, or 1,536-well microtiter plate formats. Coated micro beads containing scintillation fluid are used to capture the radiolabeled target molecules. Upon binding to the surface of the beads, they trigger the fluid in the beads to emit light, which can be detected with scintillation counters or CCD imagers. Another advantage of SPA technology is that it enables the use of low-energy radioisotopes ($^{3}H$, $^{14}C$, and $^{33}P$) as labels due to their short-range electron emission. Many different types of coated beads are commercially available; in the case for DENV MTase, streptavidin-coated beads can be used [5, 10, 11]. The RNA substrates are either tagged with a biotin moiety (also known as vitamin B7) at the 3′ end of short GTP-capped oligonucleotides for monitoring 2′-*O* activity [10] or incorporated with biotinylated CTP in IVT GTP-capped viral RNA for measuring N7 activity [5, 11]. The flavivirus MTase, RNA substrate, and $^{3}H$-AdoMet are mixed together in a well in a 96-well plate and the reaction is allowed to proceed. After the enzyme catalyzes the transfer of the $^{3}H$-methyl group from AdoMet to the RNA substrate, SPA beads are added. The binding of biotin to streptavidin is one of the strongest non-covalent interactions known in nature, with a dissociation constant ($K_d$) of the order of $\approx 10^{-14}$ M. Thus, the biotinylated RNA substrates are strongly immobilized on the surface of the SPA beads and low levels of $^{3}H$-methylated RNA can easily be detected. Due to this high sensitivity, experiments can be conducted to determine the steady-state kinetic parameters of the flavivirus MTase activities, its processivity, and catalytic mechanism. The enzyme has been shown to have random, bi-bi kinetic mechanism [5, 10, 11]. A schematic representation of the SPA technology for detecting flavivirus MTase activities is shown in Fig. 2.

### 1.3 Competitive FP Immuno-Assay

Fluorescence polarization is a well-accepted and frequently used technology for high-throughput assays [13]. FP takes advantage of the inverse relationship between rotational speed of fluorescent molecules in solution and the size of the labeled molecule or complex (for review *see* ref. 14). Briefly, the fluorophore reaches the excited state upon absorption of a photon. The duration of the excited state of the fluorophore depends on the identity of the particular fluorophore. For fluorescein, the half-life of the excited state is 9 ns. In the excited state, the fluorophore releases a photon that is of a longer wavelength than the excitation light and the fluorophore returns to the ground state. During the excited state, the molecule rotates or tumbles in solution. Small molecules rotate and release light at a different angle than the absorbed light. Larger molecules do not rotate quickly, and the light becomes polarized; the angle of release becomes more acute and eventually goes towards zero as molecular mass increases. This polarization is quantified using the ratio of fluorescence measured in the parallel and perpendicular planes relative to the excited plane. This ratio value is unitless and is referred to as polarization value (P) or commonly is multiplied by 1,000 to give millipolarization (mP) values. In general high mP values indicate that

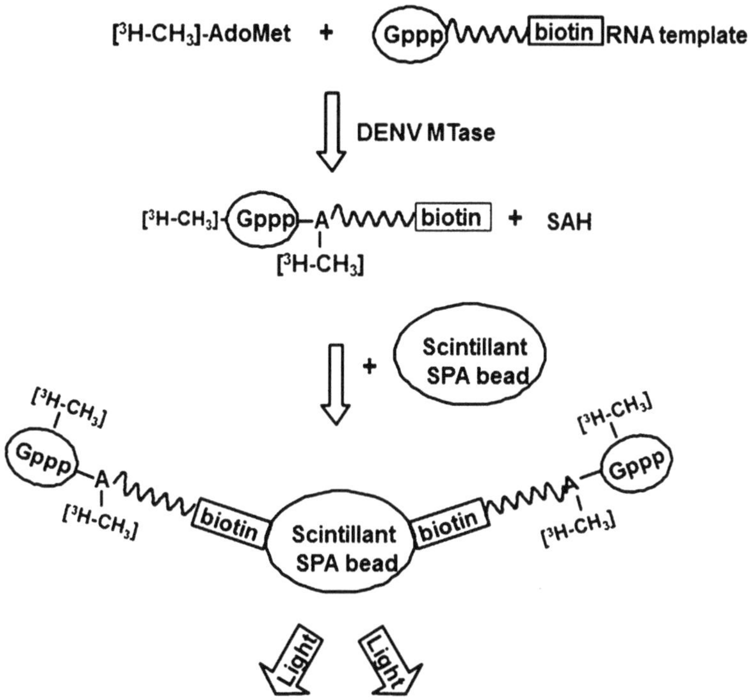


**Fig. 2** Schematic representation of the DENV MTase SPA assay. DENV MTase transfers [$^3$H-methyl] group from [$^3$H-methyl]-AdoMet to GTP-capped RNA template incorporated with biotin (as a tag at the 3'end in short RNA oligonucleotides for detection of 2'-*O* MTase activity or as CMP-biotin in long viral RNA oligonucleotides for detection of N7 MTase activity), resulting in the generation of SAH as by-product. The methylated RNA templates are captured on SPA beads and the proximity of the tritium isotope to the scintillant in the beads stimulates light emission

the fluorophore has a high apparent molecular weight, whereas low mP values indicate a low apparent molecular mass. Since FP is a ratiometric fluorescence technique, it is less subject to variability than other non-ratiometric assays.

Competitive FP immuno-assay was first described by Dandliker and coworkers [15] and incorporated the use of an antibody to increase the apparent size of a small analyte analog. The method was modified by Graves et al. [12] to detect the COMT activity using a commercial kit from Abbott. In principle, the concentration of unlabeled analytes (such as the product of the methyl-transferase reaction, SAH) can be determined by their ability to compete with fluorescently labeled analogs, commonly known as the tracer (which is a fluorescein-conjugated SAH), for antibody binding. In the absence of product (SAH), the tracer is fully bound to the antibody, dramatically increasing its apparent mass and decelerating its rotational motion. This reduced molecular rotation results in an

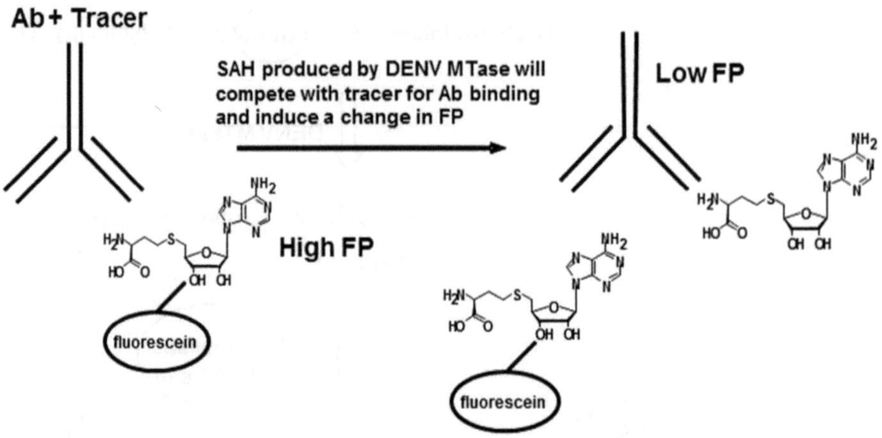


**Fig. 3** Schematic representation of the fluorescence polarization (FP) assay. SAH-fluorescein tracer binds to anti-SAH antibody (Ab) resulting in high FP. Following DENV MTase-mediated transfer of the methyl group from AdoMet to the RNA template, SAH is produced which competes with the tracer for Ab binding. As more SAH is generated and binds to the Ab, increasing amounts of tracer becomes unbound and generates a low FP

increase in FP indicated by high mP values. SAH which is generated from DENV MTase N7 or 2′-*O* reactions competes with the tracer for antibody binding, resulting in some of the tracer population being bound and others being free. The latter have greater rotational motion, resulting in a decrease in FP. As more SAH is produced, more tracers are competed out for binding to the antibody, resulting in progressive decline in FP. In this way, the activity of DENV MTase can be determined (Fig. 3).

## 2 Materials

### 2.1 Common Materials

1. IVT MEGAshortscript kit (Cat. No. AM1354; Life Technologies).
2. *Pfu Turbo* DNA polymerase (Cat. No. 600255, Stratagene).
3. 10 mM dNTP mix.
4. Ultrapure phenol:chloroform:isoamyl alcohol (25:24:1, v/v) (Life Technologies).
5. 32 mM *S*-Adenosyl-L-methionine solution (Cat. No. B9003; NEB).
6. Absolute ethanol (Molecular Biological Grade).
7. Sephadex G-25 microspin columns (Cat. No. 27-5325-01; GE Healthcare).
8. Sodium chloride.
9. Magnesium chloride.

10. Manganese chloride.
11. Potassium chloride.
12. TRIZMA base.
13. EDTA.
14. DL-dithiothreitol (DTT).
15. CHAPS.
16. Ammonium acetate ($NH_4OAc$).
17. 32 mM *S*-Adenosyl-L-methionine solution (NEB).
18. 100 mM rGTP (Promega).
19. Recombinant RNasin® Ribonuclease Inhibitor (40 U/μL) (Promega).
20. NanoDrop spectrophotometer (Thermo Scientific).

### 2.2 TLC

1. Vaccinia ScriptCap m$^7$G Capping System (Cat. No. SCCE0610; Epicenter Biotechnologies).
2. ScriptCap™ 2′-*O*-methyltransferase (Cat. No. SCMT0610; Epicenter Biotechnologies).
3. GTP, [α-$^{33}$P], 3,000 Ci/mmol, 10 mCi/mL (Cat. No. NEG606H; PerkinElmer).
4. Inorganic pyrophosphatase (Cat. No. 70950; USB Corp.).
5. Nuclease P1 (US Biological).
6. Polyethyleneimine cellulose TLC plates (Cat. No. 4473-00; JT Baker).
7. $^{33}$P-GpppA, $^{33}$P-m$^7$GpppA, and $^{33}$P-m$^7$GpppAm markers (generated by vaccinia virus ScriptCap enzymes).
8. N-7 MTase buffer: 50 mM Tris–HCl, pH 6.0, 50 mM NaCl, and 2 mM DTT.
9. 2′-*O* MTase buffer: 50 mM Tris–HCl, pH 9.0, 1 mM $MgCl_2$, and 2 mM DTT.
10. TLS solvent: 0.65 M LiCl.
11. Storage phosphor screens (Amersham Biosciences).
12. Typhoon™ 9410 variable mode imager (GE Healthcare).

### 2.3 SPA

1. *S*-Adenosyl-L-[methyl-$^3$H]-methionine, 60–85 Ci/mmol (Cat. No. TRK-865; GE Healthcare).
2. G(5′)ppp(5′)A (Cat. No. S1406L; New England Biolabs).
3. m$^7$G(5′)ppp(5′)A (Cat. No. S1405L; New England Biolabs).
4. Biotin-14-CTP (Cat. No. 19519-016; Invitrogen).
5. RNA template: 5′GpppAGAACCUG-biotin-TEG-3′ custom synthesized (Trilink Biotechnologies).

6. N-7 MTase buffer: 50 mM Tris–HCl, pH 7.5, 20 mM NaCl, and 2 mM DTT, 0.05 % CHAPS.
7. 2′-*O* MTase buffer: 50 mM Tris–HCl, pH 7.5, 10 mM KCl, 2 mM $MgCl_2$, and 2 mM DTT 0.05 % CHAPS.
8. Sealing film (Axygen).
9. Streptavidin-coated SPA beads (Cat. No. RPNQ0009, GE Healthcare).
10. 96 well ½ area white opaque micro-titer plate (Corning Costar).
11. Plate shaker.
12. Tabletop centrifuge.
13. Trilux microbeta counter (PerkinElmer).

### 2.4 FP Assay

1. N7-methylated GTP (Cat. No. M6133; Sigma).
2. RNA template (5′GpppAGAACCUG-biotin-TEG-3′), Trilink Biotechnologies, custom synthesized.
3. 96 well ½ area black opaque micro-titer plate (Cat. No. 3694; Corning Costar).
4. Anti-SAH antibody, from IMx homocysteine FPIA kit (Cat. No. 7D29-20; Abbott Laboratories).
5. Fluorescein–SAH conjugate tracer, from IMx homocysteine FPIA kit (Cat. No. D29-20; Abbott Laboratories).
6. *S*-(5′-deoxyadenosine-5′)-L-homocysteine (Cat. No. A9384, Sigma).
7. 2′-*O* MTase buffer (50 mM Tris–HCl, pH 7.5, 10 mM KCl, 2 mM $MgCl_2$, and 2 mM DTT 0.05 % CHAPS).
8. Tecan Safire2 plate reader (Männedorf, Switzerland).

## 3 Methods

### 3.1 TLC

#### 3.1.1 Preparation of T7proA2.5-DENV2-5 UTR PCR Template

1. Obtain DENV2 5′UTR nt 1–110 or 1–211 by PCR amplification from the plasmid template, pSuper-COS-FL TSV01 (harbors the infectious cDNA of DENV2 TSV01 strain; described in Lim et al. [11]), with primer-pair DV2-5′UTR (nt 1–22)-FOR and DV2-5′-UTR (nt 91–110)-REV or DV2-5′UTR (nt 190–211)-REV, respectively (listed in Table 1). The PCR reaction and conditions are given in Tables 2 and 3.
2. Link the T7 class II phi 2.5 promoter upstream of the DENV 5′ UTR sequence by PCR amplification with the primer-pair DV2-T7proPhi2.5-FOR and DV2-5′-UTR (nt 91–110)-REV or DV2-5′UTR (nt 190–211)-REV (listed in Table 1), using the same PCR reaction and conditions.

**Table 1**
**PCR primers**

| Primer name | 5′–3′ |
|---|---|
| DV2-T7proPhi2.5-FOR (T7 promoter phi 2.5 italicized) | CCCGGCGGCCGC*TAATACGACTCACTATT* AGTTGTTAGTCTACGTGG |
| DV2-5′UTR (nt 91-110)-REV | CGTTGGTTATTCATCAGAG |
| DV2-5′UTR (nt 190–211)-REV | GCAGCATTCCAAGTGAGAATC |
| D4-T7pro-211-FOR | ATATGGCGGCCGCTAATACGACTCACTA |
| D4 nt 1-110 Rev | TGGTTCATTTTTCCAGAGATCTGC |
| D4-5′UTR-L-3′UTR-REV | AGAACCTGTTGGATCAACAACACC |
| DV2-5′UTR (nt 1–22)-FOR | AGTTGTTAGTCTACGTGGACCG |

**Table 2**
**PCR reaction**

| Components | Final concentration | Volume per reaction (μL) |
|---|---|---|
| Distilled water | – | 40 |
| 10× cloned *Pfu Turbo* reaction buffer | 1× | 5 |
| 10 mM dNTP mix | 0.2 mM | 1 |
| Forward primer (100 ng/μL) | 2 ng/μL | 1 |
| Reverse primer (100 ng/μL) | 2 ng/μL | 1 |
| Plasmid pSUPER-COS-FL TSV01 or DENV2 5′ UTR PCR product (100 ng/μL) | 2 ng/μL | 1 |
| *PfuTurbo* DNA polymerase (2.5 U/μL) | | 1 |
| Total reaction voiume | | 50 |

**Table 3**
**PCR conditions**

| Segment | Number of cycles | Temperature (°C) | Duration |
|---|---|---|---|
| 1 | 1 | 95 | 2 min |
| 2 | 40 | 95<br>50<br>72 | 40 s<br>40 s<br>40 s |
| 3 | 1 | 72 | 40 s |

3. After the PCR reaction, verify the quality of the PCR product by electrophoresis through a 2 % agarose gel.
4. To purify the PCR product through a Sephadex G-25 microspin column, first invert the column several times by hand to resuspend the sephadex slurry. Then break off the tip of the column and transfer to an empty 2 mL microfuge tube (without lid). Loosen the cap slightly and spin the column at 2.9 K rpm for 1 min. Discard the tube and its contents (*see* **Note 1**).
5. Transfer the column to a new microfuge tube and slowly load the PCR reaction (maximum volume loaded to G-25 column is 100 μL) into the center of the sephadex matrix and spin the tube at 470 × *g* for 2 min (*see* **Note 1**). Discard the column. Transfer the eluate containing the PCR product into a new Eppendorf tube (with lid).
6. Add 140 μL of 5 M $NH_4OAc$ and top up the volume to 350 μL by adding sterile water. Add 350 μL phenol:chloroform:isoamyl alcohol, vortex for 5 s, and centrifuge the tube at 11,000 × *g* for 3 min.
7. Transfer the upper aqueous phase to a new microfuge tube and add 1,000 μL of 100 % ethanol. Add 2 μL of glycogen to the tube to precipitate the DNA. Vortex the tube for 10 s and incubate at –80 °C for 45 min.
8. Spin the tube at 11,000 × *g* for 20 min at 4 °C. Remove the ethanol and add 800 μL of 70 % ethanol to wash the pellet. Spin the tube for 3 min and remove the 70 % ethanol. Spin the tube again at 11,000 × *g* for 1 min at 4 °C and remove the residual ethanol.
9. Air-dry the pellet for 5 min and resuspend the pellet in 10 μL of RNAse-free water.
10. Quantify PCR products by $A_{260}$ with the NanoDrop spectrophotometer.
11. One PCR reaction with a volume of 50 μL typically yields approximately 1 μg of PCR product. Analyze PCR product by gel electrophoresis as in **step 3**.

#### *3.1.2 Preparation of In Vitro-Transcribed RNA Substrates*

1. Set up the in vitro transcription reaction for the PCR product generated in Subheading 3.1.1 using the IVT MEGAshortscript kit (Table 4).
2. Thaw all the components at room temperature (RT) and add them sequentially into a microfuge tube, at RT. The enzyme is added last.
3. Mix the tube gently by hand, and spin briefly at 11,000 × *g* for 15 s.
4. Incubate the tube at 37 °C for 6 h.

**Table 4**
**IVT reaction of uncapped DENV2 5' UTR**

| Components | Final concentration | Volume per reaction (μL) |
|---|---|---|
| Sterile distilled water | – | 6 |
| 10× T7 buffer | 1× | 2 |
| 75 mM ATP | 7.5 mM | 2 |
| 75 mM CTP | 7.5 mM | 2 |
| 75 mM GTP | 7.5 mM | 2 |
| 75 mM UTP | 7.5 mM | 2 |
| DENV2 5′ UTR PCR product (0.5 μg/μL) | 0.05 μg/μL | 2 |
| T7 enzyme mix | | 2 |
| Total reaction volume | | 20 |

5. Add 1 μL of Turbo DNase I (2U/μL) and incubate tube at 37 °C for 20 min.
6. Add 30 μL of RNase-free water to 50 μL.
7. Purify the IVT RNA through a Sephadex G-25 microspin column, as described in Subheading 3.1.1.
8. Resuspend the RNA pellet in 40 μL of RNAse-free water and quantify the yield by $A_{260}$ with a NanoDrop spectrophotometer.
9. One IVT reaction with a volume of 20 μL typically yields approximately 15 μg of IVT RNA.
10. Analyze the IVT RNA by gel electrophoresis with 2 % agarose gel. Use autoclaved running buffer and freshly washed gel apparatus.

*3.1.3 Preparation of Radiolabeled Capped IVT RNA*

1. Use the ScriptCap $m^7G$ capping system with the vaccinia virus-capping enzymes to cap the IVT RNA prepared in Subheading 3.1.2. Set up the reactions in sterile microfuge tubes. To add cap1 or cap0 RNA structure, AdoMet is included or excluded from the reaction mixture, respectively (Table 5).
2. Incubate tubes at 37 °C for 1 h.
3. Add 30 μL of sterile water to each tube and purify twice through the Sephadex G-25 microspin column as described in Subheading 3.1.1.
4. Resuspend air-dried pellet with 40 μL of RNase-free water.

**Table 5**
**Capping IVT RNA with radiolabeled $^{33}$P-GTP**

| Components | Final concentration | Volume per reaction | |
|---|---|---|---|
| | | $m^7$G*ppp-RNA (μL) | G*ppp-RNA (μL) |
| Sterile distilled water | – | 7 | 8 |
| 2 μg/μL pppA-IVT RNA | 0.25 μg/μL | 2.5 | 2.5 |
| 10× ScriptCap Capping Buffer | 1× | 2 | 2 |
| 2 mM SAM | 0.1 mM | 1 | 0 |
| 10 μCi/μL α-$^{33}$P-GTP (3,000 Ci/mmol) | 2.5 μCi/μL | 5 | 5 |
| ScriptCap capping enzyme (10 U/μL) | 0.5 U/μL | 1 | 1 |
| ScriptGuard RNase inhibitor (40 U/μL) | 1 U/μL | 0.5 | 0.5 |
| Inorganic pyrophosphatase (0.004 U/μL) | 0.0002 U/μL | 1 | 1 |
| Total reaction volume | | 20 | 20 |

*Note*: The asterisk indicates that the phosphate is $^{33}$P labeled

#### 3.1.4 DENV N7 and 2′-O MTase Assays

1. Set up the N7 or 2′-*O* MTase reactions, by first adding $m^7$GTP- or GTP-capped RNA templates and enzyme in N7 or 2′-*O* buffer, in sterile microfuge tubes, according to Table 6. Initiate methylation by adding AdoMet (*see* **Note 2**). Incubate the reactions at RT for 5 and 60 min for N7 and 2′-*O* methylation, respectively (*see* **Note 3**).
2. Prepare the stop buffer comprising a mixture of 400 μL of 5 M $NH_4OAc$, 500 μL of water, and 7.5 μL of glycogen in a microfuge tube.
3. Mix the stop buffer well and incubate the tube for 5 min at RT.
4. Add 180 μL stop solution to the tubes containing the MTase reaction mixtures and then add 200 μL phenol: chloroform: isoamyl alcohol.
5. Vortex the tubes and spin them at 11,000 × *g* for 2 min.
6. Transfer the upper phases to new microfuge tubes containing 600 μL of 100 % ethanol.
7. Mix the contents well and incubate the tubes at −20 °C for 1 h or overnight.
8. Spin the tubes at 11,000 × *g* for 20 min.
9. Remove supernatants and wash the pellets with 800 μL of 70 % ethanol.

**Table 6**
**DENV MTase reactions for TLC assay**

| | Final concentration | Final volume per well (μL) |
|---|---|---|
| Sterile distilled water | – | 11.15 |
| 1 M Tris–HCl, pH 7.5 | 50 mM | 1 |
| 250 mM NaCl | 20 mM | 1.6 |
| 1 % CHAPS (1 g/100 mL water) | 0.05 % | 1 |
| 50 mM DTT | 2 mM | 1 |
| 40 U/μL RNasin inhibitor | 0.25 U/μL | 0.25 |
| 50 μM AdoMet | 2.5 μM | 1 |
| $m^7$G*ppp- or G*ppp-RNA substrate (3–5 × $10^5$ CPM) | ~25 nM | 2 |
| 4 μM DENV2 MTase domain | 400 nM | 1 |
| Total reaction volume | | 20 |

10. Re-spin the tubes for 5 min at 11,000 × *g*.
11. Discard the supernatants and air-dry the pellets at RT.
12. Resuspend the pellets in 8 μL of sterile water. Add 1 μL of 200 mM Tris–HCl, pH 7.5 (final 20 mM), and 1 μL of nuclease P1 (5 μg/μL).
13. Incubate the tubes at 37 °C for at least 5 h or overnight.

*3.1.5 Separation of Methylation Products on TLC Plates*

1. Prepare 0.65 M of $LiCl_2$ chromatography solvent solution by dissolving 2.755 g of $LiCl_2$ powder into 100 mL of sterile water.
2. Spot 1 or 1.5 μL of the nuclease P1-digested reaction mixture onto the polyethyleneimine cellulose TLC plate, about 2 cm above the lower edge.
3. Spot the controls $m^7$G*pppAm, $m^7$G*pppA, and G*pppA adjacent to the samples.
4. Air-dry the plates for at least 20 min in a fume hood.
5. Pour 50 mL of 0.65 M $LiCl_2$ into a 2,000 mL beaker, and place the TLC plate into the beaker. Cover the top of the beaker with a glass plate or a Saran wrap.
6. Place the beaker behind a 1-cm-thick Perspex shield inside the fume hood.
7. Let the solvent run for about 30 min or until the solvent front is 1 cm from the top of the TLC plate.
8. Remove the TLC plate, blot the excess solvent with a paper towel, and dry the plate in the fume hood for 20 min.

**Table 7**
**IVT reaction of capped DENV2 5′ UTR with biotinylated CTP**

| Component amount per reaction | Final concentration | Volume per reaction (μL) |
|---|---|---|
| Sterile distilled water | – | 1.5 |
| 10× T7 buffer | 1× | 2 |
| 75 mM ATP | 2.5 mM | 0.67 |
| 75 mM CTP | 5 mM | 1.33 |
| 75 mM GTP | 7.5 mM | 2 |
| 75 mM UTP | 7.5 mM | 2 |
| 10 mM Biotin-14-CTP (Biotin-14-CTP: CTP=1:2 ) | 7.5 mM | 5 |
| 40 mM GpppA or $m^7$GpppA (GpppA or $m^7$GpppA: ATP=2:1) | 5 mM | 2.5 |
| DENV2 5′ UTR PCR product (0.5 mg/μL) | 0.05 μg/μL | 1 |
| T7 enzyme mix | – | 2 |
| Total reaction volume | | 20 |

9. Wrap the TLC plate with Saran wrap and expose it to a storage phosphor screen for 2–4 h.
10. Develop the storage phosphor screen on a Typhoon imager.
11. The migration rates of the methylation products are $m^7$G*pppA > $m^7$G*pppAm > G*pppA.

### 3.2 SPA

#### 3.2.1 Preparation of Biotinylated IVT RNA Substrates

Amplify PCR products comprising T7promoter phi 2.5 upstream of the DENV2 5′UTR nt 1–110 or 1–211 as described in Subheading 3.1.1. Use the PCR products to prepare the capped IVT RNA with biotinylated-CTP as indicated in Table 7. Purify the IVT capped RNA according to Subheading 3.1.2.

#### 3.2.2 Measurement of N7 or 2-O Methyl-Transferase Activity

1. Use the RNA substrates, Gppp-DENV 110 or 211 nt, to set up the DENV N7 MTase reaction. Use the RNA substrates, Gppp-8 nt RNA or $m^7$Gppp-DENV 110 or 211 nt, to set up the DENV 2′-*O* MTase reaction.
2. Add the RNA template and enzyme in N7 or 2′-*O* buffer, in duplicate wells, in a 96-well ½ area, white opaque plate (Tables 8 and 9; *see* **Notes 3** and **4**).
3. Seal the plate with sealing film, briefly shake at 35 × *g* at RT, and centrifuge for 30 s at 90 × *g*.
4. Remove the sealing film and add [methyl-$^3$H]-AdoMet to initiate the reaction.

**Table 8**
**DENV N7 MTase reaction for SPA assay**

| | Final concentration | Final volume per well (µL) |
|---|---|---|
| Sterile distilled water | – | 16.25 |
| 1 M Tris–HCl, pH 7.5 | 50 mM | 1.25 |
| 200 mM NaCl | 20 mM | 1 |
| 1 % CHAPS (1 g/100 mL water) | 0.05 % | 1.25 |
| 50 mM DTT | 2 mM | 1 |
| 40 U/µL RNasin inhibitor | 0.25 U/µL | 0.25 |
| 1 µCi/µL [methyl-$^3$H]-AdoMet (64 Ci/mmol) | 0.02 µCi/µL (317 nM) | 0.5 |
| 2.5 µM Gppp-DENV 110 or 211 nt RNA substrate | 250 nM | 2.5 |
| 2.5 µM DENV2 MTase domain | 100 nM | 1 |
| Total reaction volume | | 25 |

**Table 9**
**DENV 2'-*O* MTase reaction for SPA assay**

| | Final concentration | Final volume (µL) |
|---|---|---|
| Sterile distilled water | – | 11.25 |
| 1 M Tris–HCl, pH 7.0 | 50 mM | 1.25 |
| 250 mM KCl | 10 mM | 1 |
| 50 mM $MgCl_2$ | 2 mM | 1 |
| 50 mM $MnCl_2$ | 2 mM | 1 |
| 1 % CHAPS (1 g/100 mL water) | 0.05 % | 1.25 |
| 50 mM DTT | 2 mM | 1 |
| 40 U/µL RNasin inhibitor | 0.25 U/µL | 0.25 |
| 1 µCi/µL [methyl-$^3$H]-AdoMet (64 Ci/mmol; 15.87 µM) | 0.04 µCi/µL (635 nM) | 1 |
| 200 nM Gppp-8 nt RNA or 1.6 µM m$^7$Gppp-DENV 110 or 211 nt RNA substrate | 40 or 320 nM | 5 |
| 625 nM DENV2 MTase domain | 25 nM | 1 |
| Total reaction volume | | 25 |

**Table 10**
**2× stop buffer for SPA assay**

| | Final concentration | Final volume |
|---|---|---|
| Sterile distilled water | – | 535 μL |
| 1 M Tris–HCl, pH 7.0 | 100 mM | 100 μL |
| 0.5 M EDTA | 50 mM | 100 μL |
| 5 M NaCl | 300 mM | 60 μL |
| 0.5 mM SAM | 62.5 μM | 125 μL |
| 100 mg/mL SPA beads | 8 mg/mL | 80 μL |
| Total reaction volume | | 1 mL |

5. Reseal the plate, shake, and centrifuge as in **step 3**.
6. Incubate the plate at RT for 15 min.
7. Remove the sealing film and add 25 μL of 2× stop solution (Table 10).
8. Reseal the plate, shake for 20 min at 35 × *g* at RT, and centrifuge for 2 min at 90 × *g* at RT.
9. Read the plate in a microbeta counter with a counting time of 1 min/well.
10. To measure the background CPM generated by [methyl-$^3$H]-AdoMet in the absence of enzymatic activity, first aliquot 25 μL of 2× stop solution into the wells, and then add the buffer, enzyme, and substrates as described in **steps 2–5**.

#### 3.2.3 Calculation of [$^3$H]-Methyl Transferred to RNA Template

The signal or counts per min (CPM) from each test well is subtracted from the CPM obtained in the negative control well and is proportional to the level of [methyl-$^3$H] incorporated to the RNA template by the DENV MTase. The final nanomolar product formed (N; [$^3$H]-methyl group transfer) is calculated as

$$P = [(Y - B)/\mathrm{E}] \times F$$

where $Y$ is counts/min for each experiment, $B$ is counts/min of negative control well, E is counting efficiency of the microbeta counter for $^3$H (0.57 for Trilux microbeta counter), and $F$ is counts/min/nM free [$^3$H]-AdoMet.

Product formed per min is obtained as $N/T$, where $T$ is the time of reaction (15 min). The rate of product formation ($v$, nM/min) is plotted against substrate concentration and fitted into one site binding equation as follows:

$$v = \frac{V_{\max}^{app}\ [S]}{K_m^{app} + [S]}$$

**Table 11**
**DENV 2'-*O* MTase reaction for FP assay**

| | Final concentration | Final volume (μL) |
|---|---|---|
| Sterile distilled water | – | 9.65 |
| 1 M Tris–HCl, pH 7.0 | 50 mM | 1.25 |
| 250 mM KCl | 10 mM | 1 |
| 50 mM $MgCl_2$ | 2 mM | 1 |
| 50 mM $MnCl_2$ | 2 mM | 1 |
| 1 % CHAPS (1 g/100 mL water) | 0.05 % | 1.25 |
| 50 mM DTT | 2 mM | 1 |
| 40 U/μL RNasin inhibitor | 0.25 U/μL | 0.25 |
| 200 nM Gppp-8 nt RNA | 40 nM | 5 |
| 10 μM SAM | 0.64 μM | 1.6 |
| 625 nM DENV2 MTase domain | 50 nM | 2 |
| Total reaction volume | | 25 |

where ${V_{max}}^{app}$ corresponds to maximal velocity and ${K_m}^{app}$ is the substrate concentration needed to reach half-maximal velocity. From ${V_{max}}^{app}$, turnover number ($k_{cat}$) can be calculated as $k_{cat} = V_{max}/[Et]$, where *Et* defines the total concentration of enzyme.

### 3.3 FP

#### 3.3.1 Measurement of 2'-O Methyl-Transferase Activity

1. Set up the DENV 2′-*O* MTase reactions in 96-well black opaque plates according to Table 11 (*see* **Notes 4** and **5**).
2. Mix the enzyme with the RNA substrate in assay buffer, leaving out $MgCl_2$ (*see* **Note 2**).
3. Seal the plate with sealing film, briefly shake at 35 × *g* at RT, and centrifuge for 30 s at 90 × *g*.
4. Remove the sealing film and add AdoMet (*see* **Note 1**) and $MgCl_2$ to start the reaction.
5. Reseal the plate with sealing film, briefly shake at 35 × *g* at RT, and centrifuge for 30 s at 90 × *g*.
6. Incubate the plate for 10 min at RT.
7. Remove the sealing film and stop the reaction with 6 μL of 5× stop buffer (Table 12).
8. Reseal the plate with sealing film, and shake the plate at 35 × *g* at RT for 1 h.
9. Read on the Tecan plate reader using the setting below (*see* **Note 6**):

**Table 12**
**5× stop buffer for FP assay**

| | Final concentration | Final volume |
|---|---|---|
| Sterile distilled water | – | 201 μL |
| 1 M Tris–HCl, pH 7.0 | 500 mM | 500 μL |
| 0.5 M EDTA | 125 mM | 24 μL |
| 5 M NaCl | 750 mM | 150 μL |
| Fluorescein-SAHC tracer | 16× diluted | 62.5 μL |
| Anti-SAHC antibody | 16× diluted | 62.5 μL |
| Total reaction volume | | 1 mL |

- Plate reader: Tecan Safire2
- Measurement mode: Polarization
- Excitation wavelength: 470 nm
- Emission wavelength: 535 nm
- Emission bandwidth: 10.0 nm
- Gain (manual): 121
- Z-Position (manual): 10,200 μm
- Number of excitation flashes: 3
- Lag time: 100 ms
- No. of readings: 20
- G-factor: 1.0411
- Valid temperature range: 25 °C

10. To measure the background FP, set up negative control wells as indicated in Table 11, but without the addition of anti-SAHC antibody in the stop buffer (total free tracer).
11. To measure the maximum FP, add stop buffer to the wells first, followed by the components in Table 11 (total bound tracer).

*3.3.2 Calculation of mP Values*

The mP values are calculated using the following equation:

$$mP = 1000 * \left[ \left( I_S - I_P \right) / \left( I_S + I_P \right) \right]$$

where $I_S$ is the parallel emission intensity and $I_P$ is the perpendicular emission intensity. The IP value is adjusted by a G-factor multiplier that generated an mP value of 27 for free fluorescein. Background fluorescence is subtracted from the parallel and perpendicular emission intensities using control reactions that lack tracer.

## 4 Notes

1. To obtain a good recovery of the PCR products and IVT RNA purified through G25 spin columns, it is important to follow the centrifugation speed and time as indicated in Subheading 3.1.1. In addition, after loading the PCR product or IVT RNA into the center of the packed resin bed, keep the slope of the resin oriented towards the center of the centrifuge.
2. Avoid multiple freeze–thaw cycles with AdoMet as it is not very stable. To ensure reproducible results for protocols in Subheadings 3.1 and 3.3, care must be taken to select a reliable source of AdoMet with high purity. It should be stored in small aliquots in −20 °C for no more than three rounds of freeze–thaw cycles.
3. The N7 and 2′-*O* buffer conditions differ in protocols in Subheadings 3.1 and 3.2 and are optimized for each method. This is because the enzyme and substrate concentrations used in both methods are not the same. In addition, as the SPA method is optimized for compound screening, the buffer conditions are specifically optimized for measureable MTase activity at neutral pH (pH 7.5).
4. To verify that the DENV N7 and/or 2′-*O* MTase activity has occurred in the SPA or FP assays, digest the RNA template after DENV MTase reaction with tobacco acid pyrophosphatase (TAP), followed by binding to SPA beads. Briefly, purify the RNA product through phenol–chloroform extraction and ethanol precipitation. Digest RNA with 1U TAP (Epicenter Biotechnologies, USA) in a volume of 25 μL, according to the manufacturer's protocol. Transfer the reaction to a 96-well white opaque plate and add equal volume of 2× stop buffer containing SPA beads. Shake the plate and read it on the microbeta counter as described in Subheading 3.2.2. Alternatively, check the reaction products by TLC.
5. The Gppp-110 nt or 211 nt RNA substrates are not used in this assay as they interfere with the FP.
6. Settings stated for Tecan plate reader (in particular gain, Z-height, and G-factor) are machine dependent and should be redetermined when another multiplate reader is used.

## References


1. Chambers TJ, Hahn CS, Galler R et al (1990) Flavivirus Genome Organization, Expression, and Replication. Annu Rev Microbiol 44: 649–688
2. Ackermann M, Padmanabhan R (2001) De novo synthesis of RNA by the dengue virus RNA-dependent RNA polymerase exhibits temperature dependence at the initiation but not elongation phase. J Biol Chem 276:39926–39937
3. Ray D, Shah A, Tilgner M et al (2006) West Nile virus 5′-cap structure is formed by

sequential guanine N-7 and ribose 2′-*O* methylations by nonstructural protein 5. J Virol 80:8362–8370
4. Egloff MP, Benarroch D, Selisko B et al (2002) An RNA cap (nucleoside-2′-*O*-)-methyltransferase in the flavivirus RNA polymerase NS5: crystal structure and functional characterization. EMBO J 21:2757–2768
5. Chung KY, Dong H, Chao AT et al (2010) Higher catalytic efficiency of N7-methylation is responsible for processive N7 and 2′-O methyl-transferase activity in dengue virus. Virology 402(1):52–60
6. Dong H, Chang DC, Hua MH et al (2012) PLoS Pathog 8(4):e1002642
7. Zhou Y, Ray D, Zhao Y et al (2007) Structure and function of flavivirus NS5 methyltransferase. J Virol 81:3891–3903
8. Dong H, Ren S, Zhang B et al (2008) West Nile virus methyltransferase catalyzes two methylations of the viral RNA cap through a substrate-repositioning mechanism. J Virol 82:4295–4307
9. Dong H, Chang DC, Xie X et al (2010) Biochemical and genetic characterization of dengue virus methyltransferase. Virology 405(2):568–578
10. Lim SP, Wen D, Yap TL et al (2008) A scintillation proximity assay for dengue virus NS5 2′-O-methyltransferase–kinetic and inhibition analyses. Antiviral Res 80:360–369
11. Lim SP, Sonntag L, Noble C et al (2011) Small-molecule inhibitors that selectively block flavivirus methyltransferase. J Biol Chem 286(8):6233–6240
12. Graves TL, Zhang Y, Scott JE (2008) A universal competitive fluorescence polarization activity assay for S-adenosylmethionine utilizing methyltransferases. Anal Biochem 373(2):296–306
13. Owicki JC (2000) Fluorescence polarization and anisotropy in high throughput screening: perspectives and primer. J Biomol Screening 340:336–340
14. Jameson DM, Croney JC (2003) Past, present, and future. Comb Chem High Throughput Screen 6:167–173
15. Dandliker WB, Kelly RJ, Dandliker J et al (1973) Fluorescence polarization immunoassay: theory, and experimental method. Immunohistochemistry 10:219–227

# Chapter 21

# Testing Antiviral Compounds in a Dengue Mouse Model

Wouter Schul, Andy Yip, and Pei-Yong Shi


## Abstract

Dengue fever is an emerging mosquito-borne flaviviral disease that threatens 2.5 billion people worldwide. No clinically approved vaccine and antiviral therapy are currently available to prevent or treat dengue virus (DENV) infection. Vertebrate animals other than primates are not normally infectable with DENV; however, a small animal dengue infection model would greatly facilitate the development of a vaccine or an antiviral therapy. To this end, a rodent model for DENV infection has been established in IFN-$\alpha/\beta$ and IFN-$\gamma$ receptor-deficient (AG129) mice. This chapter describes the protocol for the DENV infection model in AG129 mice and testing of antiviral compounds by oral gavage or parenteral injection.


**Key words** Flavivirus, Dengue virus, Animal model, Antiviral testing

## 1 Introduction

Dengue virus (DENV) belongs to the Flavivirus genus in the *Flaviviridae* family and consists of four serotypes. DENV is an enveloped virus, containing a single-stranded positive-sense RNA genome. The virus causes dengue fever (DF) which is a major worldwide health concern with significant economic cost. Two-fifths of the world's population lives in areas that are endemic to DENV, with an estimated 50 million infections per year. Approximately 500,000 cases progress to a more severe form of the disease called dengue hemorrhagic fever (DHF), often requiring hospitalization. Without proper treatment DHF can progress to dengue shock syndrome (DSS), which is often fatal [1–3].

Humans and *Aedes* mosquitoes are the main DENV host. Nonhuman primates (NHP) also serve as a reservoir of sylvatic DENV [4]. However, there is no laboratory animal model for dengue infection that faithfully reflects dengue disease at present. Rodents are not normally susceptible to DENV infection, but through the use of genetically modified mice several mouse infection models for dengue virus have been established. All these models rely on circumventing part of the murine innate immune system

Edwin Yunhao Gong (ed.), *Antiviral Methods and Protocols*, Methods in Molecular Biology, vol. 1030,
DOI 10.1007/978-1-62703-484-5_21, 

in some way, and the most reliable mouse infection models rely on a partial or a complete deficiency in the interferon response pathways.

A common genetically modified mouse strain used to establish DENV infection is the AG129 line which lacks both IFN-α/β and IFN-γ receptors in the 129/Sv background [5]. Johnson and Roehrig [6] were the first to report that DENV infection could be established in the AG129 mice and many research labs have subsequently used this system for further studies of dengue infection in vivo, including testing antiviral compounds. Other knockout mouse lines that have been shown to be infectable with DENV are A129 mice (deficient in the IFN-α/β receptors only) [7], and STAT1−/− mice [8]. It should be noted that even in the AG129 mice, which completely lack the ability to respond to type I and II interferons, not every strain of DENV can establish infection [9]. In our laboratory less than 10 % of the DENV serotype 2 (DENV2) strains could reproducibly establish an infection in the AG129 mice, and an even lower percentage was found for the three other serotypes (unpublished data). A DENV strain that can reliably infect AG129 mice can either be obtained through systematically testing all strains in a laboratory's collection (starting with the DENV2 strains) and/or by requesting a strain known to infect AG129 mice (e.g., strains TSV01, PL046, New Guinea C, WestPac74) from other laboratories.

Most strains known to infect the AG129 mice give rise to viremia, viral load in tissues, and inflammatory responses, but do not lead to overt signs of disease and animals will clear the infection within 4–7 days. However, a few strains have been generated through repeated adaptive selection that can cause severe symptoms and death after infection of AG129 mice: the D2S10 [10] and plaque-purified S221 clone [11], and the D220 strain [12]. Tan et al. [13] found that non-adapted DENV2 strain D2Y98P can also cause pathogenic and lethal infection in AG129 mice, and showed that this property is caused by a specific mutation in the viral NS4B protein which can be conferred to a nonpathogenic infectious strain by site-directed mutagenesis [14].

An additional phenomenon associated with DENV infection is known as antibody-dependent enhancement (ADE). In humans, immunity to one of the four DENV serotypes can increase disease severity upon subsequent infection with another serotype, caused by low-affinity cross-reactive antibodies. In 2010, Balsitis et al. [15] described ADE infection of AG129 mice with enhanced disease. This was achieved by injecting the AG129 mice with a low dose of cross-reactive polyclonal or monoclonal antibodies one day before DENV infection. Here we describe a well-established AG129 mouse model for testing the in vivo efficacy of antiviral compounds by oral gavage or parenteral injection.

## 2 Materials

### 2.1 Virus Production

1. C6/36 cells: ATCC (CRL-1660).
2. Virus: TSV01 (DENV2), NGC (DENV2; ATCC, VR-1255), Westpac74 (DENV1), D2S10 (mouse-adapted DENV2), D2Y98P (DENV2) (*see* **Note 1**).
3. Culture medium for C6/36 cells: RPMI 1640 medium (Gibco), 5 % fetal bovine serum (FBS), 1 % penicillin and streptomycin (P/S).

### 2.2 Mouse Infection

1. AG129 mice (*see* **Note 2**): Obtained from B&K Universal Ltd, Hull, UK, to establish a breeding colony. Male and female mice between 7 and 12 weeks of age.
2. Syringes: 0.5 mL insulin syringes with attached 29G needles (BD).

### 2.3 Dosing

Both stainless steel reusable or disposable straight needles can be used, depending on personal preference. Since mice between 7 and 12 weeks were used for infection, there is a wide range of body weights to be considered when choosing a suitable gavage needle. Younger mice usually weigh less than their older siblings.

1. Gavage needles for mice between 15 and 20 g body weights: 22G, 1 to 1.5 in. in length, 1.25 mm ball diameter.
2. Gavage needles for mice between 20 and 25 g body weights: 20G, 1–3 in. in length, 1.90–2.25 mm ball diameter.

### 2.4 Sampling

1. Anesthesia: Isoflurane (2-chloro-2-(difluoromethoxy)-1,1,1-trifluoro-ethane) used in a calibrated isoflurane anesthesia machine suitable for small animals (e.g., VetEquip Mobile Laboratory Animal Anesthesia System).
2. Collection: Microhematocrit Capillary Tubes (Fisher, Cat. No. 22-362-574), Microtainer EDTA tubes (BD, Cat. No. 365974).

### 2.5 Sample Analysis

#### 2.5.1 Plaque Assay

1. BHK-21 cells: ATCC (CRL-10).
2. Culture medium for BHK-21 cells: RPMI 1640 medium supplemented with 10 % FBS, 1 % P/S.
3. Plaque assay overlay: 0.8 % methyl cellulose with 2 % FBS:
   (a) Prepare 500 mL of 1.6 % methyl cellulose by dissolving methyl cellulose powder in water (slowly adding powder to water while stirring), autoclave, store overnight at 4 °C for hydration before use.
   (b) Prepare 500 mL of 2× RPMI 1640 by dissolving one bag of RPMI 1640 powder (for 1 L) in Millipore water. Add

20 mL FBS, 10 mL P/S, 5 mL $NaHCO_3$ (7.5 %), 25 mL of 1 M HEPES buffer. Filter through a 0.2 μm membrane.

(c) Mix 500 mL of 2× RPMI 1640 with 500 mL of 1.6 % methyl celllulose, then add 5 mL DMSO. Mix well, aliquot, and store at 4 °C until use.

4. 1 % (w/v) crystal violet solution: Dissolve 5 g crystal violet in 500 mL water, stir overnight until dissolved.

#### 2.5.2 RT-PCR

1. SuperScript III Reverse Transcriptase (Invitrogen, Cat. No. 18080-044).
2. RNasin (40 U/μL) (Promega).
3. FastStart DNA Master$^{PLUS}$ SYBR Green Kit (Roche Diagnostics, Cat. No. 03515869).
4. Glass capillaries (Roche Diagnostics).
5. QIAamp Viral RNA Mini Kit (Qiagen).

#### 2.5.3 NS1 ELISA

Platelia Dengue NS1 Ag Kit (Bio-Rad, Cat. No. 72830).

#### 2.5.4 Mouse Cytokines

1. Cytometric Bead Array (CBA) Mouse Inflammation Kit (BD, Cat. No. 552364).
2. A dual-laser flow cytometer equipped with a 488- or 532-nm and a 633- or 635-nm laser capable of distinguishing 576-, 660-, and >680-nm fluorescence.

## 3 Methods

### 3.1 Preparing Virus Stocks (See Note 3)

1. Grow C6/36 cells in RPMI 1640 medium + 5 % FBS at 28 °C, to 70–90 % confluence in one or more tissue culture flasks. A cell count can be performed to get an estimate of the number of cells per flask.
2. Take one vial of DENV stock with known titer from −80 °C freezer and thaw rapidly. Dilute the virus in RPMI 1640 medium + 5 % FBS to result in a multiplicity of infection (MOI) of 0.01–0.1 plaque-forming units (PFU)/cell (i.e., a flask containing $10^7$ cells receives 1 mL of $10^5$–$10^6$ PFU of virus).
3. Incubate the infected cells at 28 °C for 3–7 days, depending on the growth characteristics of the DENV strain. Check the signs of increased cell death (cytopathic effect) daily as indication of viral proliferation although this may not be apparent for every DENV strain.
4. After 3–7-day incubation remove the culture medium from the infected cells and centrifuge at 5,000 × *g* for 5 min. Aliquot

the supernatant into cryovials, quickly freeze in liquid nitrogen, and store at −80 °C.

5. Thaw one vial to determine the viral titer by a plaque assay (*see* Subheading 3.5.1).

### 3.2 Infection of Mice by Parenteral Injection (See Note 4)

1. Transfer AG129 mice into cages according to their experimental group.
2. Take one or more vials of DENV stock from the −80 °C and thaw rapidly. Dilute the virus to an appropriate titer in RPMI1640 medium and let it warm to room temperature.

#### 3.2.1 Intraperitoneal Injection of DENV

Intraperitoneal injection of 0.25 mL at $5 \times 10^6$ PFU/mouse or higher of a nonpathogenic DENV strain usually gives reproducible viremic infection. Therefore, the virus should be diluted to a titer of $2 \times 10^7$ PFU/mL.

The non-adapted pathogenic DENV strain D2Y98P can be injected intraperitoneally as low as $10^4$ PFU/mouse to still give 100 % mortality although injection with higher titers gives a more robust and viremic infection [13].

#### 3.2.2 Intravenous Injection of DENV

Mouse-adapted pathogenic DENV strains like D2S10 are typically injected intravenously in the tail vein. For the D2S10 strain the amount to be injected should be $10^7$ PFU/mouse or higher (*see* **Note 5**).

### 3.3 Compound Dosing (See Note 6)

Compounds to be tested for their in vivo activities can be dosed either orally or by injection.

#### 3.3.1 Injection of Compound

Completely dissolve the test compound in the vehicle fluid which is typically an aqueous buffer (e.g., PBS), or contains organic excipients allowed under animal testing protocols. The advantage of dosing by injection is that 100 % of the compound is delivered into the body allowing for maximal exposure; however, it will not reflect the oral bioavailability of a compound. Injection is typically by intravenous, subcutaneous, or intraperitoneal route.

#### 3.3.2 Oral Dosing of Compound

1. Oral dosing of a compound is performed by oral gavage and takes gastrointestinal stability and permeability, extended period of absorption, and first-pass metabolism through the liver into account which is particularly relevant if ultimately an oral dengue drug is intended.
2. For oral dosing a compound can be either in solution or suspension.
3. Fill a 1 mL syringe with gavage needle with drug solution/suspension and carefully orally dose a specific volume (typically 0.1–0.2 mL) into the stomach of a mouse.

#### 3.3.3 Dosing Schedule

The compound dosing schedule can vary from a single dose to multiple doses per day over several days, starting directly after DENV infection or one or several days after or before infection. Dosing levels are dependent on the expected exposure and in vitro potency of the compound but typically range from 3 to 300 mg/kg (i.e., milligram of compound per kilogram of mouse body weight) (*see* **Note 7**).

### 3.4 Sampling

1. Infected mice can be sampled daily for several days by retro-orbital puncture. Mice are anesthetized by isoflurane gas inhalation. A glass capillary is used to draw a small amount (0.1–0.2 mL) of blood from the retro-orbital plexus of the anesthetized mouse. The blood is collected into an EDTA tube. Alternatively, blood can be collected in tubes containing sodium citrate solution to give a final concentration of 0.4 % (*see* **Note 8**).
2. After sampling, the mouse is placed back into a cage. Make sure that it does not get hypothermic, e.g., by placing the cage on a warm pad or by wrapping the mouse with a tissue paper where it can recover from the effects of anesthesia.
3. Perform terminal bleeding similarly by retro-orbital puncture followed by cervical dislocation under isoflurane anesthesia.
4. Perform sample processing of anti-coagulated blood by immediate centrifugation in an Eppendorf centrifuge for 3 min at 7,000 × *g*. Remove the supernatant plasma carefully and transfer to a labeled cryo-tube. Subsequently, take 12 μL of plasma and add it to another labeled cryo-tube containing 588 μL of RPMI 1640 medium (effectively diluting 50×). Both tubes are then snap frozen in liquid nitrogen and stored at −80 °C for later analysis.

### 3.5 Sample Analysis

#### 3.5.1 Plaque Assay

1. Using serum-free RPMI 1640 as diluent, perform a fivefold serial dilution on the 1:50 pre-diluted mouse plasma sample (*see* Subheading 3.4, **step 4**) in a sterile 96-well microtiter plate, resulting in 4 serially diluted samples at 1:50, 1:250, 1:1,250, and 1:6,250.
2. Carefully overlay 200 μL of each diluted sample onto BHK21 cells in a 24-well plate. Repeat for the remaining samples. Swirl the plate to ensure adequate mixing and incubate the plate at 37 °C, 5 % $CO_2$ for 1 h, briefly shaking the plate gently at the 30-min mark.
3. Remove virus inoculums with vacuum aspiration and overlay with 500 μL of 0.8 % methyl cellulose medium with 2 % FBS, taking care not to disturb the cell monolayer.
4. Incubate at 37 °C, 5 % $CO_2$ for 4 to 6 days depending on the strain of dengue virus used.

5. Gently rinse the plates several times in 3.7 % formaldehyde, and then soak for 1 h at room temperature. Shake plates (wells facing down) a few times to remove the methyl-cellulose layer.
6. Rinse plates in copious volumes of water in a large container. Shake plate (wells facing down) to remove excess water from the wells.
7. Add several drops of 1 % crystal violet into each well, enough to cover the cell monolayer, and set aside for 1 min.
8. Invert plate on layers of paper towels to remove excess crystal violet.
9. Rinse plates in copious volumes of water in a beaker. Shake plates (wells facing down) to remove excess water.
10. Dry plates in a 50 °C oven and count the number of plaques on the plates to determine the viral titer of each sample.

#### 3.5.2 RT-PCR

1. Perform viral RNA isolation from the mouse plasma samples and appropriate standards (virus samples with known titer) with the QIAamp Viral RNA Mini Kit according to the manufacturer's instructions.
2. Elute RNA in 60 μL of buffer AVE (supplied with the QIAGEN kit).
3. Synthesis of cDNA is performed using SuperScript III reverse transcriptase. The reaction mixture consists of the following:

| | |
|---|---|
| RNase-free water | 5 μL |
| Random hexamers (250 ng/μL) | 1 μL |
| dNTPs (10 μM) | 1 μL |
| RNA sample | 6 μL |

4. Heat the mixture to 65 °C for 5 min and place on ice immediately. Then add the following:

| | |
|---|---|
| 5× first-strand buffer | 4 μL |
| DTT (100 mM) | 1 μL |
| RNasin (40 U/μL) | 1 μL |
| Superscript III enzyme | 1 μL |

5. Incubate the reaction mixture at 25 °C for 5 min and then at 42 °C for 60 min.
6. Stop the reaction by heating the mixture to 70 °C for 15 min.
7. Perform real-time PCR on a Roche LightCycler 2.0 system using the FastStart DNA MasterPLUS SYBR Green I kit.

Pipette the following reaction mixture into an Eppendorf tube for each sample:

| | |
|---|---|
| SYBR Green I | 4 μL |
| cDNA | 2 μL |
| Primer 1 (10 μM) (*see* **Note 9**) | 1 μL |
| Primer 2 (10 μM) (*see* **Note 9**) | 1 μL |
| PCR-grade water | 12 μL |

8. Carefully pipette the reaction mixture into the glass capillaries.
9. Fit capillaries into precooled adapters and centrifuge at 700 × *g* for 5 s.
10. Transfer the capillaries from the rotor into the Lightcycler and begin the amplification using the following LightCycler PCR program:

    Denaturing and enzyme activation:

    10 min at 95 °C, slope 20 °C/s, acquisition mode: none

    Amplification:

    40 cycles of:

    5 s at 95 °C, slope 20 °C/s, acquisition mode: none

    5 s at 60 °C, slope 20 °C/s, acquisition mode: none

    20 s at 72 °C, slope 20 °C/s, acquisition mode: single

    Melting curve:

    1 s at 95 °C, slope 20 °C/s, acquisition mode: none

    5 s at 65 °C, slope 20 °C/s, acquisition mode: none

    0 s at 95 °C, slope 0.1 °C/s, acquisition mode: continuous
11. Use the cycle crossing point (Cp) values from the standards to generate a standard curve. The log-linear correlation between the Cp value and the virus concentration holds true over a range of $10^2$ to $10^5$ PFU/mL for all four dengue serotypes. With this information, the viral titer of the mouse samples can be determined.

#### 3.5.3 NS1 ELISA

1. Carefully plan the plate layout for controls and samples.
2. Take the carrier tray and the strips out of the protective pouch, and leave at room temperature for a few minutes to allow temperature equilibration.
3. For the test samples, the dilution factor is 1:500, prepared by diluting 5 μL plasma in 45 μL PBS, mix well to get a 1:10

dilution, then add 1.2 μL from this 1:10 dilution mixture to 58.8 μL of R7 diluent, and mix well to get a 1:500 dilution.

4. For the control samples, use undiluted (neat), strictly following the indicated distribution sequence, and distribute successively in the wells:

   50 μL of diluent

   50 μL of samples (or controls)

   100 μL of diluted conjugate

5. Cover the reaction microplate with the supplied adhesive sealing film, pressing firmly onto the plate to ensure complete sealing.
6. Incubate the microplate in a 37°C incubator for 90 min.
7. Prepare the dilution of the washing solution.
8. At the end of the incubation period, carefully remove the adhesive sealing film. Remove the contents of all wells into a container for biohazard waste via vacuum aspiration. Wash the microplate 6 times with washing solution. Aspirate the washing solution. Leave the last round of washing solution in the wells.
9. Prepare the R8 + R9 solution as described in the user guide. Remove the remaining washing solution and place 160 μL of the R8 + R9 solution into each well. Allow reaction to develop at room temperature for 5 min.
10. Add 100 μL of the stop solution to each well by using the same sequence and rate of distribution as for the development solution.
11. Carefully wipe the plate bottom with lint-free tissue paper and read the optical density at 450/620 nm using an ELISA plate reader within 30 min after stopping the reaction.

#### 3.5.4 Detection of Mouse Cytokines

A brief protocol is attached below; for more details check the user guide from the manufacturer.

1. Reconstitute mouse inflammation standards in assay diluent.
2. Dilute standards by serial dilutions using the assay diluent.
3. Mix 10 μL/test of each mouse inflammation capture bead suspension (vortex before aliquoting).
4. Transfer 50 μL of mixed beads to each assay tube.
5. Add standard dilutions and test samples to the appropriate sample tubes (50 μL/tube).
6. Add PE detection reagent (50 μL/tube) and incubate for 2 h at room temperature.
7. In the meantime, perform the cytometer setup bead procedure.

8. Add Cytometer setup beads (vortex before adding) to set up tubes A, B, and C (50 μL/tube).
9. Add 50 μL of FITC positive control to tube B and 50 μL of PE positive control to tube C.
10. Incubate for 30 min at room temperature, protected from light.
11. Add 400 μL of wash buffer to tubes B and C.
12. Add 450 μL of wash buffer to tube A.
13. Use tubes A, B, and C for cytometer setup as described in more detail in the supplied user guide.
14. After 2-h incubation, wash samples with 1 mL wash buffer and centrifuge at 200 × *g* for 5 min.
15. Add 300 μL of wash buffer to each assay tube and analyze samples.
16. Acquire the samples on a flow cytometer. For details, go to bdbiosciences.com/cbasetup and select the appropriate flow cytometer under CBA Kits: Instrument Setup.
17. Analyze BD CBA Mouse Inflammation Kit data using FCAP Array software. For instructions on analysis, go to bdbiosciences.com/cbasetup and see the Guide to Analyzing Data from BD CBA Kits, using FCAP Array software.

#### 3.5.5 Histopathology

If needed, mouse liver and kidney can be harvested and fixed in 10 % formalin buffered in PBS, pH 7.4 for 2 days at room temperature before being sent for histopathology studies.

## 4 Notes

1. All procedures involving DENV should be performed according to the local biosafety regulations and protocols, including importation, storage, registration, and handling of the virus. In most countries, working with DENV needs to be conducted in a class II biosafety cabinet or higher. A risk assessment of the virus, the laboratory, and the procedures may be necessary. Researchers involved in biosafety and/or animal experiments should be evaluated for additional training needs, and should be made aware of all risks involved in the work.
2. All procedures involving mice require a protocol pre-approval by the Institutional Animal Care and Use Committee (IACUC). The investigator should have an experimental plan including the following:

(a) Calculation of the number of animals required per experimental group: This depends upon the level of variation observed between experimental animals, as well as the level of difference expected between control and experimental groups. In our laboratory we typically used 6 animals per group allowing statistically significant detection of five- to tenfold changes in viremia levels.

(b) A plan that clearly defined experimental end-points: All researchers working with infected mice should be able to recognize signs of severe illness, including hunched posture, ruffled coat, weight loss, and immobility. Mice are to be humanely terminated when they show signs of severe illness (method and signs to be agreed upon with the IACUC).

3. Production of high-titer stocks of DENV takes optimization and may vary for different strains. For some strains MOIs of 0.1–1 with 6–7 days of incubation may be required, while for other strains the optimal viral titer is reached after lower MOI infection or shorter incubation.

   DENV is grown in C6/36 mosquito cells at 28 °C which can be done by adjusting a 37 °C incubator to 28 °C. However, these incubators can be prone to overshooting the set temperature by one or more degrees, especially if the set temperature is close to room temperature and due to thermal insulation of the incubator stay at this elevated level for an hour or more resulting in poor cell and viral growth. It is therefore advisable to use a dedicated refrigerated incubator that can both heat and cool the incubation chamber to ensure incubation at the proper temperature.

4. Mouse handling and experimentation techniques: The experiments described in this chapter include references to mouse experimental techniques such as intraperitoneal and subcutaneous injections, tail vein injection, oral gavage, retro-orbital bleeding, and cervical dislocation. The details of these techniques are beyond the scope of these instructions and investigators are referred to mouse experimentation instruction manuals and hands-on training to learn more about these procedures.

5. Since intravenous injection of pathogenic virus requires small volumes (0.05–0.1 mL) containing large amounts of virus ($10^7$ PFU), this requires a stock with viral titer of $10^8$ PFU/mL or higher, which can be challenging to achieve and require optimization of the viral production (*see* Subheading 3.1). Alternatively, concentrating the virus by (ultra)centrifugation can be considered although this often results in an overall loss of infectious virus up to 90 %. Therefore, a large starting volume of virus stock may be needed.

6. Pharmacokinetic compound profiling: It is advisable that, before starting dosing novel compounds to DENV-infected AG129 mice, a series of single-dose pharmacokinetic studies are conducted with novel compounds in uninfected wild-type 129/Sv mice to determine the plasma concentrations, half-life, rate of clearance, oral bioavailability, etc. In our experience, we did not observe any difference in pharmacokinetic properties of compounds between DENV-infected AG129 mice and uninfected wild-type 129/Sv mice (unpublished data). Such pharmacokinetic data is extremely valuable to optimize the dosing schedule (e.g., multiple dosing per day for a compound with a short half-life) or to interpret any (lack of) antiviral effect.
7. Dosing schedule options: Compound efficacy testing in DENV-infected AG129 mice can involve dosing several different compounds at a fixed dose for head-to-head comparison, or a dose–response experiment of a single compound at different dosing levels. In the latter case, a threefold increase over a relevant range is often applied (e.g., dosing at 3, 10, 30, 100 mg/kg or 5, 15, 50, 150 mg/kg). It is recommended to include a negative control group in each experiment that is dosed with the vehicle fluid, and a positive control group dosed with a known DENV inhibitor, e.g., NN-DNJ, BuCast [9], or NITD008 [17].
8. Do not use heparin or heparin-treated capillaries, tubes, or containers to anti-coagulate blood. Heparin can competitively prevent binding of DENV to heparan sulfates on the surface of cells blocking viral entry, interfering with viral infection assays, e.g., plaque assays.
9. PCR Primers

   For quantitative PCR detection of DENV genomes of all four serotypes, we have been using the D1 and D2 primers as described by Lanciotti et al. [16]. The primer sequences are as follows:

   D1 forward primer: 5′-TCAATATGCTGAAACGCGCGAGAAACCG-3′

   D2 reverse primer: 5′-TTGCACCAACAGTCAATGTCTTCAGGTTC-3′

## References

1. WHO (2000) WHO report on global surveilance of epidemic-prone infectious diseases-dengue and dengue haemorrhagic fever. http://www.who.int/csr/resources/publications/dengue/CSR_ISR_2000_1/en/. Accessed 10 Oct 2012
2. Gubler DJ (1998) Dengue and dengue hemorrhagic fever. Clin Microbiol Rev 11(3):480–496
3. Suaya JA, Shepard DS, Siqueira JB et al (2009) Cost of dengue cases in eight countries in the Americas and Asia: a prospective study. Am J Trop Med Hyg 80(5):846–855
4. Vasilakis N, Holmes EC, Fokam EB et al (2007) Evolutionary processes among sylvatic dengue type 2 viruses. J Virol 81(17):9591–9595

5. Yauch LE, Shresta S (2008) Mouse models of dengue virus infection and disease. Antiviral Res 80(2):87–93
6. Johnson AJ, Roehrig JT (1999) New mouse model for dengue virus vaccine testing. J Virol 73(1):783–786
7. Cassetti MC, Durbin A, Harris E et al (2010) Report of an NIAID workshop on dengue animal models. Vaccine 28(26):4229–4234
8. Chen ST, Lin YL, Huang MT et al (2008) CLEC5A is critical for dengue-virus-induced lethal disease. Nature 453:672–676
9. Schul W, Liu W, Xu HY et al (2007) A dengue fever viremia model in mice shows reduction in viral replication and suppression of the inflammatory response after treatment with antiviral drugs. J Infect Dis 195(5):665–674
10. Shresta S, Sharar KL, Prigozhin DM et al (2006) Murine model for dengue virus-induced lethal disease with increased vascular permeability. J Virol 80(20): 10208–10217
11. Yauch LE, Zellweger RM, Kotturi MF et al (2009) A protective role for dengue virus-specific CD8+ T cells. J Immunol 182(8): 4865–4873
12. Orozco S, Schmid MA, Parameswasan P et al (2012) Characterization of a model of lethal dengue virus 2 infection in C57BL/6 mice deficient in the interferon-α/β receptor. J Gen Virol 93(Pt 10):2152–2157
13. Tan GK, Ng JK, Trasti SL et al (2010) A non-mouse-adapted dengue virus strain as a new model of severe dengue infection in AG129 mice. PLoS Negl Trop Dis 4(4):e672
14. Grant D, Tan GK, Qing M et al (2011) A single amino acid in nonstructural protein NS4B confers virulence to dengue virus in AG129 mice through enhancement of viral RNA synthesis. J Virol 85(15):7775–7787
15. Balsitis SJ, Williams KL, Lachica R et al (2010) Lethal antibody enhancement of dengue disease in mice is prevented by Fc modification. PLoS Pathog 6(2):e1000790
16. Lanciotti RS, Calisher CH, Gubler DJ et al (1992) Rapid detection and typing of dengue viruses from clinical samples by using reverse transcriptase-polymerase chain reaction. J Clin Microbiol 30:545–551
17. Yin Z, Chen YL, Schul W et al (2009) An adenosine nucleoside inhibitor of dengue virus. Proc Natl Acad Sci USA 106(48):20435–20439

# Chapter 22

# Construction of Self-Replicating Subgenomic West Nile Virus Replicons for Screening Antiviral Compounds

**Sofia L. Alcaraz-Estrada, Erin Donohue Reichert, and Radhakrishnan Padmanabhan**

## Abstract

Mosquito-borne flavivirus RNA genomes encode one long open reading frame flanking 5′- and 3′-untranslated regions (5′- and 3′-UTRs) which contain *cis*-acting RNA elements playing important roles for viral RNA translation and replication. The viral RNA encodes a single polyprotein, which is processed into three structural proteins and seven nonstructural (NS) proteins. The regions coding for the seven NS proteins are sufficient for replication of the RNA. The sequences encoding the structural genes can be deleted except for two short regions. The first one encompasses 32 amino acid (aa) residues from the N-terminal coding sequence of capsid (C) and the second, 27 aa region from the C-terminus of envelope (E) protein. The deleted region can be substituted with a gene coding for a readily quantifiable reporter to give rise to a subgenomic reporter replicon. Replicons containing a variety of reporter genes and marker genes for construction of stable mammalian cell lines are valuable reagents for studying the effects of mutations in translation and/or replication in isolation from processes like the entry and assembly of the virus particles. Here we describe the construction of two West Nile virus (WNV) replicons by overlap extension PCR and standard recombinant DNA techniques. One has a *Renilla* luciferase (Rluc) reporter gene followed by an internal ribosome entry site (element) for cap-independent translation of the open reading frame encompassing the carboxy-terminal sequence of E to NS5. The second replicon has in tandem the Rluc gene, foot and mouth disease virus 2A, and neomycin phosphotransferase gene that allows establishment of a stable mammalian cell line expressing the Rluc reporter in the presence of the neomycin analog, G418. The stable replicon-expressing Vero cell line has been used for cell-based screening and determination of $EC_{50}$ values for antiviral compounds that inhibited WNV replication.

**Key words** West Nile virus, Replicon construction, Overlap extension PCR, *Renilla* luciferase reporter, Neomycin resistance selection, Foot and mouth disease virus 2A, Bicistronic expression

## 1 Introduction

The key steps in flavivirus life cycle are entry of the virus into an appropriate host cell via receptosome-mediated endocytosis followed by uncoating, translation of the positive-sense RNA genome in the endoplasmic reticulum (ER), polyprotein processing, assembly of the viral replicase complex, genome replication, packaging of

Edwin Yunhao Gong (ed.), *Antiviral Methods and Protocols*, Methods in Molecular Biology, vol. 1030,
DOI 10.1007/978-1-62703-484-5_22, © Springer Science+Business Media, LLC 2013

the progeny RNAs into virus particles, and release of the virions into the extracellular milieu. The replicons devoid of most of the coding region of structural genes but replaced by a heterologous gene encoding a readily quantifiable reporter or a selectable marker for establishment of a stable cell line expressing the replicon are very useful tools for studying several steps of the virus life cycle except the entry, assembly, and maturation of the virions [1]. Replicons and replicon-expressing cell lines are also used frequently for screening antiviral activity of small-molecule compounds which could potentially target either a viral gene or a host gene essential for flavivirus replication [2–4]. Many replicons from different *Flaviviridae* family members have been reported, including Kunjin virus [1], West Nile virus (WNV) [2, 5–9], dengue virus [3, 10–16], tick-borne encephalitis virus [17, 18], yellow fever virus [19–21], and Japanese encephalitis virus [22].

The WNV replicons described here were constructed using the overlap extension PCR method to remove the majority of the sequences that encode the structural genes and replaced them with the *Renilla* luciferase (Rluc) gene as a reporter. The replicon sequence starts with the 5′-UTR of WNV RNA. This is followed by 96 nucleotides (nt) encoding 32 aa from the N-terminal region of the capsid protein. The N-terminal capsid coding sequence contains the 5′-cyclization sequence (5′-CS) that is involved in long-distance RNA–RNA interaction with the complementary 3′-cyclization sequence located within the 3′-UTR. This interaction was shown to be important for minus-strand RNA synthesis in vitro [23] and for RNA replication in mammalian cells [7, 24, 25]. The 32 aa capsid coding sequence was fused to the *Renilla* luciferase reporter gene. In one replicon, the Rluc gene reading frame is terminated with a stop codon. This is fused to the encephalomyocarditis virus (EMCV) leader sequence that functions as an internal ribosomal entry site (IRES) for cap-independent translation of the long open reading frame (ORF). The ORF contains 81 nt (27 aa) from the C-terminus of E protein followed by the coding sequence for the precursor polyprotein encompassing seven nonstructural proteins, $NH_2$-NS1-NS2A-NS2B-NS3-NS2B-NS3-NS4A-NS4B-NS5-COOH. The 27 aa of the C-terminal E protein fused to the NS polyprotein precursor was shown to be essential for proper processing of the E-NS1 junction and folding of the polyprotein in the endoplasmic reticulum [26]. In the second replicon, the Rluc gene reading frame is fused sequentially downstream to the 20 aa foot and mouth disease virus (FMDV) 2A protease coding sequence, neomycin resistance gene ($Neo^r$), a termination codon, and the EMCV leader followed by the same ORF as in the first replicon. This replicon allowed us to establish a stable monkey kidney (Vero) cell line expressing the WNV Rluc replicon by a selective pressure by culturing the cells in the presence of G418, an analog of neomycin.

These replicons are designated as 5′-UTR-$C_{32codons}$-Rluc*-IRES-$E_{27codons}$-NSorf-3′-UTR and 5′-UTR-$C_{32codons}$-Rluc-FMDV2A-$Neo^{r}$*-IRES-$E_{27codons}$-NSorf-3′-UTR (*indicates translation termination codon) to indicate the sequential insertion of heterologous sequences into self-replicating WNV subgenomic replicons. The first replicon is useful for transient expression of the Rluc reporter under regulation of *cis*-acting RNA elements or transacting viral or host factors. The second replicon was used to establish Vero cell line stably expressing Rluc reporter for screening small-molecule compounds as antiviral drugs.

## 2 Materials

Use molecular biology-grade chemicals and reagents and Milli-Q water for preparation of all solutions. Store all reagents at room temperature (unless indicated otherwise). Follow established protocols and regulations in the Institutional Biosafety Committee manual for handling biohazard materials such as recombinant DNA and plasmid-transformed *E. coli* waste disposal.

1. PCR: For the amplification of the fragments, use about 1 μg of plasmid DNA template, 10 μM of oligonucleotide primers (e.g., A, B, C, D, E, and F in Table 1), 10× PCR buffer for Ex-Taq polymerase (Takara Bio Inc.), and four ribonucleoside triphosphate mixture (NTP; 250 μM of each).
2. Ligation: 10× ligation buffer (500 mM Tris–HCl, pH 7.5, 100 mM $MgCl_2$, 5 mM ATP, 100 mM dithiothreitol) (*see* **Note 1**), 10 mM ATP (*see* **Note 2**), 10 mg/mL acetylated

**Table 1**
**PCR primer sequences used for construction of WNV reporter replicon with $Neo^{r}$ selectable marker for stable expression in mammalian cells**

| Primer name | Primer sequence |
|---|---|
| Primer A | 5′-agtagttcgcctgtgtgagct-3′ |
| Primer B | 5′-ctttagtcctatcaaggacaat-3′ |
| Primer C | 5′-<u>gtccttgataggactaaag</u>atggcttccaaggtgtac-3′ |
| Primer D | 5′-ggtattatcgtgtttttcaaaggaa-3′ |
| Primer E | 5′-<u>cacgataataccatg</u>gcccgtgacaggtcaattgcta-3′ |
| Primer F | 5′-accggagctcttgcctgccaat-3′ |

Primer C: Underlined sequence corresponds to the N terminal sequence of the capsid that is followed by the luciferase coding sequence
Primer E: Underlined sequence corresponds to the 3′-end of the IRES sequence followed by the C-terminal of E protein coding sequence

bovine serum albumin (BSA) (New England Biolabs) (*see* **Note 3**), 0.5 μL T4 DNA ligase (New England Biolabs).

3. Plasmid clones: The plasmid clone encoding the full-length cDNA of WNV-956 strain in pBR322 vector [6] was a gift from Drs. Lewis Markoff and Li Yu (CBER, FDA, Bethesda, MD). The cDNAs encoding the dengue virus type 2 replicons are from Padmanabhan's laboratory.
4. Restriction enzyme digestions: All restriction enzymes and 10× buffers are from New England Biolabs unless otherwise indicated. Follow incubation conditions for restriction enzyme digestions recommended by the manufacturer. For digestion with *Mfe*I + *Xba*I plasmid DNA (40 μg) is used in the presence of 1× manufacturer-supplied NEBuffer 4 (20 mM Tris–acetate, 50 mM potassium acetate, 10 mM magnesium acetate, and 1 mM dithiothreitol, pH 7.9). For digestion with *Bgl*II restriction enzyme 40 μg of plasmid DNA are cut in the presence of 1× NEBuffer 3 (50 mM Tris–HCl, pH 7.9, 100 mM NaCl, 10 mM $MgCl_2$, and 1 mM dithiothreitol).
5. Bacterial transformation: *E. coli* Stbl2 competent cells (Invitrogen) are used for transformation. Prepare SOC medium by mixing 2 g of tryptone, 0.5 g yeast extract, 0.25 mL of 1 M KCl, 1 mL of 1 M NaCl, 1 mL of 1 M $MgCl_2$, and 98 mL water. Sterilize by autoclave. After it is cooled down, add 1 mL of $MgSO_4$ and 1 mL of 2 M glucose. Filter through a 0.22 μm filter and store in 5 mL aliquots.
6. In vitro transcription: Ampliscribe SP6 High Yield Transcription kit (Epicentre) is used. Ampliscribe 10× reaction buffer (final concentration: 1×; *see* **Note 4**) and the cap analog, m7G(5′)ppp(5′)G (New England Biolabs), are used following the protocol recommended by the manufacturer. Final NTP concentrations for the production of capped WNV RNA are 5 mM each of ATP, CTP, and UTP and 1 mM of GTP. Make up a stock reaction mixture containing 25 mM each of ATP, CTP, and UTP and 5 mM of GTP by adding 5 μL each of 100 mM stock solutions of ATP, CTP, and UTP and 1 μL of 100 mM GTP into 4 μL of $H_2O$, to produce 20 μL of the mixture (*see* **Note 5**). Cap analog m7G(5′)ppp(5′)G is supplied by the manufacturer in a lyophilized form. To prepare a stock solution, add 32.75 μL of $H_2O$ to have a concentration of 40 mM (*see* **Note 6**).
7. Mammalian cell electroporation: For cell electroporation, use 1× phosphate-buffered saline (PBS). For the preparation of 1× PBS buffer, add the following: 8 g NaCl, 0.2 g KCl, 1.44 g $Na_2HPO_4$, 0.24 g $KH_2PO_4$, in 800 mL of $H_2O$. Adjust the pH to 7.4 with HCl. Add $H_2O$ to 1 L. Autoclave and store at room temperature. The equipment used was Gene PulserXcell™

Eukaryotic System (Bio-Rad). Sterile electroporation cuvettes (5 × 0.4 cm).

8. Mammalian cell culture: For the preparation of complete medium, mix the following: Dulbecco's modified Eagle medium (DMEM) (Gibco), 10 % heat-inactivated fetal bovine serum (FBS) (Gibco), 4.5 g/L glucose, 2 mM glutamine, and 1× of penicillin–streptomycin. Heat inactivation of FBS: Incubate for 30 min at 56 °C, mixing every 5 min. Make 50 mL aliquots and store at −20 °C. Use 0.25 % trypsin–EDTA (1×) (Gibco) to dissociate adherent cells from the cell culture flasks or plates.
9. Mammalian cell cloning: Bel-Art cell culture paper cloning discs (Scienceware® is trademark of Bel-Art products), 5 mm size (available from Sigma-Aldrich, Cat. No. Z374458-100EA), and a sterile fine point forceps. G418: 50 mg/mL stock solution (*see* **Note 7**).
10. Renilla luciferase (Rluc) assay kit: For quantification of Rluc activity, the kit is purchased from Promega and it contains all the reagents necessary to measure the activity using a luminometer.

## 3 Methods

### 3.1 PCR for Construction of WNV Rluc Reporter Replicon

#### 3.1.1 PCR-1

1. Follow standard PCR protocols. For initial PCR, the reaction mixture (50 μL) consists of the following components in an appropriate microcentrifuge tube (0.2 or 0.5 mL capacity, depending on the capacity of the heating block in the thermal cycler used for PCR): $H_2O$ (37.5 μL), 10× PCR buffer (5 μL), 2.5 mM each of dNTP mixture (4 μL; the final concentration of each dNTP is 200 μM), 10 μM primer A (Table 1; Fig. 1a) (1 μL), 10 μM primer B (1 μL), 500 ng of pBR322 WNV-956 as DNA template (0.5 μL), 5U Ex-Taq polymerase (1 μL). Vortex well to mix, and then spin briefly in a microfuge.
2. Perform PCR-1 using the following cycle profiles: initial denaturation at 94 °C for 3 min, followed by 30–35 cycles of denaturation at 94 °C for 1 min, annealing at 55 °C for 2 min, and extension at 72 °C for 4 min (depending on product length), and final extension, incubated at 72 °C for 2 min to produce fragment 1 (Fig. 1a).
3. Perform two additional PCR reactions in parallel using conditions described above (**step 2**): In one, primers C, D, and the dengue serotype 2 replicon cDNA (5′-UTR-$C_{N\text{-ter}}$-Rluc-FMDV2A-Neor*-IRES-$E_{C\text{-ter}}$-NSorf-3′-UTR; Reichert and Padmanabhan, manuscript in preparation) as a template; in the other, perform PCR using primers E, F, and pBR322-WNV-956

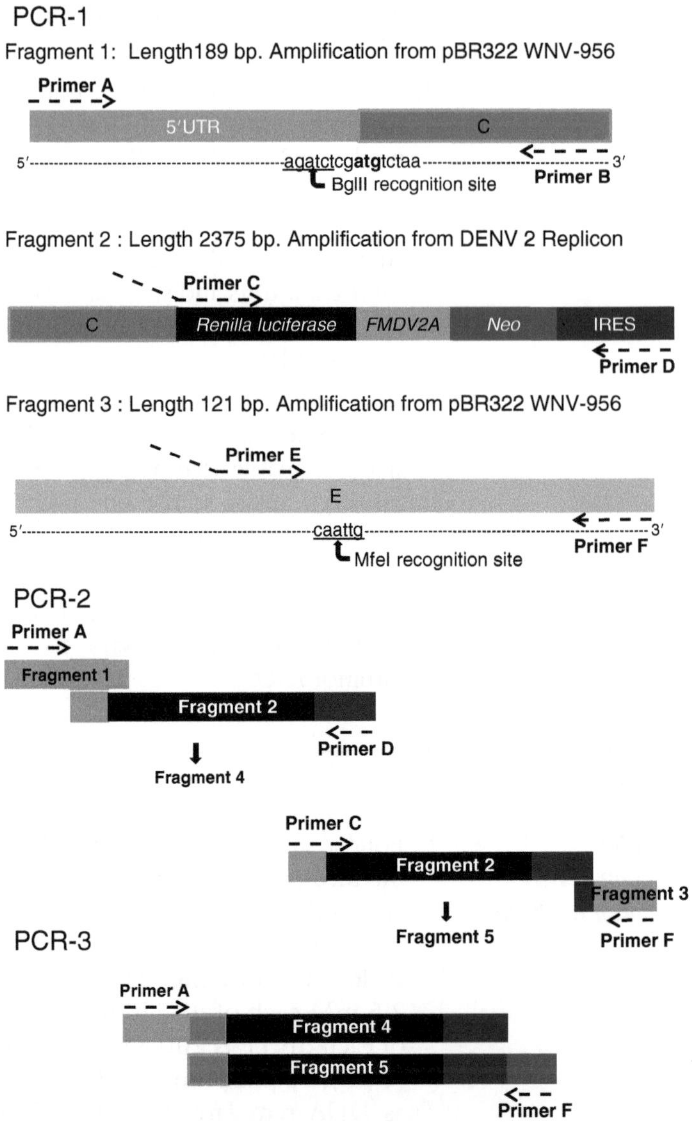


**Fig. 1** (**a**) Strategy used for construction of WNV *Renilla* luciferase reporter replicon with neomycin resistance gene as a selectable marker gene. Sequences of primers used for PCR are indicated in Tables 1 and 2. Fragments 1 and 3 contain the restriction sites (e.g., *Bgl*II and *Mfe*I) as indicated by underlined sequences. The methods used are described in detail in the text. (**b**) Strategy used for construction of WNV *Renilla* luciferase reporter replicon without a selectable marker. Sequences of the primers used for PCR are indicated in Tables 1 and 2. The methods are described in detail in the text

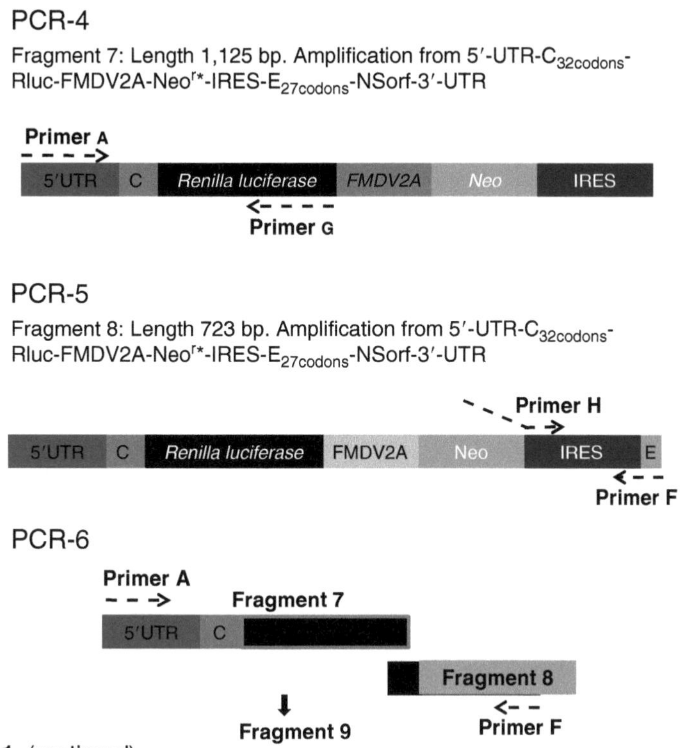


**Fig. 1** (continued)

cDNA as the template. These reactions produce PCR fragments 2 and 3, respectively. After synthesis, maintain the samples at 16 °C (*see* **Note 8**) until processed further.

4. Purify the PCR fragments using standard agarose gel electrophoresis and recovery using a kit from Zymo Research (Irvine, CA). Use a minimum volume for elution (5–10 μL) of the PCR DNA fragment adsorbed to the matrix.

*3.1.2 PCR-2*

1. In a 50 μL reaction volume, mix the following components: 10× PCR buffer (5 μL), 2.5 mM each of dNTP mix (4 μL), all of the recovered fragment 1 (10 μL), all of the recovered fragment 2 from the previous step (10 μL), 10 μM primer A (1 μL), 10 μM primer D (1 μL), and 5 U Ex-Taq polymerase (1 μL). Make up the reaction volume to 50 μL with $H_2O$ and mix well. Fragment 4 is obtained by PCR (Fig. 1a).
2. Perform another PCR reaction in parallel using gel-purified fragments 2 and 3 mixed with primers C and F and conditions described for PCR-1. Fragment 5 is obtained.

3. Purify the final PCR products by gel electrophoresis as described in Subheading 3.1.1, **step 4**. Elute the DNA fragments 4 and 5 in minimum volumes for each fragment (10 μL).

*3.1.3 PCR-3*

1. Assemble the following components in a tube: 10× PCR buffer (5 μL), 2.5 mM each of dNTP mix (4 μL), all of the recovered fragments 4 and 5 (10 μL each), 10 μM primer A (1 μL), 10 μM primer F (1 μL), and 5 U Ex-Taq polymerase (1 μL), add $H_2O$ to a volume of 50 μL, and mix.
2. Perform PCR reaction as described for PCR-1.
3. Purify the final PCR product by gel electrophoresis as described in Subheading 3.1.1, **step 4**. Clone the PCR product in a pGEMTeasy® vector (Promega) following the manufacturer's instructions. Select a few clones and verify the sequence to make sure that there are no mutations (*see* **Note 9**).

***3.2 Cloning of the WNV Rluc Reporter Replicon cDNA with Neo$^r$ Selectable Marker (5′-UTR-C$_{32codons}$-Rluc-FMDV2A-Neo$^{r*}$-IRES-E$_{27codons}$-NSorf-3′-UTR)***

1. For the first restriction enzyme digestion, mix the following components: $H_2O$ (5 μL), 40 μg of pBR322 WNV-956 plasmid (80 μL), 10× NEBuffer 4 (10 μL), and 50 U of *Mfe*I (5 μL). Incubate overnight at 37 °C.
2. Purify the restriction enzyme-digested DNA following standard protocols involving phenol:chloroform:isoamyl alcohol extraction and precipitation with ethanol.
3. To perform the next restriction enzyme digestion, add to the entire DNA that was recovered in the above step 10× NEBuffer 3 (10 μL) and 50 U of *Bgl*II (5 μL) and bring up the volume to 100 μL with $H_2O$. Incubate overnight at 37 °C.
4. In parallel, digest the recombinant plasmid containing the PCR fragment cloned into the pGEMTeasy® vector with the same restriction enzymes, *Mfe*I and *Bgl*II.
5. Purify the selected DNA fragments from the restriction enzyme digests of the pBR322 WNV-956 plasmid and the PCR product released from the pGEMTeasy® vector. Elute each fragment in 10 μL of $H_2O$, mix both purified fragments in a 1.5 μL Eppendorf tube, and dry in a vacuum centrifuge (*see* **Note 10**). Resuspend the DNA fragments in 6.5 μL of $H_2O$.
6. Perform the ligation reaction by adding 6.5 μL of the DNA from the above **step 5**, 10× ligation buffer (1 μL), 10 mM ATP (1 μL), 10 mg/mL acetylated BSA (1 μL), and 400 cohesive end units of ligase T4 (0.5 μL). Incubate the ligation reaction overnight at room temperature. Use the whole ligase reaction mixture for transformation of competent *E. coli* Stbl2 cells.
7. Grow selected bacterial colonies in LB broth + ampicillin (100 μg/mL). Plasmid DNAs representing full-length replicon cDNAs are prepared on a mini scale following standard recombinant DNA methods.

**Table 2**
**PCR primer sequences used for construction of WNV reporter replicon for transient expression in mammalian cells**

| Primer name | Primer sequence |
|---|---|
| Primer G | 5′-ctgctcgttcttcagcacgcgct-3′ |
| Primer H | 5′-gtgctgaagaacgagcagtgaccgcccctctccctcccc-3′ |

Primer H: Underlined sequence corresponds to the end of the luciferase gene followed by the stop codon, tga, and 5′-end of IRES sequence

8. Verify the sequences of the inserted DNA encoding the *Renilla* luciferase gene fused to the part of capsid gene (32 codons) at the 5′-end and the EMCV leader sequence at the 3′-end followed by 27 amino acid coding sequence of the E gene in the WNV replicon to ensure that they are in-frame and in correct orientation.
9. Prepare a medium-scale WNV replicon plasmid DNA purified from 250 mL culture of the transformed *E. coli* clone. This plasmid is designated as 5′-UTR-$C_{32codons}$-Rluc-FMDV2A-$Neo^{r}$*-IRES-$E_{27codons}$-NSorf-3′-UTR.

### 3.3 PCR for Construction of WNV Rluc Reporter Replicon for Transient Expression (5′-UTR-$C_{32codons}$-Rluc*-IRES-$E_{27codons}$-NSorf-3′-UTR)

The protocols for PCR are the same as described above under Subheading 3.1.1, PCR-1.

Perform two PCRs in parallel: In the first PCR-4, use the *primer A* as the forward primer and the reverse primer G (a new primer that contains the sequence from the 3′-end of Rluc gene); in the second PCR (PCR-5, Fig. 1b), use the new forward primer H that contains the sequence of the 3′-end of Rluc gene, stop codon, and the IRES overlapping with primer G containing the 3′-end of Rluc gene (*see* Table 2; Fig. 1b) and the reverse primer F (Table 1). Use the 5′-UTR-$C_{32codons}$-Rluc-FMDV2A-$Neo^{r}$*-IRES-$E_{27codons}$-NSorf-3′-UTR as a template for both PCRs. Fragments 7 and 8 are obtained. The overlap PCR (PCR-6) is carried as described above under section PCR-2. The overlap PCR product is purified as described above. Clone the PCR product into the pGEMTeasy vector. The clone pGEMTeasy-WNΔNeo-c6 is obtained. Verify the sequence. The pBR322 WNV-956 clone and pGEMTeasy-WNΔNeo-c6 clone are digested with the enzymes *Bgl*II + *Mfe*I. Purify the fragments and ligate to get the WNV replicon clone, 5′-UTR-$C_{32codons}$-Rluc*-IRES-$E_{27codons}$-NSorf-3′-UTR.

### 3.4 In Vitro Transcription and Expression of Replicon in Mammalian Cells

1. For the linearization of the plasmid, add the following: $H_2O$ (4 μL), 40 μg of plasmid 5′-UTR-$C_{32codons}$-Rluc*-IRES-$E_{27codons}$-NSorf-3′-UTR (80 μL), 10× NEBuffer 4 (10 μL), 10 mg/mL acetylated BSA (1 μL), and 50 U of Xbal (5 μL). Incubate overnight at 37 °C.

2. Purify linearized replicon DNA by extraction with phenol:chloroform:isoamyl alcohol and precipitation with ethanol. Resuspend the DNA in 10 μL of $H_2O$.
3. Mix the following components for the in vitro transcription reaction: $H_2O$ (5 μL), Amplicap 10× transcription buffer (2 μL), 5 mM each of NTPs (4 μL), 40 mM cap analog (2 μL; the final concentration is 4 mM), 100 mM dithiothreitol (2 μL), 2 μL of the purified linear WNV replicon cDNA, RNase inhibitor (1 μL), and SP6 DNA-dependent RNA polymerase enzyme solution (2 μL from the AmpliCap™ SP6 High Yield Message Maker kit) for a final volume of 20 μL. Incubate for 2 h at 37 °C. If desired, treat the in vitro transcription reaction with 1 μL RNase-free DNase I and incubate for 15 min at 37 °C (*see* **Note 11**). Purify the RNA by phenol:chloroform:isoamyl alcohol extraction and ethanol precipitation. Resuspend the pellet in 12 μL of $H_2O$.
4. Verify the integrity of the WNV replicon RNA produced in the in vitro transcription reaction by electrophoresis of a 0.5 μL aliquot on a 0.6 % agarose gel. Aliquot the in vitro transcript and store at −80 °C (*see* **Note 12**).
5. For expression of replicon RNA in mammalian cells and quantification of *Renilla* luciferase activity, transfect chosen mammalian cells such as Vero cells ($1 \times 10^6$ cells) by electroporation with 1.5 μg of the RNA. Immediately after electroporation, allow the cells to recover by addition of growth medium containing 10 % fetal bovine serum and kept at 37 °C. Incubate for 5 min and plate the cells into 6-well tissue culture plates. Lyse cells using the *Renilla* luciferase cell lysis solution provided in the kit at different time points (*see* **Note 13**). Determine the luciferase activity in the cell lysates collected at various time points (*see* **Note 14**).

### 3.5 Establishment of Stable Cell Line Expressing WNV Rluc Replicon

Before establishment of stable cell line, it is important to assess the concentration of G418 that would be cytotoxic to normal Vero cells but not toxic to cells replicating WNV RNA encoding $Neo^r$ gene which would inactivate G418. This protocol is described below:

#### 3.5.1 Estimation of Cytotoxicity of G418 to Vero Cells

1. Grow Vero cells from a stock vial stored in liquid nitrogen. When the cells become confluent, remove the medium and wash once with PBS. Add 2 mL of 0.25 % trypsin–EDTA to the cells and incubate for 2 min in a humidified 37 °C incubator with 5 % $CO_2$. Remove trypsin by aspiration and incubate again for 2 min in the 37 °C incubator. Resuspend cells in 10 mL of PBS and count the cells using a hemocytometer following standard cell culture techniques.
2. Plate in each well of a 6-well plate $1.25 \times 10^5$ cells in 2 mL of complete medium. 9 wells are needed.

3. After 24 h, remove medium by aspiration and add 2 mL of complete medium with different concentrations of G418 ranging from 0 to 800 μg/mL (*see* **Note 15**).
4. For the cytotoxicity curve, add the following stock solution of G418 (50 mg/mL) to each 2 mL of medium: (1) 0 μL of G418; (2) 100 μg/mL (4 μL G418); (3) 200 μg/mL (8 μL of G418); (4) 300 μg/mL (12 μL of G418); (5) 400 μg/mL (16 μL of G418); (6) 500 μg/mL (20 μL of G418); (7) 600 μg/mL (24 μL of G418); (8) 700 μg/mL (28 μL of G418); and (9) 800 μg/mL (32 μL of G418).
5. Replenish medium every other day with fresh G418.
6. After 2 weeks, select the lowest concentration of antibiotic that kills all the cells and use this concentration to create the stable cell lines.

*3.5.2 Electroporation*

1. Plate $2.5 \times 10^6$ Vero cells in a sterile tissue culture 100 mm dish. 48 h after seeding, use them for electroporation (*see* **Note 16**).
2. Trypsinize the cells as mentioned above. Count the cells, centrifuge at 600 × *g* for 8 min, and resuspend the cells in chilled PBS at 4 °C to have $1 \times 10^6$ cells per 100 μL.
3. Add 100 μL of the cells and 1.5 μg of the RNA in a 1.5 mL Eppendorf tube, mix, and transfer to the electroporation cuvette.
4. Electroporate using these settings: 450V and 50 μF, resistance ∞, and one pulse.
5. After electroporation, immediately add 1 mL of pre-warmed complete medium to the cuvette containing the cells.
6. Transfer the cells to a 100 mm sterile tissue culture dish with medium without any G418.
7. 24 h post electroporation, change the medium and add fresh medium containing G418 at the appropriate concentration (estimated in Subheading 3.5.1).
8. Every 48 h, remove the medium and add fresh10 mL of fresh medium with 60 μL of G418 (50 mg/μL) at a final concentration of 300 μg/mL.
9. After a week when the cells become confluent, trypsinize the cells, and place half of the cells in a new 100 mm sterile tissue culture dish (*see* **Note 17**).
10. Repeat the replenishment of medium every 48 h followed by trypsinization and replate half of the cells into a new 100 mm sterile tissue culture dish.
11. After 15 days, the majority of the cells that do not have replicating WNV replicon would die leaving several foci of stable clones.

#### 3.5.3 Cell Cloning

1. Identify the cell foci with an inverted microscope and mark them using a permanent ink marker.
2. Place 4 mL of trypsin in a new 100 mm sterile tissue culture dish and let the cloning discs soak.
3. In parallel, remove the medium from the plates containing the cell foci, add 10 mL of 1× PBS to wash the cells, and remove all traces of liquid by aspiration.
4. Place one cloning disc in each cell clone and incubate for 4 min in a 37 °C $CO_2$ incubator.
5. Take the cloning disc with a sterile fine-point forceps and place each cloning disc in a well of a 24-well plate that has 500 μL of medium without G418.
6. After 24-h incubation, remove the medium and cloning disc from the wells. Replenish with 500 μL fresh complete medium and add 3 μL of G418 (50 mg/μL).
7. Change medium every 48 h and add G418.
8. When cells become confluent, trypsinize the cells and plate each clone in an individual well of a 6-well plate in 1 mL of complete medium containing 300 μg/μL of G418.
9. Change medium containing G418 every 48 h.
10. When cells become confluent, trypsinize the cells and plate each clone in an individual 100 mm sterile tissue culture dish in 10 mL of complete medium with 60 μL of G418 (300 μg/μL).
11. When cells reach confluence, test for luciferase activity. Make stocks of chosen stable Vero cell clones.
12. Quantify the Rluc activity using a Centro LB 960 Microplate Luminometer (Berthold Technologies) and MikroWin Version 4.0 Software.
13. Wash transfected cells in wells of a 6-well plate using PBS and lyse by scraping into 250 μL *Renilla* luciferase assay lysis buffer. The cells are collected and vortexed vigorously. In some cases, the lysate is centrifuged in a microcentrifuge to remove any cellular debris. If the cell lysates need to be stored until all the samples are collected, store them at −20 °C.
14. Luciferase assays are performed in accordance with the manufacturer's instructions. An aliquot of the lysate (50 μL) is placed in wells of an opaque 96-well plate. For each sample, 50 μL of *Renilla* luciferase assay buffer with substrate (Promega) is injected into each well and delayed for 2 s before reading luminescence over a period of 10 s. The Rluc activities of four clones are shown in Fig. 2.

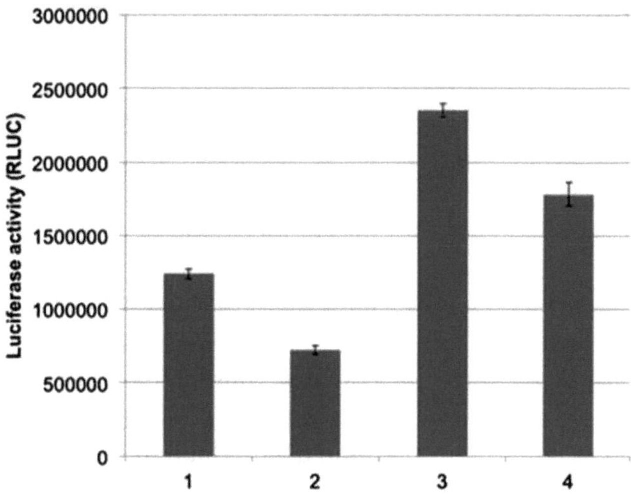


**Fig. 2** *Renilla* luciferase expression levels of four cell clones selected by G418 treatment. The isolation of stable cell clones expressing *Renilla* luciferase reporter is described in the text. The numbers in *X*-axis represent the four independent clones isolated stably expressing *Renilla* luciferase and their levels in the same amount of cell lysates

### 3.6 Antiviral Drug Screening

1. Cell-based replicon assays are usually performed in triplicates in transparent 96-well plates. The replicon-containing Vero cells are grown in T-75 flasks in complete medium. Add G418 (300 μg/mL). The medium and G418 are replaced every 3 days.
2. Each well of a 96-well plate is seeded with approximately $2.5 \times 10^4$ cells and incubated at 37 °C and 5 % $CO_2$. After 24-h incubation, small-molecule compounds to be screened are added to a final DMSO concentration of 0.5 % (*see* **Note 18**). Primary screens are performed at a single concentration of each compound (ranging from 10 to 40 μM). Cells are incubated for 24 h in the presence of the compound and the Rluc activity is measured (*see* Subheading 3.5.3, **step 14**) (*see* **Note 19**). Those compounds that show more than 50 % inhibition are tested at various concentrations (for example, concentration range from 0 to 40 μM) maintaining the DMSO concentration constant.
3. The relative luciferase unit (RLU) of the no-inhibitor control containing only 0.5 % DMSO is set as 0 % inhibition or 100 % relative response. All other RLUs are set as a percentage of the control. $EC_{50}$ values are calculated by nonlinear regression fitting (GraphPad Prism 5 or an equivalent software) of signal vs. concentration data points to the standard dose–response equation, $y = \text{bottom} + (\text{top} - \text{bottom})/(1 + 10^{(\log(\text{EC50}) - x)})$ where $x$ is the logarithm of compound concentration, $y$ is the response signal (RLU), and bottom and top refer to plateaus of the sigmoid response curve.

## 4 Notes

1. The ligation buffer has ATP and dithiothreitol that are very labile; so freeze–thaw cycles are not recommended. Store aliquots of 5 μL each at −20 °C (for short-term storage of 1–2 months) or −80 °C (for long-term storage). Once an aliquot is thawed and partly used, the remainder is discarded. The dithiothreitol in the buffer has a strong smell. Prepare aliquots in a fume hood to avoid inhalation.
2. Ligase requires ATP as a substrate. dATP forms a more stable complex, thus inhibiting the ligation reaction.
3. BSA is added to stabilize the restriction enzyme from denaturation due to dilution.
4. Thaw at room temperature. If any white precipitate is observed, incubate for 10 min at 37 °C or until it is completely dissolved.
5. The NTP mixture can be frozen and thawed once.
6. Store the cap analog stock solution in 5 μL aliquots.
7. If purchased in a powder form, verify the amount of active ingredient, e.g., if the active ingredient is 775 μg/mg, take 645.2 mg and dissolve in 10 mL PBS to prepare a stock solution of 50 mg/mL. Filter-sterilize by passing through a 0.22 μm filter fitted onto a 50 mL syringe. The antibiotic is stable for several months at 4 °C.
8. Thermal cycler machine can be programmed at the end of PCR cycles to store samples at a preset temperature. Setting the storage temperature at 16 °C for short periods of time is recommended by a technical representative of the manufacturer as (1) DNA samples are stable at that temperature and (2) it is good for optimal performance of the PCR machine.
9. We used pGEMTEasy (Promega Corporation) as a T-A vector. The sequence analysis of the insert is done using primers annealing to either T7 or SP6 promoter in the vector as well as the primers from the internal sequence. For the amplification of the overlap extension PCR, we used Ex-Taq polymerase that adds 3′-A overhangs to approximately 80 % of the PCR products. We did not see any extra addition of A in the internal sequence of the fused fragment, only at the ends that allowed the cloning to the T-A vector.
10. For purification of DNA fragments from agarose gels, there are a number of protocols available; some are based on selective adsorption to column matrices and others on glass beads. We prefer protocols based on column matrices and elution into minimum volume from the column. With regard to protocols

based on beads, it is important to ensure that no beads remain in the eluted and purified DNA which could interfere with solubility and inhibit the subsequent ligation reaction.

11. We have observed that it is not necessary to digest the DNA template by treatment with DNase I after the in vitro transcription.
12. RNA is very unstable to freeze–thaw cycles. Therefore, we recommend the in vitro-transcribed RNA to be stored as 1.5 μL aliquots at −80 °C.
13. If desired, the cell extracts can be stored in a 96-well plate at −20 °C until samples at all time points are collected and then the luciferase activity can be measured. This is recommended for optimal use of the buffers and luciferase substrate in the kit.
14. Based on published results from several laboratories including our own [15, 16], a peak of *Renilla* luciferase activity is expected in the sample collected 2–4 h post transfection which declines gradually perhaps due to a short half-life of the luciferase enzyme. From the 24–96-h post-transfection time points, a steady increase in luciferase signal has been observed due to translation of the newly replicated RNA [15, 16].
15. Cells are at ~25 % confluence at this stage.
16. Vero cells at a density of $2.5 \times 10^6$ are needed to reach confluence in a 100 mm sterile tissue culture dish in 48 h. Cells are passaged 2–3 days prior to electroporation and are at 70–85 % confluence on the day of electroporation. Higher cell densities may cause lower transfection efficiencies and increased cell death.
17. Antibiotics work best when cells are actively dividing. If the cells become too dense, the antibiotic efficiency will decrease. It is best to split cells such that they are not more than 25 % confluent.
18. DMSO is cytotoxic to mammalian cells and this cytotoxicity could vary depending on the cell line. It is preferable to test the concentration at which the cell growth is not affected prior to screening the compounds.
19. In reporter replicon-expressing stable cells, the reporter is continuously expressed due to translation of replicon RNA followed by polyprotein processing. Therefore, incubation of cells in the presence of an inhibitor compound for 24 h is sufficient whether the compound of interest has an inhibitory effect on translation, polyprotein processing, or replication. Longer incubation period may have cytotoxicity effect of the compound, if any, complicating the interpretation of the results.

## Acknowledgments

The research was supported by NIH grants R01 AI-32078 and U01 AI 54776 (R.P). We gratefully acknowledge the help and comments provided by Dr. Rosa del Angel during the preparation of the manuscript.

### References

1. Khromykh AA, Westaway EG (1997) Subgenomic replicons of the flavivirus Kunjin: construction and applications. J Virol 71: 1497–1505
2. Scholle F, Girard YA, Zhao Q et al (2004) trans-Packaged West Nile virus-like particles: infectious properties in vitro and in infected mosquito vectors. J Virol 78:11605–11614
3. Ng CY, Gu F, Phong WY et al (2007) Construction and characterization of a stable subgenomic dengue virus type 2 replicon system for antiviral compound and siRNA testing. Antiviral Res 76:222–231
4. Mueller NH, Pattabiraman N, Ansarah-Sobrinho C, Viswanathan P, Pierson TC, Padmanabhan R (2008) Identification and biochemical characterization of small-molecule inhibitors of west Nile virus serine protease by a high-throughput screen. Antimicrob Agents Chemother 52:3385–3393
5. Shi PY, Tilgner M, Lo MK (2002) Construction and characterization of subgenomic replicons of New York strain of West Nile virus. Virology 296:219–233
6. Yamshchikov VF, Wengler G, Perelygin AA et al (2001) An infectious clone of the West Nile flavivirus. Virology 281:294–304
7. Lo MK, Tilgner M, Bernard KA, Shi PY (2003) Functional analysis of mosquito-borne flavivirus conserved sequence elements within 3′ untranslated region of West Nile virus by use of a reporting replicon that differentiates between viral translation and RNA replication. J Virol 77:10004–10014
8. Rossi SL, Zhao Q, O'Donnell VK, Mason PW (2005) Adaptation of West Nile virus replicons to cells in culture and use of replicon-bearing cells to probe antiviral action. Virology 331: 457–470
9. Maeda J, Takagi H, Hashimoto S et al (2008) A PCR-based protocol for generating West Nile virus replicons. J Virol Methods 148:244–252
10. Pang X, Zhang M, Dayton AI (2001) Development of Dengue virus type 2 replicons capable of prolonged expression in host cells. BMC Microbiol 1:18
11. Alvarez DE, De Lella Ezcurra SA, Fucito L, Gamarnik AV (2005) Role of RNA structures present at the 3′UTR of dengue virus on translation, RNA synthesis, and viral replication. Virology 339:200–212
12. Jones M, Davidson A, Hibbert L et al (2005) Dengue virus inhibits alpha interferon signaling by reducing STAT2 expression. J Virol 79:5414–5420
13. Suzuki R, de Borba L, Duarte dos Santos CN, Mason PW (2007) Construction of an infectious cDNA clone for a Brazilian prototype strain of dengue virus type 1: characterization of a temperature-sensitive mutation in NS1. Virology 362:374–383
14. Mosimann AL, de Borba L, Bordignon J et al (2010) Construction and characterization of a stable subgenomic replicon system of a Brazilian dengue virus type 3 strain (BR DEN3 290-02). J Virol Methods 163:147–152
15. Alcaraz-Estrada SL, Manzano MI, Del Angel RM, Levis R, Padmanabhan R (2010) Construction of a dengue virus type 4 reporter replicon and analysis of temperature-sensitive mutations in non-structural proteins 3 and 5. J Gen Virol 91:2713–2718
16. Manzano M, Reichert ED, Polo S, Falgout B, Kasprzak W, Shapiro BA, Padmanabhan R (2011) Identification of *Cis*-acting elements in the 3′-untranslated region of the dengue virus type 2 RNA that modulate translation and replication. J Biol Chem 286:22521–22534
17. Gehrke R, Ecker M, Aberle SW et al (2003) Incorporation of tick-borne encephalitis virus replicons into virus-like particles by a packaging cell line. J Virol 77:8924–8933
18. Hayasaka D, Yoshii K, Ueki T et al (2004) Sub-genomic replicons of Tick-borne encephalitis virus. Arch Virol 149:1245–1256
19. Corver J, Lenches E, Smith K et al (2003) Fine mapping of a *cis*-acting sequence element in yellow fever virus RNA that is required for RNA replication and cyclization. J Virol 77:2265–2270
20. Molenkamp R, Kooi EA, Lucassen MA et al (2003) Yellow fever virus replicons as an expression system for hepatitis C virus structural proteins. J Virol 77:1644–1648
21. Jones CT, Patkar CG, Kuhn RJ (2005) Construction and applications of yellow fever virus replicons. Virology 331:247–259

22. Yun SI, Kim SY, Rice CM, Lee YM (2003) Development and application of a reverse genetics system for Japanese encephalitis virus. J Virol 77:6450–6465
23. You S, Padmanabhan R (1999) A novel in vitro replication system for Dengue virus. Initiation of RNA synthesis at the 3′-end of exogenous viral RNA templates requires 5′- and 3′-terminal complementary sequence motifs of the viral RNA. J Biol Chem 274:33714–33722
24. Khromykh AA, Meka H, Guyatt KJ, Westaway EG (2001) Essential role of cyclization sequences in flavivirus RNA replication. J Virol 75:6719–6728
25. Alvarez DE, Lodeiro MF, Luduena SJ et al (2005) Long-range RNA–RNA interactions circularize the dengue virus genome. J Virol 79:6631–6643
26. Falgout B, Chanock R, Lai CJ (1989) Proper processing of dengue virus nonstructural glycoprotein NS1 requires the N-terminal hydrophobic signal sequence and the downstream nonstructural protein NS2a. J Virol 63:1852–1860

# Part IV

## Herpes Viruses

# Chapter 23

# Methods for Screening and Profiling Inhibitors of Herpes Simplex Viruses


## Edwin Yunhao Gong



## Abstract

Herpes simplex viruses (HSV) establish lifelong latent infections in infected hosts that reactivate periodically and result in virus shedding and recurrent diseases, such as genital herpes. Sexually transmitted infections (STIs) caused by HSV are a major public health problem worldwide. At present, the only effective antiviral drugs for treatment of HSV are nucleoside analogues, which are incorporated into the DNA chain and terminate the chain elongation during virus replication. With increasing emergence of drug resistance, novel drugs for new viral targets are warranted. In this chapter, several screening and profiling assays including plaque reduction assays, cytopathic effect inhibition assay, and in vitro cytotoxicity assay for identifying and evaluating inhibitors of HSV are described. Assays for mode of action studies, such as virus adsorption and penetration, are also presented.



**Key words** Herpes simplex virus type 1, Herpes simplex virus type 2, Plaque assay, CPE inhibition assay, In vitro toxicity, Virus adsorption, Virus penetration


## 1 Introduction

Herpes simplex virus type 1 (HSV-1) and type 2 (HSV-2) are two closely related members of the herpesvirus family, *Herpesviridae*, which infect humans and can cause many diseases including sexually transmitted diseases, such as genital herpes. HSV-1 mainly causes nongenital infection, such as labialis or cold sores, and keratitis, while HSV-2 is more associated with genital herpes. Both HSV-1 and HSV-2 cause lifelong infection with periodic recurrent infections. Studies show that more than 80 and 30 % of adults in the world are seropositive for HSV-1 and HSV-2, respectively (for reviews *see* refs. 1, 2). Based on the information from the World Health Organization, global estimate of the prevalence of HSV-2 infection is 535.5 million (16.2 %) and the incidence of HSV-2 infection is 23.6 million (0.7 %) [3].

Both HSV-1 and HSV-2 are treatable, but incurable because of the latent infection. In the absence of effective vaccines to


Edwin Yunhao Gong (ed.), *Antiviral Methods and Protocols*, Methods in Molecular Biology, vol. 1030, DOI 10.1007/978-1-62703-484-5_23, © Springer Science+Business Media, LLC 2013

control HSV infections, the treatment of clinical HSV infections relies on the nucleoside antiviral drugs, for example, famciclovir (a prodrug of penciclovir), valaciclovir (a prodrug of acyclovir), and acyclovir. Acyclovir itself has very poor bioavailability and is mostly used for topical treatment. With increasing emergence of drug resistance, novel anti-HSV drugs with novel modes of action are warranted. In this chapter, several cell-based antiviral screening and profiling assays for identifying and evaluating inhibitors of HSV-1 and HSV-2 are described.

## 2 Materials

### *2.1 Cells, Virus, and Reagents*

1. Vero cells (African green monkey kidney cells): American Type Culture Collection (ATCC).
2. HSV-1, F strain (ATCC).
3. HSV-2, G strain (ATCC).
4. Eagle's minimum essential medium (MEM) without L-glutamine.
5. 10× MEM without L-glutamine and sodium bicarbonate (Invitrogen).
6. Fetal calf serum (FCS).
7. 200 mM L-glutamine (100×).
8. 50 mg/mL gentamicin.
9. 0.05 % trypsin–EDTA.
10. Vero cell growth medium (MEM/10 % FCS): Eagle's MEM supplemented with 10 % FCS, 2 mM L-glutamine, and 0.02 mg/mL gentamicin.
11. Vero cell maintenance medium (MEM/2 % FCS): Same as the growth medium but supplemented with 2 % FCS.
12. 2× MEM/2 % FCS: 2× MEM diluted from 10× MEM without L-glutamine and sodium bicarbonate (**item 5**) and supplemented with 4 % FCS, 2×2 mM L-glutamine, 2×7.5 % $NaHCO_3$, 0.04 mg/mL gentamicin. Adjust to the final volume with autoclaved deionized $H_2O$.
13. Dulbecco's phosphate-buffered saline (D-PBS), no calcium chloride, no magnesium chloride (Gibco).
14. 10 % formalin in D-PBS.
15. 0.5 % crystal violet solution (Sigma-Aldrich).
16. 1.0 % (w/v) methylcellulose in water: Weigh 1.0 g of methylcellulose in an Erlenmeyer flask and add 100 mL of deionized $H_2O$ into the flask. Place the cap loosely on the bottle. Put a stirring bar into the flask and place on a magnetic stirrer. Heat the flask under stirring until boiling, and then allow the flask to cool to room temperature. Continue stirring the flask overnight.

17. 0.01 % (w/v) neutral red solution: Dissolve 0.01 g of neutral red (Sigma-Aldrich) in 100 mL of D-PBS (*see* **Note 1**).
18. 1 % glacial acetic acid/50 % ethanol: 1 % glacial acetic acid, 50 % ethanol, and 49 % deionized $H_2O$.
19. Acidic glycine-(HCl): 0.1M glycine in saline, pH 3.0 (sterile filter). Store at room temperature.
20. Reference compounds:
    (a) Acyclovir (Sigma-Aldrich).
    (b) Penciclovir (Sigma-Aldrich).

### 2.2 Consumables and Equipment

1. 6-well, flat-bottom plates: Corning® Costar® cell culture plates.
2. 96-well cell culture plates.
3. Nunc® TripleFlasks (F8542, Sigma-Aldrich).
4. Absorbent pad.
5. Multichannel pipettes.
6. Multidrop combi liquid dispenser (Thermo Scientific) (optional).
7. Cell counter (Coulter Electronics LTD).
8. Magnetic stirrer.
9. Light-box.
10. Vacuum pump with liquid waste container (optional).
11. Shaker.
12. ELISA reader.

## 3 Methods

Perform all procedures in a biosafety cabinet unless otherwise specified. All incubations at 37 °C for culturing cells described in this chapter are performed using a humidified 5 % $CO_2$ incubator.

### 3.1 Anti-HSV Efficacy Testing Using a Plaque Reduction Assay

The plaque reduction assay is a susceptibility test for anti-HSV compounds [4–6]. It is considered to be the "gold standard" for HSV susceptibility tests and is performed according to the guidelines of National Committee for Clinical Laboratory Standards (NCCLS). The test compound is used to treat the cells after virus infection. The antiviral effect is then determined by reduction of viral plaques.

#### 3.1.1 Antiviral Efficacy Testing of Compounds on Infected Cells (See **Note 2**)

1. Seed 6-well plates (Corning® Costar® cell culture plates, *see* **Note 3**) with $6.0 \times 10^5$ Vero cells per well in a volume of 3 mL MEM/10 % FCS and incubate at 37 °C overnight. The cell monolayer should be confluent by the following day.

2. On the following day, check the cell confluence with an inverted microscope. If the monolayer is confluent, continue to **step 3**. If not, wait and use them the next day if confluent.
3. Perform tenfold or fivefold serial dilutions to prepare compound solutions at 2× the final testing concentrations desired using 2× MEM/2 % FCS. Duplicate wells are required for each testing concentration.
4. Thaw one vial of either HSV-1 or HSV-2 stock quickly in a 37 °C water-bath with shaking. Wipe the vial surface with 70 % ethanol for disinfection. Prepare a virus dilution of 100 plaque-forming units (PFU)/mL with MEM/2 % FCS.
5. Remove the old culture medium from the 6-well plates by inverting the plate over an absorbent pad or aspirating out with a plastic pipette.
6. Rinse the cells once with D-PBS.
7. Add 1 mL of diluted virus to each well (100 PFU per well).
8. Incubate the plates at 37 °C for 1 h, swirling the plates periodically to evenly spread the virus over the cell monolayer.
9. During incubation period, prepare methylcellulose overlay by mixing 1 % methylcellulose with 2× final testing concentrations of test compound (**step 3**) in 1:1 ratio (1.5 mL of 1 % methylcellulose in water + 1.5 mL of 2× final testing concentrations of test compound) (*see* **Note 4**). For the control wells, mix 1 % methylcellulose with 2× MEM/2 % FCS in 1:1 ratio.
10. After 1-h incubation, discard viral inoculum from the plates by inverting the plates over an absorbent pad (*see* **Note 5**).
11. Add 3 mL of methylcellulose overlay containing various concentrations of test compound to the corresponding wells. Add 3 mL of methylcellulose overlay containing no compound to the control wells.
12. Incubate the plates at 37 °C for 2 days for HSV-2 and 3 days for HSV-1.
13. At the end of incubation period, check the plaque size under a microscope. When the plaques are of sufficient size to count, proceed to **step 14**.
14. Discard the methylcellulose overlay from plates by inverting the plate over an absorbent pad (*see* **Note 5**).
15. Add 2 mL of 10 % formalin to each well to fix the cells. Incubate at room temperature for 10 min.
16. Discard formalin from the plates.
17. Stain fixed cells by adding 2–3 drops of 0.5 % crystal violet to each well. Leave at room temperature for 5 min.
18. Remove excess of crystal violet by rinsing the plates in a gentle stream of tap water.

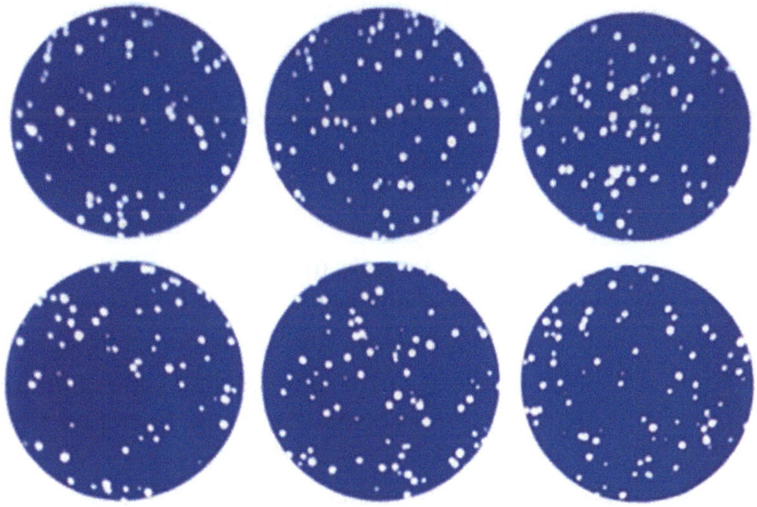

**Fig. 1** An image of crystal violet-stained HSV-2 plaques using Vero cells in a 6-well plate. The plaques appear as *white dots* and the *blue background* is the crystal violet-stained cell monolayer

19. Air-dry the plates by putting upside down over an absorbent pad.
20. Put the plates upside down on a light-box. Count the plaques in each well using a colony-counting pen. An image of crystal violet-stained HSV-2 plaques using Vero cells in a 6-well plate is shown in Fig. 1.
21. Enter the plaque counts to an Excel sheet or a GraphPad data sheet. For each testing drug concentration, average the plaque counts from the duplicate wells and calculate the percentage of inhibition. The $EC_{50}$ (effective concentration achieving 50 % of inhibition of viral plaques) values of test compounds are calculated using the GraphPad Prism software.

#### *3.1.2 Virucidal Activity of Test Compounds*

Many compounds are able to directly kill or inactivate virus particles outside of the cells, for example, the detergent-like compounds. This class of compounds is particularly useful as microbicides. The following protocol is designed to test the virucidal activity of compounds: a known amount of virus particles (PFU) are mixed with different concentrations of test compound and incubated for either a single duration (i.e., 30 min or 1 h) or a time-course (e.g., 0, 15, 30, and 60 min). The mixture is then diluted to a concentration at which the compound should no longer be active. The live virus in the compound-treated mixture can then be quantified using a plaque reduction assay.

1. Seed 6-well plates with Vero cells as described in Subheading 3.1.1, **steps 1** and **2**.
2. Dilute the virus stock of either HSV-1 or HSV-2 to generate $2.0 \times 10^5$ PFU/mL in 1 mL MEM/2 % FCS (the final virus input is 100 PFU per well).

3. Prepare 2× final concentrations of test compound by either tenfold or fivefold serial dilutions using MEM/2 % FCS.
4. Mix equal volumes of diluted virus with each drug dilution (2× final concentrations) in a 5-mL tube. Include a control containing no compound. Incubate the compound–virus mixture in a 37 °C water-bath for either one time-point, i.e., 1 h, or a time-course, i.e., 0, 15, 30, and 60 min.
5. At each time-point, perform a 1 in 1,000 dilution of the compound-treated virus mixture with MEM/2 % FCS.
6. Remove the old culture medium from the plates.
7. Add 1.0 mL of diluted compound-treated virus mixture to the corresponding wells. Duplicate wells are required for each drug concentration.
8. Incubate the plates at 37 °C for 1 h, swirling fluid on plate periodically to allow viral adsorption evenly.
9. After 1-h incubation, discard the viral inoculum.
10. Wash the infected cells once with D-PBS.
11. Prepare methylcellulose overlay by mixing 2× MEM/2 % FCS with 1 % methylcellulose (ratio 1:1). Add 3 mL of methylcellulose overlay to each well.
12. Incubate the plates at 37 °C for 2 days for HSV-2 and 3 days for HSV-1.
13. Fix and stain the cells as described under Subheading 3.1.1, **steps 13–21**.

### *3.2 Screening and Profiling Anti-HSV Compounds by a CPE Inhibition Assay*

1. Seed 96-well plates with $2.0 \times 10^4$ Vero cells per well in 100 μL of MEM/10 % FCS and incubate at 37 °C overnight. The cell monolayer should be 100 % confluent on the following day.
2. On day 2, warm the culture medium in a 37 °C water-bath.
3. Prepare fourfold or fivefold serial dilutions of test compounds in MEM/2 % FCS including positive controls, acyclovir and penciclovir. Duplicate or triplicate wells are required for each drug concentration (*see* **Note 6**).
4. Thaw one vial of either HSV-1 or HSV-2 stock quickly in a 37 °C water-bath with shaking. Prepare enough volume of virus dilution for a multiplicity of infection (MOI) of 0.01 PFU/cell in MEM/2 % FCS.
5. Aspirate out the old culture medium from the overnight culture using a glass pipette attaching to a vacuum pump (*see* **Note 7**) or a multichannel pipette.
6. Add 50 μL of either HSV-1 or HSV-2 dilution (0.01 MOI of each) to all wells of a 96-well plate except the wells of uninfected cell control (column 12). For uninfected cell control wells, add 50 μL of MEM/2 % FCS.

7. Incubate the cells at 37 °C for 1 h, swirling the inoculum on plates periodically to allow viral infection evenly on cells.
8. After 1 h of incubation, aspirate out the viral inoculum using a glass pipette attaching to a vacuum pump.
9. Add 100 μL of diluted test compounds and reference compounds to the corresponding duplicate or triplicate wells of columns 1–9. Add 100 μL of MEM/2 % FCS to columns 10 and 11 (virus control containing no compound) and column 12 (uninfected cell control containing no compound).
10. Incubate plates at 37 °C for 2–3 days until the cells in the virus control wells show 100 % cytopathic effect (CPE).
11. Once the cells in the virus control wells display 100 % CPE, perform neutral red dye uptake assay to assess the antiviral effect (see below from **steps 12–19**).
12. Remove the culture medium from the plates.
13. Pipette 100 μL of 0.01 % neutral red solution to each well using a multichannel pipette.
14. Incubate the plates in a 37 °C incubator for 30 min.
15. Remove the neutral red solution from plates by inverting plate over an absorbent pad.
16. Wash all wells with 200 μL of D-PBS twice.
17. Pipette 100 μL of 1 % glacial acetic acid/50 % ethanol to each well using a multichannel pipette.
18. Shake the plate at 100–150 rpm on a shaker for 15 min.
19. Measure the absorbance at $OD_{550nm}$ using an ELISA reader.
20. Calculate the $EC_{50}$ of test compounds: Using an Excel sheet or a GraphPad data sheet, average the absorbance values of the duplicate or the triplicate wells and calculate the percentage of inhibition for each drug concentration. Then calculate the $EC_{50}$ values using computer software, such as GraphPad Prism.

### 3.3 Evaluation of In Vitro Toxicity of Compounds Using a Dye Uptake Assay

1. Seed $2.0 \times 10^4$ Vero cells per well in 100 μL of MEM/10 % FCS to the columns 1–11 of a 96-well plate. Pipette 100 μL of MEM/10 % FCS into the wells of column 12 (medium control). Incubate the plates at 37 °C overnight.
2. On the following day, prepare fourfold or fivefold serial dilutions of a test compound including acyclovir, a positive control, in MEM/2 % FCS. Duplicate or triplicate wells are recommended for each drug concentration (*see* **Note 8**).
3. Remove the culture medium of Vero cells from 96-well plates.
4. Add 100 μL of serially diluted compounds to the corresponding wells (columns 1–10) of the plates. The cell control (containing no compound, column 11) and the medium control (column 12) wells receive 100 μL of MEM/2 % FCS.

5. Incubate the plates at 37 °C for desired time intervals, i.e., 3 days.
6. At the end of incubation period, perform neutral red staining as described in Subheading 3.2, **steps 12–19**.
7. Average the absorbance values of the duplicate or the triplicate wells and calculate the 50 % cytotoxicity ($CC_{50}$) using a computer software, such as GraphPad Prism.

### 3.4 Antiviral Effect of a Compound on Virus Adsorption

The inhibitory effect of a compound on HSV binding to cells can be examined by measuring the adsorption curve in the presence of several doses of test compound [5, 6]. In this experiment, susceptible cells are prechilled to 4 °C. The virus binding to cells is performed at 4 °C in the presence of test compound. At this temperature, virions can bind to their cellular receptors, but cannot penetrate the plasma membrane. Unbound virions blocked by the test compounds are removed by washing. Bound virions are taken up by cells when the temperature is elevated to 37 °C and subsequently visualized by a plaque assay.

1. Seed 6-well plates with $6.0 \times 10^5$ Vero cells per well in 3 mL of MEM/10 % FCS and incubate at 37 °C overnight.
2. Cool the culture medium (MEM/2 % FCS) and D-PBS to 4 °C.
3. On the following day, check the cell confluence on 6-well plates under a microscope. If the cells are confluent, continue to **step 4**.
4. Pre-warm MEM/2 % FCS, 2× MEM/2 % FCS, and 1 % methylcellulose in a 37 °C water-bath.
5. Prepare serial dilutions (fivefold or tenfold) of test compounds with *cold* MEM/2 % FCS and store the tubes in 4 °C until needed.
6. Label the 6-well plates with virus type, compound name, compound concentration, and time-point, for example, 0, 15, 30, 45, 60, 90, and 120 min.
7. Place the plates in a 4 °C refrigerator or a cold room for at least 30 min to cool down the cells.
8. After approximately 30-min incubation at 4 °C, bring the 6-well plates to a biosafety cabinet and put on ice.
9. Remove the old culture medium.
10. Add 1 mL of *cold* diluted compound to the corresponding wells while keeping the plates on ice. Add 1 mL of prechilled medium to the control wells (no compound). Duplicate wells are required for each drug concentration.
11. Return the plates to 4 °C and incubate for 1 h.
12. Prepare a virus dilution of $1 \times 10^4$ PFU/mL with cold MEM/2 % FCS. While keeping the plates on ice, infect the cells by adding 20 μL of diluted virus (200 PFU/well) to each well.

13. Incubate the plates at 4 °C for desired time-points, such as 0, 15, 30, 45, 60, 90, and 120 min. Swirl the plates periodically. For time-point 0, after adding the virus, immediately proceed with **step 14**.
14. At each time-point, remove the appropriate plate from 4 °C and wash the cells twice with *cold* D-PBS and once with *cold* 1× MEM without fetal bovine serum to remove the unadsorbed virus.
15. Immediately add 1 mL of *warm* MEM/2 % FCS to the wells and incubate the plates at 37 °C for 90 min.
16. Remove the medium by inverting the plates over an absorbent pad and then cover the cells with 3 mL of warm methylcellulose overlay (mix equal volumes of 2× MEM/2 % FCS with 1 % methylcellulose).
17. Incubate the plates at 37 °C for 2 days for HSV-2 and 3 days for HSV-1.
18. Once viral plaques are clearly visible, remove the methylcellulose overlay. Fix and stain the cells as described in **steps 14–19** under Subheading 3.1.1.
19. Count the plaques on a light-box. Using Excel sheet, plot the mean plaque count (PFU) of the duplicate wells (*y*-axis) vs. time (*x*-axis) to generate the adsorption curves.

### 3.5 Antiviral Effect of Compounds on Virus Penetration

The inhibitory effect of a test compound on HSV penetration of cells can be tested by measuring the penetration rate of HSV in the presence of test compound [5, 6]. In this experiment, the virus binding to cells is performed at 4 °C. Unbound virions are removed by washing with cold D-PBS. Penetration of virus occurs after addition of warm medium containing various concentrations of test compound and when the temperature is increased to 37 °C. Virions that have bound, but not internalized, are stripped off by a short exposure to acidic glycine.

1. Seed 6-well plates with $6.0 \times 10^5$ Vero cells per well in 3 mL of MEM/10 % FCS and incubate at 37 °C overnight.
2. On the following day, check the cell confluence under a microscope. If the monolayer is confluent, continue to next step.
3. Cool the cells on 6-well plates in a 4 °C cold room or a 4 °C refrigerator for at least 30 min.
4. Prepare a virus dilution to result in 200 PFU/mL using *cold* MEM/2 % FCS and put the dilution on ice.
5. Remove the culture medium from the prechilled plates.
6. Aliquot 1 mL of prechilled virus dilution (200 PFU) into each well while keeping the plate on ice.
7. Incubate the plates at 4 °C for 2 h to allow maximum virus binding, swirling the plates periodically.

8. After 2-h incubation, remove the viral inoculum from the plates.
9. Wash the cells twice with cold D-PBS.
10. Add 2 mL of pre-warmed serially diluted compound to the corresponding wells. Duplicate wells are needed for each drug concentration. For the no-compound control wells, add 2 mL of MEM/2 % FCS.
11. Incubate the plates at 37 °C for desired length of time.
12. At each time-point (0, 15, 30, 45, 60, 90, 120, and 150 min), remove the culture medium and wash the cells twice with 1 mL of warm D-PBS.
13. Discard D-PBS, and treat the cells with 1 mL of acidic glycine (pH 3) for 1 min.
14. Discard the acidic glycine, and wash the cells with 1 mL of warm 1× MEM (no serum).
15. Immediately add 3 mL of methylcellulose overlay to each well as described in Subheading 3.4, **steps 16–18**.
16. Count the plaques. Using Excel sheet, plot the mean plaque count (PFU) of the duplicate wells (*y*-axis) vs. time (*x*-axis) to generate the penetration curves.

## 4 Notes

1. Neutral red dye is a hazardous chemical and should be handled carefully with gloves.
2. This plaque reduction assay protocol can also be used for titrating HSV virus stocks. The only difference is that the methylcellulose overlay contains no compound.
3. Different brands of 6-well plates can affect the quality of plaque reduction assay. The Corning® Costar® 6-well flat-bottom cell culture plate (*see* Subheading 2.2, **item 1**) is recommended for the HSV plaque assay.
4. The mixture of methylcellulose overlay with test compounds can be placed in a 37 °C water-bath or incubator to reduce the viscosity of methylcellulose.
5. The discarded medium contains live HSV. Use an appropriate disinfectant to decontaminate the virus-contaminated medium such as Virkon or bleach tablets.
6. Alternatively, the CPE inhibition assay can be performed in the premade 96-well plates, in which the test compounds are serially diluted and plated in the corresponding wells (columns 1–9) in a volume of 25 μL. Columns 10 and 11 (virus control) and column 12 (cell control) receive 25 μL of medium with

DMSO. In this setting, add 50 μL of cell suspension to all wells of the 96-well plate using a multidrop combi liquid dispenser (select the 96-well program) or a multiple channel pipette. Then add 25 μL (an MOI of 0.01) of either HSV-1 or HSV-2 to the wells of columns 1–11 and 25 μL of medium to the wells of column 12. Incubate the plates at 37 °C until the CPE in virus control wells reaches 100 % (3 days). The neutral red dye uptake assay can be performed as in **steps 12–20** under Subheading 3.2.

7. Make sure not to aspirate the cells off from the plates.
8. Alternatively, the in vitro cytotoxicity assay can also be performed in the premade 96-well plates, in which the test compounds are serially diluted and plated in the corresponding wells (columns 1–9) in a volume of 25 μL. Columns 10 and 11 (cell control) and column 12 (medium control) receive 25 μL of medium with DMSO. In this setting, 75 μL of cell suspension can be dispensed to the wells of columns 1–11 and 75 μL of medium can be added to the wells of column 12 by a multidrop combi liquid dispenser or a multiple channel pipette. The plates are incubated at 37 °C for 3 days. The neutral red dye uptake assay can be performed as in **steps 12–19** under Subheading 3.2.

## References


1. Coen D, Schaffer P (2003) Antiherpesvirus drugs: a promising spectrum of new drugs and drug targets. Nature Rev Drug Discovery 2:278–288
2. Roizman R, Knipe D, Whitley R (2007) Herpes simplex viruses. In: Knipe D, Howley P et al (eds) Fields virology, 5th edn. Lippincott Williams & Wilkins, New York, NY, pp 2501–2601
3. Looker K, Garnett G, Schmid G (2008) An estimate of the global prevalence and incidence of herpes simplex virus type 2 infection. Bulletin of the World Health Organization 86 (10): 805–812
4. Gong Y, Wen A, Cheung D et al (2001) Preclinical evaluation of docusate as protective agent from herpes simplex viruses. Antiviral Res 52(1):25–32
5. Gong Y, Matthews B, Cheung D et al (2002) Evidence of dual sites of action of dendrimers: SPL-2999 inhibits both virus entry and late stages of herpes simplex virus replication. Antiviral Res 55(2):319–329
6. Gong E, Matthews B, McCarthy T et al (2005) Evaluation of dendrimer SPL7013, a lead microbicide candidate against herpes simplex viruses. Antiviral Res 68(3):139–146

# Chapter 24

# In Vivo Evaluation of Antiviral Efficacy Against Genital Herpes Using Mouse and Guinea Pig Models

**Frances Valencia, Ronald L. Veselenak, and Nigel Bourne**

## Abstract

Both the guinea pig and mouse are important animal models for the study of genital herpes. The murine model has been used extensively to evaluate vaccines and antiviral agents by measuring the incidence of infection and the magnitude of viral replication; however, this model is limited with regard to distinguishing between candidate vaccines or treatments. In contrast, the guinea pig closely mimics human infection and provides an excellent model of both primary and recurrent genital herpes disease. This animal model is especially important in the study of viral transmission through the evaluation of latent viral reactivation and virus shedding into the genital tract.

Here, we describe methodologies to determine viral infection, severity of primary disease, and quantification of primary viral replication in the genital tract for both the guinea pig and murine models of genital herpes. Additionally, we detail the evaluation of the onset of primary disease and progression to the day of death in the mouse model. Further, we summarize methods to assess the frequency of recurrences, frequency and magnitude of virus shedding, and latent viral load in the sensory nerve ganglia of the guinea pig.

**Key words** Herpes simplex viruses, HSV-2, Genital herpes, Animal model, Mouse, Guinea pig

## 1 Introduction

Herpes simplex virus type 2 (HSV-2) is the primary cause of genital herpes worldwide, although HSV-1 is becoming increasingly important as a cause of genital herpes [1, 2]. HSV is typically transmitted via the genital tract where it infects and passes through the mucosal epithelium before establishing a lifelong latent infection in the dorsal root ganglia (DRG) of the innervating nerves [3]. A number of antiviral agents have been developed that are effective in reducing the severity of primary and recurrent disease. Although such antivirals have generally been used episodically to reduce disease, recent studies indicate that at least in the short term, suppressive antiviral therapy can reduce HSV viral shedding into the genital tract and so potentially transmission to uninfected partners [4, 5]. However, none of the current antiviral agents are able to

Edwin Yunhao Gong (ed.), *Antiviral Methods and Protocols*, Methods in Molecular Biology, vol. 1030,
DOI 10.1007/978-1-62703-484-5_24, © Springer Science+Business Media, LLC 2013

clear an established virus infection. Thus, there is a clear need to develop additional treatments for prophylactic or therapeutic use to control this disease of global importance.

The mouse and guinea pig models of genital herpes are well established and have been used extensively to evaluate the efficacy of new antiviral agents. Mice can undergo vaginal virus infection resulting in viral replication in the vaginal mucosa and the development of primary genital skin disease. However, in order to achieve uniform infection the mouse requires hormonal pretreatment. Further, mice that do not receive effective prophylactic or therapeutic treatment fail to resolve the primary disease resulting in progressive neurological involvement and a high incidence of mortality. Mice that do survive do not experience spontaneous recurrent disease. Thus, the mouse provides a good model for initial studies examining the impact of antiviral efficacy on infection, vaginal virus replication, and primary genital herpes disease. Once antiviral treatment proves to be effective additional endpoints of disease outcome not possible in mice can be evaluated in the guinea pig. Guinea pigs, unlike mice, do not need hormone pretreatment and provide a model of genital herpes that more closely mimics human disease. Following intravaginal virus challenge, guinea pigs experience virus replication in the genital mucosa and develop a self-limiting vesiculo-ulcerative primary disease that leads to the establishment of a latent infection in the DRG. Episodic reactivation of virus from these latently infected ganglia results in viral shedding into the genital tract and the potential for the development of recurrent skin lesions, similar to human disease.

The mouse can be treated with antivirals prior to infection to determine the efficacy of treatment on the incidence of infection and clinical signs of disease (CSD) or treated during primary disease to evaluate the efficacy of antivirals on the day of death. While the guinea pig can also be treated prior to infection to determine antiviral efficacy on the incidence of infection, severity of primary disease, and subsequent recurrent disease and during primary disease to determine antiviral efficacy on severity of primary disease and subsequent recurrent disease, this model also allows for treatment during recurrent disease to examine the suppressive effects of antiviral interventions on recurrent lesions and vaginal viral shedding.

HSV-1 and HSV-2 can both be used for vaginal studies in mice and guinea pigs. HSV-1 typically gives a lower incidence of disease than HSV-2 in the murine model. In the guinea pig HSV-1 and HSV-2 produce comparable primary genital skin disease; however as in humans, HSV-1 produces fewer recurrences and less viral shedding than HSV-2.

## 2 Materials

### 2.1 Reagents

1. Medium 199 supplemented with Earle's salts, L-glutamine, and 2.2 g/L sodium bicarbonate.
2. FBS: Fetal bovine serum heat inactivated (56 °C, 30 min).
3. Glutamine: 200 mM L-glutamine.
4. Pen–Strep: Penicillin–streptomycin (10,000 U/mL penicillin; 10,000 μg/mL streptomycin).
5. 2 % Media: Sterile filtered Medium 199 supplemented with 1 % glutamine (v/v), 1 % pen–strep (v/v), and 2 % FBS (v/v).
6. 2 % agarose overlay: Dilute a 2 % agarose stock 1:40 using 2 % media pre-warmed to 37 °C.
7. 10 % Media: Sterile filtered Medium 199 supplemented with 1 % glutamine (v/v), 1 % pen–strep (v/v), and 10 % FBS (v/v).
8. 2.3 % crystal violet solution.
9. DNeasy DNA isolation kit (Qiagen) (*see* **Note 1**).
10. iQ Supermix (Bio-Rad).

### 2.2 Equipment

1. 37 °C incubator with 5 % $CO_2$ in air for determination of infection and quantification of viral replication.
2. Inverted binocular microscope for cytopathic effect (CPE) observation of infected Vero cells.
3. Platform rocker for incubating 24-well plates at room temperature.
4. Real-time PCR detection system for DNA quantification.
5. Ultracentrifuge.

### 2.3 Animals

1. Female inbred or outbred mice (typically 6–8 weeks old; *see* **Note 2**).
2. Female Hartley guinea pigs weighing 250–350 g (typically 5 weeks old; *see* **Note 2**).

### 2.4 Cell Line and Viruses

1. Vero (African green monkey kidney) cells (ATCC, CCL-B1).
2. Both HSV-1 and HSV-2 strains can be used for antiviral studies. Our group typically uses HSV-1 strain 17 syn+ (for mouse and guinea pig studies), HSV-2 strain 186 (for mouse studies), and HSV-2 strains MS or HSV-2333 (for guinea pig studies). To prepare virus stocks for use in animal studies *see* **Note 3**. Virus titer in stocks is determined by plaque assay using tenfold dilutions ($10^{-1}$ to $10^{-8}$) of stock virus (*see* **Note 4**).

## 3 Methods

### *3.1 Measurement of Antiviral Activity Against Genital Herpes in Mice*

A high-titer inoculum ($10^4$ PFU) is typically used to ensure uniform infection in all mice and to produce a high incidence of CSD in the control animals. Initial evaluations of antiviral efficacy generally require at least 10 animals per group. Antivirals may be administered subcutaneously, intraperitoneally, orally (in food and drinking water or by gavage), or intravaginally. Intramuscular treatment is also possible although not recommended because it can be difficult particularly for repeated administrations. The timing of antiviral treatment relative to virus inoculation should also be considered during study design. The model allows a number of virologic and clinical endpoints to be evaluated including the incidence of infection, the incidence, magnitude and duration of vaginal virus replication, incidence of CSD, time to onset of CSD, and day of death (*see* **Note 5**). A generalized timeline for genital herpes studies in the mouse is shown in Fig. 1.

#### *3.1.1 Intravaginal Inoculation of Mice with HSV-2*

1. Seven days prior to virus inoculation, administer 2 mg/mouse of medroxy-progesterone acetate in sterile saline by subcutaneous injection (final volume of 100 μL; *see* **Note 6**).
2. Dilute virus stock to a final concentration of $6.7 \times 10^5$ PFU/mL immediately prior to use and maintain on ice.
3. Swab the vaginal vault with a swab moistened in Media 199 and then instill 15 μL ($1 \times 10^4$ PFU) of virus inoculum into the vagina. For ease of handling during virus inoculation, mice can be anesthetized if desired.

#### *3.1.2 Evaluation of HSV-2 Infection and Vaginal Viral Replication*

1. To determine the incidence of infection and clearance of virus from the vagina, collect vaginal swabs from mice daily from days 1 to 7 post inoculation. Place each swab in 1 mL of cold 2 % media, maintain on ice, and immediately add sample to Vero cell monolayers to determine viral infection of cells using a CPE assay (*see* **Note 7**). After plating freeze the remainder of swab samples at −80 °C.

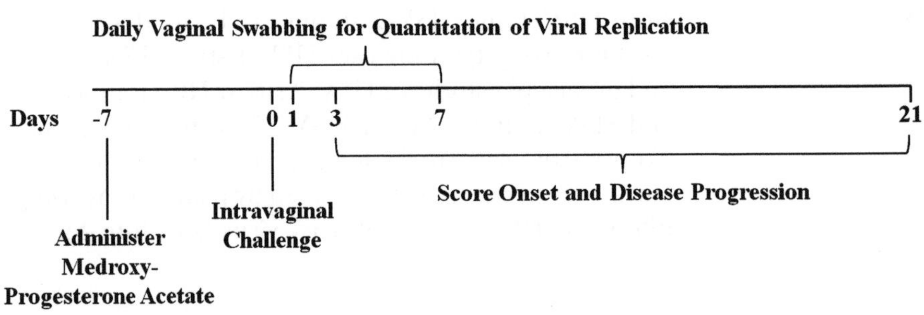


**Fig. 1** Timeline for antiviral evaluation of genital HSV-2 in the mouse model

2. To determine vaginal virus titers in infected mice (those animals corresponding to a positive result from **step 1** above), conduct a plaque assay using tenfold dilutions ($10^{-1}$ to $10^{-3}$) of sample (*see* **Note 4**).

##### *3.1.3 Evaluation of Genital Herpes Disease*

1. Beginning on day 3 post inoculation, observe mice daily for signs of genital herpes disease. In the mouse these include erythema (swelling and redness around the perineum), hair loss, and chronic wetness in the genital region. Mice do not develop vesiculo-ulcerative lesions. Many investigators use reduction in the incidence of clinical signs as the chief measure of antiviral efficacy, although a numerical score scale may also be used to quantify disease severity [6, 7].
2. As the disease progresses, animals can develop additional neurologic signs including hind leg paralysis (HLP). Animals that develop HLP rarely recover and should be euthanized to prevent suffering. Such animals are considered dead the next day for data analysis.

### *3.2 Measurement of Antiviral Activity Against Genital Herpes in the Guinea Pig*

Guinea pigs develop a self-limiting primary genital skin disease; however, like humans, not all infected animals will develop clinical signs of disease. Additionally, there can be considerable variability in the number of spontaneous recurrences observed between control animals. For these reasons, fairly large groups (typically 12–15) will be needed to provide adequate statistical power to antiviral efficacy studies. Antiviral treatments are typically administered subcutaneously, intraperitoneally, intramuscularly, intravaginally, by oral gavage, or in food and drinking water. Intravenous treatment is possible although not recommended for repeated treatments. As with the mouse, the timing of antiviral treatment should also be considered during study design. Prophylactic treatments allow for a broader range of endpoints to be evaluated, including the incidence of infection and primary CSD, vaginal virus replication and clearance during primary disease, onset of CSD, severity of primary disease, frequency of recurrent vesiculo-ulcerative lesions, frequency and amount of vaginal virus shedding, and latent viral load. However, unlike the mouse model, the guinea pig experiences recurrent genital herpes disease and viral shedding into the genital tract, allowing the evaluation of therapeutic interventions that can be administered during primary infection or during recurrent disease at multiple time points. Treatment during primary infection allows for evaluation of severity of primary disease, frequency of recurrent vesiculo-ulcerative lesions, frequency and magnitude of vaginal virus shedding, and latent viral load. Treatment during recurrent disease allows for evaluation of antiviral efficacy on frequency of recurrent vesiculo-ulcerative lesions, frequency and amount of vaginal virus shedding, and latent viral load. A generalized timeline for genital herpes studies in the guinea pig is shown in Fig. 2.

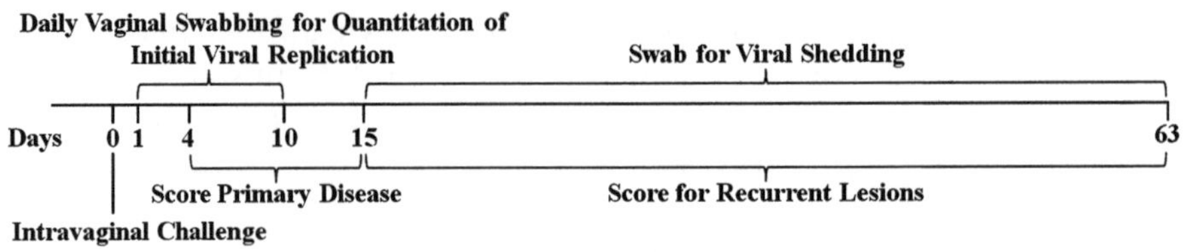


**Fig. 2** Timeline for antiviral evaluation of genital HSV-2 in the guinea pig model

*3.2.1 Infection of Guinea Pig with HSV-2*

1. Dilute virus stock to a final concentration of $1 \times 10^7$ PFU/mL immediately prior to use and maintain on ice.
2. Guinea pigs possess a vaginal membrane that must be ruptured prior to inoculation. This is done during initial pre-swabbing of the vaginal lumen with a swab moistened in medium 199 followed by a second swab of the lumen using a non-moistened swab.
3. Instill 100 μL ($1 \times 10^6$ PFU) of prepared virus inoculum into the vagina.

*3.2.2 Evaluation of HSV-2 Infection and Vaginal Viral Replication During Primary Disease*

1. To determine the incidence of infection and to quantify the magnitude and duration of vaginal virus replication, collect vaginal swabs from the guinea pigs daily from day 1 to 10 post inoculation. Place the swab in 1 mL of cold 2 % media, keep on ice, and immediately add an aliquot (typically 0.1–0.2 mL) to Vero cell monolayers to determine virus infectivity of cells by CPE assay (*see* **Note 7**). Freeze the remainder of the swab samples at −80 °C.
2. To quantify virus titer of infected guinea pigs (those animals corresponding to a positive result from **step 1** above) conduct a plaque assay using tenfold dilutions ($10^{-1}$ to $10^{-6}$) of the swab sample (*see* **Note 4**).

*3.2.3 Evaluation of Primary Genital Herpes Disease*

1. To evaluate the impact of antivirals on primary disease severity, individual animals are evaluated and a score given daily from onset (typically about 4–5 days post inoculation) to resolution (typically about 12–14 days post inoculation). The cumulative daily lesion score will provide a quantitative measure of primary disease severity. Typically, a lesion score scale ranging from 0 (no disease) to 4 (severe vesiculo-ulcerative disease of the perineum) that utilizes increments of 0.5 is used [8]. Briefly, animals with no disease are scored as 0; those with redness or swelling of the perineum as 0.5; animals with scores above 0.5 have vesicular lesions with the score reflecting the number and confluence of the vesicles on the perineum. Primary genital skin disease in guinea pigs typically peaks in severity about 8–9 days post inoculation (*see* **Note 8**). In addition to genital skin disease, some animals may develop other

complications during the primary infection (*see* **Note 9**). The ability of an antiviral to reduce the incidence of these complications provides another measure of efficacy.

2. As part of daily examination during primary disease it is important that the bladder be gently palpated to check for the occurrence of urinary retention and that the urine be expressed if retention is present (*see* **Note 10**).

#### 3.2.4 Evaluation of Recurrent Genital Herpes Disease and HSV-2 Shedding into the Genital Tract

1. To evaluate the impact of antiviral treatment on the frequency of spontaneous recurrent disease, animals are observed daily for recurrent lesions following resolution of primary disease. These evaluations typically start at day 15 post inoculation and continue for several weeks. Beyond approximately day 63 post inoculation, spontaneous recurrences in untreated animals decline in rate to an extent that makes measuring reductions due to antiviral treatment problematic. Recurrent lesions normally last only 1–2 days and are scored as 0 (no disease), 0.5 (localized redness but no vesicle), or 1.0 (vesicular lesion present).
2. To evaluate the impact of antiviral treatment on frequency and magnitude of HSV-2 shedding into the genital tract during recurrent disease (between days 15 and 63 post inoculation), vaginal swab samples can be collected daily. The timing and length of the antiviral treatments will determine the period over which vaginal swabs should be collected. Typically daily assessment of shedding for a 21-day period will be sufficient to allow reductions in incidence and magnitude to be determined. Swab tips should be placed in 1 mL of cold 2 % media. Samples can be frozen at −80 °C and batched for processing or processed immediately. Processing will include DNA extraction and quantitative PCR (qPCR).

#### 3.2.5 Evaluation of Latent Viral Burden in the DRG

To determine the impact of antiviral treatment on the magnitude of latent virus infection in the DRG, at the completion of the study, the DRG are harvested from the humanely euthanized animals and placed into sterile tubes. The samples can be processed immediately or stored at −80 °C until they are needed for DNA extraction and real-time PCR quantitation.

#### 3.2.6 DNA Extraction and qPCR for Viral Shedding and Latent Viral Burden

1. If samples have been frozen for storage, thaw before proceeding to DNA extraction.
2. To extract DNA from the swab samples, follow the protocol provided by the kit manufacturer. For vaginal swabs, see "Purification of Total DNA from Animal Blood or Cells (DNeasy 96 Protocol)" (Qiagen). Our laboratory uses the protocol "Purification of Total DNA from Animal Tissues (Spin-Column Protocol)" (Qiagen) for DRG extractions.

Extracted DNA is stable at 4 °C for several days; however evaporation can be an issue. For long-term storage, place extracted samples at −20 °C until qPCR can be conducted.

3. Prepare separate PCR master mixes for each target, HSV and GAPDH. The final volume of reagents to add to the master mix will be determined by the number of samples to be assayed (*see* **Note 11**).
   (a) Add 12.5 μL 1× iQ Supermix (or similar reagent) to a sterile pipet basin.
   (b) Add 1 μL each of forward and reverse primers (final concentration 5 pmol each) specific for the target, HSV (forward: 5′-CGC ATC AAG ACC ACC TCC TC-3′; reverse: 5′-GCT CGC ACC ACG CGA-3′) or GAPDH (forward: 5′-AAT GGG AAG CTC ACA GGT ATG G-3′; reverse: 5′-ATG TCA TCG TAT TTG GCC GGT-3′), to the pipet basin.
   (c) Add 0.5 μL TaqMan probe (final concentration 2.5 pmol) specific for the target, HSV-2 (5′-FAM-CGG CGA TGC GCC CCA G-BHQ1-3′) or GAPDH (5′-TET-TCC AGG CGG CAG GTC AGA TCC ACA-BHQ1-3′), to the basin.
   (d) Finally, add 5 μL of DNase/RNase-free water to the basin.
   (e) Mix the reagents in the basin by gently rocking back and forth.
4. Aliquot 20 μL of the master mix to all wells of a 96-well PCR plate that will contain sample DNA. Due to differences in the PCR protocols, a separate PCR plate will be needed for each target, HSV and GAPDH.
5. Prepare quantitation standards by diluting cloned amplimers specific for the target run (HSV or GAPDH) using tenfold dilutions (five dilutions, $10^6$–$10^2$).
6. Add 5 μL/well DNA template or quantitation standards to appropriate wells. Negative control reactions lacking template (e.g., water only) should also be included to assess contamination.
7. Seal the plate using optical grade sealing film.
8. Centrifuge the plate for ~60 s at 5800 × *g* to concentrate liquid at the bottom of the wells and remove bubbles that may have formed during the pipetting process.
9. Run the appropriate PCR protocol for each target, HSV (Table 1) or GAPDH (Table 2).
10. For data analysis, HSV-2 values should be normalized to account for cellular load using the results of GAPDH qPCR run.
    (a) Calculate the average GAPDH value from all treatment groups (study GADPH).

**Table 1**
**PCR program protocol for the determination of HSV-2 load**

| Cycle | Repeats | Step | Time (min) | Hold | Temp (°C) | PCR acquisition |
|---|---|---|---|---|---|---|
| 1 | 1 | | | | | |
| | | 1 | 2:00 | No | 95.0 | |
| 2 | 45 | | | | | |
| | | 1 | 0:20 | No | 95.0 | |
| | | 2 | 1:00 | No | 60.0 | Real time |
| 3 | 1 | | | | | |
| | | 1 | | Yes | 4.0 | |

**Table 2**
**PCR program protocol for the determination of cellular GAPDH load**

| Cycle | Repeats | Step | Time (min) | Hold | Temp (°C) | PCR acquisition |
|---|---|---|---|---|---|---|
| 1 | 1 | | | | | |
| | | 1 | 1:30 | No | 95.0 | |
| 2 | 40 | | | | | |
| | | 1 | 0:30 | No | 95.0 | |
| | | 2 | 0:30 | No | 55.0 | |
| | | 3 | 0:45 | No | 72.0 | Real time |
| 3 | 1 | | | | | |
| | | 1 | | Yes | 4.0 | |

(b) Calculate the adjusted GAPDH by dividing the GAPDH qPCR value for each individual sample by the average GAPDH (e.g., sample GAPDH/study GAPDH=adjusted GADPH). This will account for the cellular load in each sample using the average GAPDH obtained from all samples.

(c) Finally, divide the HSV-2 qPCR value for each sample by the adjusted GAPDH value (sample HSV-2/adjusted GAPDH). This will normalize the HSV-2 values for each sample to account for variability in the amount of cellular material collected during sampling and provide for better comparisons of treatment groups.

## 4 Notes

1. A number of commercial kits for DNA extraction are available. Our group has had success using the DNeasy 96 Blood & Tissue kit to analyze vaginal swab samples and the DNeasy Blood & Tissue spin-column protocol for analysis of DRG samples.
2. Animals are typically allowed to acclimate to the vivarium for 7 days prior to initiation of studies. This allows physiological and psychological stabilization before any experimental manipulation occurs.
3. For virus stock production, infect confluent Vero cell monolayers at a multiplicity of infection of 0.01. Observe the cell monolayers daily using standard light microscopy until ~50–75 % of the cells show CPE, typically between 3 and 4 days post inoculation. Transfer cells and media to 50 mL conical tubes and centrifuge at low speed (800 × *g*, 10 min at 4 °C). Discard supernatant and resuspend the pellet in 1 mL of 2 % media and freeze (−80 °C). Thaw and refreeze cells a total of 3 times. After the final thaw, remove the remaining cells by low-speed centrifugation and decant supernatant containing virus, aliquot to cryovials, and store frozen (−80 °C).
4. To determine virus titers, plaque assays are conducted using a 24-well plate with a confluent Vero cell monolayer. Serially dilute samples and add 200 μL/well in duplicate. Rock the plate gently for 1 h at room temperature. Add agarose overlay (600 μL/well) and incubate the plates at 37 °C for 2 days. Following incubation, aspirate the wells and stain with 1 mL of crystal violet. Count the number of plaques per well to determine the titer of the virus. Only wells containing between 10 and 100 plaques should be counted to reduce error when calculating the titer.
5. Although the day of death is of limited clinical relevance for herpes antivirals, it can provide useful information about the ability of antivirals to impact disease progression in this model.
6. Medroxy-progesterone pretreatment is necessary to induce diestrus in the animals, thinning the vaginal epithelium to allow a uniform infection with HSV-2.
7. CPE assays are typically conducted using a 24-well plate with a confluent Vero cell monolayer. Add 200 μL of sample per well. Rock the plate gently for 1 h at room temperature. Add warm media (600 μL/well) and incubate the plates at 37 °C for 5 days. Begin observing wells for CPE at day 3 post inoculation as vaginal swabs also collect bacteria present in the vaginal lumen that can kill the cells and result in false positives.

Following incubation, aspirate the wells and stain with 1 mL of crystal violet.

8. Guinea pigs can develop open sores as a result of scratching or biting at lesions during primary disease. In addition, secondary infections of the perineum can develop resulting in extensive scarring. If either of these events occurs so that the animal cannot be evaluated accurately for primary disease severity or the presence of recurrent lesions later in the course of the study, the animal should be excluded from the data analysis.
9. Guinea pigs can experience a number of neurologic complications during primary infection. These include urinary retention, incontinence, and unilateral or bilateral hind limb paralysis. In the majority of animals all of these signs will resolve spontaneously within a few days. Animals that do not resolve hind limb paralysis within 3–4 days, or which show signs of self-mutilation of the limbs, should be humanely euthanized. Animals that do not resolve incontinence but remain healthy and do not develop secondary infections of the perineum can remain in the study. In addition, some animals have reduced food and water intake during primary disease. Supplemental fluids may be administered subcutaneously during this period as needed.
10. To check for urinary retention, gently massage the lower abdomen to determine if the bladder is round and distended. If retention is found, gently express the bladder. This process is very important; failure to express the bladder if there is retention can result in rupture.
11. Two PCR runs will be needed for each sample: one to quantify the amount of HSV-2 DNA present and the second to quantify the housekeeping gene GAPDH. This allows normalization of the HSV load to the cellular load present in each sample. For each run, separate plates and master mixes must be prepared for the HSV and GAPDH targets. To determine the final volume of each reagent to be added, multiply the number of samples by the indicated amount of each reagent. For example, 10 samples would require 120 μL iQ supermix, 10 μL of each forward and reverse primer, 5 μL of probe, and 50 μL of water. The PCR assay described here for the HSV target is specific to HSV-2.

## References


1. Nieuwenhuis RF, van Doornum GJ, Mulder PG et al (2006) Importance of herpes simplex virus type-1 (HSV-1) in primary genital herpes. Acta Derm Venereol 86(2):129–134
2. Xu F et al (2006) Trends in herpes simplex virus type 1 and type 2 seroprevalence in the United States. JAMA 296(8):964–973
3. Corey L (2005) Herpes simplex virus. In: Mandell G, Bennett J, Dolin R (eds) Practices of infectious diseases, 6th edn. Elsevier, Philadelphia, PA, pp 1762–1780
4. Corey L et al (2004) Once-daily valacyclovir to reduce the risk of transmission of genital herpes. N Engl J Med 350(1):11–20

5. Schiffer JT, Magaret A, Selke S, Corey L, Wald A (2011) Detailed analysis of mucosal herpes simplex virus-2 replication kinetics with and without antiviral therapy. J Antimicrob Chemother 66(11):2593–2600
6. Harandi AM, Svennerholm B, Holmgren J, Eriksson K (2001) Differential roles of B cells and IFN-gamma-secreting CD4(+) T cells in innate and adaptive immune control of genital herpes simplex virus type 2 infection in mice. J Gen Virol 82(4):845–853
7. Gillgrass AE, Tang VA, Towarnicki KM et al (2005) Protection against genital herpes infection in mice immunized under different hormonal conditions correlates with induction of vagina-associated lymphoid tissue. J Virol 79(5):3117–3126
8. Stanberry LR, Kern ER, Richards JT et al (1982) Genital herpes in guinea pigs: pathogenesis of the primary infection and description of recurrent disease. J Infect Dis 146(3): 397–404

# Chapter 25

# A Fluorescence-Based High-Throughput Screening Assay for Identifying Human Cytomegalovirus Inhibitors

Christel Van den Eynde, Ellen Van Damme, Tania Ivens, and Edwin Yunhao Gong

## Abstract

Human cytomegalovirus (HCMV) is a common opportunistic pathogen that can cause devastating morbidity and mortality amongst neonates and immune-compromised patients. The current standard of care for HCMV infection is limited to four antiviral compounds that have major limitations in terms of long-term use, toxicity, and use during pregnancy. To provide patients with alternative treatment options to decrease HCMV-related morbidity and mortality, new drugs with novel modes of action are warranted. Here, we describe a validated high-throughput fluorescence antiviral screening assay based on infection of fibroblast cells with a fluorescently tagged reference strain of HCMV (AD169-GFP) to screen and profile HCMV inhibitors.

**Key words** Cytomegalovirus, High-throughput screening, Fluorescence-based screening assay, HCMV inhibitor

## 1 Introduction

Human cytomegalovirus (HCMV) is a common opportunistic pathogen that is transmitted via body fluids such as urine, saliva, vaginal secretions, semen, or breast milk [1]. It is a double-stranded DNA virus belonging to the *Betaherpesvirinae* subfamily that primarily infects fibroblasts, endothelial cells, epithelial cells, and myloid cells such as dendritic cells, monocytes, and macrophages. As all herpesviruses, HCMV can establish latency but unlike alpha- or gammaherpesviruses, HCMV specifically targets blood monocytes and $CD34^+$ stem cells as sites of latency. Upon allogeneic stimulation or differentiation, the virus can reactivate and engage in lytic replication. In this part of the life cycle, the full HCMV gene array, categorized in immediate–early, early, and late transcripts, is transcribed and translated, which eventually results in the production of infectious virus [2–4].

Edwin Yunhao Gong (ed.), *Antiviral Methods and Protocols*, Methods in Molecular Biology, vol. 1030, DOI 10.1007/978-1-62703-484-5_25, 

Worldwide, the prevalence of HCMV is between 45 and 100 % depending on age, location, gender, and socioeconomic status [5]. While acute HCMV infection remains without symptoms or causes a mononucleosis-like disease in healthy persons [6], in neonates and immune-compromised individuals, such as HIV and transplant patients, the virus can result in severe morbidity and mortality [7–10].

The present licensed antiviral drugs to treat HCMV infections are ganciclovir, cidofovir, foscarnet, and fomivirsen (discontinued). Although these drugs provide treatment options, they have major limitations in terms of hemotoxicity, carcinogenicity, kidney problems, and hepatotoxicity. In addition, because of potential teratogenic effects none of these compounds is suitable to treat neonatal HCMV [6]. New drugs such as CMX001 (the hexadecyloxypropyl prodrug of cidofovir) and AIC246 (Letermovir) are currently under development and have passed phase IIb clinical trials [11, 12]. Nevertheless, potential resistance issues of the drugs on the market and the need for drugs with novel mechanisms of action justify the discovery of novel antiviral drugs to provide patients with alternative treatment options to decrease HCMV morbidity and mortality.

In this chapter we describe a fluorescence-based high-throughput screening assay for evaluating cytomegalovirus inhibitors. This robust assay provides the advantage of being homogeneous (mix and measure), fast, and easy, and also enables high-throughput screening of compound libraries.

## 2 Materials

### 2.1 Virus, Cells, Media, and Reagents

1. Cytomegalovirus: Laboratory-adapted HCMV strain AD169-GFP licensed from Dr. M. Marschall, Universität Erlangen-Nürnberg, Germany [13] (*see* **Note 1**).
2. Hel-299 cells: American Type Culture Collection (ATCC, CCL-137).
3. Hel-299 cell growth medium (MEM/10 % FCS): Eagle's minimum essential medium (MEM) supplemented with 10 % fetal calf serum (FCS), 1.0 mM sodium pyruvate, 0.1 mM nonessential amino acids, 2 mM L-glutamine, and 0.02 mg/mL gentamicin.
4. Hel-299 cell maintenance medium (MEM/2 % FCS): Same as the growth medium but supplemented with 2 % FCS.
5. Freezing medium for HCMV stocks (freshly prepared): FCS supplemented with 10 % DMSO.
6. Reference compound: Ganciclovir (Sigma-Aldrich) [14] (*see* **Note 2**).

7. Resazurin (7-hydroxy-3H-phenoxazin-3-one-10-oxide) (Sigma-Aldrich) (*see* **Note 3**).
8. Potassium ferricyanide $K_3Fe(CN)_6$ (Sigma-Aldrich) (*see* **Note 4**).
9. Potassium ferrocyanide $K_4Fe(CN)_6{\cdot}3H_2O$ (Sigma-Aldrich) (*see* **Note 4**).
10. Preparation of 0.1 M potassium phosphate buffer (pH7.4) (PPB): Weigh 2.72 g of $KH_2PO_4$ and 13.86 g $K_2HPO_4$ (in a beaker) and dissolve in 1 L deionized water. Check the pH and correct if it is not equal to pH7.4. Sterilize the PPB through a 0.22 μm filter and store at room temperature.
11. Preparation of 20× resazurin stock solution:
    (a) Prepare a fresh stock of 3 g/L resazurin in PPB (PPB-A).
    (b) Make a 30 mM potassium ferricyanide stock by dissolving 9.87 g/L $K_3Fe(CN)_6$ in PPB (PPB-B).
    (c) Weigh 12.66 g $K_4Fe(CN)_6{\cdot}3H_2O$ and dissolve in 1 L of PPB to prepare a stock of 30 mM potassium ferrocyanide (PPB-C).
    (d) Mix equal volumes of PPB-A, PPB-B, and PPB-C to obtain a 20× resazurin stock solution.
12. Preparation of the final resazurin working solution: Make a 20-fold dilution of the 20× resazurin stock solution in PPB and filter-sterilize. This working solution can be stored at 4 °C for a maximum of 2 weeks.

### 2.2 Disposables and Equipment

1. 96-well black-view plates (Costar).
2. 96-well V-bottom transparent plates.
3. Centrifuge.
4. Multichannel pipettes.
5. Automatic fluorescence microscope (or fluorescence spectrophotometer).
6. Multidrop combi liquid dispenser (Thermo Scientific).
7. Fluorescence microscope.

## 3 Methods

All incubations at 37 °C for Hel-299 cells described in this chapter are performed using a humidified 5 % $CO_2$ incubator unless otherwise stated.

### 3.1 Preparation of Cell-Associated HCMV-AD169-GFP Stocks

1. Seed Hel-299 cells in 180 $cm^2$ flasks using MEM/10 % FCS and incubate at 37°C overnight. The cells should reach ~80–90 % confluence on the day of infection (*see* **Note 5**).
2. On the following day, remove the culture medium from the flasks.

3. Wash the cells with 20 mL PBS (Dulbecco's phosphate-buffered saline, no calcium chloride and no magnesium chloride).
4. Discard the PBS.
5. Add AD169-GFP virus at a multiplicity of infection (MOI) of 0.01 $TCID_{50}$/cell (the 50 % tissue culture infective dose per cell) (*see* Subheading 3.2) in a volume of 15 mL MEM/2 % FCS (*see* **Note 6**).
6. Incubate the cells at 37 °C for 2 h with occasional tilting to spread the virus over the whole-cell monolayer.
7. After incubation, remove the virus inoculum (*see* **Note 7**).
8. Add 18 mL of MEM/2% FCS to the flask and incubate at 37 °C.
9. From day 3 post infection, check the infected cells daily for cytopathic effect (CPE) under an inverted microscope or for fluorescent foci under a fluorescent microscope.
10. When the cells reach ~100 % CPE or ~100 % of the cells are fluorescent, discard the culture supernatants and trypsinize the infected cells with 0.05 % trypsin–EDTA after washing with PBS (*see* **Note 6**). Resuspend the infected cells in 15 mL of MEM/2% FCS.
11. Collect the infected cells into one or more sterile Falcon™ conical tubes and centrifuge at 500×*g* for 5 min to pellet the infected cells.
12. Discard the supernatant and resuspend the infected cells in an appropriate amount (4 or 5 mL for each 180 $cm^2$ flask) of freezing medium. Pool the infected cells and aliquot 0.5 mL into each cryovial.
13. Freeze the vials at −80°C overnight and subsequently store the cryovials in liquid nitrogen.

### 3.2 Virus Titration

1. Prepare a 96-well titration plate by seeding Hel-299 cells at a density of 25,000 cells/75 μL/well in MEM/2 % FCS.
2. To make serial dilutions of virus stocks, take a new plate and add 90 μL MEM/2 % FCS to all wells except in the first column. Add undiluted virus to all wells of column 1 (8 wells) and make a tenfold serial dilution by transferring 10 μL from well to well until column 11. Column 12 is used as an uninfected cell control and receives medium only (*see* **Note 8**).
3. Transfer 25 μL of the virus dilution series to the corresponding wells of the titration plate with cells (the final volume in the titration plate is 100 μL per well). Incubate the plate at 37 °C for 4 to 5 days. After incubation, check the CPE under an inverted microscope or assess the fluorescent foci under a fluorescent microscope. The endpoint is when the undiluted virus in column 1 gives full CPE (or all the cells are fluorescent).

4. Count the number of wells with CPE per virus dilution or count the number of wells with fluorescent foci under a fluorescent microscope and calculate the 50 % endpoint ($TCID_{50}$/mL) according to the method of Reed and Muench [15].

### 3.3 Fluorescence-Based Screening Assay

The anti-HCMV compound screening assay described below is performed using a 96-well plate format, but it can be easily adapted to a 384-well plate format.

#### 3.3.1 Preparation of Compound Test Plates

1. Make a serial dilution (four- or fivefold) of the test compounds to prepare 4× final test concentrations in a 96-well V-bottom transparent plate in MEM/2 % FCS. The DMSO concentration in each well is 2 % (*see* **Note 9**). Include a reference compound ganciclovir as a positive dose–response control. The test compounds can be plated in different formats: (1) one compound per row with 9 dilutions; (2) two compounds per row with 5 dilutions per compound; (3) one compound with 9 dilutions in two row (duplicate wells for each dilution); and (4) two compounds with 5 dilutions of each compound in two rows (duplicate wells for each dilution).
2. Transfer 25 μL of this dilution series to the corresponding wells of a 96-well black-view plate. If performing 9 dilutions per compound, columns 10–12 receive 25 μL of MEM/2 % FCS containing 2 % DMSO. If performing 5 dilutions per compound, columns 11 and 12 receive 25 μL of MEM/2 % FCS containing 2 % DMSO.
3. For each set of test compounds, prepare duplicate plates: one for activity testing and one for toxicity testing.

#### 3.3.2 Antiviral Screening of HCMV Inhibitors

1. Prepare a Hel-299 cell suspension at $5 \times 10^5$ cells/mL in MEM/2 % FCS (prepare 5.5 mL for each 96-well plate). To keep the cell suspension homogenized, put a sterile stirring bar into the cell suspension and leave the cells gently stirring on a magnetic stirrer.
2. Thaw one or more vials of AD169-GFP virus stocks (depending on the number of plates to be tested) in a 37 °C water-bath with shaking. Make an appropriate virus dilution to result in an MOI of 0.01 $TCID_{50}$/cell (*see* **Note 6**) in MEM/2 % FCS (prepare 3 mL of virus dilution for each 96-well plate).
3. Use a multidrop combi liquid dispenser or a multichannel pipette to dispense 50 μL of cell suspension to all wells of compound test plates (*see* Subheading 3.3.1). The final cell input is 25,000 cells per well.
4. Dispense 25 μL of virus suspension to columns 1–11. Columns 10 and 11 serve as virus controls if 9 dilutions per compound are prepared, and column 11 serves as a virus control if 5 dilutions per compound with two compounds per row are made.

Column 12 serves as cell control and receives 25 μL of MEM/2 % FCS.

5. Incubate the plates at 37 °C for 5 days.
6. At the end of incubation, measure the fluorescent intensity of each well using an automatic fluorescence microscope at a wavelength of 488 nm or using any fluorescence spectrophotometer.
7. Calculate the 50 and 90 % effective concentrations ($EC_{50}$ and $EC_{90}$) of the test compounds based on the reduction of the fluorescence intensity relative to the untreated virus controls using computer software, such as GraphPad Prism. The $EC_{50}$ and $EC_{90}$ are defined, respectively, as the concentration of a compound achieving 50 and 90 % inhibition of the fluorescent signals.

### 3.4 Determination of the Cytotoxicity of Test Compounds

1. Prepare a Hel-299 cell suspension at a density of $5 \times 10^5$ cells/mL in MEM/2 % FCS (5.5 mL per 96-well plate). To keep the cell suspension homogenized, leave the cells gently stirring on a magnetic stirrer with a sterile stirring bar.
2. Using a multidrop combi liquid dispenser or a multichannel pipette, dispense 50 μL of cell suspension (final cell density is 25,000 cells per well) into columns 1–11 of a 96-well plate containing 25 μL of test compounds or medium (*see* Subheading 3.3.1).
3. Add 25 μL of MEM/2 % FCS into all wells of columns 1–11 and 75 μL of MEM/2 % FCS into the wells of column 12. Column 12 is used as medium control and contains no cells and no compound.
4. Incubate the plates at 37 °C for 5 days.
5. At the end of incubation, add 30 μL of resazurin working solution (*see* **Note 10**) to each well using a multichannel pipette or a multidrop combi liquid dispenser.
6. Incubate the plates at 37 °C for 6 h. Leave the plates in the incubator for a longer time if the blue resazurin is not yet converted to the fluorescent pink resorufin.
7. Measure the fluorescent resorufin signal using an automatic fluorescence microscope at a wavelength of 514 nm. Alternatively, a fluorescence spectrophotometer can be used to measure the change from blue to pink at excitation of 530 nm and emission of 580 nm.
8. Calculate the 50 % cytotoxic concentration ($CC_{50}$) for each compound using GraphPad or comparable software.
9. Calculate the selectivity index (SI): $SI = CC_{50}/EC_{50}$. SI can be used to select compound hits.

## 4 Notes

1. According to the Belgian biosafety regulation, the handling and manipulation of HCMV are restricted in a biosafety level two laboratory.
2. Beware that ganciclovir, although approved as a drug, can be harmful to the unborn child and fertility. Always wear proper personal protection, such as gloves, glasses, and a mouth mask, when working with this compound.
3. Resazurin may be toxic when inhaled, ingested, or when it comes in contact with the skin. Always wear proper personal protection, such as gloves, glasses, and a mouth mask, when working with this compound.
4. Potassium ferricyanide and potassium ferrocyanide are toxic when ingested, inhaled, or through skin contact. These substances can permeate the skin; thus always wear proper personal protection equipment.
5. For subculture, the Hel-299 cells are split every 3 or 4 days and cultured in MEM/10 % FCS. The cells are split 1:4 for a 3-day culture and 1:6 for a 4-day culture.
6. The HCMV AD169 reference strain is mostly cell associated, meaning that the majority of the virus particles are associated with cells or cell membranes instead of secreted into the culture medium. Therefore, the term MOI used in this chapter for HCMV is the number of infected foci (or fluorescent cells), rather than the actual number of infectious virions per cell. Each foci may contain more than one virus particles.
7. Use an appropriate disinfectant to decontaminate the virus-contaminated medium such as Virkon.
8. To avoid carryover of virus, transfer the virus suspension from the highest concentration to the first dilution with minimal touching of the medium present in the well. Discard the tip and use a new tip to mix medium and virus before transferring the suspension to the next dilution. Use this procedure for every dilution step.
9. The final concentration of DMSO should not exceed 0.5 % because of the toxicity on cells. The volume of the compounds in each well of the plate is 25 μL and the total volume of each well is 100 μL. Therefore, the DMSO concentration for the compound dilutions and the control wells is 2 %.
10. The resazurin assay for cytotoxicity is based on the reduction of the blue resazurin by NADH, produced by the cells, into the pink fluorescent product, resorufin. The formation of the pink fluorescent resorufin is directly related to the number of viable cells in the culture.

## Acknowledgments

The authors wish to thank Dr. M. Marschall (Universität Erlangen-Nürnberg, Germany) for providing the AD169-GFP virus used in this chapter.

### References

1. Mocarski ES, Shenk T, Pass RF (2001) Cytomegaloviruses. In: Knipe D, Howley P et al (eds) Fields Virology, 5th edn. Lippincott Williams & Wilkins, Philadelphia, pp 2701–2772
2. Reeves M, Sissons P, Sinclair J (2005) Reactivation of human cytomegalovirus in dendritic cells. Discov Med 5:170–174
3. Reeves MB, Lehner PJ, Sissons JG, Sinclair JH (2005) An in vitro model for the regulation of human cytomegalovirus latency and reactivation in dendritic cells by chromatin remodelling. J Gen Virol 86:2949–2954
4. Reeves MB, MacAry PA, Lehner PJ et al (2005) Latency, chromatin remodeling, and reactivation of human cytomegalovirus in the dendritic cells of healthy carriers. Proc Natl Acad Sci USA 102:4140–4145
5. Cannon MJ, Schmid DS, Hyde TB (2010) Review of cytomegalovirus seroprevalence and demographic characteristics associated with infection. Rev Med Virol 20:202–213
6. Boeckh M, Geballe AP (2011) Cytomegalovirus: pathogen, paradigm, and puzzle. J Clin Invest 121:1673–1680
7. Pass RF, Fowler KB, Boppana SB et al (2006) Congenital cytomegalovirus infection following first trimester maternal infection: symptoms at birth and outcome. J Clin Virol 35:216–220
8. Kenneson A, Cannon MJ (2007) Review and meta-analysis of the epidemiology of congenital cytomegalovirus (CMV) infection. Rev Med Virol 17:253–276
9. Griffiths PD (2006) CMV as a cofactor enhancing progression of AIDS. J Clin Virol 35:489–492
10. Crough T, Khanna R (2009) Immunobiology of human cytomegalovirus: from bench to bedside. Clin Microbiol Rev 22:76–98
11. Aicuris (2012) Press release. http://wwwaicuriscom/10d71/News_Publicationshtm. Accessed 29 May 2012
12. Chimerix (2012) Clinical trials. http://wwwchimerixcom/therapeutic-programs/category/clinical-trials-patient-information/. Accessed 29 May 2012
13. Marschall M, Freitag M, Weiler S et al (2000) Recombinant green fluorescent protein-expressing human cytomegalovirus as a tool for screening antiviral agents. Antimicrob Agents Chemother 44:1588–1597
14. Matthews T, Boehme R (1988) Antiviral activity and mechanism of action of ganciclovir. Rev Infect Dis 10:S490–S494
15. Reed LJ, Muench H (1938) A simple method of estimating fifty percent endpoints. Am J Hygiene 27:493–497

# Part V

## Respiratory Viruses

# Chapter 26

# A Fluorescence-Based High-Throughput Antiviral Compound Screening Assay Against Respiratory Syncytial Virus

**Leen Kwanten, Ben De Clerck, and Dirk Roymans**

## Abstract

Respiratory syncytial virus (RSV) is a common virus that infects people of all ages and causes cold-like symptoms in most cases. However, more serious infections occur in the younger and older extremities of the population causing severe lung infections such as bronchiolitis and pneumonia. The current standard of care is mostly limited to supportive treatment, although prophylaxis by passive immunization with the humanized monoclonal antibody palivizumab and therapeutic intervention with aerosolized ribavirin are available. Unfortunately, administration of palivizumab is restricted to at-risk infants up to the age of two and is associated with high cost, while ribavirin treatment is hindered by questionable efficacy and safety reasons. Consequently, the development of novel specific RSV antiviral drugs is needed to help decrease RSV-related morbidity and mortality. We describe here a fluorescence-based high-throughput screening assay to discover RSV inhibitors which is based on the infection of HeLa cells with a recombinant RSV strain that contains an enhanced green fluorescent protein coding sequence in its viral genome.

**Key words** Respiratory syncytial virus, High-throughput screening, Fluorescence, RSV inhibitors

## 1 Introduction

Respiratory syncytial virus (RSV) is a pneumovirus belonging to the *Paramyxoviridae* family. In cell culture, the virus induces characteristic cytopathogenic effects like syncytia formation, hence explaining its name. The virus is composed of a phospholipid envelope surrounding the helical nucleocapsid that contains a negative-sense single-stranded RNA genome. The genome encodes 11 viral proteins, 9 of which are structural. Three proteins are located on the surface of the virion, i.e., the fusion (F), the attachment (G), and the small hydrophobic (SH) protein. Proteins F and G mediate virus attachment and fusion, respectively [1, 2]. The virus appears to spread via large droplets or through fomite contamination. Infection requires either close contact with infected individuals or contact of contaminated hands with nasal or conjunctival mucosa [3]. In temperate

Edwin Yunhao Gong (ed.), *Antiviral Methods and Protocols*, Methods in Molecular Biology vol. 1030,
DOI 10.1007/978-1-62703-484-5_26, © Springer Science+Business Media, LLC 2013

climates, outbreaks of RSV infection usually start in the fall and end in the spring, while in subtropical regions RSV persists all year long [3, 4]. The virus is a predominant cause of acute lower respiratory tract infections with symptoms ranging from severe pneumonia and bronchiolitis to much milder infections similar to the common cold. Though RSV does not pose a threat to healthy adults, it represents a major disease burden in infants, young children, elderly, and immunocompromised people [5, 6]. The virus is responsible for a significant amount of morbidity and mortality each year. Current treatment options are limited to supportive care, administration of ribavirin to severely ill patients, and prophylactic administration of a monoclonal antibody, palivizumab, to high-risk children [7]. However, both treatments have several limitations. Treatment with ribavirin is limited because of its problematic route of aerosol administration, questionable efficacy, and teratogenic effects, while prophylaxis with palivizumab is restricted to infants till the age of 2 that are at high risk to develop severe RSV infection and by high cost.

Consequently, the medical need for the development of specific RSV antiviral drugs is high.

We describe here a cell-based antiviral screening assay using a recombinant RSV strain that contains an enhanced green fluorescent protein (eGFP) coding sequence in its viral genome [8]. To discover potentially interesting RSV antiviral compounds, the recombinant virus is incubated together with test compounds and HeLa cells. Since the level of eGFP expression in infected cells correlates with the level of viral replication, the antiviral activity of test compounds can be assessed by measuring the level of eGFP fluorescence in infected cells exposed to compounds. A major advantage of this type of assay is that compounds inhibiting either viral or cellular targets important for viral replication can be identified in one assay.

Since the reported assay is cell based, potential nonspecific, compound-related cytotoxicity needs to be assessed in parallel. The method described here uses a commercially available luminescence-based assay to measure the adenosine triphosphate (ATP) levels in the cells. Since ATP is an essential energy carrier needed for proper functioning of cells, cytotoxic effects of test compounds result in a rapid decrease of cytoplasmic ATP levels. Both the activity and toxicity assays should be performed in parallel to assess the RSV antiviral potential of compounds.

## 2 Materials

### 2.1 Cells, Virus, Media, and Reagents

1. HeLa cells: American Type Culture Collection (ATCC number: CCL-2™).
2. RSV-eGFP virus (rgRSV 224): Licensed from the National Institutes of Health (NIH) (for description *see* Subheading 3.1.1 and refs. 8, 9) (*see* **Note 1**).

3. HeLa cell culture medium (RPMI/10 % FCS): RPMI 1640 (Gibco) supplemented with 10 % heat-inactivated fetal calf serum (FCS-HI) and 0.04 % gentamicin (*see* **Note 2**).
4. HeLa cell maintenance medium (RPMI/2 % FCS): RPMI 1640 supplemented with 2 % FCS-HI and 0.04 % gentamicin.
5. 40 % (w/v) sucrose solution in RPMI/10 % FCS.
6. 0.05 % of Trypsin–EDTA (1×).
7. Dulbecco's phosphate-buffered saline (DPBS), without calcium and magnesium.
8. Dimethylsulfoxide (DMSO) (pharma solvent grade).
9. Liquid nitrogen.
10. Disinfectant solutions: 70 % ethanol, 1 % Virkon (*see* **Note 3**).
11. ATPLite™ 1-Step kit (PerkinElmer).

### 2.2 Consumables and Equipment

1. 175 $cm^2$ culture flasks (tissue culture treated).
2. Autoclaved multidrop cassettes.
3. Cell scraper.
4. 384-well sterile black-view and white microtiter plates.
5. 96-well sterile V-bottom microtiter plates.
6. Viewlux™ (PerkinElmer) or any other luminescence reader.
7. Fluorescence microscope.
8. Fluorescence reader (*see* **Note 4**).
9. Multidrop combi liquid dispenser (Thermo Scientific).
10. Centrifuge.
11. Incubator (humidified, 5 % $CO_2$, 37 °C).

## 3 Methods

### 3.1 Preparation of the RSV-eGFP

#### 3.1.1 Recombinant RSV-eGFP

For in vitro RSV antiviral experiments, recombinant human RSV rgRSV224 is used. The rgRSV224 containing an eGFP coding sequence was generated through a reverse genetic approach by inserting an eGFP gene between the 3′UTR and the first viral gene NS1 of the RSV genome [8, 9].

#### 3.1.2 Preparation of Virus Stocks

1. Grow the HeLa cells in 175 $cm^2$ culture flasks in RPMI/10 % FCS till approximately 80 % confluence.
2. Remove the medium and rinse the cells once with DPBS to remove the dead cells.
3. Thaw a vial of RSV-eGFP. Dilute the virus with warm RPMI/2 % FCS to generate a multiplicity of infection (MOI) of 1 plaque-forming unit (PFU)/cell.

4. Add 20 mL of virus dilution to the cells and incubate the flasks on an automated shaker for 1 to 2 h between 20 and 37 °C.
5. After incubation, remove the viral inoculum and wash the infected cells with DPBS once. Add 20 mL of warm RPMI/2 % FCS to the infected cells.
6. Incubate the flasks at 37 °C, 5 % $CO_2$. From day 2 post infection, check the infected cells under an inverted microscope for cytopathogenic effect (CPE) and under a fluorescent microscope for occurrence of fluorescence every day. Harvest the virus culture when approximately 60 % of CPE is observed (*see* **Note 5**).
7. Collect the supernatant from the culture flasks in sterile Falcon™ conical tubes.
8. Detach the remaining cells by scraping the cells with a cell scraper and add them to the supernatant.
9. Centrifuge the supernatant for 10 min at 2,095 ×*g* and 4 °C.
10. Transfer the supernatant to a new sterile tube.
11. To release intracellular virus particles from the cell pellet, submit the cell pellet to 3 cycles of snap freezing/thawing (submerge the cell pellet in liquid nitrogen for 10 s, thaw the pellet using running tap water). Centrifuge the frozen/thawed pellet again for 5 min at 2,095 ×*g* and pool the supernatants.
12. Add an appropriate volume of sucrose (40 % stock solution) to a final concentration of 25 % to the pooled supernatants (*see* **Note 6**).
13. Aliquot the virus in cryovials and store at −80 °C.
14. Determine the virus titer by a standard plaque assay.

### 3.2 RSV Antiviral and Toxicity Assays

The optimal virus input in the RSV antiviral screening assay is an MOI of 1 PFU/cell. At this MOI, sufficient fluorescent signal is produced to distinguish a viral infection control (positive/high control) from an uninfected cell control (negative/low control).

#### 3.2.1 Preparation of Compound Plates

1. Prepare fourfold serial compound dilutions in a V-bottom 96-well microtiter plate in RPMI/10 % FCS. Each row contains a different compound. The DMSO concentration in each well is 2 %. The highest compound concentration should be in column 1, and next 8 dilutions are plated through columns 2–9. Columns 10, 11, and 12 are control wells and receive cell culture medium/DMSO only.
2. Transfer 10 μL of each compound dilution to each of the corresponding quadruplet wells of a black 384-well clear bottom microtiter plate (for activity testing) and a white 384-well microtiter plate (for toxicity testing). For example, well A1 in a 96-well plate is transferred to the wells A1, A2, B1, and B2 of a 384-well plate (*see* **Note 7**).

3. Prepare reference compound plates for both activity and toxicity tests in every experiment. The reference compounds have known antiviral activities against RSV which should remain constant over different experiments. The fluorescence values obtained from the positive and negative controls can serve as quality evaluation of generated data and assay performance. Compounds with a reported antiviral activity against RSV can be used as reference compounds [10–13].

*3.2.2 Preparation of Cell Suspension*

1. Culture the HeLa cells in RPMI/10 % FCS in 175 $cm^2$ culture flasks till approximately 80 % confluence.
2. Before starting the experiment check the cells with a microscope for yeast, fungi, or bacterial contamination.
3. Remove the culture medium and wash the cells once with DPBS.
4. Add 5 mL of 0.05 % trypsin–EDTA to each flask. Incubate for 1 min at room temperature and aspirate out the trypsin–EDTA solution.
5. Place the cells in an incubator for a few minutes until detached. If the cells are not detached, tap the flask gently until all cells detached.
6. Resuspend the cells in RPMI/10 % FCS and take a sample to count the cells.
7. Prepare HeLa cell suspensions (the final density is 3,000 cells per well):
   - For activity plates, prepare 8 mL of cell suspension per plate at a density of 150,000 cells/mL.
   - For toxicity plates, prepare 12 mL of cell suspension per plate at a density of 100,000 cells/mL.

*3.2.3 Preparation of Virus Suspension*

1. Thaw an appropriate volume of RSV-eGFP stock solution according to the number of plates to be tested.
2. Prepare 4 mL per test plate of RSV-eGFP virus dilution in RPMI/10 % FCS to result in an MOI of 1 PFU/cell.

*3.2.4 Antiviral Screening and Cytotoxicity Assays*

The cell suspension, virus dilution, and medium (for control wells) are dispensed into the wells of the test plates using a multidrop combi liquid dispenser. In the activity plates columns 19–22 serve as virus controls (VC), and columns 23–24 serve as cell controls (CC). In the toxicity plates, columns 19–22 are cell controls, and columns 23–24 are medium controls (MC).

1. Use a multidrop combi liquid dispenser to dispense 10 μL of RPMI/10 % FCS into the CC wells of the activity plates.
2. Transfer 30 μL of RPMI/10 % FCS into the MC wells of the toxicity plates.

3. Dispense 10 μL/well of virus suspension into columns 1–22 of the activity plates.
4. Use a fresh multidrop cassette (*see* **Note 8**) to dispense 20 μL/well of the activity HeLa cell suspension (density of 150,000 cells/mL) to all wells of the activity plates (whole plate). The cell suspension should be gently stirred on a magnet stirrer in order to keep the cells homogeneous.
5. Dispense 30 μL/well of the toxicity cell suspension (density of 100,000 cells/mL) to columns 1–22 of the toxicity plates.
6. Incubate the plates at 37 °C and 5 % $CO_2$ for 3 days.

#### *3.2.5 Reading the Test Plates*

Three days post virus infection and compound treatment, the activity plates are read with an in-house-developed mega screening microscope coupled to a twister robot system for automation. The fluorescence intensity is measured at a wavelength of 488 nm. Viral replication is quantified by measuring the eGFP expression in the HeLa cells.

In parallel, the toxicity plates are read by measuring the ATP levels of the live cells using an ATPLite™ 1-Step kit according to the manufacturer's instructions. Briefly, equilibrate all reagents to room temperature before use. Next, add an appropriate volume of buffer to the substrate bottle, mix by inversion of the bottle, and let equilibrate for 5 min. Then, add 40 μL of reconstituted substrate to each well of the toxicity plates and shake the plates for 2 min. Measure the luminescence using a Viewlux™ machine. An exposure time of 0.1 s and 3 times binning are used.

#### *3.2.6 Data Analysis*

To evaluate the antiviral effect of test compounds, the 50 and 90 % effective concentrations ($EC_{50}$ and $EC_{90}$) are calculated. $EC_{50}$ and $EC_{90}$ values correspond to the compound concentrations at which 50 and 90 % of viral replication is inhibited (reduction of fluorescent intensity), respectively. The cytotoxicity of a test compound is expressed by the 50 % cytotoxic concentration ($CC_{50}$), which is the concentration of compound causing 50 % decrease of luminescence in the compound-treated cells. The $CC_{50}/EC_{50}$ ratio defines the selectivity index (SI), which allows the ranking of compounds with respect to the highest specificity/safety margin. The $EC_{50}$, $EC_{90}$, and $CC_{50}$ can be calculated by using different types of software (e.g., Graphpad Prism) allowing generation of dose–response curves by a nonlinear regression.

## 4 Notes

1. According to the Belgian biosafety regulations, the handling and manipulation of RSV are restricted to a biosafety level two laboratory. Carry out all manipulations with appropriate

personal protective clothes and gloves; and take into account all safety procedures for working with biohazardous materials.

2. To heat inactivate the FCS, warm the serum to 56 °C for 30 min.
3. Use an appropriate disinfectant. One can use 70 % ethanol to decontaminate surfaces. Virus-contaminated medium can be inactivated by adding 1 % Virkon.
4. We use an automated custom-made fluorescence reader. Alternative fluorescent spectrophotometers are commercially available, e.g., Argus microscope and Acumen explorer. The most important specification is that focusing on the fluorescent cells is possible.
5. We recommend not waiting for full CPE for harvesting the virus since the progeny RSV released from the cells is not stable at 37 °C.
6. RSV is a thermolabile virus. For long-term storage at −80 °C, a stabilizer should be added to the virus stock to prevent the decrease of virus titer [14]. A final concentration of 25 % sucrose can be used as a stabilizer for storage.
7. Downscaling of the RSV antiviral assay from a 384-well format to a 96-well format is possible if, for example, the automation of the assay is not possible. Instead of quadruplet measurements for each test concentration in one 384-well plate, only a single measurement can be performed per concentration for a 96-well format.
8. Cassette tubing should be changed to avoid viral contamination of the cell controls.

## Acknowledgments

The authors wish to thank Koen Van Acker and Pascale Holemans who assisted in the development of the original RSV fluorescence-based assay in a 96-well plate format.

### References

1. Krusat T, Streckert HJ (1997) Heparin-dependent attachment of respiratory syncytial virus (RSV) to host cells. Arch Virol 142:1247–1254
2. Lamb RA (1993) Paramyxovirus fusion: a hypothesis for changes. Virology 197:1–11
3. Martin AJ, Gardner PS, McQuillin J (1978) Epidemiology of respiratory viral infection among paediatric inpatients over a six-year period in north-east England. Lancet 2:1035–1038
4. Weber MW, Mulholland EK, Greenwood BM (1998) Respiratory syncytial virus infection in tropical and developing countries. Trop Med Int Health 268–280
5. Nair H, Nokes DJ, Gessner BD et al (2010) Global burden of acute lower respiratory infections due to respiratory syncytial virus in

young children: a systematic review and meta-analysis. Lancet 375:1545–1555
6. Falsey AR, Hennessey PA, Formica MA et al (2005) Respiratory syncytial virus infection in elderly and high-risk adults. New England J Medicine 352:1749–1759
7. Collins PL, Crowe JE Jr (2007) Respiratory syncytial virus and metapneumovirus. In: Knipe DM, Howley PM et al (eds) Fields virology. Lippincott Williams & Wilkins, Philadelphia, PA, pp 1601–1646
8. Hallak LK, Spillmann D, Collins PL, Peeples ME (2000) Glycosaminoglycan sulfation requirements for respiratory syncytial virus infection. J Virol 74:10508–10513
9. Hallak LK, Collins PL, Knudson W, Peeples ME (2000) Iduronic acid-containing glycosaminoglycans on target cells are required for efficient respiratory syncytial virus infection. Virology 271:264–275
10. Cianci C, Genovesi EV, Lamb L et al (2004) Oral efficacy of a respiratory syncytial virus inhibitor in rodent models of infection. Antimicrob Agents Chemother 48:2448–2454
11. Bonfanti JF, Meyer C, Doublet F et al (2008) Selection of a respiratory syncytial virus fusion inhibitor clinical candidate. 2. Discovery of a morpholinopropylaminobenzimidazole derivative (TMC353121). J Medicinal Chem 51:875–896
12. Ventre K, Randolph AG (2007) Ribavirin for respiratory syncytial virus infection of the lower respiratory tract in infants and young children. Cochrane Database Syst Rev 1:CD000181
13. Chapman J, Abbott E, Alber DG et al (2007) RSV604, a novel inhibitor of respiratory syncytial virus replication. Antimicrob Agents Chemother 51:3346–3353
14. Gupta CK, Leszczynski J, Gupta RK, Siber GR (1996) Stabilization of respiratory syncytial virus (RSV) against thermal inactivation and freeze-thaw cycles for development and control of RSV vaccines and immune globulin. Vaccine 14:1417–1420

# Chapter 27

# Screening and Evaluation of Anti-respiratory Syncytial Virus Compounds in Cultured Cells

Anna Lundin, Tomas Bergström, and Edward Trybala


## Abstract

Respiratory syncytial virus (RSV) is a highly contagious pathogen that infects mainly ciliated cells of respiratory epithelium and type 1 pneumocytes in the alveoli frequently causing serious respiratory disease in infants, elderly, and immunocompromised patients. At present, prevention/treatment of RSV infection is limited to the use of specific anti-RSV antibody or an aerosol formulation of ribavirin, a drug of suboptimal efficacy and low safety profile. There is an urgent need for development of novel anti-RSV drugs and virucides. Here we describe the cell culture-based methods used in our laboratory in identification of novel inhibitors of RSV including the P13 fusion inhibitor, and the PG545 virucide. Protocols for antiviral screening, evaluation of anti-RSV potency, and elucidation of mode of antiviral activity of test compounds are described.


**Key words** Respiratory syncytial virus, Antiviral screening, Mode of activity, Antiviral drugs, Drug-resistant virus variants, Virucidal activity

## 1 Introduction

To search for novel anti-respiratory syncytial virus (RSV) hit compounds we screened several diversity collections of up to 17,000 compounds for antiviral activity in cultured cells. This assay was performed in 384-well plate format and identification of hit compounds was based on their capability to protect cells from the virus-induced cytopathic effect. Using this approach several novel anti-RSV compounds including P13 and C15 were identified [1]. In another approach we studied whether specific modifications of a sulfated oligosaccharide structure would improve anti-RSV activity. Sulfated oligo- and polysaccharides are mimetics of cell surface glycosaminoglycan receptors that are utilized by RSV, and many other viruses, for initial binding to cells [2, 3]. We found that conjugation of cholestanol to the reducing end of a sulfated oligosaccharide enhanced anti-RSV [4], anti-herpes simplex virus [5],

Edwin Yunhao Gong (ed.), *Antiviral Methods and Protocols*, Methods in Molecular Biology, vol. 1030,
DOI 10.1007/978-1-62703-484-5_27, © Springer Science+Business Media, LLC 2013

and anti-HIV [6] activity of the native oligosaccharide. By using several assays, designed to elucidate the mode of antiviral activity, we found that both P13 and C15 inhibited RSV-induced cell fusion [1] while the cholestanol-sulfated oligosaccharide conjugate (PG545) manifested the RSV-inactivating (virucidal) activity, an antiviral mode of action absent in native sulfated oligosaccharides of muparfostat [4]. Here we present protocols for methods used in the screening and evaluation of anti-RSV compounds.

## 2 Materials

### *2.1 Cells, Viruses, and Media*

1. Human laryngeal epidermoid carcinoma (HEp-2) cells (ATCC, CCL-23).
2. RSV strain A2 [7, 8] (ATCC, VR-1540) (*see* **Note 1**).
3. Eagle's minimum essential medium (EMEM).
4. Dulbecco's modified Eagle's medium (DMEM).
5. DMEM methionine- and cystine-free (Invitrogen).
6. Fetal bovine serum (FBS). Inactivated by incubation for 30 min at 56 °C water bath (HI-FBS).
7. Penicillin/streptomycin (PEST) stock: 10 mg/mL of streptomycin and 6 mg/mL of penicillin in Hank's balanced salt solution. Sterilize by filtration.
8. Amphotericin B stock: 250 μg/mL in deionized water.
9. L-Glutamine stock: 29.2 mg/mL in deionized water. Sterilize by filtration.
10. Phosphate-buffered saline (PBS): 137 mM NaCl, 2.7 mM KCl, 8.1 mM $Na_2HPO_4$, 1.5 mM $KH_2PO_4$, 1 mM $CaCl_2$, and 0.5 mM $MgCl_2$ in deionized water.
11. PBS without calcium and magnesium ions (PBS-A). Autoclave at 120 °C for 20 min, and store at 4 °C.
12. Bovine serum albumin (BSA), globulin-free.
13. HEPES stock: 0.4 M in EMEM. Sterilize by filtration.
14. 50 % Sucrose solution for stabilization of RSV infectivity: Dissolve 50 g of sucrose in PBS to give a final volume of 100 mL (w/v). Sterilize by filtration. 30, 40, and 50 % sucrose solutions for ultracentrifugation are prepared in the same manner (w/v).
15. HEp-2 cell growth medium: DMEM supplemented with 10 % FBS, 1 % L-glutamine stock, and 1 % PEST stock (DMEM-G).
16. HEp-2 cell maintenance medium: DMEM supplemented with 2 % HI-FBS, 1 % L-glutamine stock, 1 % PEST stock, and 10 mM HEPES (DMEM-M).

17. HEp-2 cell screening medium: DMEM supplemented with 2 % HI-FBS, 1 % L-glutamine stock, 2 % PEST stock, 0.2 % amphotericin B stock, and 10 mM HEPES (DMEM-S).
18. Methyl cellulose (Methocel® MC, Sigma, 64625) stock solution: 1.5 g of methyl cellulose in 100 mL of Hank's balanced salt solution. Autoclave at 120 °C for 20 min, and store at 4 °C.
19. Methyl cellulose overlay medium: Supplement 100 mL of methyl cellulose stock solution with 100 mL of DMEM, 4 mL of HI-FBS, 2 mL of L-glutamine stock, 2 mL of PEST stock, and 10 mM HEPES.

### 2.2 Chemicals and Screening Equipment

1. Diversity collection of compounds such as the ChemBioNet collection of 16,671 compounds (Leibniz Institute for Molecular Pharmacology, Germany) supplied at a concentration of 10 mM in 5 μL of DMSO in 384-well plate format. The plates are sealed with DMSO-resistant aluminum tape, and stored at −20 °C (*see* **Note 1**).
2. L-($^{35}$S) methionine/L-($^{35}$S) cysteine (Expre$^{35}$S$^{35}$S Protein Labelling Mix, PerkinElmer, NEG072007MC) (*see* **Note 1**).
3. Monoclonal antibody 131/G2 (Lifespan Biosciences, LS-C56536) specific for the G protein of RSV A2 strain.
4. Peroxidase-conjugated, affinity pure F(ab′)$_2$ fragment goat anti-mouse IgG (Jackson ImmunoResearch Laboratories, 115-036-062).
5. Peroxidase substrate: Dissolve 10 mg of 4-chloro-1-naphthol in 1 mL of ethanol and then dilute it in 100 mL of PBS-A. Add 20 μL of 30 % $H_2O_2$ immediately prior to use (*see* **Note 1**).
6. Glutaraldehyde (Sigma, G7651) (*see* **Note 1**).
7. Ribavirin (Sigma, R9644) stock: 10 mM in sterile deionized water (*see* **Note 1**). Store at −20 °C.
8. 1 % Crystal violet solution: In a fume hood dissolve 1 g of crystal violet in 87 mL of 70 % ethanol, and supplement with 8.7 mL of formalin and 4.3 mL of acetic acid (*see* **Note 1**).
9. CellTiter 96® Aqueous One Solution reagent (Promega, G3580) for cytotoxicity measurement (*see* **Note 1**).
10. Dimethylsulfoxide (DMSO) (*see* **Note 1**).
11. DMSO-resistant aluminum sealing tape for 384-well plates (Thermowell™ Sealing Tape, Costar-Corning).
12. Pipettes, 16-channels, adjustable volume of 5–50 μL (Finnpipette, Thermo Electron), 0.5–10 μL (CAPP, CTKSL10-16AZ), and 0.2–2 μL (Anachem, SL02-16AZ). Sterile tips recommended for these pipettes by manufacturers.
13. Cluster 384-well plates for culturing of cells (Corning, CLS-3701), and for storage of library compounds or DMSO solvent (Corning, CSL-3657).

14. Ultracentrifugation tubes Ultra-Clear™, 16 × 102 mm (Beckman-Coulter).
15. Multiskan FC microplate reader (Thermo Fisher Scientific).

## 3 Methods

### 3.1 Screening of Diversity Collection of Compounds for Anti-RSV Activity

We identify anti-RSV hits based on the capability of a test compound to protect HEp-2 cells from the virus-induced cytopathic effect (CPE). The assay permits the preliminary discrimination between the specific antiviral activity and the adverse effects of test compound on cells. Prior to this assay the infectivity of the virus stock is titrated in HEp-2 cells in 384-well plate format, i.e., under conditions of the screening assay.

#### 3.1.1 Titration of RSV Infectivity in a 384-Well Plate

1. Adjust suspension of HEp-2 cells with DMEM-G to contain $1.2 \times 10^5$ cells/mL and pour it into a sterile square petri dish (*see* **Note 2**). Unless otherwise stated this concentration of HEp-2 cells is used for plating the cells in other protocols described in this chapter. Transfer 50 μL of cell suspension into each well of a 384-well cell culture plate using a 16-channel pipette. Incubate at 37 °C for 24 h in a humidified 5 % $CO_2$ incubator. Unless otherwise stated all incubations at 37 °C described in this report are performed using the $CO_2$ incubator.
2. On the following day, inspect the cells under an inverted microscope. The cell monolayer should be ~60–90 % confluent. Remove growth medium supernatant from cells by inverting the plate and then shaking out the medium in a waste container. Add 25 μL of warm DMEM-S.
3. Using a 16-channel pipette (Anachem), transfer 0.8 μL of 25 % DMSO (solvent for library compounds) solution in deionized water from a control library plate to cells in each well of a 384-well plate (*see* **Note 3**).
4. Place the plate on a rocker platform and shake for 1 min at 750 rotation cycles/min to redistribute the DMSO solvent.
5. Perform serial tenfold dilutions of RSV stock (*see* **Note 4**) in DMEM-S and transfer 25 μL volumes of each virus dilution to cells growing in two columns of wells (32 wells) of 384-well plate.
6. Centrifuge the plate at 100 × *g* for 1 min to remove any air bubbles.
7. Shake the plate for 1 min at 750 rotation cycles/min to redistribute the virus inoculum.
8. Incubate the cells at 37 °C for 4 days.

9. Inspect the cells under an inverted microscope for the presence of the virus-induced CPE. Use low magnification (10×4) under which an entire bottom area of a single well of the 384-well plate is visible. Record the wells in which viral CPE is present and calculate the number of the tissue culture infectious doses ($TCID_{50}$) in RSV stock using any statistical method for computation of 50 % end-point (e.g., Reed-Muench method).

*3.1.2 Anti-RSV Screening Assay*

The ChemBioNet diversity library of compounds is provided in 384-well plate format. Each well in columns 1–22 comprises different test compounds at 10 mM concentrations in 5 μL of DMSO. Each well in columns 23 and 24 comprises 5 μL of DMSO solvent to serve as controls.

1. Thaw the library compounds by placing the 384-well plate in a 37 °C incubator (no $CO_2$) for 20–30 min. Shake the plate for 1 min at 750 rotation cycles/min, and then centrifuge at 100×*g* for 1 min.
2. Make a copy of library compounds at a dilution of 1:4 (*see* **Note 3**) as follows: In a fume hood, remove the aluminum sealing tape and using a 16-channel pipette transfer 1 μL of compounds to an empty 384-well polypropylene storage plate (Corning, CSL-3657). Add 3 μL of deionized water to each well. Seal both the original and a copy library plate with the DMSO-resistant aluminum tape and store at −20 °C.
3. Prepare the HEp-2 cells in 384-well plates as described under Subheading 3.1.1, **step 1**.
4. On the following day, thaw the diluted library copy as described in **step 1**.
5. Remove the medium from cells growing in 384-well plates and add fresh medium as described under Subheading 3.1.1, **step 2**.
6. Work in a fume hood. Using a 16-channel pipette transfer 0.8 μL of test compounds to corresponding wells with cells in a 384-well plate. Shake the plate for 1 min at 750 rotation cycles/min to redistribute test compounds and DMSO (*see* **Note 5**).
7. Dilute the virus stock in DMEM-S to contain 1,000 $TCID_{50}$/25 μL, and pour it in a sterile square petri dish. Using a 16-channel pipette transfer 25 μL of virus inoculum to cells in each well of columns 1–23. Cells in each well of column 24 receive 25 μL of DMEM-S only (*see* **Note 6**).
8. Centrifuge the plate at 100×*g* for 1 min to remove any air bubbles. Shake the plate for 1 min at 750 rotation cycles/min, and incubate at 37 °C for 4 days.
9. Inspect the cells under an inverted microscope at low magnification (10×4) under which an entire bottom area of a single

well of 384-well plate is visible. Begin to inspect cells for their protection against RSV when cells in all wells of column 23, which received the virus inoculum but no test compound, show extensive RSV-induced CPE. To verify the possibility that the cell-protective activity is not due to adverse effects of hit compound on cells, compare morphology of HEp-2 cells incubated in the presence of hit in question with mock-treated (DMSO) uninfected cells in column 24 (*see* **Note 7**).

### 3.2 Effect of a Test Compound on RSV Infectivity and Cell Viability

#### 3.2.1 Anti-RSV Potency of a Test Compound-Plaque Number Reduction Assay

1. Add 2 mL of HEp-2 cell suspension ($1.2 \times 10^5$ cells/mL) in DMEM-G to each well of a 12-well cell culture plate.
2. On the following day, rinse the cells twice with 1 mL of DMEM-M.
3. Perform serial fivefold dilutions of a test compound in 1 mL of DMEM-M in a blank 24-well plate. Include a control sample with 1 mL of DMEM-M and no test compound. Perform corresponding dilutions of the solvent used for the test compound (*see* **Note 8**). Add 100 μL of DMEM-M comprising 200 PFU of RSV to each dilution of test compound, solvent, and control samples. Shake the plate gently to redistribute the viral inoculum.
4. Remove the medium from the HEp-2 cells, and add 0.5 mL of the virus–compound mixture to each well (in duplicate). Incubate the cells at 37 °C for 3 h.
5. Remove the virus–compound mixture from cells, rinse the cells once with 1 mL of DMEM-M, and add 1 mL of warm methyl cellulose (MC) overlay medium supplemented with the same concentrations of a test compound or a solvent as in original dilutions (*see* **Note 9**). Incubate cells at 37 °C for 3 days.
6. To visualize the viral plaques, remove, in a fume hood, the MC overlay medium and add 200 μL of 1 % solution of crystal violet. Spread the dye by rocking the plate gently, and stain the cells for 2 min at room temperature. Rinse the cell monolayer thrice with 1 mL of tap water. Dry the cells and count viral plaques under an inverted microscope.
7. Alternatively, the viral plaques can be visualized by immunostaining with antibodies specific for the surface components of RSV, such as monoclonal antibody 131/G2 specific for the G protein of RSV A2 strain, as follows:
   (a) Rinse the cell monolayer twice with 1 mL of PBS.
   (b) Fix the cells with 0.25 % glutaraldehyde in 0.5 mL PBS for 15 min at room temperature, and then rinse the cells thrice with 1 mL of PBS.
   (c) Add monoclonal anti-G protein antibody diluted in 0.5 mL of PBS supplemented with 2 % BSA, and incubate for 90 min at room temperature.

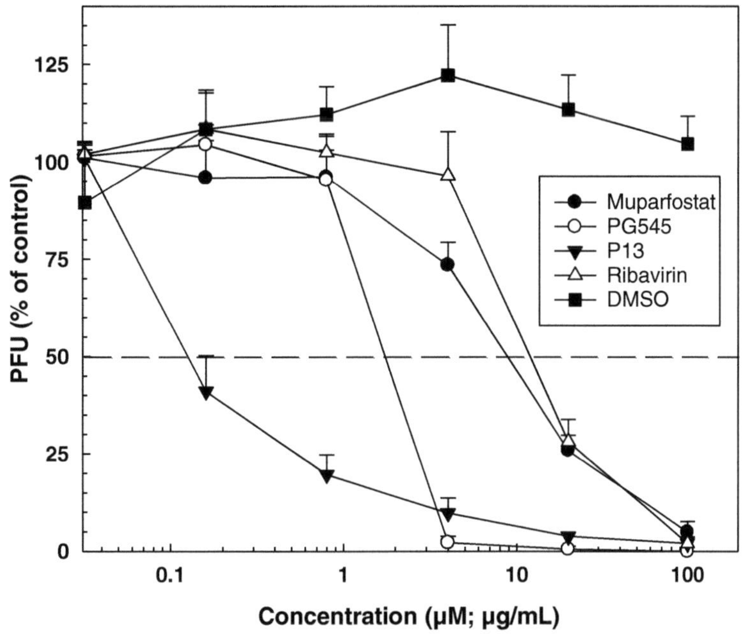


**Fig. 1** Anti-RSV activities of the test compounds as determined by the viral plaque number reduction assay. The test compounds muparfostat, PG545 (0.16–100 µg/mL), P13, and ribavirin (0.16–100 µM) were mixed with 200 PFU of RSV, added to HEp-2 cells, and incubated for 3 days for the development of viral plaques. The solvents, i.e., deionized water for muparfostat, PG545, and ribavirin, or DMSO for P13 at concentrations (0.0016–1 %) corresponding to those present in respective dilutions of test compounds, were processed in identical manner. Data are expressed as a percentage of the number of RSV-induced plaques found in cells incubated with test compound or solvent relative to those developed in cells in the absence of test compound or solvent. The concentration of test compound that reduced the number of RSV plaques by 50 % ($IC_{50}$) can be interpolated from the dose–response curve. Deionized water (not shown) and DMSO did not reduce the plaque forming activity of RSV. Sulfated oligosaccharides (muparfostat) and their conjugates (PG545) have no exact molecular weight and therefore their concentration is expressed in µg/mL

(d) Rinse the cells thrice with 1 mL of PBS, and add peroxidase-conjugated goat anti-mouse IgG at 1:1,000 dilution in 0.5 mL of PBS supplemented with 2 % BSA.

(e) After further incubation for 90 min at room temperature, rinse the cells thrice with PBS-A, and add 0.5 ml of the substrate solution (*see* Subheading 2.2, item 5).

(f) When the viral plaques are visible, remove the substrate and add 1 mL of PBS-A (Fig. 1).

#### *3.2.2 Anti-RSV Potency of a Test Compound-Plaque Size Reduction Assay*

1. Prepare the HEp-2 cells in 12-well plates as described under Subheading 3.2.1, **steps 1** and **2**.
2. Inoculate cells in each well with 100 PFU of RSV A2 strain in 0.5 mL of DMEM-M. Incubate cells at 37 °C for 3 h.

3. Prepare serial fivefold dilutions of a test compound or a solvent in MC overlay medium as described under Subheading 3.2.1, **step 5** (*see* **Note 9**). Include control samples that contain no test compound.
4. Remove the virus inoculum from HEp-2 cells, rinse the cells once with 1 mL of DMEM-M, and add 1 mL of MC overlay medium prepared in **step 3**. Incubate cells at 37 °C for 3 days.
5. Visualize viral plaques as described under Subheading 3.2.1, **step 6** or 7.
6. Capture images of 20 neighboring viral plaques for each concentration of test compound, solvent, and mock-treated infected cells. We use a DC300 digital camera (Leica, Heerbrugg, Switzerland) attached to a Diavert microscope (Leitz-Wetzlar, Germany) for image capture, and IM500 image software (Leica, Cambridge, UK) for measurement of the area of viral plaque images (*see* **Note 10**).

#### *3.2.3 Cytostatic/Cytotoxic Activity of a Test Compound-Cell Viability Assay*

We investigate the effect of a test compound on viability of cells by using the CellTiter 96® Aqueous One Solution reagent that contains a novel tetrazolium compound MTS (Promega).

1. Add 200 μL of HEp-2 cell suspension in DMEM-G into each well of a 96-well cell culture plate. Incubate at 37 °C for 24 h.
2. On the following day, perform serial fivefold dilutions of a test compound in warm DMEM-M in a sterile blank 96-well plate. Include control samples that contain no test compound in DMEM-M. Carry out identical dilutions for solvent of the test compound.
3. Remove the medium from HEp-2 cells. Rinse the cells once with 200 μL of the DMEM-M medium, and add 100 μL of the working dilutions of a test compound, solvent, or control sample to duplicate wells. Incubate cells at 37 °C for 3 days. During the incubation period inspect the cells under a microscope daily. Record any visible changes in cell morphology and in confluence of cell monolayer as related to mock-treated cells.
4. At the end of incubation period add, according to the manufacturer's instruction, 20 μL of the CellTiter 96® Aqueous One Solution reagent to each well. Shake the plate for 30 s at 500 rotation cycles/min, and incubate cells at 37 °C for 1–2 h. Record the absorbance at 490 nm against the background of 650 nm using a microplate reader.
5. We express the results as a percentage of absorbance value found in cells incubated with the test compound or the solvent relative to that developed in cells in the absence of the test compound or the solvent. Interpolate the concentration of test compound that reduced the cell viability by 50 % ($CC_{50}$) from

the dose–response curve. Complement these data with description of any visible changes in cell morphology and proliferation (*see* **step 3**).

### 3.3 Mode of Anti-RSV Activity of a Test Compound

#### 3.3.1 The Time-of-Addition Assay

To identify which step of the virus life cycle is being targeted by a test compound, the compound is added to cells at different time-points, i.e., at −1, 0, 2, 4, 6, and 8 h counted relative to the beginning of inoculation of cells with RSV (0 h).

1. Prepare HEp-2 cells in a 12-well plate as described under Subheading 3.2.1, **step 1**. Duplicate wells are required for each time-point, and quadruplicate wells are necessary for control purposes.
2. On the following day, rinse the cells once with 1 mL of warm DMEM-M, and add 445 μL of the same medium to all wells.
3. To ensure good anti-RSV activity and lack of adverse effects on cell viability, select a specific concentration of a test compound (delimited by $IC_{50}$ and $CC_{50}$ values) to be used in the assay, and if necessary pre-dilute the test compound stock.
4. Add 5 μL of the test compound to cells (in duplicate wells) at the time-point −1 h counted relative to the beginning of inoculation of cells with RSV (0 h). Rock the plate gently to redistribute the compound.
5. Add 5 μL of the test compound to cells (in duplicate wells) at the time-point 0 h, and then inoculate cells in all wells including quadruplicate control wells with RSV in 50 μL of DMEM-M at the multiplicity of infection (MOI) of 1 (time-point 0 h). Rock the plate gently to redistribute the compound and the virus.
6. At the time-point 2 h, remove the medium from all wells. Rinse the cells once with 1 mL of warm DMEM-M, and add 495 μL of the same medium. Add 5 μL of the test compound to cells in duplicate wells (time-point 2 h). Because test compound added at time-points −1 and 0 h was removed together with non-adsorbed virus, resupply 5 μL of test compound to these wells. Rock the plate gently to redistribute the compound.
7. At the time-point 2 h, scrape and harvest the cells and culture medium from two control wells. Subject the cell suspension to three cycles of vortexing for 5 s each, mix with an equal volume of 50 % sucrose, and store at −80 °C (*see* **Note 11**).
8. Add 5 μL of the test compound to cells in duplicate wells at the time-points 4, 6, and 8 h. Rock the plate gently to redistribute the compound.
9. Harvest the cells and the medium (*see* **step 7**) at time-point 24 h. Test these samples for residual infectivity by the viral plaque assay (Fig. 2), and for the presence of viral RNA by qPCR assay.

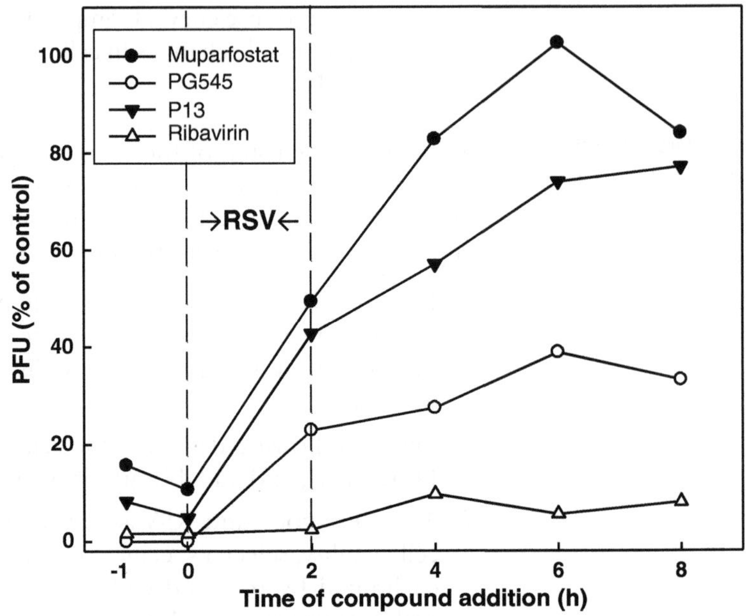


**Fig. 2** Mode of anti-RSV activity of the test compounds as determined by the time-of-addition assay. Muparfostat (20 μg/mL), PG545 (10 μg/mL), P13 (10 μM), and ribavirin (20 μM) were added to HEp-2 cells at different time-points relative to the beginning of inoculation of cells with RSV (0 h). The virus inoculum was incubated with cells for 2 h. The compounds remained on cells from the time of their addition until harvesting of infectious culture media and cells at the time-point 24 h. Two separate experiments were carried out in duplicate for each compound. The results are expressed as a percentage of an infectious virus titer found in samples incubated in the presence of test compound relative to that determined in the mock-treated infectious material. Muparfostat, PG545, and P13 exhibited the greatest reduction of RSV infectivity when added at the time-point 0 h suggesting that the virus attachment to and/or entry into the cells could be targeted by these compounds. In contrast, ribavirin blocked RSV infectivity even when added 8 h post infection suggesting that some post-entry events are affected

*3.3.2 The Virucidal Activity of a Test Compound*

Some compounds may interact with virus particles in a virucidal manner that results in permanent inactivation of viral infectivity with or without disintegration of virions. The principle of this assay is to incubate the test compound with the virus and then investigate whether the virus remains infectious when the compound in the virus–compound mixture is diluted out to non-inhibitory concentrations. Here, we describe the assay used in evaluation of compound PG545 which, in serum-free medium, inhibited RSV infectivity with an $IC_{50}$ value of 0.4 μg/mL [4]. PG545 is a conjugate of sulfated tetrasaccharide and cholestanol. The tetrasaccharide component targets the G protein of RSV and inhibits virus attachment to cells while cholestanol is likely to interact with the viral lipid envelope, thus causing permanent inactivation of virion infectivity [4].

1. Calculate the amount of virus and the concentration of test compound to be mixed and incubated in the assay. The number of viral PFU units should be at least 100 times greater than the highest selected number of $IC_{50}$ doses of test compound, so it is possible to determine viral infectivity after dilution of the virus–compound mixture to non-inhibitory (<1 $IC_{50}$ dose) concentration of test compound. In the case of PG545 we incubate $10^5$ PFU of RSV with compound at 100, 10, and 1 μg/mL that corresponds to 250, 25, and 2.5 $IC_{50}$ doses/mL, respectively.
2. Prepare HEp-2 cells in 12-well plates as described under Subheading 3.2.1, **step 1**.
3. Dilute the stock of test compound in its solvent to 10, 1, and 0.1 mg/mL concentrations.
4. Dilute the RSV A2 stock in DMEM-M without FBS, to contain $10^5$ PFU/0.5 mL. In this step we use serum-free media to avoid possible interference of serum components with lipophilic PG545.
5. Transfer 0.5 mL of the virus dilution to four separate 2-mL microtubes. Add 5 μL of the test compound dilutions to achieve final concentrations of 100, 10, and 1 μg/mL. To control sample add 5 μL of solvent. Close the tubes and mix their content thoroughly by inverting the tube ten times.
6. Place the tubes in a 37 °C water bath and incubate for 15 min.
7. Meanwhile, prepare material required for titration of virus infectivity as this task should be performed without delay at the end of the virus–compound incubation period. To this end, remove the growth medium from HEp-2 cells in 12-well plates and add 0.45 mL of DMEM-M. Add 180 μL of DMEM-M to several wells of a blank 96-well plate.
8. Add 200 μL of the virus–compound mixture to an empty well in a 96-well plate and perform serial tenfold dilutions ($10^0$–$10^{-3}$) by transferring 20 μL of sample to 180 μL of DMEM-M. Add 50 μL of each dilution to cells in duplicate wells of 12-well plate comprising 0.45 mL of DMEM-M, and incubate for 3 h at 37 °C. Remove the inoculum from cells and add 1.5 mL of MC overlay medium. Incubate for 3 days at 37 °C. Visualize viral plaques as described under Subheading 3.2.1, **step 6** or 7.
9. Calculate the final concentration of test compound (number of $IC_{50}$ doses) present in cell culture supernatant (*see* **step 8**) after inoculation with all dilutions of the virus–compound mixture (Table 1). Reduced number or absence of viral plaques developed in the presence of non-inhibitory concentrations of compound (i.e., <1 $IC_{50}$) as related to the solvent-treated RSV implies virucidal activity of test compound.

**Table 1**
**The RSV inactivating (virucidal) activity of test compounds[a]**

| | | PG545 | | | P13 | | |
|---|---|---|---|---|---|---|---|
| | | | RSV plaque number[c] | | | RSV plaque number[c] | |
| Concentration (μg/mL or μM[b]) | Mixture dilution (neg. $\log_{10}$) | $IC_{50}$ doses | Compound | Solvent[d] | $IC_{50}$ doses | Compound | Solvent[d] |
| 100 | 0 | 25 | 0.0 | | 77 | | |
| | 1 | 2.5 | 0.0 | | 7.7 | | |
| | 2 | **0.25** | **0.0** | 46.3 | **0.77** | **60.5** | 63.0 |
| 10 | 0 | 2.5 | 0.0 | | 7.7 | | |
| | 1 | **0.25** | **0.0** | | **0.77** | | |
| | 2 | **0.025** | **0.0** | 46.3 | **0.077** | **65.8** | 63.0 |
| 1 | 0 | **0.25** | | | **0.77** | | |
| | 1 | **0.025** | **118.0** | | **0.077** | | |
| | 2 | **0.0025** | | 46.3 | **0.0077** | **66.0** | 63.0 |

[a]The test compound at indicated concentrations (see column 1) was mixed with $10^5$ PFU of the virus and incubated for 15 min at 37 °C. The mixture was then diluted (see column 2), added to cells, and the number of $IC_{50}$ doses of test compounds present during the development of viral plaques is specified (see columns 3 and 6). Reduced number or an absence of viral plaques developed at <1 $IC_{50}$ dose (in boldface) as related to the solvent control implies virucidal properties of test compound

[b]Refers to PG545 (μg/mL) and P13 (μM)

[c]Mean values from two separate experiments performed in duplicate

[d]PG545 and P13 were solubilized in deionized water and DMSO, respectively. Because the test compounds were prediluted in their respective solvents prior to the addition to the virus, the same control of the solvent is valid for all concentrations of test compound specified in column 1

### *3.3.3 Does the Test Compound Target the Virus Binding to Cells?*

The most straightforward method of testing this possibility is to study the effect of a test compound on the binding of purified radiolabeled virus to cells. It can be difficult to obtain preparations of well-purified RSV due to the variability in shape and size of RSV virions including spherical and filamentous forms [9], their instability [10], and aggregation with cellular components. Techaarpornkul et al. [11] and Kwilas et al. [12] purified RSV by two rounds of centrifugation over a linear 25–55 % sucrose gradient. The cell binding activity of this preparation of purified virus was substantially reduced by soluble heparan sulfate [11], a well-known inhibitor of RSV attachment to cell surface glycosaminoglycan receptors. In our laboratory we purify herpes simplex virus by centrifugation over a discontinuous 30, 40, and 50 % sucrose gradient [13] where the virus particles concentrate and form a band at the 40/50 % sucrose interphase position. When the same protocol was applied to RSV we noted that the 40/50 % interphase fraction exhibited higher infectivity and radioactivity than other fractions of sucrose gradient. In line with other protocols of RSV purification [11, 12] we subject the 40/50 % fraction to the second round of centrifugation over a discontinuous 30/40/50 % sucrose gradient.

#### Purification of Radiolabeled RSV

1. Prepare HEp-2 cells in two 162 $cm^2$ flasks by plating 50 mL of cell suspension in DMEM-G to each bottle.
2. On the following day, remove the medium and rinse the cells once with 15 mL of DMEM-M. Add 15 mL of the same medium, and inoculate the cells with RSV at MOI of 0.2. Incubate for 16 h at 37 °C.
3. Remove the medium and add 18 mL of fresh medium composed of methionine- and cysteine-free DMEM (90 %) and normal DMEM (10 %), supplemented with 2 % HI-FBS, 1 % L-glutamine, 1 % PEST stocks, 10 mM HEPES, and Expre L-($^{35}$S) methionine/L-($^{35}$S) cysteine Protein Labelling Mix (50 μCi/mL).
4. Incubate the cells for 24 h at 37 °C. Close the bottles tightly as volatile $^{35}$S-containing compounds may form.
5. Scrape the cells into the medium and collect into a 50-mL tube. Pipette the cells and then vortex three times for 5 s (*see* **Note 11**).
6. Centrifuge the suspension at 300 × *g* for 5 min at 4 °C.
7. Collect the supernatant and clarify it by centrifugation at 4,000 × *g* for 5 min at 4 °C.
8. Perform the three-step discontinuous sucrose gradient in four Ultra-Clear 16 mm × 102 mm tubes, each consisting of 2 mL of 50, 40, and 30 % sucrose. Mark the position of the 40/50 % interphase on the tube wall.
9. Load 8.5 mL of the virus supernatant (*see* **step 7**) onto the top of the gradient and centrifuge at 120,000 × *g* ($r_{max}$) (25,000 rpm) for 2 h at 4 °C using the SW 28.1 bucket rotor and Optima LE-80K ultracentrifuge (Beckman). Select the slow acceleration for the start and no breaks option at the end of centrifugation period.
10. Puncture the tube with a 21 G needle attached to a 1 mL syringe just below the marked position of a 40/50 % interphase, and slowly collect a 1-mL fraction (*see* **Note 12**).
11. Combine the 40/50 % sucrose interphase fractions, obtained from four tubes, and dilute in 13 mL of DMEM supplemented with 10 mM HEPES.
12. Prepare new gradients in two Ultra-Clear tubes, load the virus material, centrifuge, and collect the 40/50 % interphase fraction as described in **steps 8–10**.
13. Combine the 40/50 % fractions and quantify their radioactivity and infectivity. Aliquot and store the preparation at −80 °C.

#### Assay of Inhibition of Virus Attachment to Cells

1. Plate 1 mL of HEp-2 cell suspension ($1.2 \times 10^5$ cells/mL) in DMEM-G in each well of a 24-well cell culture plate and propagate the cells for 2 days at 37 °C.
2. Incubate the cells for 30 min in a cold (4–8 °C) room.

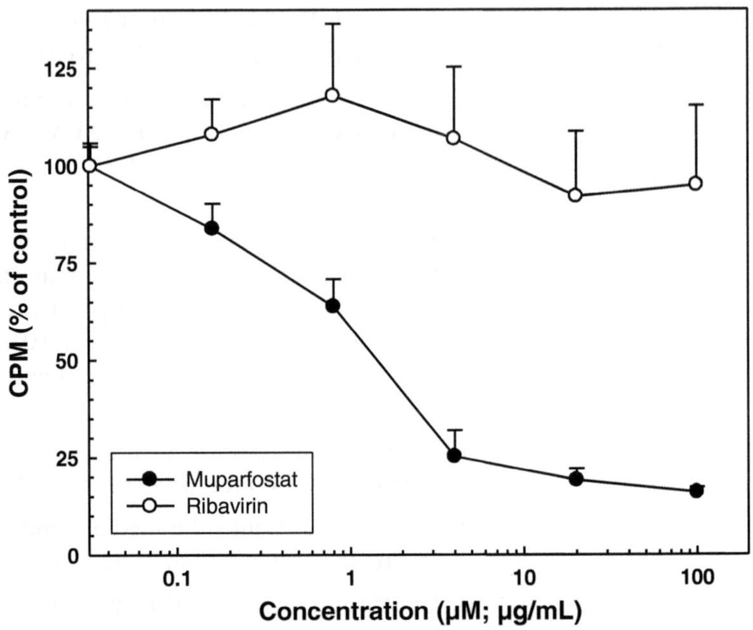


**Fig. 3** The RSV cell-binding activity as target for antiviral compounds. Sulfated oligosaccharide muparfostat (0.16–100 μg/mL), a known inhibitor of RSV binding to cells [4], and anti-RSV drug ribavirin (0.16–100 μM) were mixed with ~70,000 cpm of purified radiolabeled RSV, added to precooled HEp-2 cells, and incubated for 2 h at 4 °C to allow the virus to bind to cells. The cells with any attached RSV were lysed with 5 % SDS and subjected to quantification of radioactivity. The results are expressed as a percentage of a number of cpm that attached to cells in the presence of test compound relative to that bound to cells in the absence of test compound

3. Meanwhile, using a blank 24-well plate, perform serial fivefold dilutions of test compound or its solvent (if other than deionized water) in cold EMEM supplemented with 10 mM HEPES (EMEM-H). Include control samples that contain no test compound or its solvent.
4. Dilute the stock of purified radiolabeled virus with EMEM-H supplemented with 1 % BSA to contain $10^6$ cpm/mL (*see* **Note 13**).
5. Add 70 μL of diluted virus stock to 600 μL of fivefold dilutions of test compound. Mix and incubate at cold room for 5 min.
6. Rinse the cells twice with 0.5 mL of cold EMEM-H medium. Add 320 μL of the virus–compound mixture to cells in duplicate wells. Place the plate on a rocker platform in cold room, and incubate for 2 h at 75 rotation cycles/min.
7. Remove the unbound viral inoculum, and rinse the cells thrice with 500 μL of EMEM.
8. Lyse the cells with 200 μL of 5 % SDS, and transfer the lysate to scintillation vials for quantification of radioactivity (Fig. 3).

*3.3.4 Identification of the RSV Gene(s) Targeted by a Test Compound by Selection of the Drug-Resistant Viral Variants*

Choosing the specific concentration of a test compound to be applied during the virus passage in cells is of importance for successful selection of the drug-resistant viral variants. We choose this concentration based on the dose–response curves used in the determination of $IC_{50}$ value (Fig. 1). For compounds such as P13, which demonstrate no complete inhibition of RSV infectivity at concentrations much higher than the $IC_{50}$ value, we choose a relatively high concentration from a range of 5–100 $IC_{50}$ doses. For compounds such as PG545, which show complete virus inhibition at concentrations higher than $IC_{50}$ value, we begin passaging of RSV in cells in the presence of relatively low concentrations (2 $IC_{50}$) and then increase this concentration progressively in subsequent passages.

1. Prior to the drug-selection experiment prepare the stock of plaque-purified virus (*see* **Note 14**). To this end prepare HEp-2 cells in 6-well plates by adding 4 mL of cell suspension in DMEM-G.
2. On the following day, perform serial tenfold dilutions of the stock of RSV A2 strain in the DMEM-M and add 1 mL of these dilutions to cells.
3. After 3 h of incubation at 37 °C remove the inoculum and add 3 mL of the MC overlay medium. After 3 days of incubation, collect several plaques from wells inoculated with the limiting virus dilution where relatively few (<10) well-separated viral plaques developed. Collect the viral plaque directly through the MC overlay medium in a total volume of 5 μL and transfer it to a new cell monolayer containing 1 mL of DMEM-M. Shake the plate and repeat the procedure described in **step 3** two times. Prepare stock of plaque-purified virus.
4. In a 10-mL tube mix $3 \times 10^5$ PFU of plaque-purified virus with 5 mL of DMEM-M comprising test compound at relatively high concentration. Choose this concentration from a range corresponding to 5–100 $IC_{50}$ doses. This concentration must be at least two times lesser than the $CC_{50}$ value. Use this concentration of test compound in all subsequent virus passages in cells. In a control tube mix the virus with corresponding concentration of the test compound solvent.
5. Remove the growth medium from HEp-2 cells and rinse them once with 2 mL of DMEM-M. Add 1 mL portions of the virus–compound mixture to cells in separate wells. Add 1 mL portions of the virus–solvent mixture to cells in a separate 6-well plate.
6. After 3 h of incubation at 37 °C, remove the inoculum, rinse the cells once with 2 mL of DMEM-M, and add 2 mL of DMEM-M comprising the same concentrations of test compounds or solvent.

7. When the RSV-induced CPE is advanced, scrape the cells into culture supernatant. Combine supernatants in a single 10-mL tube and centrifuge at 300×*g* for 5 min. Remove the supernatant except for the bottom 2 mL, resuspend the cell pellet by pipetting, and vortex it thrice for 5 s each. Pellet the cells by centrifugation at 1,000×*g* for 5 min, mix the supernatant with an equal volume of 50 % sucrose (*see* **Note 11**), and store in aliquots at −80 °C.
8. While an extensive RSV-induced CPE in solvent-treated cells is expected to occur after 2 days of passage 1, little or no viral CPE may be seen at this time-point in the presence of test compound. Therefore incubate the cells with test compounds for 5 days. Harvest the cells, even in the absence of detectable CPE, as described in **step** 7. Do not freeze an entire volume of harvested material. Instead, collect 1 mL of the virus harvest, prior to its mixing with 50 % sucrose, add to 4 mL of DMEM-M with test compound, and immediately transfer 1 mL portions of this solution to cells. In control sample mix 20 μL of the passage 1 stock with the solvent solution in 5 mL of DMEM-E. Repeat **steps** 7 and **8** for passages 2 and 3.
9. When the CPE, induced by the drug-exposed virus, tends to be increasingly advanced over the first three passages, titrate this virus and the mock-passaged virus to use the same amounts of these viruses in preferred volume of 5–20 μL of the stocks for mixing with 5 mL volumes of the compound or the solvent and distributing 1 mL of the mixture to cells in 6-well plate for passage 4 (*see* **step 4**).
10. Titrate the virus stocks and continue with serial virus passages in cells (*see* **step 9**) until the progression of viral CPE in the presence of test compound and its solvent is similar.
11. Prepare stocks of viruses passaged in the presence of test compounds or their solvent (*see* **step** 7). Determine sensitivity of these viruses to test compound as described under Subheading 3.2.1.
12. If the virus passaged in the presence of a test compound shows decreased sensitivity to the compound as related to original and mock-passaged RSV then subject these viruses to plaque purification (*see* **steps 1–3**). Purify at least three viral plaques from each virus.
13. Determine sensitivity to the test compound of an original plaque-purified RSV (**step 3**) and viruses specified in **step 12** as described under Subheading 3.2.1.
14. Sequence an entire genome of each plaque-purified virus variant specified in **steps 3** and **12**. Identification of the same nucleotide change in all the drug-resistant viral plaque variants suggests

contribution of this alteration to the reduced virus sensitivity to the test compound.

15. When the virus-induced CPE is not detectable after passage 3, repeat **steps 4–8** using the concentration of test compound corresponding to 2 $IC_{50}$ doses. Passage the virus with this dose of test compound until the progression of viral CPE is similar to that observed in control (solvent-treated) cells. Subsequently, passage this virus in the presence of progressively increasing concentrations of test compound. Repeat **steps 11–14**.

## 4 Notes

1. Follow the local rules of good laboratory practice while working with respiratory syncytial virus, radioactive isotopes, and potentially toxic chemicals.
2. Shake the cell suspension and pour a small volume into a petri dish (~25 mL) to limit cell sedimentation prior to the transfer to 384-well plates.
3. Addition of library compounds in undiluted DMSO to cells may result in a rapid sedimentation (sinking) of this solvent on cells and partial destruction of the cell monolayer. Handle DMSO with caution and avoid contact with the skin as this solvent can be quickly absorbed from the skin surface together with any solubilized compound.
4. We prepare RSV stock as described by Hallak et al. [14], and stabilize it with 25 % sucrose [10]. Briefly, inoculate HEp-2 cells in two 162 $cm^2$ flasks with 15 mL of DMEM-M comprising ~$5 \times 10^6$ PFU. After incubation for 4 h at 37 °C, remove the inoculum and add 25 mL of fresh DMEM-M to each bottle. When CPE is advanced (usually after 40–48 h), release the cells from the bottom with the cell-scraper and collect the cells and the medium from two bottles into a 50-mL tube. Centrifuge for 5 min at $500 \times g$. Remove the supernatant except for the bottom 5 mL and resuspend the cell pellet by pipetting. Subject the cell suspension to three cycles of vortexing for 5 s each, and then centrifuge for 5 min at $1{,}000 \times g$. Collect the supernatant, mix it with an equal volume of 50 % sucrose, and store at −80 °C. The titer of the RSV A2 stock, prepared in this manner, is usually $1–3 \times 10^7$ PFU/mL.
5. The collections of compounds may not always be sterile, and despite the use of the screening medium with double concentration of PEST (DMEM-S), microbial contamination may occasionally occur in some wells with cells.

6. We recommend using a relatively high input dose of RSV (1,000 $TCID_{50}$) so that the viral CPE spreads rapidly and is clearly distinguishable from mock-infected control cells in column 24.
7. Cytotoxic/cytostatic activity of some test compounds can be manifested as cell lysis, shrinking of cells, presence of large poorly growing cells, altered cell shape (fibroblast-like) with clear edges, and others. Particles of insoluble compound in the form of needles or large dark aggregates may occasionally be seen. Time consideration: We test four separate 384-well plates with library compounds (1,408 compounds) in a single screening experiment. Whole experiment including transfer of library compounds to cell culture plates takes approximately 1.5–2 h. The reading of these four plates under a microscope takes approximately 1–1.5 h.
8. Calculate the final concentrations of solvents present in diluted solutions of the test compound. We found that 1 % deionized water or 1 % DMSO (present at 100 μM solutions of our test compounds) had no inhibitory effect on infection of HEp-2 cells by RSV.
9. Do not carry out serial dilutions of test compound in MC overlay medium. Instead, perform serial fivefold dilutions of test compound in DMEM-M at concentrations ten times higher than those used in the assay. Add 0.25 mL of these dilutions to tubes containing 2.25 mL of MC overlay medium and mix by vortexing.
10. Due to natural variability in size and shape of RSV-induced plaques, do not select random viral plaques for the area measurement. Instead, capture images and measure area of all plaques present, e.g., in the upper-left quarter of the well.
11. RSV virions are known to occur in spherical and filamentous forms [9] that frequently remain associated with the cell surface. These cell-associated virions can be released from the cell surface by vortexing of infected cells [14]. Gupta et al. [10] reported that the presence of 25 % sucrose in the RSV preparation stabilizes its infectivity in different temperatures and during the freezing and thawing of the virus.
12. Comparative analysis of different fractions collected from this discontinuous sucrose gradient revealed that the 40/50 % interphase fraction exhibited the greatest number of both infectious units of RSV and the $^{35}S$ cpm.
13. Tachaarpornkul et al. [11] diluted purified virus with a medium comprising 1 % BSA, a component that may block nonspecific interactions of RSV particles.
14. Use of the plaque-purified virus in selection experiments may facilitate comparative analysis of genomes of original virus and the drug-resistant viral variant(s).

## Acknowledgments

This work was supported by grants from the Torsten and Ragnar Södeberg Foundation, the Mizutani Foundation for Glycoscience, and the Sahlgren's University Hospital Läkarutbildningsavtal.

### References

1. Lundin A, Bergstrom T, Bendrioua L, Kann N, Adamiak B, Trybala E (2010) Two novel fusion inhibitors of human respiratory syncytial virus. Antivir Res 88:317–324
2. Vaheri A (1964) Heparin and related polyionic substances as virus inhibitors. Acta Pathol Microbiol Scand Suppl 171:1–98
3. Hallak LK, Kwilas SA, Peeples ME (2007) Interaction between respiratory syncytial virus and glycosaminoglycans, including heparan sulfate. Meth Mol Biol 379:15–34
4. Lundin A, Bergstrom T, Andrighetti-Fröhner CR, Bendrioua L, Ferro V, Trybala E (2012) Potent anti-respiratory syncytial virus activity of a cholestanol-sulfated tetrasaccharide conjugate. Antivir Res 93:101–109
5. Ekblad M, Adamiak B, Bergstrom T, Johnstone KD, Karoli T, Liu L, Ferro V, Trybala E (2010) A highly lipophilic sulfated tetrasaccharide glycoside related to muparfostat (PI-88) exhibits virucidal activity against herpes simplex virus. Antivir Res 86:196–203
6. Said J, Trybala E, Andersson E et al (2010) Lipophile-conjugated sulfated oligosaccharides as novel microbicides against HIV-1. Antivir Res 86:286–295
7. Lewis FA, Rae ML, Lehmann NI, Ferris AA (1961) A syncytial virus associated with epidemic disease of the lower respiratory tract in infants and young children. Med J Aust 2:932–933
8. Cameron R, Buck C, Merrill D, Luttick A (2003) Identification of contaminating adenovirus type 1 in the ATCC reference strain of respiratory syncytial virus A2 (VR-1302). Virus Res 92:151–156
9. Norrby E, Marusyk H, Orvell C (1970) Morphogenesis of respiratory syncytial virus in a green monkey kidney cell line (Vero). J Virol 6:237–242
10. Gupta CK, Leszczynski J, Gupta RK, Siber GR (1996) Stabilization of respiratory syncytial virus (RSV) against thermal inactivation and freeze-thaw cycles for development and control of RSV vaccines and immune globulin. Vaccine 14:1417–1420
11. Techaarpornkul S, Barretto N, Peeples ME (2001) Functional analysis of recombinant respiratory syncytial virus deletion mutants lacking the small hydrophobic and/or attachment glycoprotein gene. J Virol 75:6825–6834
12. Kwilas S, Liesman RM, Zhang L et al (2009) Respiratory syncytial virus grown in Vero cells contains a truncated attachment protein that alters its infectivity and dependence on glycosaminoglycans. J Virol 83:10710–10718
13. Trybala E, Liljeqvist JA, Svennerholm B, Bergström T (2000) Herpes simplex virus types 1 and 2 differ in their interaction with heparan sulfate. J Virol 74:9106–9114
14. Hallak LK, Collins PL, Knudson W, Peeples ME (2000) Iduronic acid-containing glycosaminoglycans on target cells are required for efficient respiratory syncytial virus infection. Virology 271:264–275

# Chapter 28

# Evaluation of Antiviral Efficacy Against Human Respiratory Syncytial Virus Using Cotton Rat and Mouse Models

Joke Van den Berg, Leen Kwanten, and Dirk Roymans

## Abstract

Infection with human respiratory syncytial virus (hRSV) causes a wide spectrum of respiratory disease in infants, young children, and elderly persons. No vaccine is available today and hRSV treatment options are limited. As a consequence, the treatment of hRSV infection remains largely supportive and new therapeutic options are needed to treat severe lower respiratory tract hRSV disease. Several animal models have been developed to study hRSV disease and evaluate novel therapies or preventive measures such as vaccines. However, each of these models reproduces different aspects of hRSV disease, and therefore, an appropriate model should be selected on the basis of the scientific question under investigation. In this chapter, we describe how cotton rats and Balb/c mice are used in our laboratory to test the in vivo efficacy of small-molecule inhibitors against hRSV.

**Key words** Human respiratory syncytial virus, Cotton rat, Balb/c mice, hRSV inhibitors

## 1 Introduction

hRSV was first isolated in 1957 and has since emerged as an important human respiratory pathogen [1–4]. The virus contains a negative-sense, single-stranded RNA genome and belongs to the family *Paramyxoviridae*, subfamily *Pneumovirinae* [5]. In healthy adults, the infection is generally restricted to the upper respiratory tract and presents with symptoms similar to the common cold. However, progression to a more severe lower respiratory tract infection, often associated with long-term pathology and/or death, is observed in infants and children younger than 5 years, in immunocompromised patients and in the elderly [6–8]. hRSV is therefore considered to be the most important pathogen causing lower respiratory tract disease such as bronchiolitis and pneumonia in infants and young children [6, 9, 10], and in the hospitalized elderly or severely immunocompromised adults with hRSV

Edwin Yunhao Gong (ed.), *Antiviral Methods and Protocols*, Methods in Molecular Biology, vol. 1030,
DOI 10.1007/978-1-62703-484-5_28, © Springer Science+Business Media, LLC 2013

pneumonia, mortality rates can reach 20 and 70 % respectively [11, 12]. hRSV treatment options remain largely supportive and are limited to prophylactic passive immunization with the monoclonal antibody palivizumab or to treatment with ribavirin [13, 14]. Palivizumab is associated with weight-based dosing limitations as a consequence of the high cost of the drug and therefore is only indicated for use in high risk infants up to the age of 2. Treatment with ribavirin is limited due to its problematic mode of aerosolic administration, questionable efficacy, and poor safety profile. Consequently, new therapeutic strategies are needed to reduce hRSV-associated morbidity and mortality.

Several animal models have been developed to better understand hRSV pathogenesis and to support the development of new treatment options or preventive measures [15, 16]. However, no single animal model covers the complete spectrum of disease as observed in humans, and therefore, selection of the right animal model should be driven by the scientific hypothesis under investigation. To enable the selection of appropriate small-molecule hRSV inhibitors as candidates for clinical drug development, it is imperative to evaluate their in vivo antiviral efficacies against the virus. For this purpose, pharmaceutical companies often use rodent models, especially cotton rats and Balb/c mice, because of limited associated cost, and well-characterized standards of housing and veterinary care. However, it is important to note that the infected animals do not express any obvious disease signs of lower respiratory infection like coughing or wheezing, and that the amount of virus recovered from the lungs is directly proportional to the amount of challenge virus. Despite the fact that the rodent models are only semi-permissive to hRSV infection, the predictive quality to assess antiviral activity in research related to respiratory viruses is high. For example, antibodies RespiGam® and palivizumab advanced to clinical trials on the basis of efficacy and safety data generated in cotton rats, bypassing the need for testing in primates [17]. Moreover, the model accurately predicted the dose of palivizumab currently used in the at-risk infants.

We used the cotton rat and Balb/c mouse models to demonstrate the in vivo efficacy of RSV inhibitors [18–21]. The protocol reported here describes a prophylactic drug administration schedule, but can easily be adopted to test different dosing regimens and different routes of administration. The efficacy of compounds is measured at maximal pulmonary RSV titer, which is reached after 4 days of viral challenge. Lungs are lavaged with physiological water and the infectious virus titer in broncho-alveolar lavage fluid (BALF) and lavaged lung tissue is determined by a standard plaque assay and quantitative RT-PCR.

## 2 Materials

### 2.1 Animals

1. Cotton rats: Harlan Laboratories B.V., The Netherlands (*see* **Note 1**).
2. BALB/c mice: Janvier, Belgium (*see* **Note 2**).

### 2.2 Cells, Virus, Media, and Reagents

1. Long strain of hRSV: ATCC (American Type Culture Collection, Cat. No. VR-26™) (*see* **Note 3**).
2. Vero cells (ATCC, Cat. No. CCL-81™).
3. Dulbecco's phosphate buffered saline, without calcium and magnesium (DPBS).
4. Cell culture medium: Dulbecco's minimum essential medium (DMEM) supplemented with 10 % fetal calf serum (FCS) and 0.04 % gentamicin.
5. Disinfectant solutions.
6. 2-Chloro-2-(difluoromethoxy)-1,1,1-trifluoro-ethane (isoflurane).
7. Plaque medium: 100 mL of 10 × Dulbecco's modified Eagle medium (DMEM) (Gibco 042-90132M) supplemented with 333 mL aqua bidest (distilled water), 40 mL sodium bicarbonate, 12 mL penicillin–streptomycin (peni-strept), 12 mL L-glutamine, 60 mL of heat inactivated FCS and adjust to pH 7.4 with 1 M NaOH.
8. SeaPlaque agarose medium: 3 % SeaPlaque agarose (Lonza 50100) solution in aqua bidest, mix and autoclave.
9. Agarose growth medium: heat the agarose solution in microwave until agarose dissolved, then mix with equal volume of the plaque medium and keep the solution at 42 °C in a water-bath.
10. +99 % DL-1,4-dithiothreitol (DTT), DNase, RNase, and Protease free (Acros).
11. NucliSens extraction lysis buffer (bioMérieux).
12. NucliSens buffer 3 (bioMérieux).
13. NucliSens magnetic beads (bioMérieux).
14. RNase-free water.
15. Yeast tRNA.
16. One-step RT qPCR Master Mix (Invitrogen, RT-QPRT-032X).
17. Primers and probes (*see* Table 1).

### 2.3 Consumables and Equipment

1. 175 $cm^2$ Culture flasks (tissue culture treated).
2. Cell scraper.

**Table 1**
**qRT-PCR primers and probes**

| Name | Sequence (5′–3′) |
|---|---|
| RSV-A-F-RV732 | AGTTACACCTGCATTAACACTAAATTCC |
| RSV-A-F-FW656 | CTGTGATAGARTTCCAACAAAAGAACA |
| IEC-RSV-B-N-RV508 | TGGTTATTACAAGAGCAGCTATACACAGT |
| IEC-RSV-B-N-FW435 | GGCTCCAGAATATAGGCATGATTC |
| Probe RSVA | FAM-CAGACTACTAGAGATTACC-NFQ-MGB |
| Probe IEC-RSV-B | FAM-TACCGTACTCTAGCCTA-3′-NFQ-MGB |

3. Centrifuge.
4. Barré anesthesia machine.
5. GentleMACS™ Dissociator and M tubes.
6. Syringes.
7. PE90-tubing adapter.
8. Animal dissection set: scissor, scalpel, etc.
9. Incubator (humidified, 5 % $CO_2$, 37 °C).
10. 6-Well Falcon plates (BD).
11. EasyMAG (bioMérieux).
12. Biomek laboratory automation workstation (Beckman/Coulter).

## 3 Methods

### *3.1 Animal Infection with Long hRSV*

1. Thaw an appropriate amount of Long hRSV stock and prepare viral inoculum in cell culture medium: $10^6$ PFU (plaque forming units) per 0.05 mL for mice and $10^6$ PFU per 0.1 mL for cotton rats.
2. Next, anesthetize the animals with isoflurane (*see* **Note 4**).
3. Inoculate dropwise by the intranasal route.

### *3.2 Animal Dosing with hRSV Inhibitors*

To test the in vivo antiviral efficacy of compounds, use different dosing groups. Each group contains at least five animals, varying from low to high doses depending on the activity of the compound in vitro, e.g., 1, 4, 10, 40, and 100 mg/kg. Also include an additional group of untreated controls to monitor the viral replication efficiency. Different routes of administration are possible, e.g., oral or intravenous. For comparison, an extra group of animals can be included, in which a compound with known antiviral activity against hRSV is administered [22–24].

Make sure that the formulation of test compounds is compatible with the administration route. In some of our experiments, animals are treated 1 h before infection with different doses of compound. Alternatively, a multiple dosing regimen is also possible. Depending on the experimental setup, prophylactic or therapeutic, animals are treated either 60 min before or just after infection. Animals can be treated once but multiple dosing is also possible.

### 3.3 Harvesting Lung and BALF Samples

1. Euthanize all test animals 4 days after hRSV inoculation by $CO_2$ gas asphyxiation (*see* **Note 5**).
2. Fixate the animals for further processing so that trachea and chest areas are easy accessible.
3. Expose the trachea (*see* **Note 6**).
4. Make a small incision in the trachea and ensure that the region is free of blood.
5. Fill a syringe with 2.0 mL DPBS for cotton rats and 1.0 mL for mice and attach to a needle tubing adapter on which polyethylene tubing is fixed (*see* **Note 7**).
6. Introduce the free side of the tubing into the trachea.
7. Slowly instill the DPBS into the lungs.
8. Dwell time of the instilled DPBS is 15 s.
9. Slowly aspirate the DPBS/BALF and store the BALF samples in 15 mL Falcon™ tubes.
10. Record the volume of collected BALF.
11. Aseptically remove the lung.
12. Record the lung weight and place in M tubes filled with 2.5 mL DPBS.
13. Discard the animal.
14. Homogenize the lung samples in the GentleMACS™ Dissociator.
15. Centrifuge lung and BALF samples at 1,500 × *g* for 10 min.
16. Collect 50 μL of lung and BALF supernatant into cryovials for qRT-PCR analysis. The remainder of the supernatant is used for plaque titration (*see* **Note 8**).

### 3.4 Determination of hRSV Titers

Plaque titration of the lung and BALF samples is performed via a standard plaque assay. Briefly:

#### 3.4.1 Plaque Assay

1. Add tenfold serially diluted lung/BALF supernatant to Vero cell monolayers in 6-well Falcon plates.
2. Incubate at room temperature for 1 h.

3. Remove the viral inoculum.
4. Wash once with DPBS.
5. Add 3 mL of agarose growth medium.
6. Incubate at 37 °C for 7 days.
7. After 7 days, fix the cells with 4 % formaldehyde.
8. Stain the cells with 0.4 % methylene blue solution and count the plaques.
9. Express the viral titer in $\log_{10}$ PFU/mL.

#### 3.4.2 RNA Extraction and qRT-PCR Assay

The RNA from the samples is extracted using the NucliSens easyMAG system of bioMérieux, which is an automated system for total nucleic acid extraction from a variety of samples and volumes using magnetic beads (*see* **Note 9**).

1. Add 2.15 mL of DTT to the qRT-PCR samples.
2. Add 4 mL of NucliSens extraction lysis buffer to each sample.
3. Incubate for 10 min.
4. Transfer 3 mL of lysed samples to an EasyMAG extraction apparatus.
5. Add 100 μL of internal extraction control (IEC) to each sample and perform the extraction according to the manufacturer's instructions (*see* **Note 10**).
6. Elute the RNA in 110 μL of elution buffer.
7. Prepare two real-time RT-PCR mixtures (one mixture with RSV-A primers/probe for the samples and another mixture for the IEC-B control in a 96-well PCR plate): 10 μL of extracted RNA, 0.360 μL yeast tRNA, 150 μL EuroScript reaction buffer, 0.15 μL Euroscript RT, 0.1 μM probe (probe RSV-A or probe IEC-RSV-B), 0.9 μM of each primer (RSV-A or IEC-RSV-B), 1.5 mM $MgCl_2$, and RNase/DNase-free water to a final volume of 30 μL. Also add eight tenfold serial dilutions of an RSV RNA standard on each qRT-PCR plate with the same reaction mixture used to process the samples, which is used to generate a standard curve for absolute quantitation of RNA copy numbers.
8. Transfer the 96-well PCR plates to a laboratory automation workstation, e.g., Biomek. Combine both the RSV-A and the IEC-RSV-B plates to a 384-well format and seal the plate with an ABI seal.
9. Perform real-time RT-PCR on an ABI 7900HT. The cycling conditions are: reverse transcription of RNA at 48 °C for 30 min, then denaturing at 95 °C for 10 min, followed by 45 cycles of 95 °C for 15 s, and 60 °C for 1 min.
10. Use the SDS software program of Applied Biosystems for data analysis.

## 4 Notes

1. Both male and female cotton rats can be used. All animals should be housed individually under controlled conditions: specific pathogen free, 23 °C, 60 % humidity, and a normal light–dark cycle. Food and water are available ad libitum. Make all efforts to minimize animal discomfort and limit the number of animals (Declaration of Helsinki 86/609/EEC). All experimental work with animals must be performed under anesthesia (isoflurane).
2. Keep the BALB/c mice also under controlled conditions. These animals can be housed in groups.
3. Carry out all manipulations with appropriate personal protective clothes and gloves; and take into account all safety procedures for working with biohazardous materials. According to the Belgian biosafety regulations, both in vitro and in vivo experiments for hRSV are restricted to a biosafety level two laboratory (L2).
4. Make sure that the animals are properly anesthetized by checking the corneal eye reflex and look for reaction with the toe pinch. This should be done after every anesthesia procedure.
5. Four days post-infection viral replication is peaked.
6. For exposing the trachea, limit the use of scalpels and scissors. We recommend tearing the tissue surrounding the trachea. Otherwise blood influx will occur that may complicate the BALF sampling.
7. Use prechilled DPBS. The length of the polyethylene tubing should be chosen so that it can get into the lungs, but not too long because this will penetrate the lungs and the BALF will be lost.
8. Determining the virus yield with the plaque assay should be performed within 2 h after sample collection. Otherwise the virus titers may decrease. Store the RNA samples for extraction at −80 °C.
9. The volumes of reagents added to the extraction were optimized to accommodate RNA extractions from different origins (e.g., sputum, nasal wash, BAL) and donor species (e.g., human, monkey, rodent) according to a single protocol, but these can be adjusted.
10. The extraction control consists of 50 μL of NucliSens magnetic beads, 10 μL of internal extraction control (IEC) RNA, and 40 μL of NucliSens Extraction Buffer 3.

## Acknowledgments

The authors wish to thank Koen Andries and Tom Gevers for the design of some of the cotton rat protocols described here.

### References

1. Chanock R, Roizman B, Myers R (1957) Recovery from infants with respiratory illness of a virus related to chimpanzee coryza agent (CCA). I. Isolation, properties and characterization. Am J Hyg 66:281–290
2. Chanock R, Finberg L (1957) Recovery from infants with respiratory illness of a virus related to chimpanzee coryza agent (CCA). II. Epidemiologic aspects of infection in infants and young children. Am J Hyg 66:291–300
3. Ruuskanen O, Lahti E, Jennings LC et al (2011) Viral pneumonia. Lancet 377:1264–1275
4. Nair H, Nokes JD, Gessner BD et al (2010) Global burden of acute lower respiratory infections due to respiratory syncytial virus in young children: a systematic review and meta-analysis. Lancet 375:1545–1555
5. Collins PL, Crowe JE Jr (2007) Respiratory syncytial virus and metapneumovirus. In: Knipe DM, Howley PM (eds) Fields virology, 5th edn. Lippincott Williams & Wilkins, Philadelphia, pp 1601–1646
6. Hall CB, Weinberg GA, Iwane MK et al (2009) The burden of respiratory syncytial virus in young children. New Engl J Med 360:588–598
7. Couch RB, Englund JA, Whimbey E (1997) Respiratory viral infections in immunocompetent and immunocompromised persons. Am J Med 102:2–9
8. Falsey AR, Hennessey PA, Formica MA et al (2005) Respiratory syncytial virus infection in the elderly and high-risk adults. New Engl J Med 352:1749–1759
9. Rudan I, Boschi-Pinto C, Biloglav Z et al (2008) Epidemiology and etiology of childhood pneumonia. Bull World Health Organ 86:408–416
10. Simoes EAF (1999) Respiratory syncytial virus infection. Lancet 354:847–852
11. Thompson WW, Shay DK, Weintraub E et al (2003) Mortality associated with influenza and respiratory syncytial virus in the United States. J Am Med Assoc 289:179–186
12. Falsey AR, Walsh EE (2000) Respiratory syncytial virus infection in adults. Clin Microbiol Rev 13:371–384
13. Roymans D, Koul A (2011) Treatment of respiratory syncytial virus infection: past, present and future. In: Resch B (ed) Human respiratory syncytial virus infection. InTech Europe, Rijeka, pp 197–234
14. Roymans D, Koul A (2010) Respiratory syncytial virus: a prioritized or neglected target? Future Med Chem 2:1523–1527
15. Byrd LG, Prince GA (1997) Animal models of respiratory syncytial virus infection. Clin Infect Dis 25:1363–1368
16. Bem RA, Domachowske JB, Rosenberg HF (2011) Animal models of human respiratory syncytial virus disease. Am J Physiol Lung Cell Mol Physiol 301:L148–L156
17. Boukhvalova MS, Prince GA, Blanco JCG (2009) The cotton rat model of respiratory viral infections. Biologicals 37:152–159
18. Rouan M-C, Gevers T, Roymans D et al (2010) Pharmacokinetics-pharmacodynamics of a respiratory syncytial virus fusion inhibitor in the cotton rat model. Antimicrob Agents Chemother 54:4534–4539
19. Olszewska W, Ispas G, Schnoeller C et al (2011) Antiviral and lung protective activity of a novel respiratory syncytial virus fusion inhibitor in a mouse model. Eur Respir J 38:401–408
20. Wyde PR, Chetty SN, Timmerman P et al (2003) Short duration aerosols of JNJ 2408068 (R170591) administered prophylactically or therapeutically protect cotton rats from experimental respiratory syncytial virus infection. Antiviral Res 60:221–231
21. Bonfanti JF, Roymans D (2009) Prospects for the development of fusion inhibitors to treat human respiratory syncytial virus infection. Curr Opin Drug Discov Devel 12:479–487
22. Cianci C, Genovesi EV, Lamb L et al (2004) Oral efficacy of a respiratory syncytial virus inhibitor in rodent models of infection. Antimicrob Agents Chemother 48:2448–2454
23. Bonfanti JF, Meyer C, Doublet F et al (2008) Selection of a respiratory syncytial virus fusion inhibitor clinical candidate. 2. Discovery of a morpholinopropylaminobenzimidazole derivative (TMC353121). J Med Chem 51:875–896
24. Ventre K, Randolph AG (2007) Ribavirin for respiratory syncytial virus infection of the lower respiratory tract in infants and young children. Cochrane Database Syst Rev 1:CD000181

# Chapter 29

# In Vivo Evaluation of Antiviral Compounds on Respiratory Syncytial Virus Using a Juvenile Vervet Monkey (*Chlorocebus pygerythrus*) Infection Model

**Lieselot Houspie, Hans Stevens, Maina Ngotho, Els Keyaerts, Gabriela Ispas, René Verloes, Marc Van Ranst, and Piet Maes**

## Abstract

Several animal models with varying susceptibilities to respiratory syncytial virus (RSV) infection have been developed to study the specific aspects of RSV disease. Many of these models are used for testing antiviral compounds or in vaccine efficacy studies during preclinical evaluation. In this chapter, we describe the study design of an efficacy study of an RSV inhibitor, performed in a juvenile vervet monkey model for RSV.

**Key words** Respiratory syncytial virus, Monkey model, RSV inhibitor, Animal infection model

## 1 Introduction

The development of different animal models for human RSV infection resulted in a better understanding of the immunopathology and virology associated with RSV [1]. Replication of RSV has been demonstrated in BALB/c mice [2], cotton rats [3], nonhuman primates [4, 5], lambs [6], chinchillas [7], and other animals. Each individual model can recapitulate some of the key features of RSV disease. The preference for a certain model depends on the aspect of RSV disease under investigation [8]. Rodent models are valued for their immunological response being similar to humans and are the primary models for vaccine and antiviral efficacy testing. However, due to the semi-permissiveness to RSV infection of these models, the results of efficacy assessments of vaccines and antiviral compounds in rodent models need to be interpreted with care [9]. Nonhuman primate models are genetically closely related to humans and the respiratory tract is anatomically similar. RSV replication can be induced in several nonhuman primate species

Edwin Yunhao Gong (ed.), *Antiviral Methods and Protocols*, Methods in Molecular Biology, vol. 1030,
DOI 10.1007/978-1-62703-484-5_29, 

and the virus dynamics in these models show similarity to virus replication in human adults in terms of incubation period, viral load, and shedding dynamics [10]. Nonetheless, the chimpanzee is the only nonhuman primate species with symptoms that clinically resemble disease in seronegative human subjects [8]. The African green monkey, one of the most extensively studied species among nonhuman primates, is considered a valuable model for RSV vaccine efficacy and safety testing because this species reliably reproduces the vaccine enhanced disease after vaccination with formalin inactivated RSV vaccines [11–13]. This primate model has previously been applied to evaluate the efficacy of a small molecule fusion inhibitor RFI-641 [14, 15] in which the viral load reduction was taken as a measure for the evaluation of efficacy. The potency of this fusion inhibitor was evaluated in three different animal models and demonstrated antiviral activity when used in a prophylactic regimen in mice, cotton rats and African green monkeys. In African green monkeys, an antiviral effect after therapeutic administration was shown, in addition to the previous observed prophylactic efficacy. This observation underlines the importance of efficacy testing in different animal models that individually exhibit different aspects of RSV disease and have different permissiveness for RSV infection. In our study, the efficacy of a novel small molecule fusion inhibitor with a benzimidazole structure was studied in a juvenile African green monkey model [9, 16]. The design of the efficacy study is described.

## 2 Materials

### 2.1 Animals

1. Juvenile vervet monkeys (*Chlorocebus pygerythrus*) between 12 and 24 months of age (*see* **Note 1**).
2. Isolated single stainless steel cages measuring 90 cm × 60 cm × 60 cm for animal housing in a BSL-2 animal facility at room temperatures of between 23 and 25 °C (*see* **Note 2**).
3. RSV IgG ELISA (Cat. No. 5114-8; Diagnostic Automation, Inc.).

### 2.2 Implantation of DST-nano-T Temperature Loggers

1. Gas sterilized DST nano-T temperature logger devices (Star Oddi).
2. Scalpel to make an incision in the peritoneal cavity.
3. Razor to shave the fur of the vervet.
4. Iso-Betadine solution for disinfecting the incision site.
5. Surgical suture and needles.
6. Disinfecting Iso-Betadine cream to avoid inflammation at the incision site.

### 2.3 Anti-RSV Compound

1. Test compound dissolved in the appropriate solvent. Storage conditions of the compound need to be checked.
2. Vehicle for the non-treated, infection-control group.
3. A 3 mL BD Luer-Lock syringe with needle for intravenous injection of the compound.
4. Alcohol swab for the injection site.

### 2.4 Infection with RSV Clinical Isolate

1. Individual vials of 2 mL RSV-A virus stock at a titer of 5 log PFU/mL [17] (*see* **Note 3**).
2. Pediatric nasogastric feeding tube (Flocare, Nutricia).
3. Laryngoscope.
4. 3-mL Luer-Lock syringes without needle.

### 2.5 Nasal and Throat Swab Collection

1. Viral transport swabs, carrying a virus transport medium soaked sponge in a container tube (Cat. No. 328C; COPAN Diagnostics, Inc.).
2. Laryngoscope.

### 2.6 RNA Extraction

QIAamp viral RNA mini kit (Cat. No. 52906; Qiagen) (*see* **Note 4**).

### 2.7 qRT-PCR

1. One-step reverse transcriptase qPCR kit (Cat. No. RT-QPRT-032X; Eurogentec).
2. 200 nM of RSV-A F gene specific primers:

   Forward primer: 5′-CTGTGATAGARTTCCAACAAAAGAA CA-3′.

   Reverse primer: 5′-AGTTACACCTGCATTAACACTAAATT CC-3′.
3. 100 nM of RSV-A MGB probe: FAM-5′-CAGACTA CTAGAGATTACC-3′-NFQ-MGB (Applied Biosystems).
4. cRNA standards, constructed using the MEGAshortscript T7 kit (Ambion).
5. Optical 96-well fast thermal cycling plate (Applied Biosystems).
6. ABI 7500 Fast Real Time PCR system (Applied Biosystems).

### 2.8 RSV Plaque Assay

1. HEp-2 cells: European Collection of Cell Culture (Cat. No. 85011412).
2. Cell culture 6-well plates.
3. Eagle's minimum essential medium supplemented with 2 % FBS (MEM/2 % FBS).
4. PBS.
5. SeaPlaque agarose (Lonza).
6. Neutral red.

## 3 Methods

All experimental steps and animal handling require wearing personal protections, i.e., a gown or Tyvek suit, two pairs of gloves, sleeves, double eye protection (safety glasses and face shield), and an FFP2 face mask.

### *3.1 Animals and Animal Welfare*

1. All experiments with nonhuman primates must be carried out in accordance with the local animal welfare and ethics laws. The local ethical board needs to review and approve the protocol before start of the study. Take all possible actions to minimize the suffering of animals and perform all surgery under anesthesia.
2. Animals are subjected to a 90-day quarantine period during which they are screened for tuberculosis and checked for general clinical health. All animals must be clinically healthy and free of tuberculosis. Allow the monkeys to habituate to laboratory conditions for over 8 months [18]. As the animals may have or develop worms in their feces, they need to be dewormed regularly with anthelmintics before being recruited into the experiment.
3. Check the existing IgG levels of RSV infections for all animals by RSV IgG ELISA kit according to the manufacturer's instructions. IgG positive animals should be excluded from the study population.
4. Randomly divide animals of both sexes into groups and provide isolated housing in separate facilities for the different groups.

### *3.2 Study Design*

1. Divide the vervet monkeys into groups following a prophylactic, therapeutic, or combined prophylactic–therapeutic regimen.
2. Include a non-treated, infected control group, receiving only the vehicle.
3. Each group consists of a minimum of five animals.
4. Collect nasal and throat swabs during a minimum of 14 days, every other day after infection under anesthesia (*see* **Note 5**).

### *3.3 Implantation of the Temperature Logger Devices*

1. Program the temperature logger (DST nano-T) using the Mercury software (Star Oddi). Set the temperature-sampling interval at 5 min. The measurements start at least 14 days before infection with RSV and end at least 7 days after the last sampling day.
2. Sterilize the temperature logger devices and individually pack them in a sterile package.

3. Anesthetize the animals with the appropriate anesthetic agent.
4. Implant the temperature logger device 3 weeks prior to the start of the experiment in the abdomen of the vervet (*see* **Note 6**).
5. For the readout of temperature, remove the logger from the abdomen, approximately 7 days after the last sampling day and connect the logger to the communication box interface. Temperature data can be exported using the Mercury software.

### 3.4 Infection with RSV

1. Fill a 3 mL Luer tip syringe with the correct amount of inoculum.
2. Anesthetize the animal with the appropriate anesthetic agent prior to inoculation.
3. Once under anesthesia, place the animal in dorsal recumbence, so the nostrils point to the ceiling. Place the tip of the syringe into the nostril and slowly press the plunger, drop by drop. Inject 0.25 mL of virus inoculum into the nasal passage of one nostril.
4. Repeat the procedure for the second nostril.
5. Leave the animal's head tilted back for about 20 s. This will allow the inoculums to flow into the nasal passage.
6. An assistant opens the animal's mouth. With the aid of the laryngoscope, locate and open the epiglottis.
7. Introduce the small end of the nasogastric tube into the glottis. Attach the syringe containing 1 mL RSV inoculums to the sterile catheter tube. Once in place, inject the inoculum into the trachea. Before inserting the syringe with the inoculum on the catheter, pull back the syringe allowing 2 cc of air into the syringe. This air will push all of the inoculums through the catheter. Fill 5 cc of air into a 5 cc slip tip syringe and blow through the catheter. Remove the catheter from the trachea and observe the animal until recovered from the anesthetic. Use new or sterilized equipment for each animal or disinfect laryngoscope blades or tubes between animals using isopropanol.
8. Administer atropine intramuscularly if necessary following the dosage.

### 3.5 Nasal and Throat Swab Collection

1. Label the transport swab tubes at a minimum with animal number, date of collection, type of sample, and type of medium.
2. Anesthetize the animal with the appropriate anesthetic agent, prior to inoculation. Once under anesthesia, place the animal in dorsal recumbence, so the nostrils point to the ceiling.
3. Insert the cotton-tipped applicator, which has been pre-wetted with viral culture medium from inside the tube, into the

animal's nose. During nasal swab collection, rotate and rub the cotton tip against the inner wall of the nasal passage. Swab both nostrils using one swab. Place the tip of the swab into the COPAN tube.

4. An assistant opens the animal's mouth, and remove excess saliva and food particles.
5. Insert the cotton-tipped applicator, pre-wetted with viral culture medium from inside the tube, into the animal's mouth. Make sure not to touch any surface such as the buccal mucosa (cheek), teeth, hard palate (roof of the mouth), or tongue until the tip of the swab is past the base of the tongue. Once the tip of the swab is past the base of the tongue, press the tip against the dorsal surface of the pharyngeal area and roll, making sure not to touch the teeth, tongue, and buccal mucosa. Place the tip of the swab into the COPAN tube.
6. Place samples on ice until they can be stored at −80 °C.

### 3.6 Injection of RSV Inhibitor

1. Formulate the compound in accordance with the route of administration.
2. Administer the slow bolus injection through the appropriate route: via aerosol, oral, subcutaneous, intramuscularly or intravenously.

### 3.7 RNA Extraction

1. Resuspend the nasal and throat swabs in 1 mL of fresh viral transport medium.
2. Transfer the swabs and aspirate the viral transport medium into a 2 mL tube.
3. Isolate the viral RNA using a QIAamp viral RNA mini kit according to the manufacturer's instructions.
4. For each extraction, include a negative control so that the sample cross-contamination can be excluded.

### 3.8 qRT-PCR

1. Construct cRNA standards by inserting a T7 promoter sequence into the 5′ end of the forward primer of the qRT-PCR. Perform a PCR reaction using the modified forward primer and reverse primer and quantify the cDNA concentration spectrophotometrically at a wavelength of 260 nm. Perform an in vitro transcription using the MEGAshortscript overnight at 37 °C following the manufacturer's instructions. Quantify the cRNA spectrophotometrically at 260 nm.
2. Use the Eurogentec One-Step Reverse Transcriptase qPCR kit for viral load quantitation. A final volume of 25 μL consists of 5 μL extracted viral RNA, 12.5 μL of Eurogentec One-Step Reverse Transcriptase qPCR Master Mix with ROX as a passive

reference, 0.125 μL Euroscript+RT & RNase inhibitor (Eurogentec), 200 nM of RSV-A F gene specific forward and reverse primers, and 100 nM of RSV-A MGB probe. The qRT-PCR program is initiated at 48 °C for 30 min and followed by 10 min at 95 °C. Subsequently, 45 cycles of 95 °C for 15 s, and 60 °C for 1 min using the ABI 7500 Fast Real Time PCR system. Each quantitation is performed in duplicate.

### 3.9 RSV Plaque Assay

1. Prepare tenfold serial dilutions of the virus in MEM/2 % FBS.
2. Inoculate 200 μL of each dilution to the duplicate wells of a 6-well plate. Include a non-infected cell control and an infected positive control.
3. Allow the virus to absorb for 2–4 h.
4. Afterwards, remove the inoculum and wash the cell monolayer twice with PBS.
5. Prepare the plaque overlay by melting 1 % SeaPlaque agarose in MEM/2 % FBS.
6. When the agarose cooled to approximately 37 °C, add 1 mL of plaque overlay to each well. Allow the agarose to be solidified at room temperature.
7. Add 1 mL of MEM/2 % FBS to each well and incubate the plates at 37 °C [19].
8. Discard the medium after 7 days and fix the cells overnight with 1 % formaldehyde in 0.9 % NaCl.
9. Remove the plaque overlay by using a spatula and stain the plaques with 0.05 % neutral red solution for 1 h.
10. Rinse the wells with water and dry the wells on paper towels.
11. Count the plaques macroscopically in the lowest dilution where a maximum of 50 plaque forming units (PFU) are present. Calculate the mean number of PFU/mL, taking into account the dilution factor of the inoculum.

### 3.10 Evaluation of Antiviral Efficacy

Viral load parameters can be statistically analyzed for significance using statistical software.

1. Use the viral load data to determine the viral replication over time for all groups. Calculate the area under the curve for the different groups.
2. Assess the reduction at the peak viral load in comparison with the non-treated, infected control group.
3. Compare the virus dynamics observed for the control group to the treated groups. A longer incubation, lower peak of viral load, more rapid clearance of virus replication may be indicative of reduction in virus replication (*see* **Note 7**).

## 4 Notes

1. The juvenile vervet monkeys were provided by the Institute for Primate Research in Nairobi, Kenya. This species (*Chlorocebus pygerythrus*) is common in Kenya.
2. Each animal must be housed separately in isolated cages. The direct contact between the animals must be prohibited. Animals must be housed per treatment group in a separate room to prevent virus transmission among animals.
3. Inoculate a fresh clinical specimen, found positive for RSV, in HEp-2 cells during 7 days to cultivate the RSV strain. To obtain sufficiently high infectious titers and virus stock, several rounds of replication are necessary in the cells. Cytopathic effect should be checked daily using a light microscope. To determine the infectious dose of the inoculum, a plaque assay should be performed. Here, we used an infectious dose of 5 log PFU/mL, targeting 100 % infection of the animals.
4. We opted for the QiAmp viral RNA minikit supplied by Qiagen, but other kits for viral RNA extraction can also be used.
5. A different sample type that involves a more invasive collection procedure for the animal can be chosen, such as a nasal wash, oro-nasopharyngeal aspirate or broncho-alveolar lavage (BAL). The nasal and throat swabs may be of lower quality for viral load quantification in comparison with a nasal wash or an aspirate. The acquisition of BAL may also allow the quantitation of immunological markers for RSV infection.
6. The implantation site of the temperature logger can be chosen randomly. The temperature monitored using the DST-nano-T in this model demonstrated that temperature recording is less influenced by external factors (e.g., day–night rhythm) when the logger is implanted in the peritoneal cavity.
7. To allow further interpretation of the antiviral activity, the pharmacokinetic properties of the compound should be investigated. In addition, necropsy performed at certain time points during the experiment can be used to assess the histological differences in the lung between treated and untreated groups. Evaluation of the clinical symptoms (respiratory rate, fever, oxygenation, etc.) in relation to the treated or untreated groups may elucidate the influence of the compound on clinical symptoms.

## Acknowledgments

This research was supported in part by a grant of the Institute for the Promotion of Innovation by Science and Technology in Flanders (IWT), project number 070005. Piet Maes is supported by a postdoctoral grant from the "Fonds voor Wetenschappelijk Onderzoek (FWO)-Vlaanderen". We gratefully acknowledge Dr. Thomas Karuiki and the staff at the Institute of Primate Research, Nairobi, Kenya for the excellent technical assistance. We would also like to thank our colleagues of the Laboratory of Clinical Virology and R&D of Janssen Infectious Diseases for their contribution to the study design and review of the manuscript.

### References

1. Bem RA, Domachowske JB, Rosenberg HF (2011) Animal models of human respiratory syncytial virus disease. Am J Physiol Lung Cell Mol Physiol 301:148–156
2. Kong X, Hellermann GR, Patton G et al (2005) An immunocompromised BALB/c mouse model for respiratory syncytial virus infection. Virol J 2:3
3. Boukhvalova MS, Prince GA, Blanco JC (2009) The cotton rat model of respiratory viral infections. Biologicals 37:152–159
4. Prince GA, Suffin SC, Prevar DA et al (1979) Respiratory syncytial virus infection in owl monkeys: viral shedding, immunological response, and associated illness caused by wild-type virus and two temperature-sensitive mutants. Infect Immun 26:1009–1013
5. Belshe RB, Richardson LS, London WT et al (1977) Experimental respiratory syncytial virus infection of four species of primates. J Med Virol 1:157–162
6. Olivier A, Gallup J, de Macedo MM et al (2009) Human respiratory syncytial virus A2 strain replicates and induces innate immune responses by respiratory epithelia of neonatal lambs. Int J Exp Pathol 90:431–438
7. Grieves JL, Jurcisek JA, Quist B et al (2010) Mapping the anatomy of respiratory syncytial virus infection of the upper airways in chinchillas (*Chinchilla lanigera*). Comp Med 60:225–232
8. Byrd LG, Prince GA (1997) Animal models of respiratory syncytial virus infection. Clin Infect Dis 25:1363–1368
9. Bonfanti JF, Meyer C, Doublet F et al (2008) Selection of a respiratory syncytial virus fusion inhibitor clinical candidate. 2. Discovery of a morpholinopropylaminobenzimidazole derivative (TMC353121). J Med Chem 51:875–896
10. Houspie L, Ipsas G, Keyaerts E et al (2013) Juvenile vervet monkey model for human respiratory syncytial virus infection (manuscript in preparation)
11. Graham BS (2011) Biological challenges and technological opportunities for respiratory syncytial virus vaccine development. Immunol Rev 239:149–166
12. Jin H, Cheng X, Traina-Dorge VL et al (2003) Evaluation of recombinant respiratory syncytial virus gene deletion mutants in African green monkeys for their potential as live attenuated vaccine candidates. Vaccine 21:3647–3652
13. Kakuk TJ, Soike K, Brideau RJ et al (1993) A human respiratory syncytial virus (RSV) primate model of enhanced pulmonary pathology induced with a formalin-inactivated RSV vaccine but not a recombinant FG subunit vaccine. J Infect Dis 167:553–561
14. Weiss WJ, Murphy T, Lynch ME et al (2003) Inhalation efficacy of RFI-641 in an African green monkey model of RSV infection. J Med Primatol 32:82–88
15. Huntley CC, Weiss WJ, Gazumyan A et al (2002) RFI-641, a potent respiratory syncytial virus inhibitor. Antimicrob Agents Chemother 46:841–847
16. Olszewska W, Ispas G, Schnoeller C et al (2011) Antiviral and lung protective activity of a novel respiratory syncytial virus fusion inhibitor in a mouse model. Eur Respir J 38:401–408

17. Houspie L, Keyaerts E, Maes P et al (2012) Susceptibility of the PER.C6 cell line for infection with clinical human respiratory syncytial virus isolates. J Virol Methods 181:37–42
18. Kagira JM, Ngotho M, Thuita JK et al (2007) Hematological changes in vervet monkeys (*Chlorocebus aethiops*) during eight months' adaptation to captivity. Am J Primatol 69: 1053–1063
19. Ooi LS, Ho WS, Ngai KL et al (2010) Narcissus tazetta lectin shows strong inhibitory effects against respiratory syncytial virus, influenza A (H1N1, H3N2, H5N1) and B viruses. J Biosci 35:95–103

# Chapter 30

# Methods to Determine Mechanism of Action of Anti-influenza Inhibitors

**Angela Luttick, Stephanie Hamilton, and Simon P. Tucker**

## Abstract

Mechanism of action studies can be used to demonstrate an inhibitor's ability to specifically inhibit viral replication via a virus-specific or host cell target. A well-characterized mechanism of action is useful in evaluating potential off-target toxicities (e.g., a viral polymerase inhibitor may be assessed for inhibition of host polymerase) and in designing studies to monitor the development of resistance. Several methods can be used to elucidate the mechanism of action of an anti-influenza inhibitor. The first group of methods establishes that the activity of an inhibitor occurs at concentrations that do not cause cytotoxicity, investigates the selective inhibition of influenza, and indicates that inhibition is virus specific in nature. The second group of methods establishes the site of action, typically a target protein, and includes genotypic and phenotypic analysis of variants selected under inhibitor pressure. Finally, methods for measuring virion associated activities and their inhibition are described.

**Key words** Antiviral, Cytotherapeutic index, Influenza virus, Mechanism of action, Multiplicity of infection, Neuraminidase inhibition, Polymerase, Selection of resistant variants, Time of addition

## 1 Introduction

Influenza is an infection of the human respiratory tract which causes significant morbidity and mortality. The impact of influenza virus infections is estimated to run into billions of dollars worldwide with an estimated 3.5 million cases of severe illness and 250,000–500,000 deaths annually [1]. Vaccination has a major role in the prevention of influenza and its associated complications, but the limited uptake of vaccination, viral antigenic drift and the emergence of pandemic strains mean that there remains an important role for antiviral therapy in the management of influenza [2].

Influenza viruses are from the family *Orthomyxoviridae* and are classified into three antigenic types: A, B and C. Influenza A and B are considered the principal causative agents of significant human disease. Influenza A viruses are classified into subtypes based on the antigenicity of their two major surface glycoproteins: the

Edwin Yunhao Gong (ed.), *Antiviral Methods and Protocols*, Methods in Molecular Biology, vol. 1030,
DOI 10.1007/978-1-62703-484-5_30, © Springer Science+Business Media, LLC 2013

hemagglutinin (HA) and the neuraminidase (NA). To date, 16 HA subtypes (H1–H16) and nine NA subtypes (N1–N9) have been described, but only a limited number of subtypes have become established in humans [3]. The 2009 pandemic was caused by the H1N1 swine virus, and seasonal influenza epidemics have in recent years been caused by influenza A virus subtypes H1N1 and H3N2 together with influenza B.

Unlike influenza A, which has a large animal (principally avian) reservoir and can be considered a zoonotic disease, influenza B viruses are usually found only in humans. Influenza B viruses can cause morbidity and mortality, but in general are associated with less severe epidemics than influenza A viruses. Although influenza type B viruses can cause human epidemics, they have not caused pandemics [4]. Influenza type C viruses cause mild illness in humans and are yet to be the causative agent of epidemics or pandemics [4].

Mechanism of action studies can be guided in part by what is known about the "druggable" molecular targets that have yielded effective antivirals. Presently, there are two classes of anti-influenza drugs licensed worldwide for therapeutic and prophylactic use: the M2 ion channel (adamantane) inhibitors and the neuraminidase inhibitors. The M2 ion channel inhibitors, amantadine and rimantadine, act upon the M2 ion channel of influenza A viruses [5, 6] and are therefore influenza A selective. The adamantane inhibitors rapidly select resistant virus variants that exhibit little, if any, replicative disadvantage, and in recent years adamantane resistance has become widespread in circulating influenza A viruses, including the 2009 H1N1 pandemic viruses. Consequently, they have not been recommended as a monotherapy for the treatment or prophylaxis of influenza since the 2005 influenza season [7–9].

The second class of influenza antiviral drugs, the neuraminidase inhibitors (NAIs), bind to the NA surface glycoprotein of newly formed virus particles and prevent their efficient release from the host cell, thus inhibiting viral growth through preventing the spread of infection to uninfected cells [10, 11]. Two NAIs are licensed for human use worldwide: the inhaled drug zanamivir (Relenza®) and the oral drug oseltamivir phosphate (Tamiflu®). Both drugs require twice daily dosing for treatment. Parenterally administered neuraminidase inhibitors, intravenous oseltamivir, zanamivir and peramivir (Rapiacta®) are also in clinical development, the latter of which is approved for use in Japan and Korea. Laninamivir (marketed as Inavir® in Japan) is an inhaled NAI with a long duration of action that is administered once (for example, one administration on a single day compared to ten administrations over 5 days for oseltamivir phosphate). Laninamivir is currently approved for use in Japan [12, 13] and is under clinical development in the rest of the world. NAIs are effective inhibitors of all known NA and have antiviral activity against influenza A and B, and human, avian and animal influenza viruses. Potency is generally in the low nanomolar range in neuraminidase inhibition assays [14, 15].

Initial studies in cell culture emphasized that resistance to neuraminidase inhibitors emerges much less readily than against M2 channel inhibitors, amantadine and rimantadine, and requires extensive passaging. Furthermore, mutations were initially selected in HA, which reduced receptor binding affinity and consequently the requirement for active NA, prior to selection of mutations in NA which directly reduced inhibition of enzyme activity [16]. Thus, two mechanisms of resistance have been recognized in in-vitro studies. The first involves a reduction in the HA-binding affinity for the virus receptor. The lower affinity results in a reduction in the virus's dependence on NA activity during virus release from infected cells. Substitutions in the HA have been identified in viruses selected in the presence of NA inhibitors, although sequence analysis of the HA gene itself is a poor predictor of the drug related phenotype of the virus. The second mechanism of resistance involves the substitution of residues within NA. Such substitutions can reduce the affinity of the inhibitor for the active site. The World Health Organisation reported in November 2012 that 13 amino acid substitutions had been identified in NA from clinical specimens that were associated with reduced susceptibility to NAI in vitro [17]. Of these only one (H274Y, N2 numbering, associated with oseltamivir and peramivir resistance) was considered to be clinically relevant at that time; the others being observed infrequently and their association with clinical resistance was reported as "less clear".

The emergence of viruses with reduced antiviral susceptibility in humans has been principally associated with a few characteristic NA mutations. The frequency of resistance emergence in early clinical trials was less than 1 % for zanamivir [18] and on treatment with oseltamivir, 1–2 % in adults and 5–6 % in children [19]. Analysis of H1N1 viruses isolated in and post 2007–2008 seasons has shown a striking increase in the frequency of circulating oseltamivir resistant viruses containing the H274Y mutation [19]. More recent studies with oseltamivir have indicated that resistance can be as high as 18 % in treated children [19]. In contrast, reports of zanamivir resistance are rare and no studies to date have identified widespread resistance [15]. Studies to date have shown that laninamivir has a zanamivir-like resistance profile.

Mutations associated with clinical resistance (in vivo) show virus subtype differences: for example, R292K and to a lesser degree E119V have been identified in oseltamivir resistance of A (H3N2) viruses, and in N1 viruses, the usual mutation conferring high resistance is H274Y. There has only been one reported instance of clinical resistance to zanamivir, an R152K mutation in influenza B, with the predominant mutations identified in cell culture associated with residue 119 (E119G/A/D) [20].

In addition to the clinically proven targets of NA and M2, there are many steps in a virus life cycle that can be targeted by potential antiviral inhibitors. Inhibitors can act directly to inhibit a virus by targeting specific viral-encoded functions. For example, influenza HA mediated entry or fusion, or acting indirectly, e.g., ribavirin-an IMPDH inhibitor (amongst other mechanisms that have been described). The mechanism of action of a putative antiviral may require confirmation or elucidation depending on the source and nature of the inhibitor. A well characterized mechanism of action is useful at several levels: knowledge of the molecular target facilitates lead optimization efforts, assists in the design of off-target/selectivity screening and may inform safety assessment studies together with the design of studies to assess the development of resistance. Several methods can be used to elucidate the mechanism of action of an anti-influenza inhibitor. These include assays suitable for establishing that the antiviral effect is specific or, alternatively, mediated by perturbation of host cell processes. These assays include methods for establishing the cytotherapeutic index, assessing selectivity for anti-influenza activity and multiplicity of infection effects. Further experiments to isolate the molecular target of a putative antiviral include determining the point of interference with the viral replication cycle via time of addition studies and the identification of the gene product target through selection of resistant variants. In addition, direct effects on virus specific activities can be determined via, for example, neuraminidase inhibition or polymerase assays.

The cell-killing property of an inhibitor, independent of the mechanism of cell death, is referred to as cytotoxicity. Establishing that the activity of an inhibitor occurs at concentrations that do not cause cytotoxicity is a critical component of establishing that the measured antiviral effect is virus specific and forms the cornerstone in the selection of a potential drug candidate for further development.

The cytotoxicity of an inhibitor may be evaluated in cell culture by measuring the effect on cell viability using assays—such as the MTT assay and the ATP release assay—that exploit cellular processes associated with living cells. The MTT assay is based on the reduction of 3-(4,5-dimethylthiazol-2-yl)-2,5-diphenyltetrazolium bromide (MTT) by a mitochondrial dehydrogenases of metabolically active (viable) cells to a purple/blue formazan product. The assay is a simple, relatively cost-effective method providing a quantitative result that can be measured spectrophotometrically [21]. The ATP release assay measures the amount of adenosine triphosphate (ATP) present in cell cultures. As ATP can only be synthesized by metabolically active (viable) cells, the amount of ATP present in cell cultures is directly proportional to the number of viable cells [22]. There are, of course, many alternatives and additional methods that may be employed to measure viability. The cytotoxic effect of an inhibitor can be determined by calculating the median 50 % cellular cytotoxic concentration

($CC_{50}$), which is defined as the concentration of test inhibitor resulting in a 50 % reduction in cell viability.

It is recommended that cytotoxicity evaluations are performed in stationary and dividing cell lines of different origins to ascertain the potential for cell-cycle, species, or tissue-specific toxicities [23]. At least two different methods for measuring cell viability should be performed in order to increase the likelihood of identifying genuine cytotoxic effects. Both the MTT and ATP release assays will be described in this chapter.

It remains a possibility that anti-influenza inhibitors may also possess antiviral activity against other viruses. The specificity of the antiviral activity can be evaluated, for example, using a cytopathic effect (CPE) assay for activity against a variety of viruses, including positive single-stranded RNA viruses, e.g., human rhinovirus, negative single-stranded RNA viruses, e.g., respiratory syncytial virus and double-stranded DNA viruses, e.g., herpes simplex virus. Although it is recommended that antiviral activity against viruses other than influenza be assessed, specific methodology describing non-influenza virus assays is not described in this chapter. Unless one is specifically targeting mechanisms of action that are expected to yield broad spectrum inhibitors (e.g., targeting a host function or inhibitor classes with potential for broad spectrum activity such as nucleoside/nucleotide analogues), putative direct acting antiviral compounds which exhibit activity against many different viruses should be reviewed cautiously since their effects may be nonselective and potentially mediated via a low level cytotoxicity.

A time of addition assay can be used to identify when in the virus life cycle an inhibitor interferes, and can therefore improve understanding of the mechanism of action of an inhibitor. A reduction in titer when added shortly after infection indicates that an inhibitor is acting early in the replication cycle and may be interfering with mechanisms such as virus attachment, fusion or entry into the host cell. Amantadine is an example of an influenza A inhibitor which acts early in the replication cycle and interferes with the uncoating of the virus inside the cell [24]. Conversely, an inhibitor that acts later in the replication cycle may be affecting processes such as viral release. For example, it has been shown that neuraminidase inhibitors such as zanamivir act late in the infection cycle (Fig. 1) consistent with inhibition of neuraminidase mediated release [25–27]. Inhibitors that act in the middle of the cycle may affect one or more of the many processes involved in viral replication. For example, the polymerase inhibitor T-705 inhibits virus replication when added to viral cultures within 6 h of viral infection [28].

Control inhibitors with a known mechanism of action and defined effects on the virus replication cycle should be included in these studies.

An often defining feature of a direct acting antiviral compound is its multiplicity of infection (MOI) dependency, meaning that the

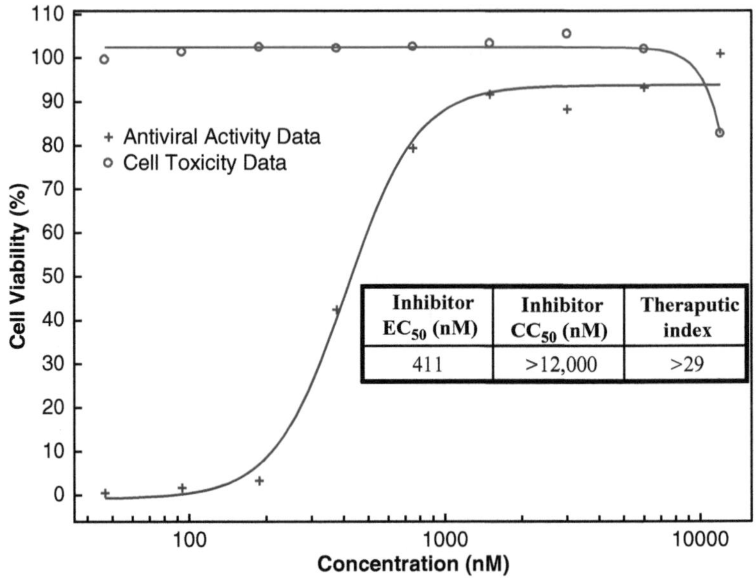


**Fig. 1** Example of $EC_{50}$ and $CC_{50}$

antiviral potency of certain inhibitors decreases as the ratio of input virus to cells increases [29]. The MOI dependency of an inhibitor can be tested by adding a range of concentrations of virus to a constant number of cells. Evaluation of the antiviral activity of the inhibitor is then usually determined by quantification of cytopathic effects. Loss of potency is reflected by an increase in $EC_{50}$ value at higher MOI. MOI-dependence does not negate the clinical potential of an antiviral inhibitor (for example, the clinically effective neuraminidase inhibitors oseltamivir and zanamivir are MOI-dependent), but may be useful for demonstrating that antiviral activity is mediated by a virus specific mechanism.

The emergence of antiviral drug resistance has been documented for virtually all antiviral inhibitors with virus-specific activity [30]. Drug resistance may be caused by single or multiple mutations in the viral genome and can emerge either through spontaneous mutation and outgrowth in the absence of drug exposure or by the natural selection of resistant variants, which are afforded a replicative advantage over the susceptible wild-type virus, once the susceptible virus has been eliminated or its replication significantly reduced by the inhibitor. In addition to providing compelling mechanism of action related data, in vitro resistance selection studies comprise an essential aspect of the antiviral drug development process [23]. Although resistance data generated in vitro are not necessarily predictive of clinical resistance, these studies can identify mutations in the viral genome that may potentially confer resistance to an investigational inhibitor.

Viruses that have a reduced susceptibility to an antiviral drug can be isolated by passage of a low initial virus inoculum in the presence of an increasing concentration of drug, with the initial concentration ideally at or near to the 50 % effective concentration ($EC_{50}$): the low multiplicity high passage method. Alternatively, a high virus inoculum may be exposed to high concentrations of the inhibitor with the goal of eradicating the wild-type virus and thereby allowing the resistant variants to grow: the high multiplicity low passage method. Various permutations and combinations of these methods have also been described. The virus variants generated in these experiments are usually characterized by genotypic analysis, which provides an understanding of the genetic basis of drug resistance, and by phenotypic analysis, which provides a quantitative assessment of drug resistance and the characteristics of the resistant variant.

Finally, the chemical structure of the putative antiviral compound may provide insights into the prospective mechanism of action which may lead one to consider evaluating known "druggable" targets at an early stage in mechanism of action studies. For example, it would be reasonable to suspect that a nucleoside/nucleotide analogue may be acting via the viral polymerase.

In this chapter, methods to elucidate the mechanism of action of an anti-influenza inhibitor will be described. In particular, methodology for establishing a cytotherapeutic index in cell lines from various origins, multiplicity of infection effects and interference with the viral replication cycle via time of addition studies will be discussed. Resistance selection based approaches to determine the mechanism of action will be delineated using methods for the selection of resistance variants to NAIs as an example. The assay of known druggable targets will be introduced through methods for the assay of neuraminidase inhibition (in this case it also serves the purpose of introducing a phenotypic characterization assay for resistant variants) and the influenza virus RNA-dependent RNA polymerase. The methods described seek to strike a balance between laboratory prepared reagents and the use of commercially available kits and reagents. The naming of specific commercial kits/reagents is for convenience only and a result of personal experience in their use. It should in no manner be interpreted as any form of endorsement and it is appreciated that there may be alternatives. Equally, the methods described are intended to be useful for a wide range of laboratories and seek to avoid wherever possible the use of highly specialized equipment, consumables or practices. It will be appreciated that many of the assays can be formatted differently, miniaturized, conducted using alternative technologies, or optimized for higher throughput, etc.

It is noted that these methods contemplate the handling of infectious agents, human and animal derived cell lines, and chemical substances together with the use of various items of equipment. Appropriate containment, training and safety precautions should be taken.

## 2 Materials

### 2.1 Cells, Virus, Media, Reagents, and Test Article

#### 2.1.1 Cells

1. Madin-Darby Canine Kidney (MDCK) cells: European Collection of Cell Cultures (ECACC: cat. no. 85011435).
2. HepG2 cells (ATCC: cat. no. HB-8065).
3. HEK293 (kidney) cells (ATCC: cat. no. CRL-1573).

#### 2.1.2 Virus

1. Stocks of previously titered influenza virus, e.g., A/Brisbane/59/07 (ECACC: cat. no. 0911081v), A/California/4/09 (ECACCC: cat. no. 0905192v), A/Mississippi/3/01 and A/Fukui/20/04 (isirv-AVG: www.isirv.org/avg) preferably titered by plaque assay.

#### 2.1.3 Media

1. MDCK and HepG2 cell growth media: minimum essential medium (MEM) supplemented with 10 % heat inactivated fetal bovine serum, 1 % L-glutamine.
2. MDCK cytotoxicity assay media and influenza virus growth media: MEM supplemented with 1 % L-glutamine, 1.2 μg/mL TPCK-trypsin, 1× insulin transferrin selenium (ITS), 0.2 % bovine serum albumin.
3. HepG2 cytotoxicity assay media: MEM supplemented with 5 % heat inactivated fetal bovine serum, 1 % L-glutamine.
4. HEK293 cell growth media: Dulbecco's modified Eagle's medium (DMEM) supplemented with 10 % fetal bovine serum.
5. HEK293 cytotoxicity assay media: DMEM supplemented with 5 % fetal bovine serum.

#### 2.1.4 Reagents

1. 3-(4,5 dimethylthiazol-2-yl)-2-5-diphenyltetrazolium bromide (MTT).
2. 2-Propanol.
3. DPBS.
4. CellTiter-Glo Luminescent Cell Viability Assay Kit (Promega).
5. NTE buffer: 100 mM NaCl, 10 mM Tris–HCl, pH 7.5, 1 mM EDTA.
6. Polyethylene glycol 6000 (if high capacity ultracentrifuge rotor is unavailable).
7. 15 and 60 % sucrose (w/v) in NTE.
8. 20 and 45 % potassium tartrate (w/v) in NTE.
9. Bradford reagent.

#### 2.1.5 Test and Control Articles

1. 10 mM stock of a putative antiviral (*see* **Note 1**).
2. 10 mM stock neuraminidase inhibitor, e.g., zanamivir (GlaxoSmithKline).
3. 10 mM stock of cytotoxicity control, e.g., ribavirin (Sigma-Aldrich).

### 2.2 Genotypic Analysis of Selected Variants

1. Qiagen QIAamp® Viral RNA Extraction Kit.
2. Specific influenza HA and NA primers designed according to the published DNA sequence of the wild-type strain.
3. Superscript III One-Step RT-PCR System with Platinum Taq DNA Polymerase kit (Invitrogen).
4. 1 % Agarose gel in 1× TAE buffer.
5. SYBR safe DNA gel stain.
6. TAE buffer (Tris pH 7.4, acetic acid, EDTA).
7. 6× Gel-loading DNA dye.
8. 1 kb DNA molecular marker.
9. QIAquick® PCR Purification Kit.
10. Big Dye Terminator Mix version 3.1.
11. Ethanol.
12. 70 % Ethanol in DEPC water.
13. 3 M sodium acetate (pH 5.2).

### 2.3 Phenotypic Analysis of Selected Variants (Neuraminidase Assay)

1. Stocks of previously titered wild-type virus and drug exposed variants, preferably titered by plaque assay.
2. NA-Star® Influenza Neuraminidase Inhibitor Resistance Detection kit (Applied Biosystems: cat. no. 4374422).

### 2.4 Transcriptase Assay

1. Transcriptase buffer: 150 mM NaCl, 50 mM Tris–HCl, pH 7, 2 mM ATP, 0.2 mM GTP, 0.4 mM UTP, and 0.4 mM CTP.
2. ApG dinucleotide (TriLink BioTechnologies).
3. NP-40.
4. $^{3}$H-UTP (Moravek Biochemicals and Radiochemicals).
5. $MgCl_2$.
6. Filter paper (~2.5 cm diameter squares or circles).
7. 10 and 5 % trichloroacetic acid (TCA) containing 20 mM sodium pyrophosphate.

### 2.5 Consumables and Equipment

1. Flat bottomed clear and white 96-well plates.
2. Flat bottomed 24-well plates.
3. 25 cm$^2$ Cell culture flasks.
4. Eppendorf tubes.

5. PCR tubes.
6. Gradient maker.
7. Centrifuge and ultracentrifuge (ideally with rotor capable of handling 1–2 L of culture fluid, plus swinging bucket rotor capable of 110,000 × *g*) plus suitable centrifuge tubes.
8. 1–2 mL pipettes or 1–2 mL syringes with large and fine gauge hypodermic needles.
9. Scintillation fluid, vials, and scintillation counter.
10. Class II biological safety cabinet.
11. Plate reader which can read luminescence.

## 3 Methods

Carry out all procedures at room temperature unless otherwise specified.

### 3.1 Cytotoxicity Assay

Growth curves of HepG2, HEK293 and MDCK cells (other cell lines may be substituted as desired) should be performed prior to the cytotoxicity assessment to ensure that cells are in the logarithmic phase of the cell cycle for the entirety of the actively dividing study, and in the stationary phase of the cell cycle for the stationary cytotoxicity assessment. This can be performed by plotting cell density over time for a 5–7 day period.

#### 3.1.1 Stationary Cytotoxicity Assay

1. Seed 100 μL of cells in 96-well plates at a concentration of 5,000 cells/well (HepG2) or 6,000 cells/well (HEK293) in cytotoxicity assay media and incubate at 37 °C in a humidified 5 % $CO_2$ atmosphere until cells are in stationary phase (incubation length predetermined in cell growth curve experiments). Cells are seeded in to clear plates for MTT viability assessment and white plates for viability assessment via ATP release.
2. Prepare a dilution series of the putative inhibitor and control compound known to cause cytotoxicity with a published literature $CC_{50}$ value (e.g., Ribavirin [31]).
3. A nine point, twofold dilution series can be prepared by adding 150 μL of cytotoxicity assay media to wells in columns 3–11, rows B-G in a 96-well plate. Add 300 μL of compound to wells B-G in column 2 and serially dilute compound 1 in 2 across the assay plate to row 10. The cytotoxicity of two compounds in three replicates (rows B, C, D and E, F, G respectively) can be assessed on each 96-well plate.
4. Take 100 μL from columns 2–11, rows B–G and add to the corresponding wells of the 96-well tissue culture plate. Column 11 serves as the cell control. Add 100 μL of media to columns 1 and 12 which serves as background when measuring

absorbance at 540/690 nm. Incubate plates at 37 °C in a humidified 5 % $CO_2$ atmosphere for 48 h before determining cell viability using either MTT or ATP release as end points.

*3.1.2 Dividing Cytotoxicity Assay*

If desired, antiviral effects can be assayed in parallel by infecting MDCK cells with minimally sufficient influenza virus to ensure destruction of the cell monolayer within 3 days (the appropriate MOI for the available virus should be determined but is generally in the range 0.01–0.001). MDCK cells are therefore included along with HepG2 and HEK293 cells, as in the method below.

1. In a 96-well plate, prepare a threefold dilution series (eight concentrations) of inhibitor in MDCK, HepG2 or HEK293 cytotoxicity assay media (*see* **Note 2**). This can be done by adding 100 μL of media to columns 2–11, rows B–G. Add 50 μL of inhibitor to wells (replicates, e.g., triplicates are recommended) in column 2 and serially dilute 50 μL 1:3 from column 2–10. Column 11 serves as the untreated cell control. Add 200 μL of media to the outside wells of the plate to eliminate plate edge effects.
2. Add 100 μL of cells (MDCK: $5 \times 10^5$ cells/mL, HepG2: $5 \times 10^4$ cells/mL, HEK293: $6 \times 10^4$ cells/mL) to columns 2–11, rows B–G and incubate for 3 days at 37 °C in a humidified 5 % $CO_2$ atmosphere.
3. Measure cell viability using the MTT assay.

*3.1.3 Determining Cell Viability and Calculating a $CC_{50}$*

1. MTT: prepare a 3 mg/mL solution of MTT in MEM and add 100 μL to all wells in the 96-well plate. Incubate for 2 h at 37 °C in a humidified 5 % $CO_2$ atmosphere. Aspirate the unreacted MTT and solubilize the formazan crystals by adding 200 μL of 2-propanol and incubation for 30 min at room temperature. Read the absorbance at 540 and 690 nm using a UV spectrophotometer (*see* **Note 3**).
2. ATP release: equilibrate the 96-well plate to room temperature and add 100 μL of CellTiter-Glo reagent to each well (*see* **Note 4**). Shake the plates for 2 min and incubate at room temperature for 10 min before reading the luminescence.
3. Determine the percentage of viable cells present at each compound concentration in comparison to the cell control. Generate a dose response curve by plotting percent viability versus compound concentration. The 50 % cellular cytotoxic concentration which leads to a 50 % reduction in cell viability ($CC_{50}$) can then be determined. If an $EC_{50}$ is also determined (please refer to MOI assay), the $CC_{50}$ and $EC_{50}$ values can be plotted and the therapeutic index calculated by dividing the $CC_{50}$ by the $EC_{50}$ (Fig. 1). The higher the therapeutic index, the more likely the compound is exerting a specific antiviral effect rather than inhibiting viral cytopathology through a cytotoxic effect on the host cell.

### 3.2 Time of Addition Assay

1. Seed MDCK cells into 24-well plates at a density of $1.5 \times 10^5$ cells per well in a final volume of 500 μL of cell growth media and incubate overnight at 37 °C in a humidified 5 % $CO_2$ atmosphere.
2. Remove media and infect cells with 100 μL of influenza virus at $4.5 \times 10^5$ PFU/mL (estimated MOI = 0.3) in virus growth media allowing the virus to attach for 1 h at 4 °C before removing and washing the plates 2× in DPBS.
3. Add 250 μL of virus growth media to wells and incubate the cells for 0–24 h at 37 °C adding inhibitor at 0, 2, 4, 6, 8, 12, 14, 16, 18, 20, and 22 h post virus attachment (*see* **Note 5**). Inhibitor is added immediately post media addition for the 0 h time point. Include a well for the virus control at each time point.
4. Harvest virus from wells by freezing–thawing and quantify the titer of virus at each time point via $TCID_{50}$ assay.

### 3.3 $TCID_{50}$ Assay

1. Seed MDCK cells in 96-well plates at a concentration of $2 \times 10^4$ cells/well in 200 μL of virus growth media and incubate overnight at 37 °C in a humidified 5 % $CO_2$ atmosphere.
2. Dilute the virus harvested from the time of addition study 1:10 or 1:100 in virus growth media (*see* **Note 6**) and add 100 μL of virus to wells in column 1 of the $TCID_{50}$ plate. Serially dilute virus threefold across the plate resulting in a total of 11 different virus concentrations (*see* **Note 7**). Wells in column 12 are virus free and are used as cell control wells.
3. Incubate plates for 3 days at 37 °C in a humidified 5 % $CO_2$ atmosphere before scoring the wells microscopically for virus induced CPE. Return the cells to the incubator and score again over the next ~1–3 days to ensure that the assay has reached its end-point.
4. Record wells showing CPE as "+" and wells lacking CPE as "–".
5. The $TCID_{50}$ of the virus suspension can then be calculated using the method of Reed and Muench [32].
6. Plot a graph of virus titer versus time to establish at which time point(s) in the virus life cycle the inhibitor reduces virus titer (Fig. 2) (*see* **Note 8**).

### 3.4 MOI Assay

1. In a 96-well plate, prepare a threefold dilution series (eight concentrations) of inhibitor in virus growth media. This can be done by adding 100 μL of virus growth media to columns 2–11, rows B–G. Add 50 μL of inhibitor to wells in column 3 and serially dilute 50 μL 1:3 across plate to column 10. Add 200 μL of virus growth media to the outside wells of the plate

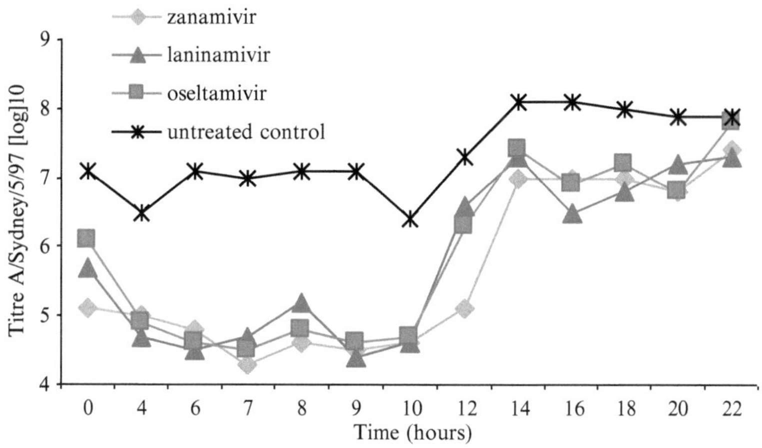


**Fig. 2** Expected reductions in virus titer by neuraminidase inhibitors using time of addition assay

to eliminate plate edge effects. The plate layout allows for three replicates of the inhibitor (rows B–D and E–G) and two different MOIs/plates. Column 2 serves as the cell control and column 11 the virus control.

2. Suspend MDCK cells at a concentration of $5 \times 10^5$ cells/mL and add sufficient influenza virus to the cell mixture to generate well mixed suspensions containing cells and virus at MOIs of 1, 0.1, 0.01, 0.001 and 0.0001. For example, add $5 \times 10^5$ plaque forming units per mL of cells to obtain an MOI of 1. Immediately add 100 μL of cell/virus mixture to columns 3–11, rows B–G, and 100 μL of virus growth media containing $5 \times 10^5$ cells/mL of uninfected MDCK cells to column 2. Incubate for 3 days at 37 °C in a humidified 5 % $CO_2$ atmosphere.
3. Prepare a 3 mg/mL solution of MTT in MEM and add 100 μL to columns 2–11, rows B–G of the 96-well plate. Incubate for 2 h at 37 °C in a humidified 5 % $CO_2$ atmosphere. Aspirate the unreacted MTT and solubilize the formazan crystals by addition of 200 μL of 2-propanol and incubation for 30 min at room temperature. Read the absorbance at 540 and 690 nm.
4. The percent cell protection achieved by the control and putative inhibitor in influenza infected cells is calculated by the method of Pauwels et al. [33]. Calculate $EC_{50}$ values using the percent cell protection results by nonlinear regression analysis using the Hill (sigmoid $E_{max}$) equation.
5. In general, an increase in $EC_{50}$ value with increasing MOI is characteristic of a direct acting antiviral (Fig. 3).

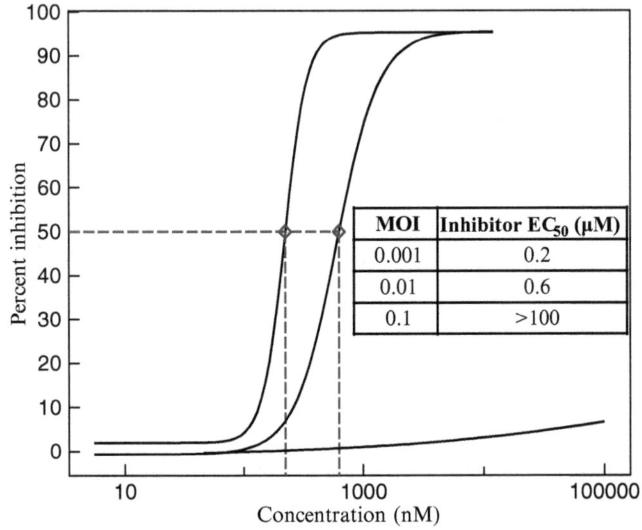


**Fig. 3** Example of MOI effect on inhibitor $EC_{50}$

### 3.5 Selection of Drug Resistant Variants

#### 3.5.1 Neuraminidase Variant Selection: Low Multiplicity High Passage Method

1. Seed MDCK cells in cell growth media at a concentration of $5 \times 10^5$ cells per 25 $cm^2$ flask. The cell monolayer should be approximately 70 % confluent after an overnight incubation at 37 °C in a humidified 5 % $CO_2$ atmosphere.
2. Remove the cell growth media and wash the flasks twice using warmed DPBS (*see* **Note 9**). Inoculate cells with 1 mL of wild-type influenza virus diluted in virus growth media (*see* **Note 10**) to an estimated MOI of 0.05 plaque forming units/cell (*see* **Note 11**).
3. After adsorption for 15 min at 37 °C in a humidified 5 % $CO_2$ atmosphere, remove the virus solution and wash the flasks twice in warmed DPBS [34] (*see* **Note 12**), then add 5–10 mL of virus growth media containing influenza neuraminidase inhibitor at the inhibitor's $EC_{50}$ (*see* **Note 13**).
4. Incubate flasks at 37 °C in a humidified 5 % $CO_2$ atmosphere until 80 % or greater CPE is observed (*see* **Note 14**). Harvest virus via one freeze-thawing cycle at −80 °C and centrifuge for 10 min at 4 °C at $4{,}500 \times g$ to pellet cell debris.
5. For each subsequent passage repeat **steps 1–4**, increasing the concentration of neuraminidase inhibitor twofold with each subsequent passage (Fig. 4) (*see* **Note 15**).
6. Assess virus from each passage for mutations in both the NA and HA regions by DNA sequencing of products amplified by RT-PCR.

#### 3.5.2 Variation of Method Favoring Hemagglutinin Mutation Selection

1. Follow **steps 1** and **2** under Subheading 3.5.1.
2. Do not remove the virus after the 15 min adsorption step. Add media containing inhibitor to the flasks and follow **steps 4–6** as described in Subheading 3.5.1.

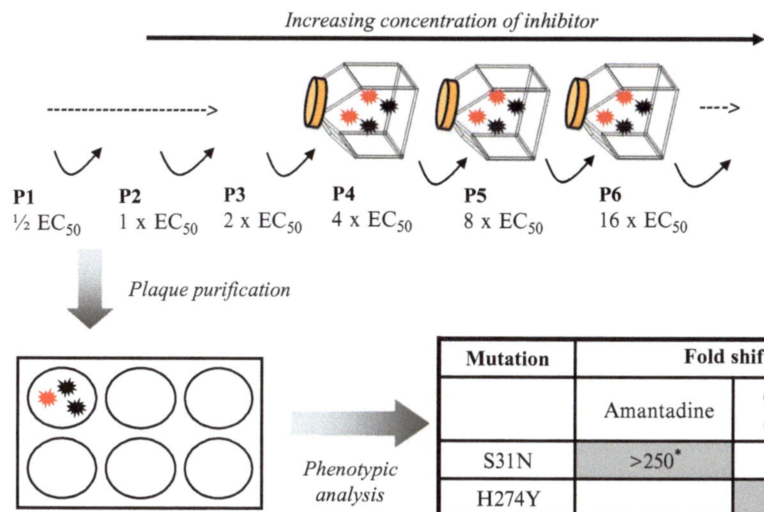


| Mutation | Fold shift in $EC_{50}$ (μM) | | |
|---|---|---|---|
| | Amantadine | Oseltamivir carboxylate | Zanamivir |
| S31N | >250* | | |
| H274Y | | 10,000# | |
| E119D | | | 7,000# |

*(38)
# Biota in-house data

**Fig. 4** Flow diagram of resistance variant selection

### *3.6 Genotypic Analysis of Selected Variants*

1. Extract nucleic acid from each passage using the QIAamp® Viral RNA Extraction Kit according to the manufacturer's instructions. The amount of thawed aliquoted virus used for the extraction can be increased from 160 to 280 μL to improve the yield of RNA.
2. Amplify RNA using HA and NA primers designed using the published sequences of the influenza isolates of interest. For example, use the Superscript III One-Step RT-PCR System with Platinum Taq DNA Polymerase kit. Add 5 μL of extracted viral RNA to each individual reaction mixture containing 25 uL of 2× reaction buffer, 1 μL of 10 μM forward primer, 1 μL of 10 μM reverse primer, 2 μL Superscript III RT Platinum Taq mix, and 16 μL of autoclaved DEPC water. Final volume is 50 μL. Amplification conditions are as follows: reactions are reverse transcribed for 60 min at 45 °C, followed by 2 min incubation at 95 °C. Reactions are then cycled five times for 15 s at 94 °C, 30 s at 45 °C, and 2 min at 68 °C before a further 30 cycles of 15 s at 94 °C, 30 s at 50 °C, and 2 min at 68 °C. Finally, the reactions should be incubated for 5 min at 68 °C (*see* **Note 16**).
3. Run 5 μL of the PCR products on a 1 % agarose gel (1 % agarose in 40 mL of TAE buffer containing 4 μL of SYBR safe DNA gel stain) and verify the size of the PCR product by comparison with a 1,000 bp size marker.
4. Purify PCR products using the QIAquick® PCR Purification Kit according to manufacturer's instructions into a final volume

of 30 μL (*see* **Note 17**). Estimate the amount of purified PCR product by electrophoresis.

5. Sequence the HA and NA regions using a system such as Big Dye Terminator Mix V3.1, with reactions comprising 3.6 pM of relevant sequencing primers and approximately 30–40 μg of PCR product in a final volume of 20 μL (*see* **Note 18**). Sequencing reaction conditions are as follows: denaturation at 96 °C for 2 min, 25 cycles of 10 s at 96 °C followed by 5 s at 50 °C and 4 min at 60 °C.
6. Remove excess dye using ethanol precipitation. Mix 20 μL of the sequencing reaction with 3 μL of 3 M sodium acetate (pH 5.2), 62 μL of 100 % ethanol, and 15 μL of water. Centrifuge samples for 20 min at 12,000 × *g* before washing 1× with 200 μL of ice cold 70 % ethanol. Centrifuge samples for a further 5 min, remove the ethanol and dry in a heat block at 70 °C. Analyze samples using electrophoretic separation and sequence determination using, for example, an Applied Biosystems 3730 Genetic Analyzer.
7. Analyze HA and NA chromatographs at each passage in comparison to the reference sequence (wild-type virus) and note any amino acid changes (*see* **Note 19**).
8. Variants containing mutations should be plaque purified and scaled up in the presence of inhibitor for further characterization. Working stocks of variants should be sequenced to reconfirm the presence of the NA or HA mutation.

### 3.7 Phenotypic Analysis of Selected Variants

Three commercially available kits can be used to determine if influenza variants have reduced susceptibility to neuraminidase inhibitors. There are currently two chemiluminescent assays kits: NA-Star® Neuraminidase and NA-XTD® Influenza Neuraminidase Assay Kit, and the fluorescent MUNANA-based neuraminidase assay kit. The next generation NA-XTD® Influenza Neuraminidase Assay Kit provides a longer-lasting chemiluminescent signal and slightly higher detection sensitivity. The MUNANA-based neuraminidase assay kit NA-Fluor™ Influenza Neuraminidase Assay Kit may offer a cost advantage over the NA-XTD® Influenza Neuraminidase Assay Kit [35, 36]. Although it is of course possible to prepare one's own reagents, it is recommended that a commercial kit be used together with appropriate controls to obviate the need for stringent quality control/quality assurance measures to ensure reagent consistency. Known influenza reference strains and neuraminidase inhibitors should be used as standard assay controls. The World Health Organization Global Influenza Surveillance Network Manual for the Laboratory Diagnosis and Virological Surveillance of Influenza is a useful source of information for assay controls and includes detailed methodology for a MUNANA-based assay which does not require a commercial kit [36].

Variant susceptibility to neuraminidase inhibitors is described below using the NA-Star® Influenza Neuraminidase Inhibitor Resistance Detection kit, following manufacturer's instructions.

1. Prepare a ten point, threefold dilution series of the inhibitor in NA-Star Assay Buffer in a dilution plate such that dilutions are 2× the final desired concentration. Prepare two dilution series with starting concentrations of 20 μM (10 μM in assay plates) and 200 nM (100 nM in assay plates). The dilution series at the higher concentration is used for assessing drug-exposed virus, assuming a phenotypic shift in the $IC_{50}$ value and the lower dilution series for generating an $IC_{50}$ using wild-type virus. Add 200 μL of NA-Star Assay buffer to wells in columns 2–12 in a 96-well dilution plate (utilize as many rows as necessary to generate the volume of diluted inhibitor required for the assay). Columns 11 and 12 will contain NA-Star Assay buffer alone and will be used for virus control and background wells. Add 300 μL of neuraminidase inhibitor at 20 μM or 200 nM to column 1 and dilute 1 in 3 by transferring 100 μL across the plate to column 10.
2. Transfer 25 μL of inhibitor dilutions and NA-Star buffer control wells from the dilution plate into corresponding columns 1–12 of the assay plate. Add 25 μL of virus diluted 1 in 10 in NA-Star buffer (*see* **Note 20**) to each well in columns 1–11, column 11 serving as the neuraminidase activity control in the absence of inhibitor and 25 μL of uninfected culture supernatant diluted in NA-Star Assay Buffer to column 12. Column 12 provides a background readout. Place the lid on the plate and shake the plate for 10 s on a plate shaker to ensure that the well contents are mixed, and incubate for 20 min at 37 °C. Using a multichannel pipette, rapidly add 10 μL of diluted NA-Star Substrate (1:500 in NA-Star Assay Buffer) to each well from column 1 to 12. Shake the plate on a plate shaker to ensure that contents are mixed and incubate at room temperature for 30 min. Add 60 μL of NA-Star Accelerator using a multichannel pipette in the same order of columns 1–12 and measure the signal for 1 s/well using a luminometer as soon as possible (if a suitable instrument is available, preferably add the NA-Star Accelerator via the luminometer's injectors and measure the signal for 1 s/well 2 s after Accelerator addition) (*see* **Note 20**).
3. Calculate the average signal intensity for each inhibitor dilution as a percentage of the signal intensity of the positive control. Use these percentages to determine the concentration of inhibitor that produces a 50 % reduction in signal ($IC_{50}$) via nonlinear regression and sigmoidal dose response curve fitting. Calculate the fold shift between wild type and variants (Fig. 4).

### 3.8 Virion Associated Enzyme Activities

Virus encoded enzymes are arguably the most compelling targets for direct acting antivirals and should be considered for evaluation as potential targets in mechanism of action studies. Virion associated activities encoded by the viral genome include an RNA-dependant RNA polymerase (or transcriptase) complex and a neuraminidase. A neuraminidase assay method is described above (*see* Subheading 3.7). The following describes an assay method for evaluation of potential inhibitors of the transcriptase complex. Purification of influenza virus is required.

#### 3.8.1 Virus Purification

1. Prepare at least 1–2 L of an influenza virus stock (grown to high titer in either embryonated eggs or tissue culture).
2. Clarify the virus stock by low speed centrifugation (1,500 × *g* for 10 min at 4 °C). All subsequent steps should be performed on ice or at 4 °C.
3. Concentrate the virus to yield 2–4 mL from 1 to 2 L of infected culture fluid:
   (a) Preferably the virus should be concentrated by centrifugation (50,000 × *g* for 90 min at 4 °C).
   (b) If a suitable large volume rotor and centrifuge are unavailable, the virus may be concentrated by polyethylene glycol precipitation:
      - Add PEG 6000 to yield 8 % (w/v) and incubate for 1 h with stirring or periodic gentle mixing.
      - Pellet the precipitated virus (10,000 × *g* for 20 min at 4 °C).
4. Discard the supernatant and resuspend the pellet in as small a volume of NTE as possible to yield an (generally) off white homogenous suspension. This is best achieved by trituration using a 1–2 mL pipette or syringe and large gauge needle, followed by passing the resuspended virus through a fine gauge needle (e.g., 25 G). Short (~1 min) periods of sonication in a water bath sonicator may also be useful to disperse aggregates (*see* **Note 21**).
5. Prepare linear 15–60 % sucrose gradients in NTE using either ~10 or ~30 mL capacity transparent ultracentrifuge tubes allowing approximately 10 mm space at the top of the tube (*see* **Note 22**).
6. Carefully layer the virus suspension on top of the sucrose gradient (*see* **Note 23**) to fill the tube.
7. Centrifuge at 110,000 × *g* for 1 h at 4 °C using a swinging bucket rotor.
8. While **step 7** is underway prepare linear 20–45 % potassium tartrate gradients in NTE using ~10 mL capacity transparent ultracentrifuge tubes as described in **step 5** (*see* **Note 24**).

9. Carefully remove the tube and visualize the virus band using a dark background and a bright light source shone from above. The milky band should appear approximately half-way down the tube. Mark the upper and lower edges of the band.
10. Collect the band using a 1 mL syringe and a bent needle (*see* **Note 25**).
11. Dilute the band 1:1 with NTE.
12. Carefully layer the virus suspension on top of the potassium tartrate gradient.
13. Centrifuge at 100,000 ×*g* overnight at 4 °C using a swinging bucket rotor.
14. Collect the band as described in **steps 9** and **10**.
15. Dilute the virus suspension at least twofold in NTE and pellet at 100,000 ×*g* for 30 min at 4 °C.
16. Resuspend in 0.5 mL NTE.
17. Measure (e.g., using a Bradford assay) and adjust the protein concentration to 5 mg/mL.
18. Store in 50 μL aliquots at −70 °C.

*3.8.2 RNA-Dependant RNA Polymerase (Transcriptase) Assay*

Although the transcriptase complex may be further purified if desired, transcriptase activity can be readily assayed in vitro using detergent disrupted, purified virus together with ribonucleotide triphosphates, divalent cations and a suitable detectable label [37]. Potential inhibitors may be assessed through addition to the following assay.

1. Incubate 0.5 mg/mL of purified virus in transcriptase buffer containing 0.5 % NP-40, 0.3 mM ApG, and 0.1 μCi/μL of $^{3}$H-UTP for 10 min at 30 °C.
2. Start the reaction by the addition of $MgCl_2$ to a final concentration of 8 mM (*see* **Note 26**).
3. Remove 10 μL aliquots at specific times (e.g., 5 min intervals), spotted on to filter paper and immediately dropped into ice cold 10 % TCA.
4. Wash the filter papers for 10 min using an orbital shaker as follows: two washes in ice cold 10 % TCA, two washes in ice-cold 5 % TCA and one wash in ethanol.
5. Allow the filters to air-dry (overnight is often convenient) and measure the bound radiolabeled RNA via a scintillation counter.

## 4 Notes

1. If the compound is prepared in an organic solvent (e.g., DMSO), the stock solution concentration should be consistent with a final concentration of the organic solvent that is

tolerable under the specified assay conditions (e.g., most cell types can tolerate ~0.2 % DMSO under the conditions described here).

2. The method described utilizes identical media to virus growth media for MDCK cells to allow direct correlation with an antiviral assay of similar format. If desired, MDCK cytotoxicity can be evaluated using MEM supplemented with 2 % fetal bovine serum, 1 % L-glutamine. The method conveniently allows the assay to be set up on a single day by seeding cells into the inhibitor dilution series and conserves plasticware by omitting a compound dilution plate. If the inhibitor interferes with cell attachment to the tissue culture plate, a false-positive result may be obtained; in which case cells may be seeded overnight and inhibitor added the next morning. Cell numbers may require adjustment to ensure a comparable number of divisions.
3. A 2-h time frame has been found to yield an acceptable correlation between cell density and MTT metabolism under the conditions described herein. If conditions are changed or a different cell line used, it is recommended that one confirm 2 h is an acceptable time frame for the cell line under consideration. The absorbance reading at 690 nm is "background" and therefore subtracted from the reading at 540 nm to eliminate the effects of nonspecific absorption.
4. Depending on the brand of plate reader it is possible to dilute the reagent tenfold without compromising signal and reproducibility.
5. The 22-h assay length represents approximately two cycles of virus replication.
6. A 1:10 or 1:100 dilution of the virus reduces the potential impact of inhibitor carryover in to the $TCID_{50}$ assay.
7. It is important to change tips after each transfer of virus across the plate. This avoids carryover of virus on tips and an overestimation of viral titer.
8. Neuraminidase inhibitors generally reduce virus titers in a time of addition experiment when added up to approximately 10 h post infection.
9. Trypsin is required for cleavage of the hemagglutinin. It is important to ensure that all traces of FBS from the cell growth media are removed so as not to inactivate the trypsin in the virus growth media.
10. TPCK-trypsin concentrations may differ depending on the supplier. An appropriate concentration of trypsin which does not cause cell detachment should be determined via titration on MDCK cells prior to use in assays.

11. The virus titer is required for calculating the MOI. Use a $TCID_{50}$ virus titration assay or plaque assay to determine the titer and calculate the MOI based on the original cell seeding density: in our hands under the conditions described there is negligible MDCK cell replication between the time of cell seeding and virus inoculation.
12. Based on the method of Blick et al. 1995 [34], adaption for growth in the presence of neuraminidase inhibitors in routine tissue culture is preferentially through mutation of the HA to yield HA with low affinity for the receptor. The wash step helps to eliminate viruses which may bind weakly through a mutation in the HA.
13. Prior to starting resistance studies, the $EC_{50}$ of the neuraminidase inhibitor against the influenza virus of interest should be determined using an appropriate cellular assay such as a plaque reduction assay.
14. This usually takes 3 days but occasionally can take up to 6 days at later passages with higher drug concentrations and increasing selection pressure. A virus, no inhibitor control flask, should also be prepared for comparison to ensure that CPE is not confused with cell death of confluent monolayers.
15. At higher concentrations of inhibitor poor influenza growth may be observed. If virus titers are low, an additional passage at the same concentration of the inhibitor may increase viral titer prior to continuing with another twofold increase in selection pressure.
16. To improve the PCR reproducibility, two sets of primers are recommended to amplify smaller sections of the NA and HA.
17. Eluting the purified DNA into DEPC-treated water rather than the elution buffer provided in the kit can improve the quality of sequencing results.
18. Ideally, the HA and NA should be sequenced in at least two consecutive sections using discrete primers. Sequencing of both DNA strands is useful for confirming mutations.
19. Sequence wild-type virus passaged in parallel with the cultures under selective pressure and confirm that amino acid changes are due to inhibitor pressure and not adaptive mutations arising from routine passaging in MDCK cells.
20. A 1:10 dilution of stock virus of titer approximately $1 \times 10^6$ PFU/mL generally performs well in the assay, yielding an acceptable signal: noise within the linear range. Otherwise the optimal dilution can be determined by experimentation using virus grown in the absence of phenol red. If several virus stocks are to be evaluated simultaneously there is a risk of well to well signal carryover from high titer viruses, particularly if

adjacent to a low titer virus. This may be controlled by using a single plate per virus. The kit protocol suggests a 1:1,000 dilution of substrate. If neuraminidase signal of a variant is low, a 1:500 dilution of the substrate will help improve signal without compromising assay reproducibility. For cost conservation purposes multiple virus samples can be evaluated by reducing the number of replicates.

21. It is important that the virus suspension be well dispersed before gradient centrifugation. If particulate material remains after resuspension, the virus preparation should be clarified by mid-speed centrifugation (10,000 × *g* for 10 min). Notably, PEG 6000 concentrated virus yields a particularly glutinous pellet that can be difficult to resuspend.
22. Preferably use a gradient maker compatible with the volume of gradient required. If a suitable gradient maker is unavailable gradients can be prepared by carefully layering equal volumes of progressive concentrations (e.g., 60, 55, 50, 45, 40, 35, 30, 25, 20, and 15 % sucrose) in a vertical tube. It is important to pipette carefully to avoid mixing the layers. The gradients should then be allowed to rest to allow diffusion in order to create an approximation of a linear gradient (overnight at 4 °C is convenient). Take care to ensure that the gradients are not subject to vibration during this period. In all cases handle gradients with care to avoid mixing.
23. Ideally, the minimum number of gradients should be used, and the virus layer should be relatively narrow and sharply defined (i.e., layered carefully to avoid mixing with the top layer of sucrose). This will ensure that the virus is not unduly diluted and assists in the formation of a reasonably tight band following rate zonal centrifugation.
24. Alternatively, but less preferably, a second identical 15–60 % sucrose gradient centrifugation step can be substituted for the tartrate equilibrium density centrifugation step.
25. The needle should be carefully bent with metal tweezers or equivalent so that liquid can be readily collected on the horizontal plane. While holding the syringe vertically insert the needle into the top of the sucrose gradient and gradually lower it to the virus band. Collect the entire band in a minimal volume of sucrose, ideally in one continuous, smooth operation. If the band is difficult to see, or if an alternative method is desired, the gradient may be fractionated and each fraction assayed for hemagglutination activity.
26. The reaction should be started using a minimal volume of $MgCl_2$ (to avoid unduly diluting the virus transcriptase assay mixture) and with immediate mixing.

## References


1. Influenza (Seasonal) Fact sheet No. 211 (2009) World Health Organization. http://www.who.int/mediacentre/factsheets/fs211/en/. Accessed 26 Sept 2012
2. Fiore AE, Uyeki TM, Broder K et al (2010) Centers for disease control. prevention and control of influenza with vaccines. Recommendations of the Advisory Committee on Immunization Practices (ACIP). Morb Mortal Wkly Rep 59:1–62
3. Wright PF, Neumann G, Kawaoka Y (2007) Orthomyxoviruses. In: Knipe DM, Howley PM, Griffin DE (eds) Fields virology, 5th edn. Lippincott Williams & Wilkins, Philadelphia, pp 1691–1740
4. Skehel JJ, Wiley DC (2000) Receptor binding and membrane fusion in virus entry: the influenza hemagglutinin. Ann Rev Biochem 69:531–569
5. Burkrinskaya AG, Vorkunova ND, Kornilayeva GV et al (1982) Influenza virus uncoating in infected cells and effect of rimantadine. J Gen Virol 60:49–59
6. Martin K, Helenius A (1991) Nuclear transport of influenza virus ribonucleoproteins: the viral matrix protein (M1) promotes export and inhibits import. Cell 67:117–130
7. Bright RA, Shay DK, Shu B et al (2006) Adamantane resistance among influenza A viruses isolated early during the 2005–2006 influenza season in the United States. J Am Med Assoc 295:891–894
8. Bright RA, Shay D, Bresee J et al (2006) High levels of adamantane resistance among influenza A (H3N2) viruses and interim guidelines for use of antiviral agents – United States, 2005–06 influenza season. Morb Mortal Wkly Rep 55:44–46
9. Dawood FS, Jain S, Finelli L et al (2009) Emergence of a novel swine-origin influenza A (H1N1) virus in humans. N Engl J Med 360:2605–2615
10. Biegel J, Bray M (2008) Current and future antiviral therapy of severe seasonal and avian influenza. Antiviral Res 78:91–102
11. Moscona A (2005) Neuraminidase inhibitors for influenza. N Engl J Med 353:1363–1373
12. Kubo S, Tomozawa T, Kakuta M et al (2010) Laninamivir prodrug CS-8958, a long-acting neuraminidase inhibitor, shows superior anti-influenza virus activity after a single administration. Antimicrob Agents Chemother 54: 1256–1264
13. Yamashita M, Tomozawa T, Kakuta M et al (2009) CS-8958, a prodrug of the new neuraminidase inhibitor R-125489, shows long-acting anti-influenza virus activity. Antimicrob Agents Chemother 53:186–192
14. Hayden F, Rollins B (1997) In vitro activity of the neuraminidase inhibitor GS4071 against influenza viruses. Antivir Res 34:86
15. Thorlund K, Awad T, Boivin G et al (2011) Systematic review of influenza resistance to the neuraminidase inhibitors. BMC Infect Dis 11:134
16. Gubareva LV, Webster RG, Hayden FG (2002) Detection of influenza virus resistance to neuraminidase inhibitors by an enzyme inhibition assay. Antiviral Res 53:47–61
17. World Health Organization: Laboratory methodologies for testing the antiviral susceptibility of influenza viruses: Neuraminidase inhibitor (NAI) (2012). http://www.who.int/influenza/gisrs_laboratory/antiviral_susceptibility/nai_overview/en/index.html. Accessed 7 May 2013
18. Collins PJ, Haire LF, Lin YP et al (2009) Structural basis for oseltamivir resistance of influenza viruses. Vaccine 27:6317–6323
19. Hurt AC, Holien JK, Parker MW et al (2009) Oseltamivir resistance and the H274Y neuraminidase mutation in seasonal pandemic and highly pathogenic influenza viruses. Drugs 69:2523–2531
20. McKimm-Breschkin J, Trivedi T, Hampson A et al (2003) Neuraminidase sequence analysis and susceptibilities of influenza virus clinical isolates to zanamivir and oseltamivir. Antimicrob Agents Chemother 47: 2264–2272
21. Mosmann T (1983) Rapid colorimetric assay for cellular growth and survival: application to proliferation and cytotoxicity assays. J Immunol Methods 65:55–63
22. Crouch SP, Kozlowski R, Slater KJ et al (1993) The use of ATP bioluminescence as a measure of cell proliferation and cytotoxicity. J Immunol Methods 160:81–88
23. Guidance for Industry: Antiviral product development – conducting and submitting virology studies to the Agency (2006) U.S. Department of Health and Human Services Food, Food and Drug Administration and Center for Drug Evaluation and Research (CDER)
24. Hay AJ, Wolstenholme AJ, Skehel JJ et al (1985) The molecular basis of the specific anti-influenza action of amantadine. EMBO J 4:3021–3024
25. Palese P, Tobita K, Ueda M et al (1974) Characterization of temperature sensitive influenza virus mutants defective in neuraminidase. Virology 61:397–410

26. Morris SJ, Price GE, Barnett JM et al (1999) Role of neuraminidase in influenza virus-induced apoptosis. J Gen Virol 80: 137–146
27. Von Itzstein M, Wu WY, Kok GB et al (1993) Rational design of potent sialidase-based inhibitors of influenza virus replication. Nature 363:418–423
28. Furuta Y, Takahashi K, Kuno-Maekawa M et al (2005) Mechanism of action of T-705 against influenza virus. Antimicrob Agents Chemother 49:981–986
29. Lischka P, Hewlett G, Wunberg T et al (2012) *In vitro* and *in vivo* activities of the novel anticytomegalovirus novel anticytomegalovirus compound AIC246. Antimicrob Agents Chemother 54:1290–1297
30. Richman D (1997) Antiviral drug resistance: issues and challenges. In: Richman DD (ed) Antiviral drug resistance. Wiley, London, pp 1–12
31. Shigeta S, Mori S, Baba M et al (1992) Antiviral activities of ribavirin, 5-ethynyl-1-beta-D-ribofuranosylimidazole-4-carboxamide, and 6'-(R)-6'-C-methylneplanocin A against several ortho- and paramyxoviruses. Antimicrob Agents Chemother 36:435–439
32. Reed LJ, Muench H (1938) A simple method of estimating fifty percent endpoints. Am J Hyg 27:493–497
33. Pauwels R, Balzarini J, Baba M et al (1988) Rapid and automated tetrazolium-based colorimetric assay for the detection of anti-HIV compounds. J Virol Methods 20:309–321
34. Blick TJ, Tiong T, Sahasrabudhe A et al (1995) Generation and characterization of an influenza virus neuraminidase variant with decreased sensitivity to the neuraminidase-specific inhibitor 4-Guanidineo-Neu5Ac2en. Virology 214:475–484
35. Wetherall NT, Trivedi T, Zeller J et al (2003) Evaluation of neuraminidase enzyme assays using different substrates to measure susceptibility of influenza virus clinical isolates to neuraminidase inhibitors: report of the neuraminidase inhibitor susceptibility network. J Clin Microbiol 41:742–750
36. WHO Global Influenza Surveillance Network (2011) Manual for the laboratory diagnosis and virological surveillance of influenza. WHO Press, World Health Organization, Geneva, Switzerland
37. McGeoch DJ, Kitron NK (1975) Influenza virion RNA-dependent RNA polymerase: stimulation by guanosine and related compounds. J Virol 15:686–695
38. Nguyen JT, Hoopes JD, Le MH et al (2010) Triple combination of amantadine, ribavirin, and oseltamivir is highly active and synergistic against drug resistant influenza virus strains in vitro. PLoS One 5(2):1–12

# Chapter 31

# Methods for Evaluation of Antiviral Efficacy Against Influenza Virus Infections in Animal Models

Donald F. Smee and Dale L. Barnard

## Abstract

Compounds undergoing preclinical development for anti-influenza virus activity require evaluation in small animal models. Laboratory mice are most commonly used for initial studies because of size, cost, and availability. Cotton rats, guinea pigs, and ferrets (particularly) have been used for more advanced studies. Each animal infection model has certain limitations relative to human influenza infections. For example, the fever response that is evident in humans only occurs with consistency in ferrets. Mice infected with mouse-adapted viruses and ferrets infected with highly pathogenic avian influenza viruses suffer severe disease, whereas cotton rats and guinea pigs manifest few symptoms. Thus, for each animal model there is a certain set of disease parameters that can be measured. Here we describe methods for assessing the efficacy of anti-influenza virus compounds in each of these animal species.

**Key words** Influenza, Antiviral, Mice, Cotton rats, Guinea pigs, Ferrets

## 1 Introduction

Infection of mice with influenza viruses has been widely used to evaluate the efficacy of antiviral agents [1, 2]. Frequently the viruses require adaptation by successive passage from animal to animal before they become lethal. Newly emerging seasonal influenza A and B viruses are not infectious to mice without adaptation, but some of the recent 2009 pandemic H1N1 do not require adaptation [3, 4]. Highly pathogenic avian H5N1 viruses are also lethal without adaptation [5]. Seasonal H3N2 viruses have been particularly difficult if not impossible to adapt to mice. The ones that cause lethal infections are few in number and are 30 or more years old. Few studies have been performed with influenza H2N2 viruses in mice. Influenza B viruses can be difficult to adapt to mice, but a few are available. Some of the influenza viruses that are mouse-adapted are commercially available, whereas other strains must be acquired from the laboratory that developed them. In this chapter we describe commonly used methods for determining antiviral

Edwin Yunhao Gong (ed.), *Antiviral Methods and Protocols*, Methods in Molecular Biology, vol. 1030,
DOI 10.1007/978-1-62703-484-5_31, © Springer Science+Business Media, LLC 2013

activities of test materials in mice infected with lethal influenza viruses. This includes a description of various parameters that can be measured and how to best statistically evaluate the data. For mouse and ferret models particularly, a number of methods can be employed, and alternative methods are presented. The field has evolved over time, and some of the methods reported in older scientific literature are not currently in use or should be replaced by more sophisticated methods, particularly those involving statistical analysis of the data. Influenza antiviral studies in cotton rats, guinea pigs, and ferrets are less common compared to mouse experiments due to cost or limitations in the models. Methodologies and assay parameters specifically used to assess antiviral activity in these species will follow those that are pertinent to mice.

## 2 Materials

### *2.1 Animals*

1. Mice are generally obtained from commercial breeders. A mouse strain frequently used is BALB/c. Swiss or Swiss Webster mice appear to be similar to BALB/c mice in terms of ability to be lethally infected, and they are considerably less expensive (*see* **Note 1**). Female mice are preferred due to aggressive behavior in males. The age of the animal influences infectivity, with older mice being less susceptible to lethal infection using seasonal virus strains. For this reason, the mice generally are infected at 6–8 weeks of age (18–20 g weight for a BALB/c mouse). Proper sorting of the mice prior to the initiation of each experiment should be done (*see* **Note 2**).
2. Cotton rats, guinea pigs, and ferrets are usually obtained from commercial sources. Some laboratories maintain colonies of cotton rats. Ferrets should be purchased as specific pathogen-free animals since they can naturally acquire influenza, and must be checked for influenza antibody status prior to use.

### *2.2 Viruses and Cells*

1. Only a certain number of seasonal influenza A H1N1, H2N2, H3N2, low pathogenic H5N1, and influenza B virus strains have been successfully adapted to mice (Table 1) (*see* **Note 3**), making them appropriate for antiviral studies. Some of the mouse-adapted virus strains can be acquired by purchase (such as from the American Type Culture Collection, Manassas, VA), others can be obtained from specific research laboratories, and new ones can be developed by mouse adaptation (*see* **Note 4**). A number of highly pathogenic avian H5N1 viruses (that do not require mouse adaptation) and a mouse-adapted H7N7 virus have been used for lethal infection studies in mice (Table 1).

**Table 1**
**Influenza virus strains lethal to mice that have been used for antiviral studies. The viruses are listed in chronological order by type**

| Virus | Strain | Literature reference |
|---|---|---|
| Influenza A (H1N1) | NWS/33 | [6] |
| | WSN/33 | [7] |
| | PR/8/34 | [8] |
| | Bayern/57/93 | [9] |
| | New Caledonia/20/99 | [10] |
| | Solomon Islands/03/2006 | [11] |
| | California/04/2009 | [12] |
| Influenza A (H2N2) | Japan/305/57 | [13] |
| | Singapore/1/57 | [14] |
| Influenza A (H3N2) | Hong Kong/1/68 | [15] |
| | Aichi/2/68 | [16] |
| | Port Chalmers/1/73 | [17] |
| | Victoria/3/75 | [6] |
| | Philippines/82 | [18] |
| Influenza A (H5N1), low pathogenic | Duck/MN/1525/81 | [19] |
| Influenza A (H5N1), high pathogenic | Chicken/Jogjakarta/BBVET/IX/04 | [5] |
| | Thailand/1(Kan-1)/04 | [5] |
| | Vietnam/1203/04 | [20] |
| | Whooper swan/Mongolia/244/05 | [5] |
| | Duck/Laos/25/06 | [5] |
| | Turkey/15/06 | [20] |
| | Turkey/65-1242/06 | [5] |
| Influenza A (H7N7) | Seal/Massachusetts/1/80 | [21] |
| | Ck/Netherlands/621557/2003 | [15] |
| | Netherlands/219/2003 | [22] |
| Influenza B | Lee/40 | [9] |
| | Hong Kong/5/72 | [6] |
| | Sichuan/379/99 | [6] |

**Table 2**
**Anesthetics useful for infecting mice intranasally with influenza viruses**

| Anesthetic | Dose (mg/kg) | Route of administration | Reference |
|---|---|---|---|
| Ether | – | Inhalation | [16] |
| Isoflurane | – | Inhalation | [12] |
| Ketamine | 100 | Intraperitoneal | [19] |
| Ketamine/xylazine | 50/5 | Intraperitoneal | [23] |

2. Cotton rats, guinea pigs, and ferrets can be infected with various non-adapted influenza virus strains. The viruses will differ in terms of their ability to cause mild to severe disease.
3. Madin-Darby canine kidney (MDCK) cells are obtained from a supplier such as the American Type Culture Collection (Manassas, VA). They are propagated in Eagle's minimum essential medium containing 5 % fetal bovine serum and 0.18–0.22 % sodium bicarbonate buffer in 5 % $CO_2$ at 37 °C. The cells are used for propagating influenza viruses and for titrating virus from infected animal lungs.

### 2.3 Anesthesia

1. Mice must be anesthetized in order to breathe in the intranasal liquid containing virus. The types of anesthetic that have been reported are presented in Table 2. Any one of these may be satisfactory to use.
2. Different types of anesthesia have been reported for cotton rats, guinea pigs, and ferrets. Isoflurane has been used for anesthetizing cotton rats. Ketamine/xylazine (30/2 mg/kg) has been given intramuscularly to guinea pigs. Ketamine/xylazine (5/0.5 mg/kg/day) effectively anesthetizes ferrets.

### 2.4 Positive Control Drugs

A positive control drug such as ribavirin, oseltamivir phosphate (Tamiflu®), or zanamivir is recommended for use in antiviral studies. Ribavirin can be purchased from Sigma (St. Louis, MO) or MP Biomedicals (Santa Ana, CA) and zanamivir from Haorui Pharma-Chem (Edison, NJ). Oseltamivir phosphate may possibly be obtained as a gift from Roche (Palo Alto, CA) or purchased from a local pharmacy (as Tamiflu®) (*see* **Note 5**).

### 2.5 Special Equipment (Optional)

1. Pulse oximeter for measuring arterial oxygen saturation in mice, such as one specifically designed for mice, MouseOx Plus (Starr Life Sciences Corp., Oakmont, PA).
2. Whole body unrestrained plethysmograph (EMKA, Paris, France) and software, with Buxco small rodent chamber (purchased through EMKA). The chamber will work for mice

and cotton rats. Guinea pigs and ferrets are too large for the chamber, although one can have custom-made chamber manufactured, which must be calibrated with the instrument.

3. Q-View® imager for reading multiplex cytokine/chemokine array microplates (both imager and microplates from Quansys Biosciences, Logan, UT) or Luminex® imaging system for magnetic bead-based cytokine/chemokine assays (both imager and beads from EMD Millipore, Billerica, MA). Alternatively, individual cytokine and chemokine kits can be purchased (e.g., R&D Systems, Minneapolis, MN), and an ELISA plate reader will be required. These kits are currently available for mice. Immunological reagents for cotton rats, guinea pigs, and ferrets cannot be performed in multiplex format. It is possible that certain individual cytokine/chemokine kits may be cross-reactive between mice and other species (which would require validation) and could be used.

## 3 Methods

### 3.1 Methods for Mice

#### 3.1.1 Virus Preparation and Titrations

1. A large stock of virus sufficient for many experiments is prepared either from a large number of mice (mouse lung homogenates) or (preferably) by a single passage of the mouse-adapted virus in MDCK cells (*see* **Note 6**) or embryonated chicken eggs. Make 0.5–1 mL aliquots of virus and freeze at −80 °C.
2. Determine in vitro virus titer of virus stock by plaque assay [24] or endpoint dilution method [25]. Express virus titer as plaque forming units per mL (PFU/mL) or 50 % cell culture infectious units per mL ($CCID_{50}$/mL).
3. Conduct a lethality titration of the virus stock in the selected strain and age of mice. Infect the mice by intranasal route when they are deeply anesthetized (*see* **Note 7**). Use several half-$log_{10}$ dilutions of virus and five or more mice per group. Determine mortality in each group over 21 days (*see* **Note 8**). Estimate the 50 % mouse lethal dose ($mLD_{50}$) from the data (*see* **Note 9**). $mLD_{50}$ calculations are determined by probit analysis or another method of linear regression. The amount of virus required to achieve the $mLD_{50}$ can then be calculated in PFU/mL or $CCID_{50}$/mL units, based upon the dilution of virus resulting in the $mLD_{50}$.

#### 3.1.2 Antiviral Experiments in Mice

1. Infect mice under anesthesia with an appropriate number of $mLD_{50}$ of virus (*see* **Note 10**). Distribute the infectious medium equally between nostrils.
2. Treat mice using the determined treatment regimen (*see* **Note 11**).

3. Allow the study to proceed 21 days, unless endpoints are achieved prior to that time. Hold some mice in each treatment group for mortality determinations. Sacrifice other animals in the group at different time intervals for determination of multiple infection parameters (*see* **Note 12**).

*3.1.3 Assay Parameters for Mice*

1. A number of parameters can be measured as part of an antiviral experiment, as described below and accompanied by an example literature citation. For a compound of unproven antiviral activity in vivo, assaying for mortality and body weight loss are sufficient for a preliminary experiment. More in-depth studies can be reserved for compounds with proven efficacy against influenza infections in mice. The statistical methods are considered the most definitive of those available and the ones we prefer. Alternative, less suitable methods are also provided. Statistical analyses are easily performed using commercially available software, such as InStat® or Prism® (preferred) (both from GraphPad, Inc., San Diego, CA).
2. Death reported using survival curves [23]. Record deaths in each group on a daily basis. The majority of mice will die from influenza between days 6–14, depending upon the virus strain. Hold the mice for 21 days to account for later mortality. Plot results as survival curves (such as Kaplan–Meier plots). Perform a log-rank test (such as Mantel–Cox) analyzing all groups for statistical significance. If significance is found, follow up with pairwise comparisons using the Gehan–Breslow–Wilcoxon test or Mantel–Cox test for pairwise comparisons with Bonferroni's corrected threshold of significance for multiple groups (*see* **Note 13**).
3. Death reported as survivors/total (or dead/total) [26]. Record deaths in each group on a daily basis. Report the results in tabular form in conjunction with mean day of death calculations (below). Analyze data for significance by chi square analysis when there are three or more groups. When the data are statistically significant, pairwise comparisons can be performed by methods such as Fisher's exact test (preferred), chi square analysis with Yates' correction, or similar test (*see* **Note 14**). These tests do not provide as much information about the mortality associated with the experiment, as do Kaplan–Meier plots analyzed by log-rank test described above.
4. Mean day of death [26]. This measurement is determined only for mice that die. Survivors are not included. To make this calculation, indicate the day of death for each mouse in the group that died. Determine the mean value ± standard deviation for the group. Present the data in tabular form (*see* **Note 14**). Mean day of death data sets are usually non-parametric, and are statistically analyzed by the two-tailed Mann–Whitney

*U*-test (pairwise comparisons), or by Kruskal–Wallis test (three or more comparisons) followed by Dunn's test for appropriate pairwise comparisons. These tests do not provide as much information about the mortality associated with the experiment, as do Kaplan–Meier plots analyzed by log-rank test described above.

5. Hazards ratio [20]. Hazard is defined as the slope of the survival curve-a measure of how rapidly subjects are dying. The hazard ratio compares two treatments. If the hazard ratio is 2.0, then the rate of deaths in one treatment group is twice the rate in the other group. We use Prism® software, which computes the hazard ratio and its confidence interval using the Mantel–Haenszel method.
6. Protection index (PI) [16]. This value estimates the degree of protection achieved by a particular treatment relative to the placebo group. It is calculated by the following equation: $PI = [(PC-1)/PC] \times 100$, where PC (the coefficient index) equals % mortality in placebo group/% mortality in the drug-treated group.
7. Body weight [27]. In order to evaluate the results statistically, mice should be weighed individually rather than as a group. In practice, every-other-day weighing may be sufficient to show and analyze trends. Analyze the results using two-way analysis of variance (ANOVA) with Tukey's multiple comparisons post-tests (*see* **Note 15**).
8. Arterial oxygen saturation [17]. Use white mice to obtain an accurate reading. Place probe on the thigh area of the animal. Wait for a consistent reading and then record the measurement. For mice that are dead on the day of reading, assign a final value of 75, which equates to the lowest oxygen saturation level of mice prior to their death (*see* **Note 16**). Analyze the results using two-way ANOVA with Tukey's multiple comparisons post-tests (*see* **Note 15**).
9. Plethysmography [28]. The plethysmograph records changes in pressure as a result of the animal's respiration while it is in a special chamber. Animals are placed in the chamber and allowed to settle down (they must not be moving inside the chamber) prior to taking measurements. Thirteen measurements are automatically recorded with the instrument indicated in **item 2** of Subheading 2.5 (*see* **Note 17**). Analyze the results using two-way ANOVA with Tukey's multiple comparisons post-tests (*see* **Note 15**).
10. Lung hemorrhage score (also inaccurately referred to in the scientific literature as lung consolidation score) [29]. Mouse lungs undergo hemorrhage (resulting in a change from pink to plum color) during severe influenza virus infection. This occurs

regionally rather than by gradual change of the entire lung from pink to plum. The measurement is made from sacrificed mice (either by cervical dislocation or chemical injection, but not using carbon dioxide since it may cause hemorrhage to the lungs). The extent of plum color in the lungs is made by visual inspection after necropsy and is estimated on a scale of 0–4 in 0.5-U increments, with a score of 4 being 100 % hemorrhaged. Because this is a subjective score, a non-parametric statistical test must be used for analysis, such as nonparametric one-way ANOVA with Dunn's multiple comparisons test (*see* **Note 18**).

11. Lung weight [29]. After scoring the lungs, place the tissue on a balance and record the weight in milligram. If there is only 1 day where samples are taken, analyze the results using one-way ANOVA with Newman–Keuls multiple comparisons test or Tukey–Kramer multiple comparisons test. If lung samples are analyzed over multiple days and day-to-day comparisons are to be made, analyze the results using two-way ANOVA with Tukey's multiple comparisons post-test (*see* **Note 15**).
12. Lung index (lung weight–body weight ratio) [16]. The weight of the lungs of the mouse is compared to whole body weight. Mice lose weight during infection and lung weights increase, changing the normal balance. Lung index is calculated by the following equation: lung weight/body weight × 100 [16]. Perform statistical analysis as described in **step 11** under Subheading 3.1.3.
13. Lung virus titer [29]. Homogenize lungs either with a mechanical device (tissue grinder or stomacher) in 1 mL of cell culture medium. Alternatively, place lungs in a stomacher bag with 1 mL cell culture medium and roll a 10-mL pipette over the tissue in the bag until the sample is completely disrupted. The latter method, although simple, is quite effective. Either freeze the samples (centrifuge at 600 × *g* to remove cellular debris) for subsequent virus titer determination or immediately plate them on MDCK cells. Titrate samples in duplicate either by plaque assay in 6-well microplates containing a monolayer of cells [24] or by endpoint dilution method [25] in 96-well microplates of confluent cells using four microwell per dilution. Dilute in tenfold dilution increments. Virus titer is expressed either as PFU/g or $CCID_{50}$/g of tissue. PFU/g is calculated by the following equation: 1/lung weight × 1/microwell inoculum volume (e.g., 0.2 mL) × 1/dilution where plaque numbers are countable × plaque count at the particular dilution. $CCID_{50}$/g is calculated by the following equation: 1/lung weight × 1/microwell inoculum volume (e.g., 0.1 mL) × endpoint dilution titer calculated by Reed–Muench method [25]. These formulas work because the tissue (regardless of its weight) is homogenized in a set volume (1 mL) of

cell culture medium (*see* **Note 19**). Perform statistical analysis as described in **step 11** under Subheading 3.1.3.

14. Cytokine and chemokine determinations [20]. Cytokines and chemokines become elevated during influenza virus infection. Treatment with antiviral agents results in more normal values compared to placebo. Cytokine and chemokine factors are determined from supernatant fluid of lung homogenates (preferred, due to strong signals) and/or from serum on different days post-infection using commercially available test kits (*see* **item 3** under Subheading 2.5). Perform statistical analysis as described in **step 11** under Subheading 3.1.3.
15. Lung histopathology assessment [30]. Lungs are removed, formalin fixed, thin sectioned, mounted on slides, and stained by standard histological methods. An individual trained in lung pathology visually evaluates the prepared slides for the degree of damage due to infection and provides a written description of the findings. If a subjective scoring system is used (e.g., mild, moderate, or severe pathology equal to scores of 1, 2, or 3), perform a nonparametric statistical evaluation as described in **step 10** under Subheading 3.1.3.
16. Drug combination studies. Use at least two doses of each compound alone and in combination. Doses of each must have been predetermined to be suboptimal so that the effects of the two in combination can be seen. A useful method for evaluating drug combinations is available on free computer software, referred to as MacSynergy [31].

### 3.2 Methods for Cotton Rats, Guinea Pigs, and Ferrets

#### 3.2.1 Methods Common Among Species

Virus preparation and conducting of antiviral experiments reported for mice are applicable to cotton rats, guinea pigs, and ferrets (*see* Subheadings 3.1.1 and 3.1.2).

#### 3.2.2 Assay Parameters for Cotton Rats

1. Cotton rats do not die from infection; thus, only a few parameters can be assessed as indicated below.
2. Body weight [32]. Perform measurements and statistical analysis as described as described in **step** 7 under Subheading 3.1.3.
3. Body temperature [32]. Body temperatures drop during the acute phase of the infection. Take daily core body temperature measurements with a digital rectal thermometer. Differences between infected and uninfected cotton rats are evident on days 1–3 of the infection. Analyze differences on these days by two-way ANOVA with Tukey's multiple comparisons post-tests.
4. Lung and nasal virus titers [32]. Cotton rats produce virus both in the upper and lower respiratory tracts. Following infection, upper respiratory tract virus is higher on days 1–2,

whereas lower respiratory tract virus is higher on days 3–4. After sacrificing the animals, collect lung and nasal tissues for homogenization by necropsy. Titrate for virus as described in **step 13** under Subheading 3.1.3, and perform statistical analysis per **step 11** under Subheading 3.1.3.

5. Lung histopathology assessment [32]. Perform this procedure as described in **step 15** under Subheading 3.1.3. Use the nonparametric statistical analysis described in **step 10** under Subheading 3.1.3.

#### 3.2.3 Assay Parameter for Guinea Pigs

Virus titers [33]. The main use of this model is assessing transmission studies from infected to uninfected animals. Guinea pigs show no clinical signs of infection, and considerably less virus is produced in lungs (requiring sacrifice of the animals) compared to virus recovered from nasal washes. For this reason, only nasal wash virus titer needs to be assessed during antiviral studies. Peak viral titers are seen on day 3 of the infection. Perform nasal washes by instilling 1 mL of PBS into the nostrils and allowing the fluid to drain onto a sterile Petri dish. Titrate for virus from the fluid samples per **step 13** under Subheading 3.1.3. Perform statistical analysis as described in **step 11** under Subheading 3.1.3.

#### 3.2.4 Assay Parameters for Ferrets

1. Influenza virus strains vary in their lethality to ferrets. Highly pathogenic avian viruses are lethal [34], whereas the pandemic influenza A/California/04/2009 (H1N1) is not [35].
2. Parameters associated with lethality [34]. These are performed and analyzed as described in **steps 2** through **6** under Subheading 3.1.3.
3. Body weight [35]. Perform measurements and statistical analysis as described in **step 7** under Subheading 3.1.3.
4. Body temperature [35]. Take daily core body temperature measurements with subcutaneous implantable temperature transponders. Differences between infected and uninfected ferrets are evident on days 1–4 of the infection. Analyze differences on these days by two-way ANOVA with Tukey's multiple comparisons post-tests.
5. Nasal wash inflammatory cell count [35]. Anesthetize ferrets with ketamine (25 mg/kg, intramuscularly) and instill 0.5 mL of sterile PBS into each nostril. Allow the fluid to drain onto a sterile Petri dish. Centrifuge samples (the same volume per sample) at 1,000 × *g* for 10 min, resuspend the pellet in PBS, and count the cells in the fluid using a hemocytometer or automated cell counter. Perform statistical analysis as described in **step 11** under Subheading 3.1.3.
6. Nasal wash protein concentration [35]. Use part of the cell-free nasal wash supernatant collected in **step 5** under Subheading 3.2.4. Perform a standard protein assay using a

commercially available kit (e.g., Bio-Rad, Hercules, CA). Perform statistical analysis as described in **step 11** under Subheading 3.1.3.

7. Nasal wash virus titer [35]. Use part of the cell-free supernatant collected in **step 5** under Subheading 3.2.4. Titrate the fluid for virus as described in **step 13** under Subheading 3.1.3. Perform statistical analysis as described in **step 11** under Subheading 3.1.3.
8. Lung or other organ virus titers [35]. This is generally not performed due to the expense of the animals, since they must be sacrificed. Perform virus titrations as described in **step 13** under Subheading 3.1.3, and statistical analysis as described in **step 11** under Subheading 3.1.3.
9. Symptoms [36]. Researchers often note symptoms, such as nasal discharge, sneezing and malaise (inactivity) with number of animals/total recorded. These parameters are evaluated similar to survivors/total data as described in **step 3** under Subheading 3.1.3, although animal numbers are often too few to derive statistical significance.

## 4 Notes

1. Mouse strains vary in their susceptibility to virus infection [37] and they also vary in temperament and ease of handling. The most sensitive mouse strain to influenza infection that we have worked with is DBA/2. Mice that are not lethal to BALB/c or Swiss Webster mice may be lethal to DBA/2 mice. However, DBA/2 mice are difficult to handle particularly when treated by oral gavage, and this may result in trauma and death unrelated to infection. C57BL/6 mice have a different susceptibility to mortality caused by influenza virus infection than BALB/c mice [38]. They may be more useful than BALB/c mice for certain antiviral studies such as those involving immunomodulatory antiviral substances, because they elicit a type 2 immune response. BALB/c mice elicit a type 1 response.
2. Animals that are received from vendors are generally unloaded from shipping containers and placed in cages by the technical staff without regard to weight. The mice will have a weight range varying by 2 or more grams from lightest to heaviest. Our method is to weigh cages of animals and make adjustments by trading lighter or heavier animals so that each cage of animals does not differ from another cage by more than 0.5 g. This is not true randomization, but will prevent executing a study where a particular cage may be significantly different in weight than the others just by chance. For the starting mg/kg/day

dose, it is important that cage weights are approximately the same. In addition, there may be occasional mice that are received from vendors that are clearly outside the specified weight range. This can be managed by purchasing five or more extra animals and not using the outliers. The extra mice can be used as normal controls (uninfected, untreated). If preferred, true randomization can be performed where animals are put into groups with no attempt to have equivalent average weights among different groups.

3. Of the viruses listed in Table 1, the H1N1 virus strain that is most commonly used is A/PR/8/34 (probably because of its availability). A number of compounds are reported to have antiviral activity against A/PR/8/34 virus infections in mice. Yet when we have tested some of these same compounds in our laboratory against other virus strains, the compounds are inactive. For this reason, the A/PR/8/34 virus strain is the least desirable of the viruses that we have used. There are several reports in the scientific literature of antiviral studies in mice using a mouse-adapted influenza A/Shangdong/09/93 (H3N2) (e.g., *see* ref. 39). It was later determined by genetic analysis that the virus was actually influenza A/Victoria/3/75 (H3N2) that had been propagated concurrently in our laboratory. Subsequent attempts to adapt influenza A/Shangdong/09/93 (H3N2) to mice were unsuccessful.

4. Mouse adaptation may be a difficult and sometimes unsuccessful endeavor. From our experience it usually takes at least seven passages or transfers from mouse to mouse to obtain a virus capable of causing severe disease and death. During the first few passages there may be very low virus titer recovered from homogenized mouse lungs. Virus titers need to improve with each subsequent passage or it could be an indication that the particular virus will never adapt. Infecting with the highest virus titer possible and harvesting lungs 3 days later to make a new virus pool is an effective strategy. Weight loss (an indication of morbidity during infection) can be monitored in mice held for this purpose. By passage 5 there may be a few deaths in a group of animals, indicating that mouse adaptation is occurring. There are a few methods that may speed up the adaptation process. One is to start with barely weaned (10–12 g) mice, since younger animals are more susceptible to infection. The age of the animal can be increased as the virus adapts. Second, intranasal treatment with mannan in conjunction with infection may speed up adaptation, particularly in the beginning of the adaptation process [40]. Mannan binds host cell collectins that otherwise neutralize influenza virus particles. Mannan should be avoided at later mouse passages so that the lethality of the infection will not be mannan dependent.

Third, seasonal H1N1 virus strains are easier to adapt to mice than seasonal H3N2 and B virus strains. Fourth, adaptation may be expedited using DBA/2 mice, which are more sensitive than others to infection. This will not guarantee that the adapted virus will be lethal to BALB/c mice, however. We prefer using inexpensive Swiss Webster mice for purposes of virus adaptation. The resulting virus should also be able to cause lethal infections in BALB/c mice. Mice can be treated daily with intranasal saline on days 1 and 2 of the infection (before harvesting lungs on day 3). This serves to spread the infection in the lungs and results in higher virus titers. Finally, it is important to infect with the highest amount of virus each time. At the onset (passage 1 in the animal), high titer virus (prepared in embryonated eggs or cell culture) is used. What usually happens is the virus recovered after the first passage in mice may be only $10^2$ or $10^3$ PFU or $CCID_{50}$ per lung. The virus may have to go back into eggs or cells to increase its titer for the second passage. Eventually the passages must go from mouse to mouse rather than back and forth between cell culture (or eggs) and mice.

5. Tamiflu® contains excipients and for this reason it is necessary to weigh out the entire capsule when preparing compound for treatment. A 75-mg capsule is based upon the liberated drug, oseltamivir carboxylate, not the prodrug form, oseltamivir phosphate. The amount of oseltamivir phosphate in the capsule is estimated to be 98.53 mg, based upon the description provided in the package insert for Tamiflu®. We have compared oseltamivir phosphate (from Roche) to Tamiflu® (from a pharmacy) for antiviral activity in mice, and found them to be nearly equivalent [27].
6. Limit the number of passages in cell culture once the virus is adapted to mice, since cell culturing of virus can lead to attenuation (loss of in vivo virulence). Cell culturing is a cost-effective way to obtain a large quantity of virus with high viral titer (as opposed to using mice to produce the virus pool). For growing virus pools, use a low multiplicity of virus infection to avoid generating defective interfering particles. A 1:100 dilution of virus stock is often sufficient for infecting cells. For most influenza viruses, better virus titer is obtained by using trypsin (10 U/mL or equivalent) in the serum-free cell culture medium. After viral cytopathology in the flask of cells reaches 100 % (about 72 h), collect the cell culture fluid, perform a low speed (600 × *g*) centrifugation, aliquot the supernatant fluid into cryovials, and freeze the vials at −80 °C. Titrate a thawed vial for virus titer determination at a later time in MDCK cells.
7. The right amount of anesthetic must be given so that the virus-containing liquid can be administered safely and the animals

will recover. Cautiously and slowly administer the intranasal liquid. Have a heating pad available to warm up animals that appear to be too deeply anesthetized. The volume of intranasal liquid containing virus is important because it can influence the degree of mortality during the infection. The volume of administered liquid influences the extent of penetration into the lungs. A lower volume provides more to the upper respiratory tract and less to the lungs [41]. We have given a 90-μL volume of liquid for better penetration into the lungs, whereas other investigators have reported using 50 μL [20].

8. Local animal care and use protocols may define mortality as a particular percent of body weight loss (such as 25 %), at which time mice are humanely euthanized [42]. This will require that each animal be ear tagged and weighed individually during the titration and subsequent antiviral study.
9. Animals under stress are more susceptible to virus infection than unstressed ones, and this will affect the resulting $mLD_{50}$. For example, mice infected 1 or 2 days after receipt from a vendor may be more susceptible to lethal infection than animals that are held for a week after receipt. Thus, it is important to perform lethality titrations and antiviral studies under the same conditions with regard to acclimation of the mice. The amount of virus used for antiviral studies that causes a particular degree of lethality is expressed as a multiple of the $mLD_{50}$ value and also is reported in terms of virus titer (PFU or $CCID_{50}$ per mouse) required to achieve the particular degree of lethality.
10. Plan experiments so that there is 90–100 % mortality in placebo-treated groups of animals. Statistical significance (for death) is more easily achieved than in studies where a greater number of placebo-treated animals survive the infection. However, experiments can be designed where high mortality is not the desired outcome, requiring other parameters of infection such as weight loss [43] as the primary indicators of efficacy.
11. A preliminary toxicity test of a new compound of unknown toxicity should be performed in uninfected mice to determine the maximum tolerated dose. This test should mimic conditions for the subsequent antiviral experiment, such as treatment vehicle, treatment route [intraperitoneal, intranasal, oral (usually by gavage rather than in the drinking water), intramuscular, or other routes]. Without knowing whether the compound has good oral bioavailability, oral gavage treatments are not recommended for the first antiviral experiment. Intraperitoneal treatments are generally equivalent to intravenous dosing, and are considerably easier to administer.

Doses can then be selected for the subsequent antiviral study, varying by twofold, half-$log_{10}$, four-, tenfold, or by other dilution scheme. The time of treatment initiation relative to virus challenge needs to be carefully chosen. To provide a new compound the maximum benefit of being active, it is recommended that that the first dose be given from 1 to 4 h prior to virus challenge or no later than 4 h after infection. Virus infections become harder to treat the longer the first treatment is delayed relative to virus challenge [10]. If a compound proves to be active under these conditions, then subsequent studies can be performed to determine how late after infection the first treatment can be initiated. Initially, twice a day treatments are recommended, with 12-h intervals preferred for efficacy over convenience treatments at 8:00 a.m. and 4:30 p.m. Compounds with short half-lives that are predicted to require several (3–4) treatments per day may do quite well with two doses per day [19]. Intranasal treatments are problematic because the intranasal liquid exacerbates the infection [44]. Less virus is required to kill mice when intranasal liquid treatments are given. The animals cannot tolerate being treated intranasally much beyond day 4 post-infection because in their weakened state they are prone to die from the anesthesia. If treatments are given by another, safer route, it is recommended that treatments be given for a minimum of 5 days. For experiments involving intranasal treatments, since these treatments exacerbate the infection, it is very important to pre-titrate the virus using the same vehicle and treatment regimen (number of doses per day and days of treatment) that will be used for the subsequent antiviral study.

12. The proper number of animals per group is important for achieving statistical significance. This will vary depending upon the tightness of the data for the parameter being measured. It is more difficult to achieve statistical differences for mortality data when survival in the placebo group is greater than 10 %, unless group sizes are larger. Ten mice per compound-treated group and 15 placebo-treated mice are adequate when mortality is 90–100 %. As survivor numbers increase, the number of mice needed for statistical analysis will also increase. For other parameters such as lung virus titers (that are tedious to perform) and lung weights, we use five mice per group titrated individually. The data are reasonably tight enough to get accurate statistical interpretations. Larger group sizes will increase probability of achieving statistical significance, however. Evaluating each animal's lung parameters then calculating mean values and standard deviations is a far better method than using a single value from lungs that are pooled from three or more mice, the data of which cannot be assessed statistically.

13. The Gehan–Breslow–Wilcoxon test takes into account numbers of dead animals, when they die, and the rate of death. When comparing many groups, the probability increases that one or more groups will be significantly different from another group purely due to chance. Since the test is a pairwise comparison, it does not take into account multiple groups. This is why an analysis of all groups must be made first by log-rank test to determine significance among all groups. To account for the individual variance that each group contributes to the entire experimental variance, the desired significance is divided by the number of comparisons made, which is called a Bonferroni's correction factor. This is calculated in the following manner. At a particular desired level of statistical significance, divide the value by the number of groups evaluated. For example in a study where there are five groups (including placebo) and the desired level of significance is $P < 0.05$, the actual value that must be achieved is $<(0.05/5)$ or $<0.01$ to be at the $P < 0.05$ level.
14. The Fisher's exact test is recommended over chi square analysis when small sample sizes are used, and even for moderate-sized samples if a computer program is available [45]. Chi square analysis with Yates' correction is more accurate than chi square analysis, although both give only an approximate $P$ value [45]. If the investigator chooses to report survivors/total data rather than survival curves, mean day of death ± standard deviation values should also be reported. Analyzing survivors/total and mean day of death is not as accurate as analyzing survival curves, but it is commonly done. Although mean day of death data are often nonparametric, they do not necessarily have to be. Generally, in the scientific literature they are reported by a nonparametric test.
15. Two-way ANOVA calculations can only be done when there are survivors in all groups. For example if by day 11 of the infection a particular group has no more survivors, then statistical interpretations can only be made through day 10. The remainder of the days (or a single day of reading) can be analyzed by one-way ANOVA, however.
16. Although measuring arterial oxygen saturation seems simple in practice, it is difficult to know when the reading is accurate. Where the probe is placed influences the reading. Arterial oxygen saturation levels begin to fall a few days prior to the time of death. Earlier readings that show declines are probably inaccurate. The lowest arterial oxygen saturation readings occur just prior to death of the animals.
17. The 13 measurements that can be automatically made with plethysmography are enhanced pause (or penh), end expiratory

pause, end inspiratory pause, peak expiratory flow, peak inspiratory flow, expiratory time, inspiratory time, tidal volume, minute volume, expired volume, breathing frequency, relaxation time, and mid expiratory flow. Of these, the ones that appear to be more useful than others for assessing respiration during an influenza virus infection in mice are penh, end expiratory pause, minute volume, tidal volume, and expired volume. These parameters should be verified for each virus infection model, because they could vary depending upon how lethal the virus infection is.

18. The nonparametric one-way ANOVA (Kruskal–Wallis) with Dunn's multiple comparisons test is extremely stringent even for data sets that obviously look significantly different from one another. It is to be used when greater than two groups are to be compared. An investigator may choose to report pairwise comparisons using the two-tailed Mann–Whitney *U*-test if only one comparison is to be made (i.e., placebo vs. one treatment). In the past the two-tailed Mann–Whitney *U*-test was primarily used. It is not appropriate to use a parametric test (e.g., Student's *t*-test) for analysis.
19. The time when maximum viral titers are detected differs from the time that lung weights and lung hemorrhage scores are at their maximum [29]. Viral titers come up early in infection (days 1–3) and are lower (but still high enough for a good measurement) by day 6. Viral titers can be quite low on day 9. In contrast, lung weights and scores are near normal on day 3, and peak just before death of the animals. Taking mice on multiple days (such as days 2, 4, and 6) will capture a curve of information. If only one time point is selected for assay, then day 6 is preferred, assuming mice in the placebo group have not died yet. Valid lung infection parameters can be measured from animals found dead in the cage on the designated day of sacrifice.

## Acknowledgment

This work was funded in part with Federal funds from the Respiratory Diseases Branch, Division of Microbiology and Infectious Diseases, National Institute of Allergy and Infectious Diseases, National Institutes of Health, Department of Health and Human Services, under Contract No. HHSN272201000039I/HHSN27200005/A37.

## References


1. Sidwell RW, Smee DF (2004) Experimental disease models of influenza virus infections: recent developments. Drug Discov Today Dis Models 1:57–63
2. Barnard DL (2009) Animal models for the study of influenza pathogenesis and therapy. Antiviral Res 82:A110–A122
3. Zhang Y, Sun H, Fan L et al (2012) Acute respiratory distress syndrome induced by a swine 2009 H1N1 variant in mice. PLoS One 7(1):e29347
4. Zhu W, Zhu Y, Qin K et al (2012) Mutations in polymerase genes enhanced the virulence of 2009 pandemic H1N1 influenza virus in mice. PLoS One 7(3):e33383
5. Govorkova EA, Ilyushina NA, McClaren JL et al (2009) Susceptibility of highly pathogenic H5N1 influenza viruses to the neuraminidase inhibitor oseltamivir differs in vitro and in a mouse model. Antimicrob Agents Chemother 53:3088–3096
6. Sidwell RW, Bailey KW, Wong M-H, Barnard DL, Smee DF (2005) In vitro and in vivo influenza virus-inhibitory effects of viramidine. Antiviral Res 68:10–17
7. Radigan KA, Urich D, Misharin AV et al (2012) The effect of rosuvastatin in a murine model of influenza a infection. PLoS One 7(4):e35788
8. Yazawa K, Kurokawa M, Obuchi M et al (2011) Anti-influenza virus activity of tricin, 4′,5,7-trihydroxy-3′,5′-dimethoxyflavone. Antivir Chem Chemother 22:1–11
9. Sidwell RW, Smee DF, Huffman JH, Barnard DL et al (2001) Influence of virus strain, challenge dose, and time of therapy initiation on the in vivo influenza inhibitory effects of RWJ-270201. Antiviral Res 51:179–187
10. Smee DF, Wong M-H, Bailey KW, Sidwell RW (2006) Activities of oseltamivir and ribavirin used alone and in combination against infections in mice with recent isolates of influenza A (H1N1) and B viruses. Antivir Chem Chemother 17:185–192
11. Murray JL, McDonald NJ, Sheng J et al (2012) Inhibition of influenza A virus replication by antagonism of a PI3K-AKT-mTOR pathway member identified by gene-trap insertional mutagenesis. Antivir Chem Chemother 22:205–215
12. Bantia S, Kellogg D, Parker C et al (2011) A single intramuscular injection of neuraminidase inhibitor peramivir demonstrates antiviral activity against novel pandemic A/California/04/2009 (H1N1) influenza virus infection in mice. Antiviral Res 90:17–21
13. Sidwell RW, Huffman JH, Call EW, Alaghamandan H, Robins RK (1986) Effect of selenazofurin on influenza A and B virus infections of mice. Antiviral Res 6:343–353
14. Ryan DM, Ticehurst J, Dempsey MH, Penn CR (1994) Inhibition of influenza virus replication in mice by GG167 (4-guanidino-2,4-dideoxy-2,3-dehydro-N-acetylneuraminic acid) is consistent with extracellular activity of viral neuraminidase (sialidase). Antimicrob Agents Chemother 38:2270–2275
15. Ekiert DC, Friesen RHE, Babha G et al (2011) A highly conserved neutralizing epitope on group 2 influenza A viruses. Science 333: 843–850
16. Galabov AS, Simeonova L, Gegova G (2006) Rimantadine and oseltamivir demonstrate synergistic combination effect in an experimental infection with type A (H3N2) influenza virus in mice. Antivir Chem Chemother 17:251–258
17. Sidwell RW, Huffman JH, Gilbert J et al (1992) Utilization of pulse oximetry for the study of the inhibitory effects of antiviral agents on influenza virus in mice. Antimicrob Agents Chemother 36:473–476
18. Yoo DG, Kim MC, Park MK et al (2012) Protective effect of ginseng polysaccharides on influenza viral infection. PLoS One 7(3):e33678
19. Sidwell RW, Barnard DL, Day CW, Smee DF et al (2007) Efficacy of orally administered T-705 on lethal avian influenza A (H5N1) virus infections in mice. Antimicrob Agents Chemother 51:845–851
20. Ilyushina NA, Hay A, Yilmaz N et al (2008) Oseltamivir-ribavirin combination therapy for highly pathogenic H5N1 influenza virus infection in mice. Antimicrob Agents Chemother 52:3889–3897
21. Kaminski MM, Ohnemus A, Cornitescu M, Staeheli P (2012) Plasmacytoid dendritic cells and Toll-like receptor 7-dependent signaling promote efficient protection of mice against highly virulent influenza A Virus. J Gen Virol 93:555–559
22. Min JY, Vogel L, Matsuoka Y, Lu B et al (2010) A live attenuated H7N7 candidate vaccine virus induces neutralizing antibody that confers protection from challenge in mice, ferrets, and monkeys. J Virol 84:11950–11960
23. Tarbet EB, Maekawa M, Furuta Y, Babu YS, Morrey JD, Smee DF (2012) Combinations of favipiravir and peramivir for the treatment of pandemic influenza A/California/04/2009 (H1N1) virus infections in mice. Antiviral Res 94:103–110
24. Appleyard G, Maber HB (1974) Plaque formation by influenza viruses in the presence of trypsin. J Gen Virol 25:351–357

25. Reed LJ, Muench H (1938) A simple method of estimating fifty per cent endpoints. Am J Hyg 27:493–498
26. Smee DF, Hurst BL, Wong M-H et al (2010) Combinations of oseltamivir and peramivir for the treatment of influenza A (H1N1) virus infections in cell culture and in mice. Antiviral Res 88:38–44
27. Smee DF, Hurst BL, Wong M-H et al (2010) Effects of the combination of favipiravir (T-705) and oseltamivir on influenza A virus infections in mice. Antimicrob Agents Chemother 54:126–133
28. Julander JG, Hagloch J, Latimer S, Motter N, Dagley A, Barnard DL, Smee DF, Morrey JD (2011) Use of plethysmography in assessing the efficacy of antivirals in a mouse model of pandemic influenza A virus. Antiviral Res 92:228–236
29. Smee DF, Bailey KW, Wong M-H et al (2008) Treatment of influenza A (H1N1) virus infections in mice and ferrets with cyanovirin-N. Antiviral Res 80:266–271
30. Verhoeven D, Teijaro JR, Farber DL (2009) Pulse-oximetry accurately predicts lung pathology and the immune response during influenza infection. Virology 390:151–156
31. Prichard MN, Shipman C Jr (1990) A three-dimensional model to analyze drug–drug interactions. Antiviral Res 14:181–205
32. Ottolini MG, Blanco JC, Eichelberger MC et al (2005) The cotton rat provides a useful small-animal model for the study of influenza virus pathogenesis. J Gen Virol 86: 2823–2830
33. Lowen AC, Mubareka S, Tumpey TM et al (2006) The guinea pig as a transmission model for human influenza viruses. Proc Natl Acad Sci USA 103:9988–9992
34. Govorkova EA, Rehg JE, Krauss S et al (2005) Lethality to ferrets of H5N1 influenza viruses isolated from humans and poultry in 2004. J Virol 79:2191–2198
35. Govorkova EA, Marathe BM, Prevost A et al (2011) Assessment of the efficacy of the neuraminidase inhibitor oseltamivir against 2009 pandemic H1N1 influenza virus in ferrets. Antiviral Res 91:81–88
36. Huang SS, Banner D, Fang Y et al (2011) Comparative analyses of pandemic H1N1 and seasonal H1N1, H3N2, and influenza B infections depict distinct clinical pictures in ferrets. PLoS One 6(11):e27512
37. Boon AC, Finkelstein D, Zheng M et al (2011) H5N1 influenza virus pathogenesis in genetically diverse mice is mediated at the level of viral load. MBio 2(5):e00171-11
38. Otte A, Sauter M, Alleva L et al (2011) Differential host determinants contribute to the pathogenesis of 2009 pandemic H1N1 and human H5N1 influenza A viruses in experimental mouse models. Am J Pathol 179:230–239
39. Sidwell RW, Smee DF, Huffman JH, Barnard DL et al (2001) In vivo influenza virus-inhibitory effects of the cyclopentane neuraminidase inhibitor RJW-270201. Antimicrob Agents Chemother 45:749–757
40. Smee DF, Wandersee MK, Wong M-H et al (2004) Treatment of mannan-enhanced influenza B virus infections in mice with oseltamivir, ribavirin and viramidine. Antivir Chem Chemother 15:261–268
41. Southam DS, Dolovich M, O'Byrne PM, Inman MD (2002) Distribution of intranasal instillations in mice: effects of volume, time, body position, and anesthesia. Am J Physiol Lung Cell Mol Physiol 282:L833–L839
42. Nguyen JT, Smee DF, Barnard DL et al (2012) Efficacy of combined therapy with amantadine, oseltamivir, and ribavirin in vivo against susceptible and amantadine-resistant influenza A viruses. PLoS One 7(1):e31006
43. Bantia S, Kellogg D, Parker CD, Babu YS (2010) Combination of peramivir and rimantadine demonstrate synergistic antiviral effects in sub-lethal influenza A (H3N2) virus mouse model. Antiviral Res 88:276–280
44. Smee DF, von Itzstein M, Bhatt B, Tarbet EB (2012) Exacerbation of influenza virus infections in mice by intranasal treatments and implications for evaluation of antiviral drugs. Antimicrob Agents Chemother 56:6328–6333
45. Motulsky H (1995) Comparing two proportions. In: Motulsky H (ed) Intuitive biostatistics. Oxford University Press, New York, pp 235–236

# Part VI

## Alphaviruses

# Chapter 32

# Development of a High-Throughput Antiviral Assay for Screening Inhibitors of Chikungunya Virus and Generation of Drug-Resistant Mutations in Cultured Cells

**Edwin Yunhao Gong, Jean-François Bonfanti, Tania Ivens, Marijke Van der Auwera, Barbara Van Kerckhove, and Guenter Kraus**

## Abstract

Chikungunya virus (CHIKV) is a mosquito-borne Alphavirus that has already infected millions of people in recent large-scale epidemics in Africa, the islands of the Indian Ocean, South and Southeast Asia, and northern Italy. The infection is still ongoing in many countries, such as India. Although the fatal rate is approximately 0.1 % in the La Réunion outbreak, it causes painful arthritis-like symptoms that can last for months or even years. Currently, neither vaccine nor approved antiviral therapy exists to protect humans from chikungunya infection. Therefore, there is an urgent unmet medical need for the development of antiviral drugs for pre-exposure prophylaxis and/or treatment of chikungunya infections. In this chapter, we describe a fully validated ATP/luminescence assay that is effective for high-throughput screening of CHIKV inhibitors. Protocols for growing CHIKV stocks and generating drug-resistant viral variants for modes of action studies of compounds are also described.

**Key words** Chikungunya virus, Antiviral, High-throughput screening, CPE inhibition, Drug resistance selection

## 1 Introduction

Chikungunya virus (CHIKV), an enveloped positive-sense single-stranded RNA virus that belongs to the *Alphavirus* genus of the *Togaviridae* family, is a reemerging arbovirus transmitted by *Aedes* mosquitoes. Geographically, it is mainly epidemic in Africa, the islands of the Indian Ocean, and Southeast Asia, in total, more than 60 countries (WHO: Guidelines for prevention and control of chikungunya fever, 2009). CHIKV was first isolated from both man and mosquitoes in Tanzania in 1952. Since then, small-scale reemergences have been documented in both Africa and Asia. In 2004, CHIKV emerged in Kenya where approximately 5,000 cases were reported. The virus continued to spread around African countries. Around 2005–2006, CHIKV crossed the Indian Ocean

Edwin Yunhao Gong (ed.), *Antiviral Methods and Protocols*, Methods in Molecular Biology, vol. 1030,
DOI 10.1007/978-1-62703-484-5_32, © Springer Science+Business Media, LLC 2013

and caused explosive outbreaks in the Indian Ocean islands. For example, in La Réunion Island (France), about 266,000 of the 775,000 inhabitants (approximately 40 % of the population) were infected with CHIKV resulting in 237 deaths. The primary vector responsible for the outbreaks in Africa is the *Aedes aegypti* mosquito. However, the virus that swiped across Réunion Island had mutation A226V in the envelope protein E1. This mutation led to not only increased transmissibility and infectivity of the virus, but also jumping vector species to the Asian tiger mosquito (*Aedes albopictus*) and, subsequently, caused explosive outbreaks in India and other Asian countries. For example, approximately 1.4 million people were infected in India during 2006–2007. Furthermore, a returned Italian traveller brought the virus to Italy after visiting India, which caused about 300 cases of CHIKV infections in Italy around 2007 (for reviews *see* refs. 1–3).

CHIKV causes a wide variety of diseases in humans. The symptoms include high fever, headache, skin rash, etc., which are indistinguishable from dengue infection in epidemic regions. CHIKV also causes illnesses such as arthritis (30–40 % of the patients can last for months and up to 5 years) and CNS (central nerve system) infection. Currently, a number of pharmaceutical and biotech companies are developing vaccines for prophylactic intervention of CHIKV infection, but all are still in preclinical stage. There are also no specific antiviral drugs approved against CHIKV infection. The treatment is purely symptomatic and is based on nonsteroidal anti-inflammatory drugs (WHO guidelines).

In the acute CHIKV infection, the viremia period lasts around 7–10 days. However, some patients continue to experience arthralgia even years later. Is this persistent chronic inflammation caused by long-term viral persistence or is it immune mediated or both? The disease mechanism is still unclear. Recent evidence showed that macrophages may act as a cellular reservoir for the persistence of CHIKV during the late stages of infection. In addition, CHIKV (RNA and proteins) was found in perivascular synovial macrophages in one chronic patient 18 months post-infection [4]. Using a nonhuman primate animal model, cynomolgus macaques (*Macaca fascicularis*), long-term CHIKV infection was observed in joints, muscles, lymphoid organs, and liver, which could explain the long-lasting CHIKV disease symptoms observed in humans [5]. However, no infectious virus was isolated in any of these studies. The understanding of the disease mechanism of the CHIKV infection is of great importance since it may impact the therapeutic targets and options.

We have developed a cell-based high-throughput screening assay that was fully validated for identifying inhibitors against CHIKV [6]. Protocols for growing CHIKV stocks and for in vitro generation of drug-resistant isolates are also presented.

## 2 Materials

### 2.1 Cells, Virus, Media, and Reagents

1. Vero cells (African green monkey kidney cells): European Collection of Cell Cultures (ECACC).
2. Huh7 cells: a human hepatoma cell line kindly provided by Dr R. Bartenschlager (University of Heidelberg, Germany).
3. CHIKV strain S27: National Collection of Pathogenic Viruses (NCPV, UK) (*see* **Note 1**).
4. Eagle's minimum essential medium without L-glutamine (MEM).
5. Dulbecco's modified Eagle's medium (DMEM).
6. Heat-inactivated fetal calf serum (FCS).
7. 200 mM L-glutamine (100×).
8. Gentamicin solution (50 mg/mL).
9. 0.05 % Trypsin–EDTA.
10. Dulbecco's phosphate-buffered saline (D-PBS), no calcium chloride, no magnesium chloride (Gibco).
11. Vero cell growth medium (MEM/10 % FCS): Eagle's minimum essential medium supplemented with 10 % FCS, 2 mM L-glutamine, and 0.02 mg/mL gentamicin.
12. Vero cell maintenance medium (MEM/2 % FCS): same as the growth medium but supplemented with 2 % FCS.
13. Huh7 cell growth medium (DMEM/10 % FCS): Dulbecco's modified Eagle's medium containing 10 % FCS, 2 mM L-glutamine, and 0.02 mg/mL gentamicin.
14. Huh7 cell maintenance medium (DMEM/2 % FCS): same as the growth medium but supplemented with 2 % FCS.
15. ATPLite™ 1-Step kit (PerkinElmer).
16. Trypan blue stain (Gibco) (optional).
17. Reference compounds:
    (a) Interferon α-2a (Roferon A, Roche).
    (b) Chloroquine (Sigma-Aldrich).
    (c) 6-Azauridine (Sigma-Aldrich).

### 2.2 Consumables

1. Nunc® TripleFlasks (F8542, Sigma-Aldrich).
2. Greiner Lumitrac 384-well plate (VWR).
3. White-view 384-well plates (white wall and transparent bottom, PerkinElmer).
4. BD Falcon™ conical tubes (15 and 50 mL).

### 2.3 Equipment

1. Centrifuge.
2. Multidrop combi liquid dispenser (Thermo Scientific).

3. ViewLux™ (PerkinElmer) or any luminescent reader.
4. Cell counter (Coulter Electronics LTD) (optional).
5. Microplate shaker.

## 3 Methods

All methods described below are performed in a biosafety level 3 laboratory (P3) according to Belgian biosafety regulation. All incubations for cells at 37 °C described in this chapter are performed using a humidified 5 % $CO_2$ incubator unless otherwise stated.

### 3.1 Preparation of CHIKV Stocks

1. Seed Vero cells in T175 flasks using the Vero cell growth medium (MEM/10 % FCS) and incubate at 37 °C overnight. The cell monolayer should be confluent on the following day (*see* **Note 2**).
2. On the following day, thaw a frozen vial of virus stock in a 37 °C water-bath quickly with shaking and prepare a viral dilution at a multiplicity of infection (MOI) of approximately 0.001 $TCID_{50}$/cell (the 50 % tissue culture infective dose per cell) in MEM/2 % FCS. Prepare 15 mL of virus dilution for each flask to be infected.
3. Remove the old culture medium from the flasks and rinse the cells once with D-PBS.
4. Add 15 mL of the virus dilution in MEM/2 % FCS to each T175 flask and swirl the flask gently to spread the virus evenly over the cell monolayer.
5. Incubate the cells at 37 °C for 1 h with occasional swirling.
6. Remove the viral inoculum.
7. Wash the infected cells once with D-PBS to remove the unbound virus.
8. Add 18 mL of fresh MEM/2 % FCS to each flask and incubate the infected cells at 37 °C for 2 days. One hundred percentage of cytopathic effect (CPE) should be observed on infected cells at day 2 post-infection using an inverted microscope.
9. Pool the culture medium containing released virus from all flasks.
10. Collect the pooled culture medium into sterile 50-mL Falcon tubes and centrifuge at 3,700 × *g* for 10 min at 4 °C to remove the cell debris.
11. Aliquot 1 mL of the supernatant to each cryovial and store in a −80 °C freezer. The titer of the virus stock can be determined as described in Subheading 3.2.

### 3.2 Titration of CHIKV Stocks

1. The day before titration, trypsinize the Vero cells from a T175 culture flask and resuspend the cells in MEM/10 % FCS.
2. Count the cells using either a cell counter or trypan blue stain and prepare a cell suspension at a density of 10,000–15,000 cells/100 μL.
3. Add 100 μL of cell suspension to each well of a 96-well plate.
4. Incubate the cells at 37 °C overnight. The cell monolayer should be confluent after overnight culture.
5. Thaw a frozen vial of virus stock in a 37 °C water-bath quickly with shaking and perform a tenfold serial dilution in MEM/2 % FCS. Prepare 1 mL for each virus dilution.
6. Remove the culture medium from the plates using a multiple-channel pipette or a sterile glass pipette attaching to a vacuum pump with a waste container (*see* **Note 3**).
7. Pipette 100 μL/well of each virus dilution to the corresponding wells of one column (eight identical wells for each virus dilution). Add 100 μL/well of undiluted virus to all wells of column 1. Columns 11–12 are uninfected cell controls and receive 100 μL/well medium only.
8. Incubate the plate at 37 °C for 2–3 days.
9. Record the wells with and without CPE for each virus dilution under an inverted microscope.
10. Calculate virus titer as the 50 % tissue culture infective dose per mL ($TCID_{50}$/mL) by the Reed–Muench method.

### 3.3 A High-Throughput Screening Assay for CHIKV Inhibitors Using a Hepatoma Cell Line

CHIKV causes extensive CPE on many cell lines, such as Vero cells and Huh7 cells (a hepatoma cell line). The high-throughput screening assay described here is based on the inhibition of viral CPE. The assay readout is luminescence. The antiviral efficacy of test compounds is determined by measuring the ATP levels in live cells protected from viral CPE by a test compound [7]. The cytotoxicity of test compounds is measured in parallel in the absence of virus infection. The hits are selected by comparing the selectivity index (SI), which is defined as the ratio of the median toxic dose ($CC_{50}$) to the median effective dose ($EC_{50}$) ($SI = CC_{50}/EC_{50}$). Our assay was fully validated using both Vero and Huh7 cells according to the FDA (Food and Drug Administration) guidelines for assay validation. However, when we screened a small library of nucleoside analogs using the Vero cells, we found that the Vero cells might not be suitable for screening the nucleoside analogs in this setting. The exact reason is unknown, but it is likely due to the phosphorylation status of the nucleoside analogs within cells in this 2-day assay. Thus, the protocol described below is based on the format using the Huh7 cells.

1. The test compounds are plated in 384-well opaque white plates (Greiner Lumitrac 384-well plate) in a volume of 0.2 μL (200 nL) per well using an Echo® liquid handling system (Labcyte). With this dispensing system, all test compounds are dissolved and diluted in 100 % DMSO. As the final volume in each well is 40 μL, the final concentration of DMSO is 0.5 %. Both cell control (absence of both compound and virus infection) and virus control (absence of compound) wells contain the same concentration of DMSO. All test compounds are four- or fivefold serially diluted to prepare 200× final test concentrations. The reference compounds are plated in a separate white-view 384-well plate (white wall with a transparent bottom) (*see* **Note 4**) in order to monitor the cells and viral CPE.

   For primary screening of compound libraries, the plate format is 80 compounds per plate with four single test concentrations for each compound. In this format, columns 21–22 are virus controls and columns 23–24 are cell controls.

   For hit confirmation and compound profiling, the plate formats are:

   (a) Eight compounds per plate: nine serially diluted test concentrations (fourfold or fivefold serial dilutions) with quadruplicate wells for each test concentration. Columns 19–22 are virus controls and 23–24 are cell controls.

   (b) 16 compounds per plate: five serially diluted test concentrations with quadruplicate wells for each test concentration. Columns 21–22 are virus controls and 23–24 are cell controls.

   All compound plates are stored in a 4 °C cold room before starting the screening experiment.

2. Three to four days before the screening experiment, seed enough Huh7 cells in T175 flasks using the Huh7 cell growth medium (DMEM/10 % FCS) and incubate at 37 °C for 3–4 days. The cells should be confluent at the day of experiment.
3. At the day of experiment, discard the old culture medium from the flasks and wash the cells with 15 mL of D-PBS.
4. Add 5 mL of 0.05 % trypsin–EDTA to each T175 flask and spread over the cell monolayer.
5. Discard excess of trypsin–EDTA and incubate the flask at room temperature for 3–5 min (*see* **Note 5**).
6. Tap the flask with the palm of the hand to detach the cells. Continue to incubate and tap until all cells are detached.
7. Add 10 mL of DMEM/2 % FCS to each flask. Rock the flask to dislodge the cells.
8. Aspirate the cells with a 10-mL plastic pipette to break into single cell suspension.

9. Count the cells using a cell counter or trypan blue solution.
10. Prepare cell suspensions in DMEM/2 % FCS (*see* **Note 6**) at a density of $2.4 \times 10^5$ cells/mL for the activity plates and a density of $1.5 \times 10^5$ cells/mL for the toxicity plates. Stir the cells gently with a sterile stirring bar on a magnetic stirrer.
11. Thaw one vial of CHIKV stock quickly in a 37 °C water-bath with shaking. Then wipe the vial surface with 70 % ethanol to avoid contamination.
12. Make an appropriate dilution of the virus in DMEM/2 % FCS: an MOI of 0.05 ($TCID_{50}$/cell) (*see* **Note 7**).
13. Using a multidrop combi liquid dispenser, dispense cells and virus into the test plates (a total volume is 40 μL per well):

    For activity plates:

    (a) Dispense 15 μL of DMEM/2 % FCS into all wells of columns 23 and 24 (cell controls).

    (b) Dispense 25 μL of cell suspension ($2.4 \times 10^5$ cells/mL) into all wells of the test plate (the final density is 6,000 cells per well).

    (c) Dispense 15 μL of virus suspension into the wells of columns 1–22.

    For toxicity plates:

    (a) Dispense 40 μL of DMEM/2 % FCS into the wells of columns 23 and 24 (medium controls).

    (b) Dispense 40 μL of cell suspension ($1.5 \times 10^5$ cells/mL) into all wells of columns 1–22 (the final density is 6,000 cells per well).
14. Incubate the plates at 37 °C for 2 days.
15. At the end of incubation, check the cells in the reference compound plate under an inverted microscope. The CPE in virus-control wells should reach ~100 %. The ATP levels in compound-treated and untreated cells are measured by a luminescent assay using an ATPLite™ 1-step kit. We use ViewLux™ to measure the luminescence. The following procedures are recommended by the manufacturer:

    (a) Equilibrate all reagents to room temperature before use.

    (b) Reconstitute the lyophilized substrate solution by adding the appropriate volume of buffer to the substrate bottle. Mix the contents by inversion and leave the solution to stand for 5 min (*see* **Note 8**).

    (c) Add 40 μL of the reconstituted reagent to each well of both activity and toxicity plates using a multidrop combi liquid dispenser.

(d) Shake the plates for 2 min at 700 rpm using an orbital microplate shaker.

(e) Program Viewlux to take a 0.1 s integrated reading of each plate.

16. The antiviral activity of a test compound is expressed as $EC_{50}$ values defined as the concentration of compound achieving 50 % inhibition of the virus-reduced luminescence signals as compared with the uninfected cell control. The 50 % cytotoxic concentration ($CC_{50}$) is determined by comparing the luminescent signal of compound-treated wells with untreated cell control wells. Both $EC_{50}$ and $CC_{50}$ are calculated using the GraphPad Prism software.

### 3.4 In Vitro Drug-Resistance Selection

1. Seed the Vero cells at a density of $5–6 \times 10^5$ cells/10 mL in a T25 culture flask with MEM/10 % FCS the day before infection. Include one flask for virus control (absence of compound).
2. Incubate the cells at 37 °C overnight.
3. On the following day, remove old culture medium and rinse the cells once with D-PBS.
4. Inoculate the cells with 3 mL of diluted CHIKV stock at an MOI of 0.001 ($TCID_{50}$/cell).
5. Incubate the flasks at 37 °C for 1 h with periodical titling.
6. Discard the viral inoculum and wash the infected cells once with D-PBS.
7. Add 10 mL of MEM/2 % FCS to the compound flask and the virus control flask. Then add 50 μL of test compound (200× final test concentration) to the compound flasks only. The final concentration of test compound is equal to its $EC_{50}$ value.
8. Incubate the flasks at 37 °C.
9. From day 2, check the viral CPE under an inverted microscope every day. Continue to incubate at 37 °C.
10. If full CPE is examined, harvest the virus by collecting the supernatant into a 15-mL Falcon™ conical tube and centrifuging at 3,700 × $g$ for 10 min to pellet the cell debris.
11. Aliquot the supernatant into five cryovials with 1.5 mL each, and store at −80 °C.
12. Continue to passage the virus by repeating **steps 1–11**. With every passage, the compound concentration is increased two-fold if 100 % CPE is observed in both virus control and compound-treated infected cells at the same time.
13. If partial CPE or no CPE is observed on the compound-treated infected cells in 4 or 5 days when compared with the virus

control, or if the CPE in the compound-treated infected cells is delayed for 2–4 days after full CPE is examined on the virus control, the compound concentration in the next passage should remain unchanged.

14. Continue to passage the virus in the presence of increased compound concentration until compound concentration has reached $2 \times CC_{50}$.
15. Purify total RNA from the virus stock of the final passage and perform a sequence analysis for the whole CHIKV genome.

## 4 Notes

1. According to the Belgian biosafety regulation, the handling and manipulation of CHIKV are carried out in a biosafety level three laboratory.
2. CHIKV is one of the fast-growing viruses and can easily grow to a high titer. Host protein synthesis is shut down after infection. Therefore, confluent cells should be used in preparing virus stocks and compound screening.
3. Make sure not to suck off cells from the wells of a 96-well plate when a glass pipette attaching to a vacuum pump is used to remove the old medium from cultured wells.
4. White-view plates (white-wall wells with transparent bottom) are used for plating reference compounds. Using this kind of plates, the viral CPE and cell growth can be monitored under a microscope.
5. When Huh7 cells are over-trypsinized in 0.05 % trypsin–EDTA at 37 °C, cell aggregates can occur and be difficult to break into single cell suspension. Therefore, we normally use room temperature to trypsinize the Huh7 cells to avoid the cell aggregates.
6. We use medium with reduced serum concentration (maintenance medium, DMEM/2 % FCS) to reduce the protein binding of some compounds.
7. This high-throughput screening assay can also be performed using the Vero cells for testing non-nucleoside analogs. In this setting, the optimal cell input is 6,000 cells per well and virus input is 0.001 MOI ($TCID_{50}$/cell). It is also a 2-day assay.
8. Reconstituted reagent is light sensitive. If a transparent bottle is used, always wrap it using aluminum foil. The leftover of reconstituted agent can be stored at −80 °C for up to 1 month.

## Acknowledgments

The assay development presented in this chapter was sponsored by an IWT (agentschap voor Innovatie door Wetenschap en Technologie: Agency for Innovation by Science and Technology) Grant from the Flemish Government, Belgium.

### References

1. Schwartz O, Albert M (2010) Biology and pathogenesis of chikungunya virus. Nat Rev Microbiol 8(7):491–500
2. Pialoux G, Gaüzère B, Jauréguiberry S, Strobel M (2007) Chikungunya, an epidemic arbovirosis. Lancet Infect Dis 7(5):319–327
3. Gould EA, Coutard B, Malet H et al (2010) Understanding the alphaviruses: recent research on important emerging pathogens and progress towards their control. Antiviral Res 87(2): 111–124
4. Hoarau J, Bandjee M et al (2010) Persistent chronic inflammation and infection by chikungunya arthritogenic alphavirus in spite of a robust host immune response. J Immunol 184: 5914–5927
5. Labadie K, Larcher T et al (2010) Chikungunya disease in nonhuman primates involves long-term viral persistence in macrophages. J Clin Invest 120(3):894–906
6. Gong E, Ivens T, Bonfanti JF, Kraus G (2012) Method for identifying inhibitors against chikungunya virus. International Publication Number: WO2012/010617 A2
7. Gong E, Ivens T, Van den Eynde C et al (2008) Development of robust antiviral assays for profiling compounds against a panel of positive-strand RNA viruses using ATP/luminescence readout. J Virol Methods 151:121–125

# Chapter 33

# A Mouse Model of Chikungunya Virus with Utility in Antiviral Studies

Ashley Dagley and Justin G. Julander

## Abstract

Chikungunya virus (CHIKV) infection generally causes a debilitating arthritis in infected patients. Infection with CHIKV is generally not life-threatening and is associated with a mortality rate <0.1 %. However, close to 100 % of those infected will develop symptoms of disease, primarily involving swelling and pain of the joints, which can last for months or even years. A model that mimics these symptoms is needed for the development of therapies to ameliorate disease and control viral infection. In this chapter, we describe the establishment of a model of CHIKV infection in mice that is nonlethal and utilizes footpad swelling and virus titer of various tissues as key disease parameters. This model was developed primarily for use in evaluating the in vivo efficacy of candidate antiviral agents, although important questions regarding basic biology and pathogenesis of the disease may also be elucidated using this system.

**Key words** Chikungunya virus, Mouse model, Antiviral, In vivo efficacy

## 1 Introduction

Chikungunya virus (CHIKV) is a mosquito-borne *Alphavirus*. It was first identified in East Africa in the early 1950s. The name was taken from the Makonde language of this region and is translated to "that which bends up," in reference to the debilitating arthritis commonly associated with infection.

Rapid dissemination of CHIKV through the body occurs after virus transmission from an infected mosquito. Local replication of the virus occurs in the skin and is disseminated through the body via the blood [1]. Initial symptoms after infection of a person with CHIKV may include a sudden onset of fever, headache and rash. The febrile phase of disease generally subsides after 2–3 days and is followed by a more severe arthralgia that may last less than a week or may persist for months or even years [2]. Swelling of joints can result in severe pain and patients are usually rendered prostrate during the initial arthritic disease.

Edwin Yunhao Gong (ed.), *Antiviral Methods and Protocols*, Methods in Molecular Biology, vol. 1030,
DOI 10.1007/978-1-62703-484-5_33, 

In the past, this disease was only recorded from a few countries within East Africa. Recently, CHIKV has spread throughout various regions of the world, increasing the need for countermeasures. An important early event included a spread across mainland Africa to Madagascar. Serious outbreaks of the virus were recorded on various islands in the Indian Ocean Region, most notably La Reunion, which resulted in the infection of approximately 1/3 of the population (~266,000 people) [3, 4]. Subsequent research identified a mutation in the E1 gene of the virus [5], which allowed infection and spread by *Aedes albopictus*, an important vector of disease that was not known to transmit CHIKV prior to this change in the virus envelope [6]. This also resulted in an increase in the observance of atypical symptoms including meningoencephalitis, maculopapular rash, renal failure, and hepatitis [7]. A mortality of 1/1,000 cases was observed during the La Reunion outbreak, although 100 % of infected patients suffered from joint swelling or pain [1].

The virus spread further, resulting in large-scale outbreaks in India and other countries in Southeast Asia [8]. In India alone, an estimated 1.39 million cases have been reported since 2006 [9]. In 2007 there was an outbreak in Italy [10], demonstrating importation by infected travelers and the risk of further spread to other areas, similar to what was observed after the West Nile virus outbreak in the USA in 1999 [11]. Other reported cases of imported CHIKV associated with infected travelers returning from affected areas occurred in North America, France, and Japan [12]. These isolated cases, however, did not result in the further spread of the virus or establishment of the virus reservoir in these areas.

To best model the natural disease associated with CHIKV infection, an animal model would include measurable disease parameters that are seen in human cases, including virus replication at relevant sites, polyarthralgia, fever, rash, and cytokine elevation. It would also be important to model the more severe form of the disease; a persistent arthralgia that lasts for months to years [13]. Severe disease in mice has been shown to be associated with age as well as immune competence [14]. Suckling mice have been used in protection studies, as challenge of neonate mice with a high virus inoculum may result in clinical signs of severe disease, including joint swelling, paralysis, and even death [15, 16]. Infection of immunocompetent adult mice may result in characteristic swelling of joints, with severity dependent on the virus strain [17]. A nonlethal model in rhesus macaques includes a rapid viremia, similar to that observed in humans, as well as a robust and rapid production of neutralizing antibody [18]. Despite available models, there is a paucity of antiviral studies in vivo [16, 19].

The rapid spread and large number of cases throughout the world are of great medical importance. There are currently no vaccine and no antiviral treatment for CHIKV, underscoring the need for research in this area. Although several models have been developed, some of these use less available strains or may be difficult to establish. We outline here a nonlethal model of CHIKV infection that includes limb swelling and arthritis in a readily available mouse strain. Our primary focus is to identify key mediators of disease and to discover compounds that are effective in controlling virus replication and ameliorating associated disease.

## 2 Materials

1. C6/36 cell line (ATCC, CRL-1660).
2. Fetal bovine serum (FBS).
3. C6/36 cell culture medium: RPMI 1640, 1× with 2.05 mM L-glutamine and 5 % FBS.
4. CHIKV strain S27 (ATCC, VR-64).
5. Incubator.
6. Cell rocker.
7. Tail vein injection platform (model IL-300 from Bioseb).
8. Isoflurane.
9. Bell jar.
10. Tuberculin syringe.
11. 12″ Digital calipers (model 01414A from Neiko).
12. Scale.
13. Surgical scissors.
14. Forceps.
15. Mortar and pestle.
16. Q-plex mouse cytokine array (Catalog #110451MS, Quansys Biosciences).
17. DBA/1J mice: female, age 4–6 weeks (The Jackson Laboratory, Sacramento, CA).
18. Vero76 cell line (ATCC CRL-1587).
19. Vero76 cell culture medium: MEM (Eagle's minimum essential medium) with 5 % FBS.
20. Gentamicin.
21. T-150 flasks.

## 3 Methods

### 3.1 Virus Infection

Perform virus inoculation within a class 2 biosafety cabinet under BSL-3 conditions. Appropriate personal protective equipment (PPE), including gloves, gown, protective eyewear, and other institutional required PPE should be worn. All animal work should be carried out under appropriate institutional animal welfare approvals and within approved animal facilities.

1. Propagate the S27 virus strain in C6/36 mosquito cell line in T-150 flasks. Approximately 18 h after seeding, add a 1/100 dilution of virus stock to cells in a 5 mL volume. After the addition of virus, place inoculated flasks on a rocker for 2 h for virus attachment. A 25 mL volume of medium (RPMI+5 % FBS) is then added to each flask, followed by incubation for 2 days at 28 °C, 5 % $CO_2$.
2. Harvest virus 2 days post-infection by centrifuging at 2,080 × *g* aliquot the supernatant, and store at −80 °C (*see* **Note 1**). Titrate the virus stock using end-point dilution method to determine the 50 % cell culture infectious dose per milliliter ($CCID_{50}$/mL), or the amount of virus required to produce 50 % cytopathic effect. For example, the virus stock used to characterize this model has a titer of $10^{9.2}$ $CCID_{50}$/0.1 mL. A 1/10 dilution of virus is used for footpad inoculation.
3. One DBA/1J mouse (*see* **Note 2**) at a time is anesthetized by placing in a Bell Jar containing a suitable amount of isoflurane. Once movement ceases and breath rate slows, the animal is removed for footpad inoculation.
4. Inject the virus dilution quickly before the mouse recovers from anesthesia. The right foot is isolated using the tail vein injection platform (*see* **Note 3**). The ventral side of the footpad should be facing up (*see* Fig. 1). Using a 0.5 mL tuberculin syringe, insert the needle bevel up into the footpad and inject 0.05 mL of virus inoculum (~$10^{8.9}$ $CCID_{50}$ of virus). A volume of 0.05 mL is then injected subcutaneously above the hock joint.
5. Inoculate sham control mice with dilution media alone (*see* **Note 4**).
6. Return mouse to caging and monitor mouse to ensure recovery from anesthesia. Also observe mice the day after infection to determine if any complications associated with footpad inoculations are observed.

### 3.2 Antiviral Treatment

This model has utility in antiviral research. Various compounds can be administered to mice at various times prior to or after virus challenge. The most important considerations are the rapid dissemination of the virus through various tissues of the mice and the

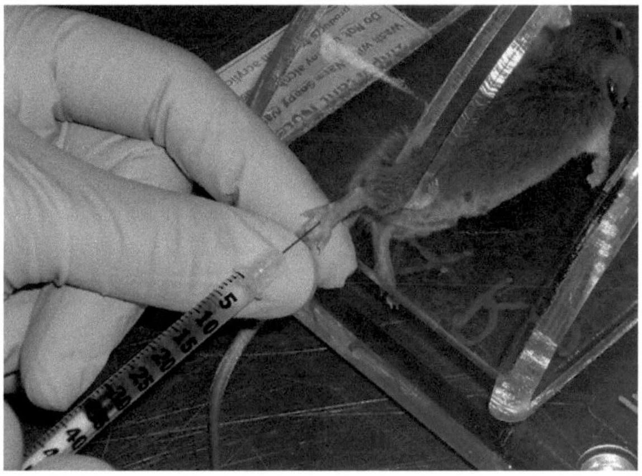

**Fig. 1** The technique for footpad injection using a tail vein injection platform to isolate the hind limb. Mice are anesthetized with isoflurane prior to injection procedure

contribution of the immune system to disease. The nature of the compound will dictate the route and timing of treatment. Below is an example of treatment with a direct acting antiviral compound.

1. Animals are properly restrained for the treatment route selected.
2. Compound is administered (*see* **Note 5**) 4 h prior to virus challenge.
3. A treatment regimen may include multiple treatments through the course of infection.
4. A toxicity control group, consisting of uninfected mice, is included. Both toxicity control and infected animals should be monitored frequently for signs of toxicity.
5. The disease parameters detailed below are used to evaluate the efficacy of compounds in the treatment of disease associated with CHIKV infection. Sufficient numbers of mice are included for statistical strength when analyzing data. We have determined that groups consisting of 7–10 animals generally provide sufficient numbers for statistical analyses.

### 3.3 Measurement of Footpad Swelling

At various times after inoculation, the footpad thickness is measured with a caliper. Peak swelling in DBA/1J mice inoculated with S27 CHIKV occurs around 6–7 days post-virus inoculation (dpi) (*see* Fig. 2).

1. Properly restrain a mouse for footpad measurement. Mice can be individually marked with an eartag or by other means for longitudinal measurements. A firm grip will reduce movement of the hind legs, but care must be taken to perform the measurements quickly to avoid undue stress to the animals (*see* **Note 6**).

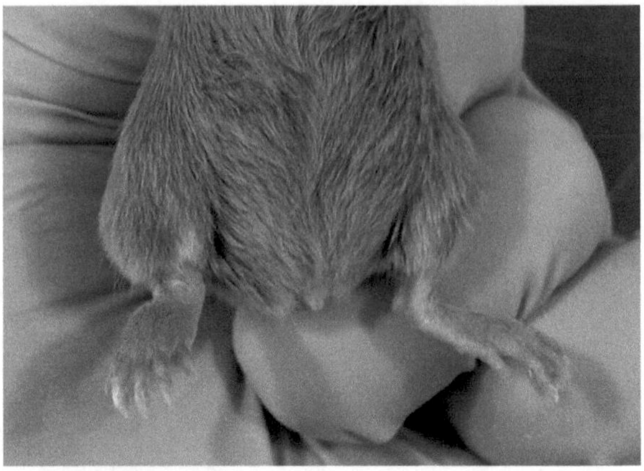

**Fig. 2** Swollen footpad of a DBA/1 J mouse on 6 days after challenge with CHIKV

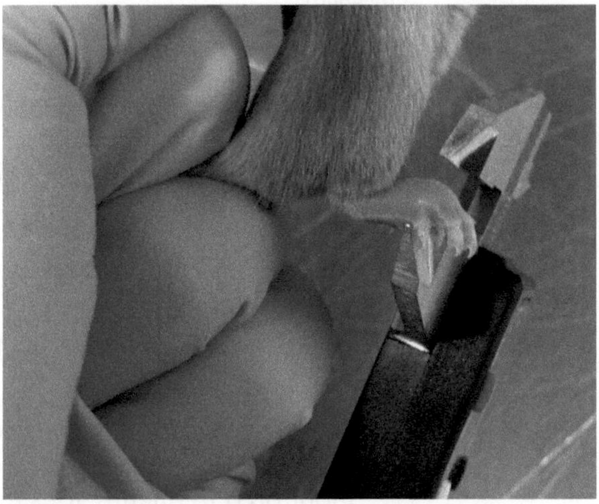

**Fig. 3** Measurement of swollen footpad using digital calipers

2. Measure the dorsoventral thickness of the footpad using a caliper (*see* Fig. 3). In mice infected with S27 CHIKV, the footpad will become obviously swollen by 5 or 6 days after virus inoculation.
3. The contralateral foot to the site of inoculation serves as an internal control for swelling. Footpad inoculation results in swelling at the site of virus injection, but does not result in measureable spread of virus or swelling in the contralateral leg. Compare measurements of the right (infected) to left (non-infected) footpad to obtain a percent increase in swelling over baseline.
4. Measure the footpad by holding the caliper perpendicular to the footpad (*see* **Note 7**). Begin with calipers open to approximately 2× the thickness of the footpad and slowly decrease the

size until calipers lightly touch the top and bottom of footpad. Care should be taken not to pinch or depress the swollen footpad. Measurements may need to be repeated to verify initial observations.

5. Record measurements for animal identification and right and left footpad thickness. It is helpful to have a second person present to help record the measurements for each animal. In addition, the mice may be ear-tagged to track longitudinal changes in swelling of individual mice.
6. Compare the right and left footpad measurements to obtain percent increase in footpad size at the site of virus inoculation. Divide the right footpad measurement by the left footpad measurement. Multiply this ratio by 100 and subtract 100 for percent increase in footpad thickness. Over several experiments, the average increase in footpad thickness in DBA/1J mice on 6 and 7 dpi ranges from 30–45 and 39–51 %, respectively (*see* **Note 8**).

### 3.4 Necropsy

Perform a necropsy on a subset of animals to acquire hind limb for quantification of viral or host disease parameters. Adequate homogenization is very important to ensure virus and cytokines are accessible for assays.

1. Euthanize group of mice by institution approved euthanasia procedure.
2. Pin all limbs except right hind leg using T-pins on a corkboard. Stretch right hind limb out using sterile tissue forceps; sever femur at midline using sterile surgical scissors. Place limb into labeled 24-well plate.
3. Place one limb into mortar.
4. Add 1 mL MEM supplemented with 50 μg/mL gentamicin.
5. Use sterile pestle to grind footpad thoroughly.
6. Collect as much liquid with pipette, taking care to avoid larger tissue fragments, and place into labeled vial (*see* **Note 9**).
7. Cold-sterilize instruments and repeat.

### 3.5 Viral Titer and Cytokine Assays

Tissues must be assayed for comparison among groups.

#### 3.5.1 Viral Titer Assay

1. Prepare plates 1 day in advance. Plate Vero cells at $4 \times 10^4$ cells per well in a 96-well plate in MEM supplemented with 5 % FBS. Cells should be 95–100 % confluent the following day.
2. Clarify homogenized limb samples by centrifugation at $2{,}500 \times g$ and 4 °C for 10 min.

3. Carefully remove medium from plates. Add 100 μL MEM with 5 % FBS to each well.
4. Dilute samples by adding 60 μL sample to 540 μL MEM with gentamicin. Serially dilute $10^{-1}$–$10^{-7}$, leave last row for cell control.
5. Transfer 100 μL of sample to the corresponding wells of a microtiter plate in four replicates, with three samples per plate.
6. Read plate for viral-induced cytopathic effect (CPE). CPE should be clearly visible by day 3 post-infection.
7. Utilize Reed–Muench method to calculate $CCID_{50}$ [20].

#### 3.5.2 Cytokine Assay

Perform cytokine assay per manufacturer's instruction. Various cytokine assays are available, but we utilized the Q-Plex Mouse Cytokine Array kit. An outline of the steps needed for this assay is briefly described below.

1. Add the appropriate dilution of sample to the plate and incubate at room temperature for 1 h. Samples must be diluted at least 1:1 with dilution buffer. If high cytokine levels are anticipated, a more dilute sample is used. An 8-point standard curve (provided with the Q-plex Array kit) is also prepared as a reference for quantification.
2. Wash the plate three times and add detection mix. Incubate as above.
3. Wash as above and add streptavidin horseradish peroxidase solution. Incubate for 15 min at room temperature.
4. Wash the plate six times and add substrate. The plate should then be imaged within 30 min of substrate addition using a supported imaging system.
5. Calculate the cytokine quantities using Q-View Software.

## 4 Notes

1. Virus titer of CHIKV strain S27 peaked 2 days after virus inoculation and declined steadily thereafter.
2. The DBA/1J mouse strain is a smaller breed of mouse. Five- to six-week-old females are used and generally weigh around 15–17 g upon arrival. We allow the animals to acclimate to the facility for 3 days prior to virus challenge. This mouse strain is commonly used in rheumatoid arthritis research.
3. The purpose of the tail vein injection platform is to isolate the foot of the mouse for footpad injection while the mouse is under anesthesia. We have found this particular model to be suitable for isolation of the hind leg for footpad inoculation, although other models may work well. The most important

consideration is to minimize stress/pain and suffering of the animals during this procedure.

4. For a true control, a 1/10 dilution of culture supernatant from C6/36 cells grown in parallel with those inoculated with CHIKV is used. During preparation of the virus stocks, a cell control flask is also propagated for the same amount of time and the supernatant is harvested to be used in studies as a negative control for swelling.
5. Depending on the compound, various treatment routes can be utilized. Common routes of treatment initiation used in our lab include intraperitoneal, subcutaneous, oral and intravenous. Since this strain of mice is smaller, a higher gage needle is recommended for injection. For continuous administration, we have found osmotic pumps (Alzet, Cupertino, CA) to be useful, which are implanted subcutaneously under the neck scruff of the animal and deliver a specific amount at a calculated rate. Since this strain of mice is smaller, a higher gage needle is recommended.
6. It may be easier to measure footpad thickness if one person restrains the animal and a second person measures the footpad thickness. The toes can be gently restrained and extended to make measurement easier. Joints of infected mice appear to have less elasticity, and may be painful for the animals. Care must be taken to reduce any undue stress or pain for the animals.
7. Measure footpads in a consistent location to ensure repeatable measurements. Topological features of the footpad can be used to consistently place calipers. The most important consideration is the comparison between right and left footpad to get an accurate indication of the increase in footpad height during peak swelling as compared with the contralateral foot.
8. After determining the percent increase in swelling of the inoculated footpad as compared with the contralateral footpad, statistical analyses of groups can be made using a one-way analysis of variance.
9. Move all solid tissue up on the side of the mortar to allow medium to pool at the bottom of mortar for ease in collection. Collected samples may be further clarified in preparation for titration in cell culture and for analysis of cytokine levels by centrifugation at 2,000 × *g*.

## Acknowledgment

This work was supported by NIH contract HHSN27220/000039I/HHSN27200005/A21.

## References


1. Dupuis-Maguiraga L, Noret M, Brun S et al (2012) Chikungunya disease: Infection-associated markers from the acute to the chronic phase of arbovirus-induced arthralgia. PLoS Negl Trop Dis 6(3):e1446
2. Couderc T, Lecuit M (2009) Focus on chikungunya pathophysiology in human and animal models. Microbes Infect 11(14–15): 1197–1205
3. Larrieu S, Balleydier E, Renault P et al (2012) Epidemiological surveillance du chikungunya on reunion island from 2005 to 2011. Med Trop Rev Corps Sante Colon 72:38–42
4. Gerardin P, Guernier V, Perrau J et al (2008) Estimating chikungunya prevalence in la reunion island outbreak by serosurveys: two methods for two critical times of the epidemic. BMC Infect Dis 8:99
5. Singh RK, Tiwari S, Mishra VK, Tiwari R, Dhole TN (2012) Molecular epidemiology of chikungunya virus: mutation in e1 gene region. J Virol Methods 185(2):213–220
6. Tsetsarkin KA, Vanlandingham DL, McGee CE, Higgs S (2007) A single mutation in chikungunya virus affects vector specificity and epidemic potential. PLoS Pathog 3(12):e201
7. Economopoulou A, Dominguez M, Helynck B et al (2009) Atypical chikungunya virus infections: clinical manifestations, mortality and risk factors for severe disease during the 2005–2006 outbreak on reunion. Epidemiol Infect 137(4):534–541
8. Kelvin AA (2011) Outbreak of chikungunya in the republic of congo and the global picture. J Infect Dev Ctries 5(6):441–444
9. Kelvin AA, Banner D, Silvi G et al (2011) Inflammatory cytokine expression is associated with chikungunya virus resolution and symptom severity. PLoS Negl Trop Dis 5(8):e1279
10. Rezza G, Nicoletti L, Angelini R et al (2007) Infection with chikungunya virus in italy: an outbreak in a temperate region. Lancet 370(9602):1840–1846
11. Savage HM, Ceianu C, Nicolescu G et al (1999) Entomologic and avian investigations of an epidemic of west Nile fever in Romania in 1996, with serologic and molecular characterization of a virus isolate from mosquitoes. Am J Trop Med Hyg 61(4):600–611
12. Gibney KB, Fischer M, Prince HE et al (2011) Chikungunya fever in the United States: a fifteen year review of cases. Clin Infect Dis 52(5): e121–e126
13. Hoarau JJ, Jaffar Bandjee MC, Krejbich Trotot P et al (2010) Persistent chronic inflammation and infection by chikungunya arthritogenic alphavirus in spite of a robust host immune response. J Immunol 184(10): 5914–5927
14. Couderc T, Chretien F, Schilte C et al (2008) A mouse model for chikungunya: young age and inefficient type-i interferon signaling are risk factors for severe disease. PLoS Pathog 4(2):e29
15. Morrison TE, Oko L, Montgomery SA et al (2011) A mouse model of chikungunya virus-induced musculoskeletal inflammatory disease: evidence of arthritis, tenosynovitis, myositis, and persistence. Am J Pathol 178(1):32–40
16. Lam S, Chen KC, Ng MM, Chu JJ (2012) Expression of plasmid-based shRNA against the E1 and nsP1 genes effectively silenced chikungunya virus replication. PLoS One 7(10): e46396
17. Gardner J, Anraku I, Le TT et al (2010) Chikungunya virus arthritis in adult wild-type mice. J Virol 84(16):8021–8032
18. Akahata W, Yang ZY, Andersen H et al (2010) A virus-like particle vaccine for epidemic chikungunya virus protects nonhuman primates against infection. Nat Med 16(3):334–338
19. Rulli NE, Rolph MS, Srikiatkhachorn A et al (2011) Protection from arthritis and myositis in a mouse model of acute chikungunya virus disease by bindarit, an inhibitor of monocyte chemotactic protein-1 synthesis. J Infect Dis 204(7):1026–1030
20. Reed LJ, Muench CH (1938) A simple method of estimating fifty percent endpoint. Am J Hyg 27:493–497

# Index



Edwin Yunhao Gong (ed.), *Antiviral Methods and Protocols*, Methods in Molecular Biology, vol. 1030, DOI 10.1007/978-1-62703-484-5, © Springer Science+Business Media, LLC 2013

MIX
Papier aus verantwortungsvollen Quellen
Paper from responsible sources
FSC® C105338

If you have any concerns about our products,
you can contact us on
**ProductSafety@springernature.com**

In case Publisher is established outside the EU,
the EU authorized representative is:
**Springer Nature Customer Service Center GmbH**
**Europaplatz 3, 69115 Heidelberg, Germany**

Printed by Libri Plureos GmbH
in Hamburg, Germany